HIGH ENERGY HADRON PHYSICS

MARTIN L. PERL

Stanford Linear Accelerator Center
Stanford University
Stanford, California

HIGH ENERGY HADRON PHYSICS

A WILEY-INTERSCIENCE PUBLICATION

JOHN WILEY & SONS, New York · London · Sydney · Toronto

Library of Congress Cataloging in Publication Data

Perl, Martin L. 1927–
High energy hadron physics.

Includes bibliographies.
1. Hadrons. 2. Nuclear reactions. I. Title.
QC793.5.H322P47 539.7'216 74-6348
ISBN 0-471-68049-4

Printed in the United States of America

10 9 8 7 6 5 4 3 2 1

I think prefaces to books such as this one are rarely read; I know that I rarely read them. But I hope that the reader will read this preface because it explains the spirit and the structure of this book. This spirit and structure, although somewhat different from other works of this type, are required, I believe, by the subject—the dynamics of hadrons at high energy and the aspects of the nature and structure of the hadrons that are important at high energy. We are primarily concerned with the strong interactions. But we are also interested in those aspects of electromagnetic and weak interactions which help us understand the dynamics and structure of the hadrons.

We are faced with a paradox in our present knowledge of strong interaction physics. We now know a great deal about the data and the phenomena of strong interaction physics. Those who have been in the field for a while acquire a broad knowledge of the regularities in these data; they acquire a general feeling for the behavior of strong interaction phenomena. But there is no complete and fundamental theory to explain these regularities and justify these feelings. There is no way to quantitatively explain—to calculate— the observed data starting from a fundamental theory. The paradox is that in strong interaction physics we know a great deal and yet we know very little. This paradox is the major problem to be faced in writing a book on this subject.

I have tried to solve the problem by the use of seven guiding principles—topic selection rules, if you like—to help me choose a path through the forests and swamps of present-day strong interaction physics. If the reader is acquainted with these topic selection rules, he will understand the material I have chosen: why I have emphasized some topics and why I have slighted or omitted other topics. The remainder of the introduction describes my seven selection rules.

An Overall View of High Energy, Strong Interaction Phenomena. There is an immense amount of data in strong interaction physics, and in it many regularities have been observed. The reader should be given an overall view of the data but should not be overwhelmed by details. Gross features of the data—elastic diffraction peaks, forward and backward peaks in inelastic scattering, the energy dependencies of cross sections, and particle multiplicities—should be emphasized. But data that appear to be of current interest only because a current theoretical point depends on them should not be emphasized. Conversely, data should not be ignored because our current theoretical understanding or interest is small. The goal is to present the reader with an unbiased view of the data, thus stimulating new ideas and new approaches.

Sources of Understanding. Our current understanding of strong interaction physics is based on many sources. We argue by analogy from classical optics, from the theory of nuclear physics, from the quantum mechanics of nonrelativistic potentials, and from quantum electrodynamics. We use quantum field theory, diagrammatic Feynman theory, and analyticity. We use group theory and very general ideas about conserved quantities. We use intuition, we make models, and we are always optimistic. The book must make clear the multiplicity and the diversity of these sources of understanding.

Recurrent Themes In some areas elementary particle physicists have not been very inventive. Some themes, some basic ideas occur again and again. Thus the absorption correction to the one-particle exchange model has risen again as Regge cuts. Thus the concept of point particles, so fruitful in atomic physics, has been resurrected in the quark and parton theories. The book should make clear these recurring themes.

How to Calculate. It is generally true in physics that one does not understand a subject unless one can make calculations in it. Thus the book attempts to show the reader by example and by exercise how calculations are made. Frequently the problem is simplified to avoid complex algebra or extensive numerical work. Knowing how to calculate is important even in an area whose theory is obsolete, such as one-particle exchange, for only by understanding clearly the problems of one-particle exchange theory can one see the need for the introduction of absorption corrections and Regge theory.

Formal Theoretical Techniques. Of course, many of the omissions from this book lie in the vast area of formal theoretical techniques. Some, such as quantized field theory and formal scattering theory, are fortunately covered in other books and can be excluded. Other formal theoretical techniques must be excluded because they are too difficult for this book, or, in my opinion, are too remote from the heart of strong interaction physics. But one must include those formal techniques and devices that are most common and that offer clarity and convenience in theoretical deductions and discussions. Examples of such techniques are rotation operators, boost operators, and the helicity formalism. Some of these techniques, particularly those related to the theory of angular momentum, appear in many nonrelativistic quantum mechanics books. It is my experience that although many students are acquainted with these techniques, they cannot use them. Therefore the techniques are reviewed and their use is encouraged.

Lower Energy Data. Strong interactions below the several-GeV energy range, the interactions of elementary particles with nuclei, the decay properties of the hadrons, and the static properties of the hadrons are all enormous subjects in themselves. To keep this book within reasonable size, it was necessary to omit the major portions of these subjects. These large omissions can be justified on the grounds that the intuition of many physicists is that the secrets of strong interaction physics are most likely to be unlocked at high energies rather than at low energies. I have not, however, neglected to introduce and discuss major concepts that have been developed in the study of the static and low energy properties of hadrons; for example, strong interaction resonance theory and $SU(3)$ theory are discussed; but a full description of the performance of calculations in these areas is not provided.

Illusions. There is no coherent theory of strong interaction physics with which one can explain quantitatively the observed phenomena. We have bits and pieces of theories–some too formal to use as a basis for calculation–some too phenomenological to permit a rigorous derivation. One must carefully avoid projecting the illusion that the theoretical situation is any better than that. Such illusions develop most readily in discussing in an introductory manner a particular theory. The introductory treatment of necessity avoids some complexities of the theory. Therefore the reader may gain the impression that the theory in its full complexity is complete and rigorous; for example, someone reading an introduction to Regge pole theory ten years ago might think that he could obtain a full understanding of two-body inelastic interactions once he had mastered all the complexities introduced by nonzero spins. This, of course, is not true. At the very least, Regge cuts are required; and there are certainly doubts about the basic validity of the use of Regge theory in elementary particle physics. Therefore, at the risk of causing some hurt feelings, the limitations and the failures of each theory have been noted and discussed.

These, then, are the guiding principles used in the writing of this book. Whether they are the correct principles to use and whether they have been truly followed we leave to the reader to judge.

This book may be used as the basic textbook for two kinds of courses on high energy hadron physics. First, it may be used for a rather full course, extending over two quarters or two semesters. In such a course the students become well acquainted with the data and phenomenology of high energy physics and with the calculational methods in the field. The instructor also has time to introduce a few special topics such as a more formal consideration of $SU(3)$ symmetry or a more detailed discussion of weak interactions.

The book may also be used for a much shorter course, such as the 10-week course I conducted. (My students had previously taken a one-quarter-long course in the fundamentals of elementary particle physics.) For the short course the book is used as follows. Primarily one eliminates from the course the chapters related to the formalism and phenomenology of particles with spin. In addition, formal material such as that in Ch. 6 is omitted. The course then consists of the following topics: total cross section and elastic scattering, (Ch. 1, Ch. 2, and Secs. 3-1 to

3-4), inelastic interactions (Ch. 7 and Ch. 8), one-particle exchange (Secs. 12-1 to 12-3 and Secs. 13-1 to 13-4), analyticity and crossing (Secs. 15-1 to 15-7), Regge theory (Ch. 17), photon-hadron interactions (Secs. 19-1 to 19-6), lepton-nucleon interactions (Ch. 20), electron-positron annihilation into hadrons (Ch. 21), and large momentum transfer collisions (Ch. 22). The book has been designed so that this sequence can be coherently followed. The elimination of particle spin related topics will not interfere with the use of the book in this way. The students will still acquire a good knowledge of, and physical insight into, high energy hadron physics. They will, of course, not have time to obtain the calculational facility that they can acquire in the longer course.

A few paragraphs about references. The literature of elementary particle physics is vast. How is one to select useful references? I have used three criteria. First, reference is made to the papers that I have found to be particularly useful in providing physical insight into some aspect of the subject under discussion. Second, I have tried to develop a feeling in the reader for the ongoing progress in the subject by giving references to the proceedings of summary conferences. Third, I have tried to avoid giving references to papers that are useful but unpublished and hard to obtain.

The textbooks to which I refer are primarily those in my personal library to which I have become accustomed. There are equally good books in quantum mechanics, in quantum electrodynamics, and in elementary particle physics with which I am not well acquainted, hence rarely use. Therefore the textbook references are simply guides to the type of book from which the reader can obtain further information.

It should be clear that the names and references given in the text must not in any way be taken as indications of those who have made the major contributions to the subject. This is true also with respect to the data used to illustrate the discussion. The reference, for the reason just given, is usually to a summary talk rather than to the experimentalists to whom we owe the data.

Similarly, I have not necessarily used the most recent graphical data to illustrate the discussion. First, that is not possible to do in a field in which experiments improve rapidly. Second, sometimes early data or a particular curve illustrate a point in the discussion with the most clarity. In that case I have used the appropriate data or curve.

Each chapter has its own references, given in the form of a number followed by two or more letters. The number refers to the chapter and the letters are the first letters in the surname of the first author. Usually two letters are sufficient, but if the first author appears several times, or in other cases of confusion, three or four letters are used. In a few cases the letters refer to an editor or to the name of a reference work.

In writing this book I am indebted to my many colleagues at the Physics Department of the University of Michigan, where I first began to practice strong interaction physics, and to my present colleagues at the Stanford Linear Accelerator Center of Stanford University. I have learned a great deal from them and I am particularly grateful to Samuel Berman, Richard Blankenbecler, James Bjorken, Stanley Brodsky, Michel Davier, Sidney Drell, Frederick Gilman, Lawrence Jones, Michael Longo, H. Pierre Noyes, Wolfgang Panofsky, David Ritson, Marc Ross, and Yung Su Tsai for discussions that have helped me gain physical insight into hadron dynamics. The theorists in this group have also been most patient in helping me with the phase problems, notation questions, and other convention problems that plague this subject.

This book was begun while I was on a sabbatical leave as a John Simon Guggenheim Fellow. That fellowship was crucial, for without it I should not have had the time to think about the book, plan it, or begin writing it. Thus it owes its existence primarily to the John Simon Guggenheim Memorial Foundation.

The sabbatical was spent at the Physics Department of Westfield College of the University of London. The hospitality provided by Professors E. H. Bellamy and E. Leader and the Physics Department were of great value to me.

Some important parts of the book were completed while I was at the Aspen Center for Physics. I am very grateful for the hospitality of Professor Loyal Durand and the Center. The many discussions I had with the physicists I met there were most useful.

MARTIN L. PERL

Palo Alto, California
June 1973

CONTENTS

CHAPTER

1

Total Cross Section and Elastic Scattering

1-1 Introduction. This book is concerned with those interactions of hadron–hadron and photon–hadron systems in which the final products are also hadrons. Among the many possible kinds of measurements on these interactions, two types of measurements are distinguished by their experimental simplicity—the total cross section and the elastic scattering differential cross section. The experimental simplicity derives not only from the relative ease with which these quantities can be measured, but also from the relative ease with which they can be described. To some extent a theoretical simplicity accompanies these quantities in that their behavior can be explained, at least superficially, using the concepts of classical optics. I have used the word "superficial" here because the complete theoretical understanding of these quantities has so far eluded us; in particular, the process of elastic scattering may be harder to understand than many inelastic processes.

Nevertheless, we begin by studying the total cross section and the elastic scattering differential cross section. The concepts of absorption and diffraction scattering, which are introduced in the first chapters, are used repeatedly in the book. More complicated and hopefully more fundamental theories of elastic scattering must reduce to the formulations we give here, because those formulations do describe the data. Also, elastic scattering at high energies is closely related to the inelastic processes at those energies; and in the calculation of those inelastic processes we use directly or indirectly our knowledge of elastic scattering. Thus the study of total cross sections and elastic scattering forms a suitable and necessary foundation for the general study of hadron–hadron interactions.

This chapter is concerned with our present knowledge of total, elastic, and inelastic cross sections in the high energy region. In the discussion of elastic scattering, the emphasis is on the small angle region usually called the diffraction peak region. The differential cross section is presented as a function of t—the square of the four-momentum transfer to the target particle. This leads to the representation of the diffraction peak differential cross section by an exponential in t. Finally, there is an introduction to the optical theorem and to the relations between the total cross section, the differential cross section, and the total inelastic cross section.

1-2 The High Energy Region and the Significance of s. High energy, strong interaction data now come from two types of accelerators (1BL): *conventional accelerators*, in which a beam of hadrons, photons, or leptons interacts with stationary nucleons, and *colliding beam accelerators* (1PE), in which two beams of particles of equal but opposite momentum interact. Existing colliding beam accelerators use the combinations of proton–proton, electron–positron, or electron–electron, but the combination of electron–proton is also feasible.

In either type of accelerator, the energy parameter of interest is the total energy in the center of mass, for this is the energy available for the production of additional hadrons. The total center of mass energy is also a measure of the potential violence of the collision; and we believe that more violent collisions may be better probes of the fundamental nature of elementary particles and strong interactions. It has become conventional to describe the total center of mass energy by its square, usually denoted by s. If two particles a and b of four-momentum P_a and P_b collide, we have

$$s = (P_a + P_b)^2 \tag{1-1}$$

Thus s does not change under Lorentz transformations and is called a *Lorentz scalar.*

For conventional accelerators, in which the beam particle, denoted by a, has mass m_a and laboratory energy E_a and the stationary target particle b has mass m_b, we write

$$s = 2m_bE_a + m_a^2 + m_b^2 \tag{1-2a}$$

For $s \gg m_a^2, s \gg m_b^2$

$$s \approx 2m_bE_a \tag{1-2b}$$

As shown in Eq. 1-2b, the crucial quantity $\sqrt{s}$ increases only as $\sqrt{E_a}$ in conventional accelerators. But in colliding beam accelerators, in which each beam has energy E, we have

$$s = 4E^2 \tag{1-3}$$

Thus in colliding beam accelerators high s can be achieved with beam energies much lower than those required for conventional accelerators.

Exercise 1-1. Review the manipulation of four-momentum vectors and the meaning of Lorentz scalars. Derive Eqs. 1-2 and 1-3.

Exercise 1.2. Derive formulas for s if the masses are unequal in a colliding beam accelerator. Also discuss the kinematical situation if the colliding beams intersect not head-on, but at a small angle, such as 15°. This is indeed the situation in some existing colliding beam accelerators (1BL, 1PE).

Having defined s, we can now say what we mean by high energies. The high energy region we emphasize is that for which $\sqrt{s}$ is much greater than a baryon mass and very much greater than a pion mass. Thus we would like $\sqrt{s}$ to be greater than 4 or 5 GeV, corresponding in conventional accelerators to beams of greater than roughly 10 GeV energy. But because of the lack in some areas of such high energy data we sometimes consider data with lower values of s.

1-3 Total Cross Section Data. In the high energy region, total cross-section measurements now exist for photons, pions, kaons, and nucleons on nucleons. The total cross-section, $\sigma_{\text{tot},ab}(s)$, for the interaction of particle a and b is a function of s. Often it is presented as a function of p_a, the magnitude of the three-momentum of particle a, taking particle b as stationary. For colliding beam data, the equivalent p_a is used. Figure 1-1 shows the total cross sections for momenta above 5 GeV/c.

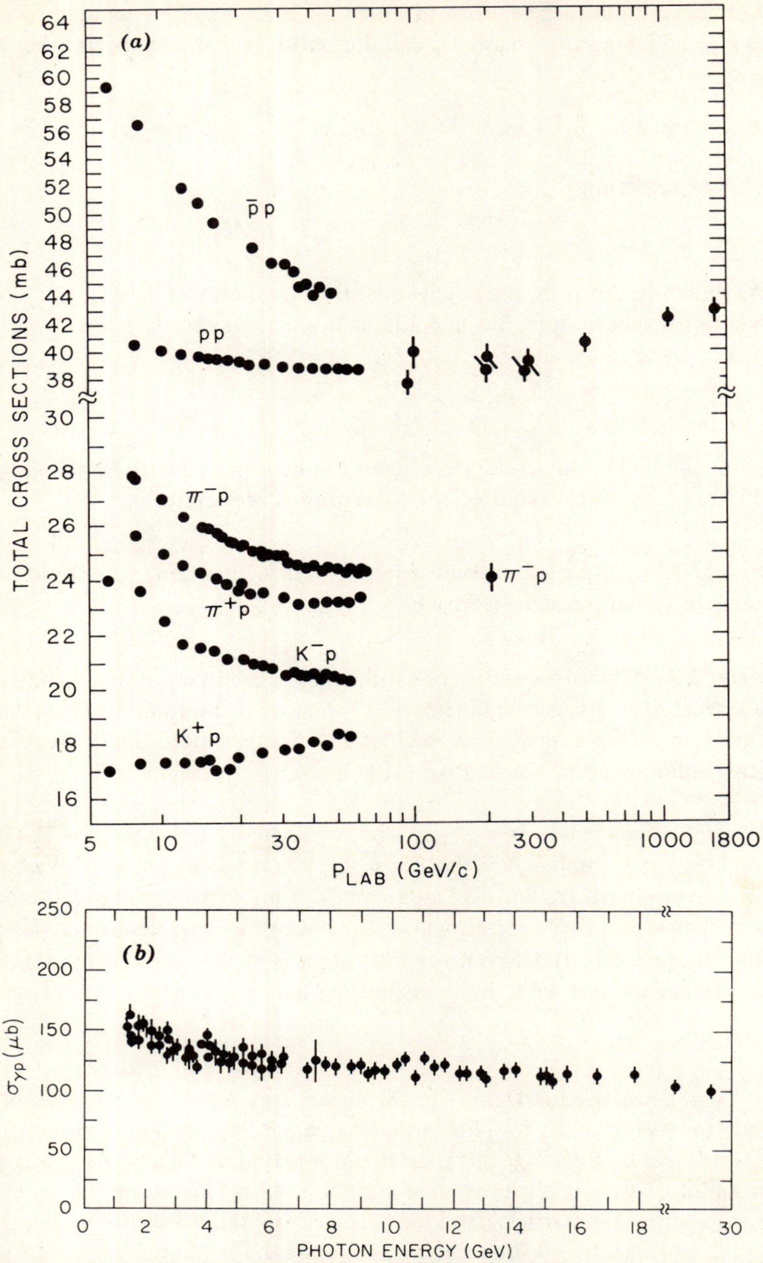

Fig. 1-1 Hadron-hadron and photon-hadron total cross sections. Note the much smaller cross section scale for the latter. Taken primarily from Ref. 1GI.

The total cross sections have the following general characteristics when the laboratory momentum of the incident particle is above 4 or 5 GeV/c.

1. As the momentum increases, σ_{tot} is either constant or changing very slowly.
2. At these higher momenta, the nucleon and antinucleon σ_{tot}'s are in the range of 40 to 60 mb and the meson σ_{tot}'s are in the range of 15 to 30 mb. The photon–proton σ_{tot} is very much lower, about 0.10 to 0.13 mb.
3. The σ_{tot}'s vary smoothly with energy; there are no sharp bumps or dips in the curves.

Thus the high energy behavior of the total cross sections shows little variety. But at low energies the different systems behave in markedly different ways. Some examples of low energy behavior are presented in Figs. 1-2 to 1-4. We see that the pion–proton total cross sections have narrow peaks. On the other hand, the proton–proton system has a smooth energy behavior; there is only a broad minimum in the total cross section. Clearly at the lower energies all sorts of total cross section energy dependencies are possible. The pion–proton total cross sections at these lower energies is discussed in Ch. 5; we only mention here that these total cross section peaks are caused by the formation of resonances.

The photon—proton system is a particularly interesting system for the photon is not a hadron. We emphasize that this total cross section includes only (1) inelastic processes in which hadrons are produced and (2) elastic photon–proton scattering. It does not include purely electromagnetic processes such as the production of electron or muon pairs. From Fig. 1-1 we see that above 2 GeV the γp total cross section is a very slowly decreasing function of the energy. Its magnitude is about 1/200 of the magnitude of the πp total cross section. Below 2 GeV there are peaks in the total cross section just like the peaks in the πp system (Fig. 1-4). Those peaks are due to the formation of some of the nucleon resonances of the πp system. Thus the entire behavior of the γp total cross section is like the behavior of the πp total cross section except that the former is two powers of 10 smaller than the latter. This phenomenon is discussed in Ch. 19.

1-4 The Kinematics of Elastic Scattering and the Significance of *t*. In elementary particle physics an elastic interaction is one in which there are two particles in the final state, each of those final state particles being identical to one of the initial state particles. The kinematic formulas that are relevant to elastic scattering are collected in App. A; here we present and review only the most important formulas. In the elastic reaction $a + b \rightarrow a + b$, the initial four-momenta are P_a and P_b and the final four-momenta are P_a' and P_b'. In the center of mass system (cms) the three-momenta and energy are designated by $\mathbf{p}_i^*$ and E_i^*, where i is any of the labels a, a', b, b'. The magnitude of the three-momentum

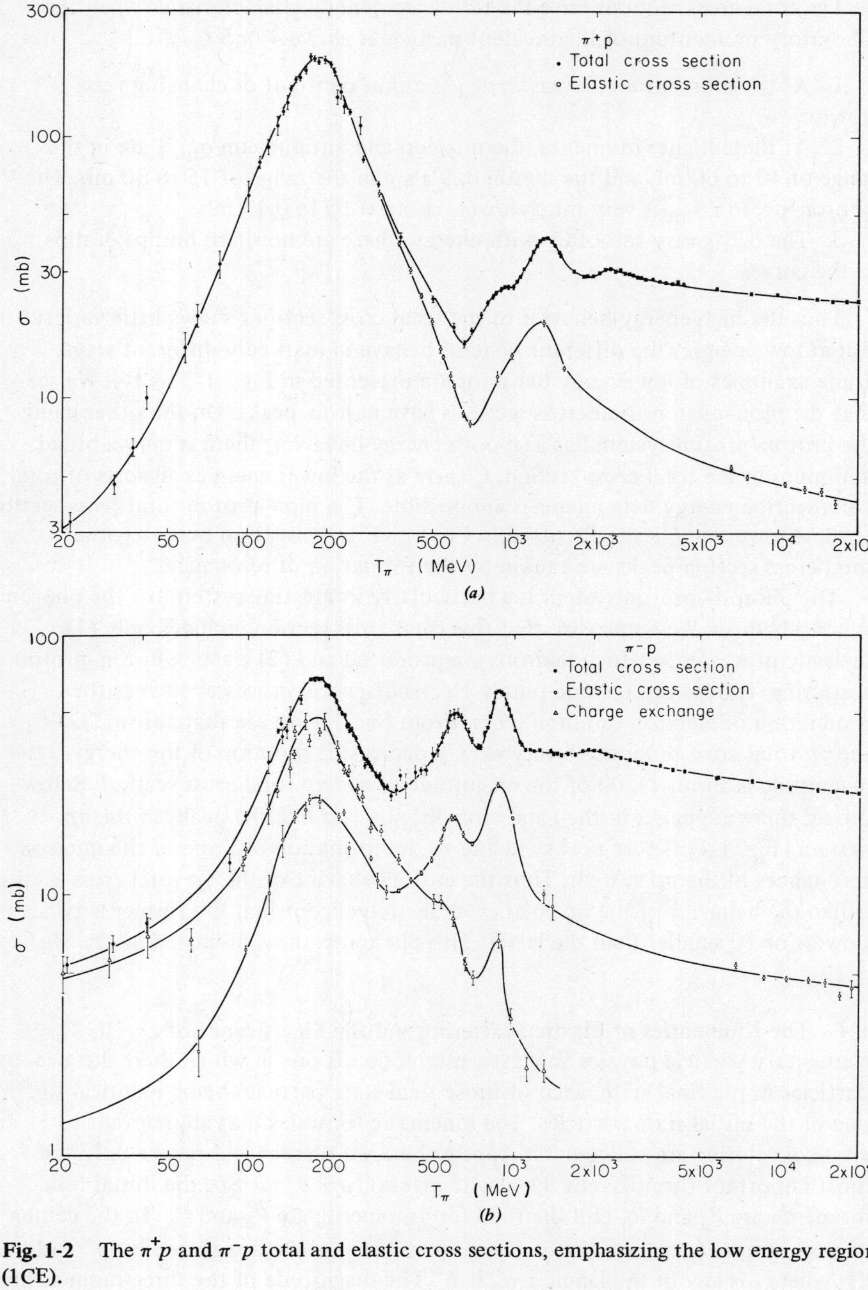

Fig. 1-2 The π^+p and π^-p total and elastic cross sections, emphasizing the low energy region (1CE).

of any of the particles in the cms is called p^*. The scattering angles in the cms are θ^* and ϕ^*, θ^* is taken between $\mathbf{p}_a^*$ and $\mathbf{p}_a'^*$; and the differential solid angle in the cms is $d\Omega^* = d\phi^* \sin\theta^*\, d\theta^*$. The differential cross section in the cms, $d\sigma/d\Omega^*$, can most generally be a function of the total energy (represented by s or p^*), of θ^*, and of ϕ^*. If all the particles are spinless, or if the measurement averages

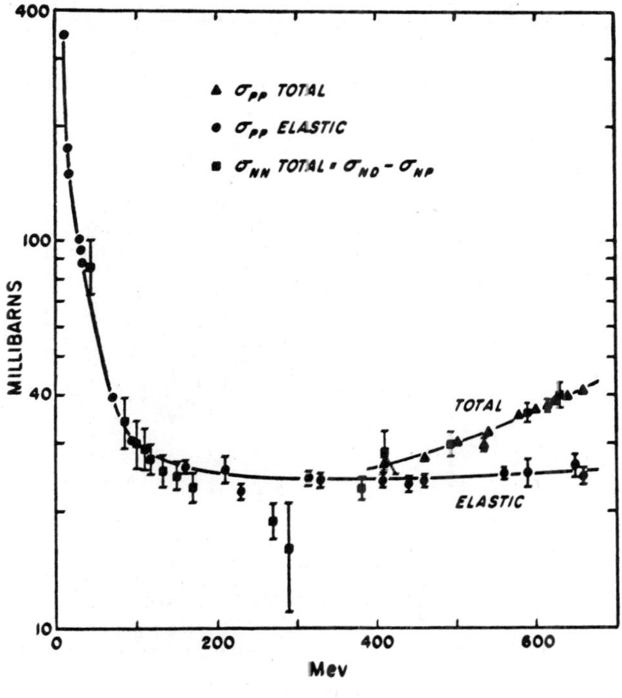

Fig. 1-3 A classic illustration of the pp total and elastic cross sections in the low energy region (1HE).

over the initial spin states and sums over the final spin states, $d\sigma/d\Omega^*$ reduces to a function only of p^* and θ^*. We shall use both $d\sigma/d\Omega^*$ and another form, $d\sigma/dt$. Here t, a Lorentz scalar, is defined by

$$t = (P_a - P_a')^2 = (P_b' - P_b)^2 \tag{1-4}$$

In the cms we have

$$t = -2p^{*2}(1 - \cos\theta^*) \tag{1-5}$$

and in the laboratory system we have the convenient relationship

$$t = -2m_b(E_b' - m_b) \tag{1-6}$$

where E_b' is the final energy of particle b in the laboratory system. We note that because of our metric, t is a negative quantity in an elastic scattering process. We use the center of mass angles θ^* and ϕ^* so often, and we use laboratory angles so seldom that for neatness and convenience I often omit the asterisk on θ^* and ϕ^* in the cms.

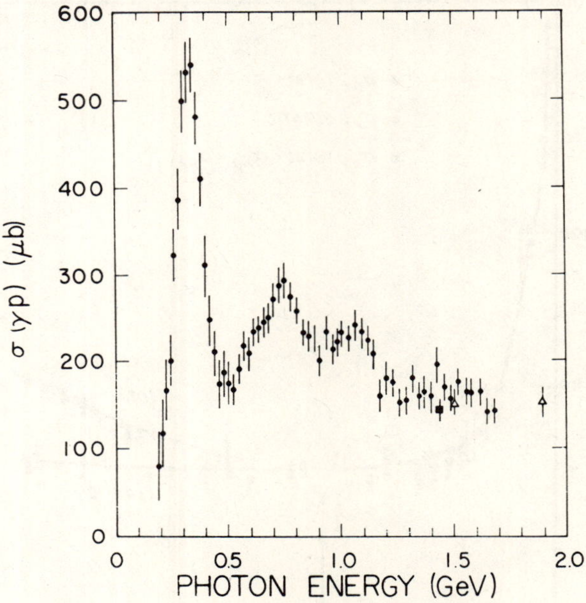

Fig. 1-4 The γp total cross section in the low energy region (1GIL).

When $d\sigma/d\Omega^*$ is independent of ϕ we calculate, disregarding a negative sign, that

$$\frac{d\sigma}{dt} = \frac{\pi}{p^{*2}} \frac{d\sigma}{d\Omega^*} \tag{1-7}$$

We usually regard $d\sigma/dt$ as a function of s and t.

The parameter t is more than just a convenient Lorentz scalar. If t is small compared with $m_b{}^2$, we can write Eq. 1-6 in its nonrelativistic form

$$t = -(\mathbf{p}_b')^2 \tag{1-8}$$

where $\mathbf{p}_b'$ is the three-momentum of particle b in the laboratory system after the scattering. Thus in this situation $\sqrt{-t}$ is the magnitude of the three-momentum transferred to the target particle. Classically a small three-momentum transfer is a glancing blow, and a large three-momentum transfer is a violent blow. Thus classically the "violence" of the collision is measured by the magnitude of t. As we develop the theory of elastic scattering, this concept is extended into the quantum mechanical realm. Through the concept of the Fourier transform (Sec. 2-4), the transfer of three-momentum is related to the spatial structure of the scattering particles. The small t collision can be regarded as a means of

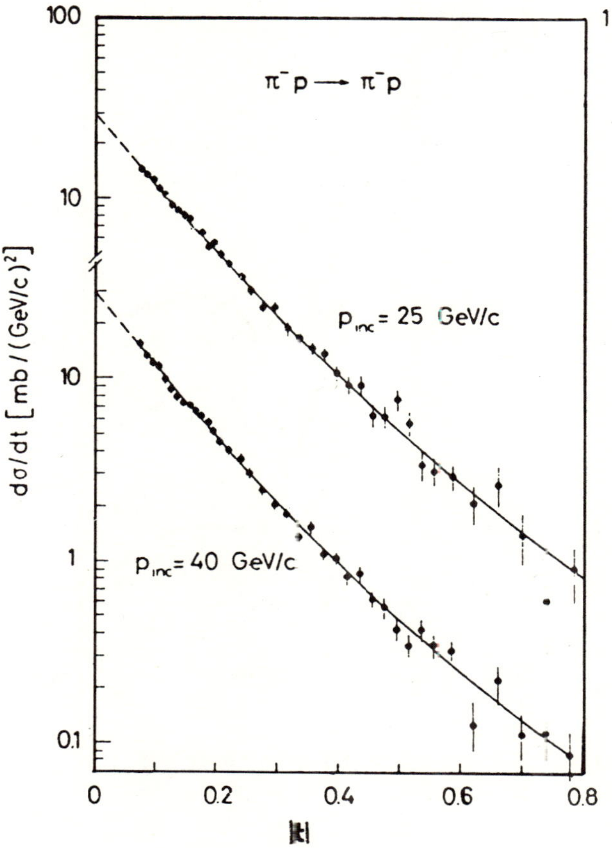

(a)

Fig. 1-5 The $d\sigma/dt$ for $\pi^- p$ elastic scattering (*a*) at small t and high energy (1GI), (*b*) over a larger t range showing a shoulder at about 0.6 $(\text{GeV}/c)^2$ and a minimum at about $3(\text{GeV}/c)^2$ (1CR), (*c*) at lower energy showing a dip at $0.6(\text{GeV}/c)^2$ (1CO), and (*d*) at still lower energy (1BU). The numbers accompanying the curves are the pion momenta.

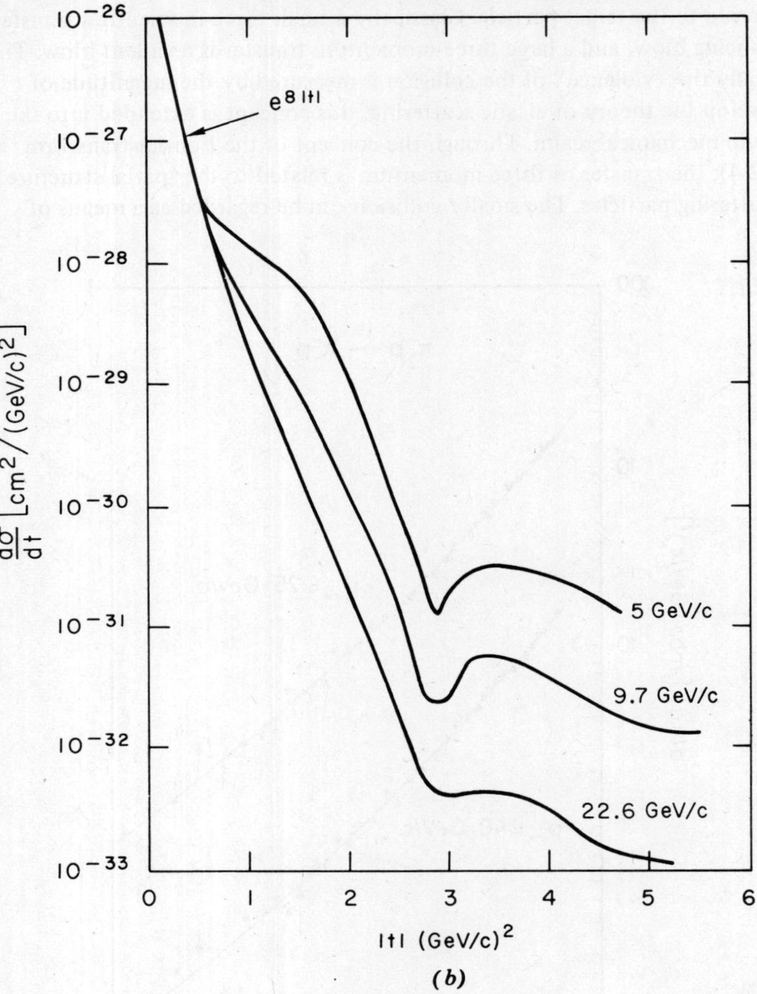

(b)

Fig. 1-5 (*Continued*)

examining no more than the overall size and shape of the interacting particles. The large t collision, on the other hand, can be regarded as a means of studying the internal structure of the particles. Thus one expects or hopes that the more violent collisions teach one more about the internal nature of elementary

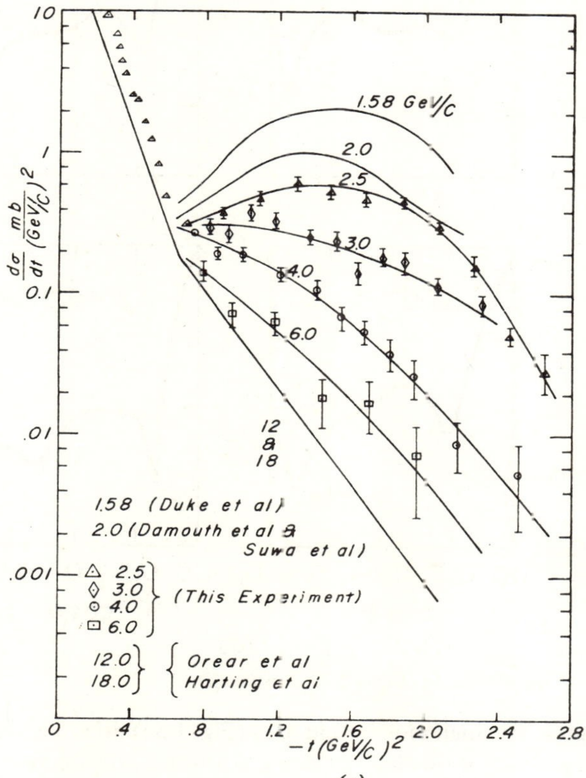

(c)

Fig. 1-5 (Continued)

particles. Unfortunately, the most secure parts of the theory are those concerning small t collisions and the theory is most obscure when t is large, say of the order of several $(GeV/c)^2$ in magnitude.

1-5 Elastic Scattering and the Diffraction Peak. To introduce the subject of [illegible] energy elastic scattering, we first look at $\pi^- p$ elastic scattering over a large

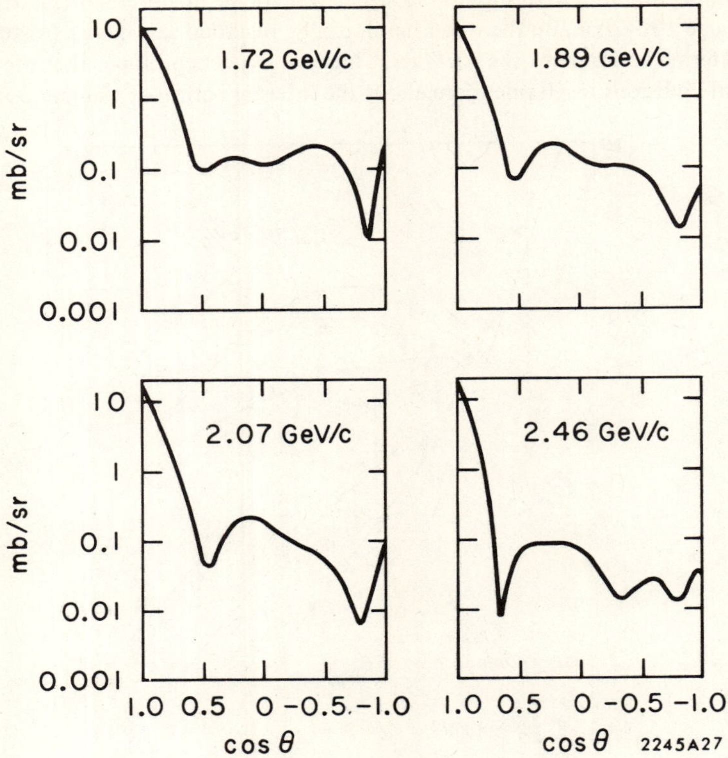

(d)

Fig. 1-5 *(Continued)*

momentum range. The high momentum data (Fig. 1-5*a*) do not extend out to large t, but they clearly show that the elastic scattering cross section is concentrated in the small t region of $|t| < 0.4$ (GeV/c)^2. In this graph and in all the graphs in Figs. 1-5 and 1-6, the quantity $\ln(d\sigma/dt)$ is plotted versus $|t|$. We observe that the rapid falloff of $d\sigma/dt$, away from $t = 0$, is roughly exponential. In fact, on these semilogarithmic plots the data can be fitted with a straight line as the cross section falls by a factor of 50 or 100 from its value at $t = 0$. Thus it has become customary to fit the small t region with the form

$$\frac{d\sigma}{dt} = A\, e^{-b|t|} \tag{1-9}$$

Such a fit is usually good for $|t| \lesssim 0.5$ (GeV/c)^2. As discussed in Sec. 1-6, at high energies b has values in the range of 5 to 13 (GeV/c)$^{-2}$. We call b the *exponential slope*.

To see what the differential cross section looks like at larger values of t, we must use the lower momentum data (Fig. 1-5b). Once again there is an exponential peak at very small t; at somewhat larger t the differential cross section still is dropping rapidly, but not with an exponential behavior. This larger t behavior is

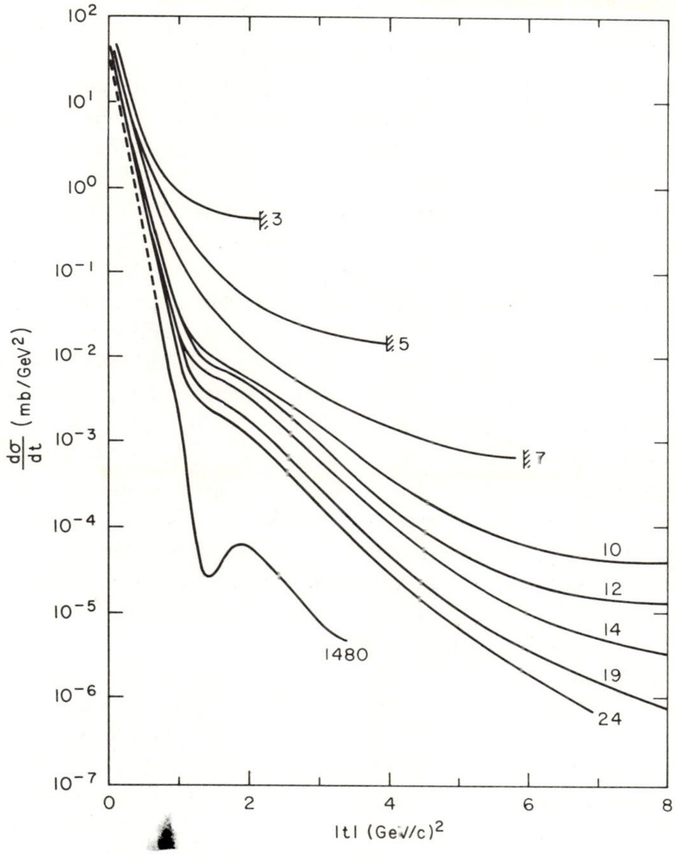

Fig. 1-6 The $d\sigma/dt$ for pp elastic scattering. The numbers accompanying the curves are the laboratory momenta of the incident protons. The highest momentum is from a pp colliding beam experiment. Taken from Ref. 1GI.

more complicated because there is a shoulder at about 0.6 (GeV/c)^2 and a dip at about 3 (GeV/c)^2. As shown in Fig. 1-5c, at still lower momenta the shoulder becomes a dip at 0.6 (GeV/c)^2, the exponential peak ending at this dip. The small t peak and the dip can be followed to yet lower momenta in Fig. 1-5d. Here the differential cross section is given versus cos θ rather than versus t. The entire angular range is covered, and a rather complicated, momentum-dependent structure is evident at the larger angles. To see the disappearance of the small t peak we must go to yet lower energies, as in Fig. 5-1.

Certainly the primary observation to be made about the differential cross sections shown in Fig. 1-5*a* through 1-5*d* is that there is a peak in the small t region. It is equally important, however, that all these peaks have almost the same shape; that is, the shape is weakly dependent on the energy at high energy. A third important observation is that at the higher energies, the peaks (in the units used here), have almost the same size; thus the curves representing different energy, small angle differential cross sections are almost superimposed. The last observation is discussed in Sec. 1-7.

The shape would not be the same at different energies if, for example, we had plotted the differential cross sections as a function of θ^*, the scattering angle in the center of mass. From Eq. 1-5, for fixed t, we see that as the energy (hence p^*) increases, the angle θ^* decreases. Therefore, if the small angle differential cross section were plotted versus θ^*, the peak would become steadily narrower as the energy increased. Now it is more than convenient that plotting the small angle differential cross sections against t yields almost an energy independent shape. This general result signifies that t is a primary variable in understanding elastic scattering and that in any theory of high energy, small angle elastic scattering, the shape must be mainly t dependent in the first approximation. Thus it becomes important to determine quantitatively just how energy independent is this shape; this question is taken up in Sec. 1-6. We emphasize that the discussion in this and the last paragraph holds only for small t. It is clear from the figures that as $|t|$ exceeds 0.5 $(\text{GeV}/c)^2$, the shape of the differential cross section becomes increasingly energy dependent.

Finally, to look at a very different system, we present a large compilation of proton-proton elastic differential cross section data in Fig. 1-6. Once again the strong small t peak is evident at all the momenta presented, and the shape of that peak changes very little with energy. However, unlike πp elastic scattering, there are no dips or shoulders in the pp differential cross section at the lower momenta. Within the statistical errors, the pp differential cross sections decrease monotonically as $|t|$ increases. However, above 8 GeV/c a shoulder and then a dip appear in the pp differential cross sections.

The high energy elastic scattering of kaons on protons, photons on protons (Fig. 2-4), neutrons on protons, and antiprotons on protons has also been measured. In all these systems the high energy elastic differential cross section is dominated by a peak at small t. This small t peak, a universal property of all hadron-hadron systems that have been studied, is ascribed to a very general process–*diffractive elastic* scattering. The small t peak is called the *diffractive peak* and that t region in elastic scattering is called the *diffraction region.* The diffractive concept is described fully in Ch. 2.

The diffraction region has no sharp t boundary. Thus the small $|t|$ region where Eq. 1-9 is a good fit is called the exponential part of the diffraction peak. As $|t|$ increases beyond the exponential region, $d\sigma/dt$ begins to flatten out—sometimes smoothly as in pp scattering (Fig. 1-6) in the 3 to 7 GeV/c range, sometimes with bumps or dips as in the π^-p data. Somewhere in the region of $|t| = 0.5$ to 1.5 (GeV/c)^2, the simple diffraction picture begins to fail. More complicated diffraction models or other models are then to be applied, and some of these models are discussed later in the book. But we are less sure of the fundamental explanation of the elastic scattering in the region where $|t|$ is of the order of 1 or more (GeV/c)^2. Therefore when we speak of the diffraction region we mean that $|t|$ is in the range of 0 to about 0.5 or 1 (GeV/c)^2, although the diffraction concept may explain scattering at larger $|t|$ values.

The other regions in the elastic differential cross section are discussed later but should be mentioned at this point. At very small angles, if both particles are charged, there will be Coulomb scattering. As shown in Sec. 10-6, the Coulomb

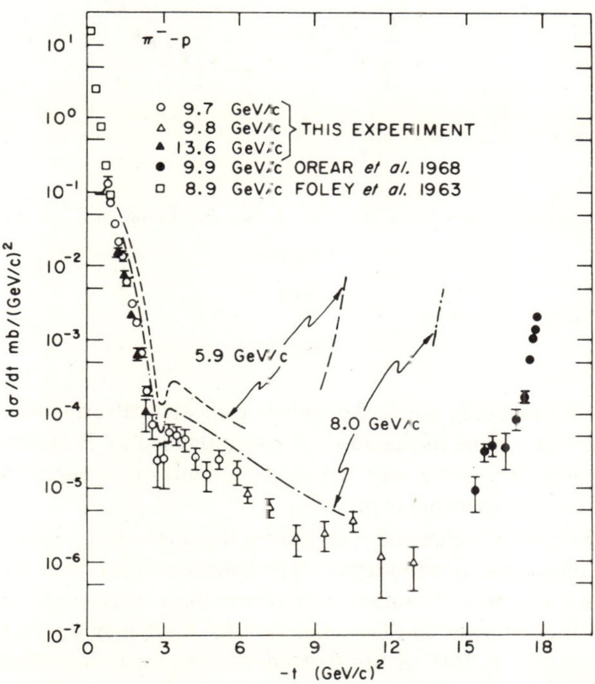

(a)

Fig. 1-7 The complete elastic differential cross section for (*a*) the π^-p system at indicated momenta (1OR), and (*b*) the π^+p system at 5.0 GeV/c (1RU).

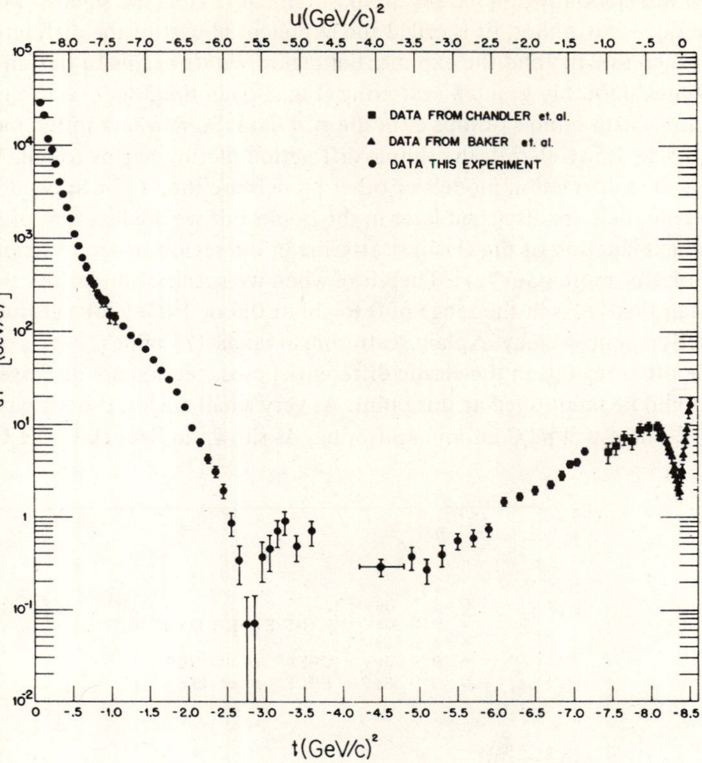

(b)

Fig. 1-7

scattering can be seen only above the hadronic elastic scattering when $|t|$ is less than 0.01 $(\mathrm{GeV/c})^2$. None of the $d\sigma/dt$ curves discussed in this section show the Coulomb scattering region because the measurements do not go to sufficiently small t values in the experiments pictured.

At the other end of the angular range from the diffraction region is the $\theta = 180°$ or backward scattering region. The behavior of $d\sigma/dt$ in the backward scattering region is extremely varied. Sometimes there is a peak at 180°, as in Fig. 1-7. For other energies and for other systems there may be smaller peaks, a dip, or a flat region at 180°. Still other systems appear to have their minimum $d\sigma/dt$ at 180°. But in no case does the $d\sigma/dt$ approach the size of the diffraction peak; it is always one or more factors of 10 smaller. In proton–proton elastic scattering, the indistinguishability of the two particles means that there is no way to separate experimentally forward scattering from backward scattering.

Therefore pp experimental differential cross sections only extend from 0 to 90° in the cms.

Between the diffraction region and the backward scattering region lies the large t region. This region may show dips or bumps in $d\sigma/dt$, or the behavior of $d\sigma/dt$ with t may be very smooth. A section of Ch. 22 is devoted to this region, which is the most difficult region to understand theoretically.

1-6 The Exponential Slope in Elastic Scattering. To return to the diffraction region, if the diffraction peak is represented by Eq. 1-9, $d\sigma/dt = A\, e^{-b|t|}$; an immediate question is, How well does Eq. 1-9 fit the data? There are no systematic studies of this question; but as we have already pointed out, as $|t|$ approaches 1 $(\mathrm{GeV/c})^2$ Eq. 1-9 does not hold. Even in the small t region, $|t| \lesssim 0.5\ (\mathrm{GeV/c})^2$, Eq. 9 is not a perfect fit. For example, in proton-proton elastic scattering (Fig. 1-8), b increases in going from the medium small t region $0.15 \lesssim |t| \lesssim 0.5$ to the very small t region $|t| < 0.1$. The increase in b is 1 or 2 $(\mathrm{GeV/c})^{-2}$, where the average value of b is about 10 $(\mathrm{GeV/c})^{-2}$. We do not

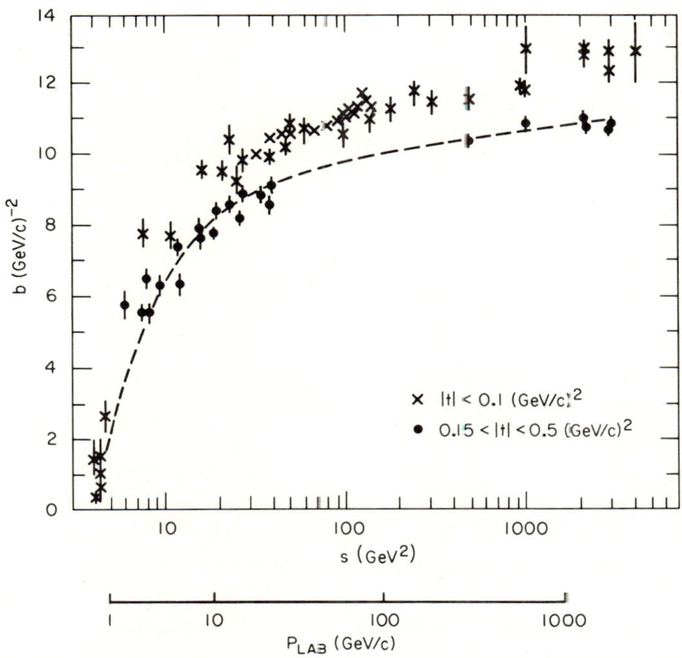

Fig. 1-8 The b from a fit of $d\sigma/dt = A\, e^{-b|t|}$ to pp elastic scattering. Taken from Ref. 1GI.

understand the significance of this small change of b in the small t region. And for the purposes of this section, and indeed for the entire book, we regard Eq. 1-9, with *b independent* of t, to be an adequate fit to the elastic diffraction peak in the region $|t| \lesssim 0.5$ (GeV/c)^2.

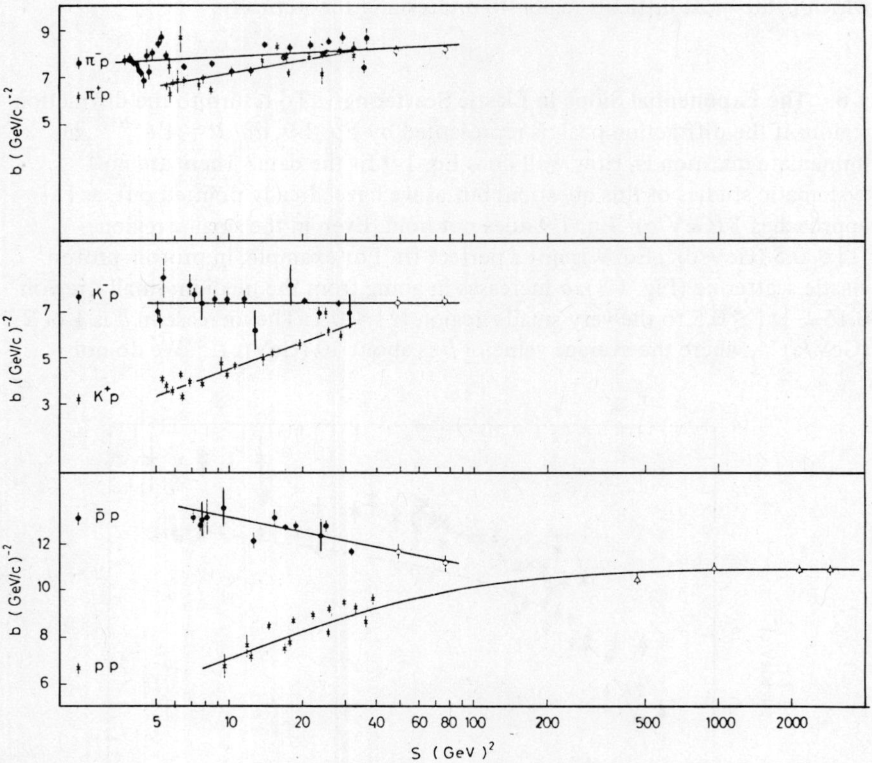

Fig. 1-9 The b from a fit of $d\sigma/dt = A\, e^{-b|t|}$ to elastic scattering in the indicated system in the range $0.1 \leqslant |t| \leqslant 0.4$ (GeV/c)^2. Taken from Ref. 1GI.

We shall now use the energy dependence of the exponential slope b to quantitatively examine the energy dependence of the shape of the diffraction peak. In Sec. 1-5 we observed that to a first approximation the diffraction peak for $|t| \lesssim 0.5$ (GeV/c)^2 has an energy-independent shape. We first turn to the pp system (Fig. 1-8), where the most extensive information lies. We observe that b increases slowly with energy, although at the highest energy b may be independent of energy. When b increases, the diffraction peak (plotted as a function of t) narrows or shrinks. This is the shrinkage phenomenon that played an important part in the initial application of Regge theory to elementary particle elastic scattering. At one time it was thought that Regge theory would require all diffraction peaks to shrink with increasing energy; but more recent applications of Regge theory do not impose this requirement. At present we do not know if shrinkage, or antishrinkage, of the diffraction peak is a very high energy phenomenon or simply an intermediate high energy phenomenon. This is discussed further in Ch. 17. Other examples of the behavior of b with energy are presented in Fig. 1-9.

1-7 Total Cross Section, Elastic Scattering, and the Inelastic Cross Section.

In this section we use the optical theorem and experimental data on the ratio of the real part to the imaginary part of the forward elastic scattering amplitude. These two subjects are discussed in detail later after the S-matrix and helicity concepts have been developed. Therefore, our introduction to these subjects is very brief (see Secs. 6-7 and 10-6).

The optical theorem relates the imaginary part of the elastic scattering amplitude at zero degrees to the total cross section. We denote the center of mass elastic scattering amplitude at zero degrees by $f(0)$. The optical theorem states

$$\text{Im}\, f(0) = \left(\frac{p^*}{4\pi}\right)\sigma_{\text{tot}} \tag{1-10}$$

where p^* is the momentum in the center of mass. The reader who finds the lack of $\hbar$ and c confusing should consult App. A. For brevity, we call $f(0)$ the forward scattering amplitude. Now the center of mass differential cross section at zero degrees is

$$\left(\frac{d\sigma}{d\Omega^*}\right)_{\theta=0} = |f(0)|^2 = [\text{Im}\, f(0)]^2 + [\text{Re}\, f(0)]^2 \tag{1-11}$$

We define the ratio

$$\alpha = \frac{\text{Re}\, f(0)}{\text{Im}\, f(0)} \tag{1-12}$$

Then using Eq. 1-10

$$\left(\frac{d\sigma}{d\Omega^*}\right)_{\theta=0} = \left[\frac{p^*\sigma_{\text{tot}}}{4\pi}\right]^2 [1 + \alpha^2] \tag{1-13}$$

Changing to the differential cross section $d\sigma/dt$ by means of Eq. 1-7, we obtain

$$\left(\frac{d\sigma}{dt}\right)_{t=0} = \left[\frac{\sigma_{\text{tot}}^2}{16\pi}\right] [1 + \alpha^2] \tag{1-14}$$

Values of α for the *pp* system are given in Fig. 1-10. Above 10 GeV/c, α lies between -0.3 and 0.0 and at least for the *pp* system is decreasing in magnitude as the energy increases. The data are not nearly as complete for the πp, Kp, and $\bar{p}p$ systems, but all measurements agree, with the $|\alpha|$ of these systems also being less than 0.3 above 10 GeV/c. Therefore, the forward elastic scattering amplitude is mostly imaginary above 10 GeV/c for all measured hadron-hadron systems. Since diffractive elastic scattering in its simplest formulation has a purely imaginary amplitude (see Ch. 2), this experimental fact is in agreement with one of the basic requirements of the diffraction model of elastic scattering.

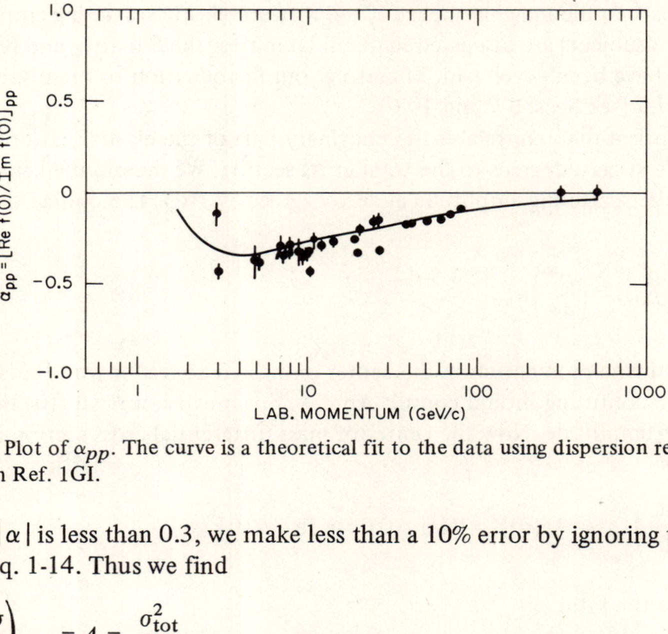

Fig. 1-10 Plot of α_{pp}. The curve is a theoretical fit to the data using dispersion relations. Taken from Ref. 1GI.

Since $|\alpha|$ is less than 0.3, we make less than a 10% error by ignoring the α term in Eq. 1-14. Thus we find

$$\left(\frac{d\sigma}{dt}\right)_{t=0} = A = \frac{\sigma_{\text{tot}}^2}{16\pi} \tag{1-15}$$

where the A parameter comes from Eq. 1-9. Now we have an immediate explanation of the observation made in Sec. 1-5 of the weak energy independence of not only the shape but also the magnitude of the small t differential cross section. The $t = 0$ differential cross section is proportional to the square of the total cross section with an approximation whose error is less than 10%. Since experimentally the total cross sections at high energy vary very slowly with energy, the same must be true for the $t = 0$ differential cross. Finally, since b also varies either slowly or not at all with energy, the entire small t diffraction peak only changes slowly with energy.

At last we write

$$\frac{d\sigma}{dt}(s, t) = \frac{\sigma_{tot}^2(s)}{16\pi} e^{-b|t|} \tag{1-16}$$

Here we have purposely emphasized that σ_{tot} is a function only of s. If we integrate Eq. 1-16 from $t = 0$ to $t = -\infty$, we only make a few percentage points of error at high energies in calculating the elastic cross section σ_{el}. Thus

$$\sigma_{el}(s) = \frac{\sigma_{tot}^2(s)}{16\pi b} \tag{1-17}$$

and the total inelastic cross section is

$$\sigma_{inel}(s) = \sigma_{tot}(s) - \sigma_{el}(s) \tag{1-18}$$

Inserting numerical values, we find

$$\sigma_{el}(s) = \frac{0.051\sigma_{tot}^2(s)}{b(s)} \tag{1-19}$$

where all the cross sections are in mb and b is in $(\text{GeV}/c)^{-2}$. Thus the πp system with a σ_{tot} of about 25 mb and a b of about 8 $(\text{GeV}/c)^{-2}$ has an elastic cross section of 4 mb and a total inelastic cross section of 21 mb.

Exercise 1-3. Derive Eq. 1-19 to ensure familiarity with the so-called elementary particle units used in this book (App. A).

Total elastic cross sections appear in Fig. 1-11. From Fig. 1-11 and from Eqs. 1-17 to 1-19, it is easy to describe the high energy behavior of σ_{inel}. For

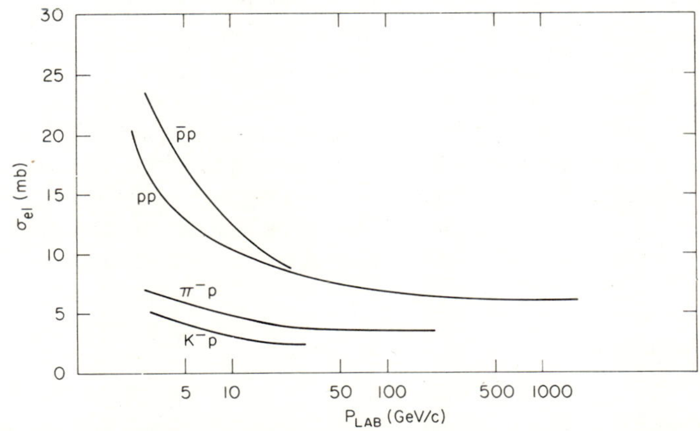

Fig. 1-11 Plot of σ_{el}, the total elastic cross section for the indicated system. Taken from Ref. 1GI.

example, at high energy σ_{tot} is made up mostly of σ_{inel}, and the change of σ_{inel} with energy will follow closely the change of σ_{tot} with energy. Another example: if b increases with energy while σ_{tot} remains constant, σ_{el} will decrease and σ_{inel} will increase. In fact, if b increases indefinitely with energy while σ_{tot} remains constant, σ_{el} will go to zero and σ_{inel} will become equal to σ_{tot}.

Other types of behavior can equally well be imagined, and Eqs. 1-16 to 1-18 provide part of the foundation for speculations on the behavior of these cross sections at very high energies. These speculations must also make use of predictions about the very high energy behavior of σ_{tot} and b. Such speculations also usually apply to what is called an asymptotic energy region. This is an energy region with no upper limit where (1) the simple quantities that describe the strong interactions—quantities like the cross sections and the exponential slope—have simple energy behaviors and (2) no new general types of phenomena occur in the strong interactions.

We close this chapter with two examples of speculations about the asymptotic energy region. First, returning to Fig. 1-1, we observe that $\sigma_{tot,pp}$ is increasing in the highest energy range. We are faced with speculative questions. Is this a temporary increase, with $\sigma_{tot,pp}$ becoming a constant at yet higher energies? Or is this an indication that $\sigma_{tot,pp}$ will continue to increase slowly with energy, perhaps as ln (s)? These questions are speculative, because we cannot obtain an experimental answer from existing accelerators, and it is very difficult to get a decisive answer from cosmic ray measurements (1HA).

Another speculation concerns the requirement for aymptotic behavior. Asymptotic behavior might occur if there is *no* physical entity or parameter that sets a *new energy scale* at very high energies. Such an entity or parameter would have to have the units of energy or mass or inverse length; of course it might be a function of several parameters. For example, the phenomena of atomic physics occur when the energy exceeds the binding energy of the atomic electron; nuclear phenomena occur when the nuclear binding energies are exceeded; and elementary particle physics phenomena occur when the energy exceeds the masses of the elementary particles. Then, if in elementary particle physics there are binding energies or masses or inverse lengths of the order of tens or hundreds of GeV in magnitude, we would not expect to have yet reached the asymptotic energy region. And if such parameters occur at all energies and play an intimate role in elementary particle dynamics, there may never be an asymptotic region.

REFERENCES

1BL *Proceedings of the Eighth International Conference on High-Energy Accelerators*, M. Hildred Blewett, Ed. (CERN, Geneva, 1971). There are of course many other books and articles on high energy accelerators; this reference provides a recent entrance to that literature.

1BU W. Busza et al., *Nuovo Cimento* **52A**, 331 (1967).

1CE R. J. Cence, *Pion-Nucleon Scattering* (Copyright © by Princeton University Press, Princeton, N.J., 1969). Figure 1-2 reprinted by permission of Princeton University Press.

1CO C. T. Coffin et al., *Phys. Rev. Lett.* **15**, 838 (1965).

1CR P. Cronillon et al., *Phys. Rev. Lett.* **30**, 403 (1973).

1GI G. Giacomelli, in *Proceedings of the Sixteenth International Conference on High Energy Physics, 1972* (National Accelerator Laboratory, Batavia, Ill., 1972).

1GIL F. J. Gilman, in *Proceedings of the Fourth International Symposium on Electron and Photon Interactions at High Energy, 1969* (Daresbury Nuclear Physics Laboratory, Daresbury, England, 1969).

1HA S. Hayakawa, *Cosmic Ray Physics* (Wiley, New York, 1969).

1HE W. N. Hess, *Rev. Mod. Phys.* **30**, 368 (1958).

1OR J. Orear et al., *Phys. Lett.* **28B**, 61 (1968).

1PE A general reference to colliding beam accelerators is C. Pellegrini, *Ann. Rev. Nucl. Sci.* **22**, 1 (1972).

1RU D. R. Rust et al., *Phys. Rev. Lett.* **24**, 1361 (1970).

CHAPTER

2

The Optical Model in the Born and Eikonal Approximations

2-1 Introduction. Since the diffraction peak occurs in all hadronic systems in the high momentum range, it cannot depend on the precise nature of the hadrons, in particular, it cannot depend on the spin of the particles. Of course the very detailed behavior of the diffraction peak, such as the existence of dips in $d\sigma/dt$ at the large $|t|$ edge of the region, will be dependent on the hadrons involved in the scattering. We are concerned here with the more general behavior and with the exponential nature of the diffraction peak. Therefore, we can simplify our work considerably by ignoring all the internal quantum numbers of the interacting particles. Specifically, we take the particles to have spin zero.

In Secs. 2-2 to 2-5 we construct a mathematical formulation of diffractive elastic scattering using the Born approximation. We then apply that formulation to the hadronic systems discussed in Ch. 1. In Sec. 2-6, using reasonable physical parameters, we reproduce all the general features of the small angle, high energy, elastic scattering in those hadronic systems.

Then an improved formulation of the optical model—the eikonal approximation—is constructed and discussed. The eikonal approximation partially takes account of multiple interactions in the elastic process. This not only yields a more realistic theory for small angle, elastic scattering but also allows extension of the optical model to larger angles.

2-2 Review of the Use of the Born Approximation in Nonrelativistic Scattering. As we do repeatedly in this book, we begin here by using nonrelativistic quantum mechanics. Consider two particles, a and b (masses m_a and m_b) with a mutual potential $V(\mathbf{r})$, $\mathbf{r}$ being the relative coordinate between the two particles. We start from the Schrödinger equation in the center of mass system. For spinless particles we have

$$[\nabla^2 + k^2]\,\psi(\mathbf{r}) = U(\mathbf{r})\psi(\mathbf{r}) \tag{2-1}$$

Here $U(\mathbf{r}) = 2mV(\mathbf{r})$, $k^2 = 2mE$, m is the reduced mass $m_a m_b/(m_a + m_b)$, E is the total kinetic energy of the system, and the initial and final three-momentum of particle a are $\mathbf{k}$ and $\mathbf{k}'$ respectively. To simplify our considerations, we assume that the potential has two properties

1. The potential falls off more rapidly than $1/r$ as r increases.
2. The potential has cylindrical symmetry about the line of motion of the incident particles.

With these assumptions, the scattering of an incident plane wave can be described by the approximate wave function

$$\psi(\mathbf{r}) = \psi_{\text{inc}}(\mathbf{r}) + \psi_{\text{scat}}(\mathbf{r}) \tag{2-2a}$$

where

$$\psi_{\text{inc}}(\mathbf{r}) = e^{i\mathbf{k}\cdot\mathbf{r}} = e^{ikz} \tag{2-2b}$$

and in the far field limit [r much larger than the dimensions of the potential $V(r)$]

$$\psi_{\text{scat}}(\mathbf{r}) = f(\theta)\frac{e^{ikr}}{r} \tag{2-3}$$

The coordinate z is along the $\mathbf{k}$ direction and θ is the center of mass scattering angle between $\mathbf{k}'$ and $\mathbf{k}$. The elastic scattering amplitude $f(\theta)$ is related to the differential cross section in the center of mass by

$$\frac{d\sigma}{d\Omega^*} = |f(\theta)|^2$$

As is discussed in many quantum mechanics or scattering theory text books (2SC, 2ME, 2LA, 2RO), the formal solution for $\psi_{\text{scat}}(\mathbf{r})$ is

$$\psi_{\text{scat}}(\mathbf{r}) = -\frac{1}{4\pi}\int d^3r' \left[U(\mathbf{r}')\psi(\mathbf{r}')\frac{e^{ikR}}{R}\right] \tag{2-4}$$

As shown in Fig. 2-1, $\mathbf{R} = \mathbf{r} - \mathbf{r}'$.

To use Eq. 2-4, an approximation must be made to $\psi(\mathbf{r})$. A simple approximation—the first Born approximation—consists simply of replacing $\psi(\mathbf{r})$ by

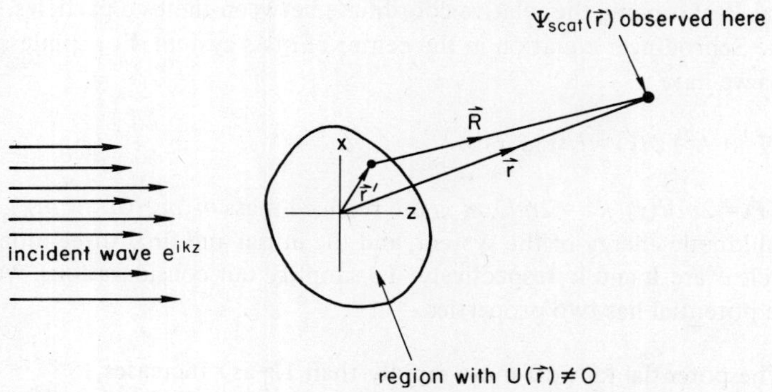

Fig. 2-1 Scattering of a plane wave by a potential.

$\psi_{\text{inc}}(\mathbf{r}) = e^{ikz}$. Thus the first Born approximation reduces Eq. 2-4 to the form

$$\psi_{\text{scat}}(\mathbf{r}) = -\frac{1}{4\pi}\int d^3r' \left[U(\mathbf{r}')\, e^{ikz'}\, \frac{e^{ikR}}{R}\right] \tag{2-5}$$

The physical meaning of the first Born approximation is illustrated in Fig. 2-1. The $\psi_{\text{scat}}(\mathbf{r})$ is simply the summation of all the spherical waves e^{ikR}/R coming from an extended source.

The validity in high energy scattering of the first Born approximation depends on two conditions. First the potential must be sufficiently "weak" so that higher order Born approximations can be neglected. In nonrelativistic quantum mechanics, if the potential has a limited range of length a, $Va/\hbar v$ should be much less than 1, where v is the velocity of the particle. We discuss this condition later when we connect the eikonal approximation with the first Born approximation. The physical significance of using just the first Born approximation is our assumption that the initial wave $\psi_{\text{inc}}(\mathbf{r})$ scatters just once on the potential and then leaves the potential region. Higher Born approximations correspond to two or more scatters on the potential.

Another condition on the validity of the first Born approximation which is particularly relevant in elementary particle physics is our assumption that the potential does not change its form during the scattering. We have written $V(\mathbf{r})$, and the absence of a time coordinate explicitly means that $V(\mathbf{r})$ does not change during the scattering process. Now in elementary particle scattering, the potential must certainly represent the force between the particles and the structures of the particles in the scattering process. By making $V(\mathbf{r})$ time independent, we have said that we can either disregard or take the time average of the variations of the force and structure during the scattering period. We are using here a form of the *impulse approximation* (2RO). We shall use the impulse approximation again in this book.

Our major concern now is to obtain a relativistic formulation of the first Born approximation, because we must in the end deal with the scattering of relativistic particles. As the reader may expect, as we become more relativistic we lose the clear meaning that a potential has in nonrelativistic theory.

2-3 The Optical Analogy. Before we become too depressed about the application of the first Born approximation to our problem of relativistic particle elastic scattering, let us obtain the same equation from a different branch of physics. Consider the scattering of a plane light wave passing through and around a disk (Fig. 2-2). Suppose the disk has radius b_0 and has a cylindrically symmetric absorption $\beta(b)$, b being the radius. Here $\beta(b)$ represents the *absorption coefficient* for the *amplitude* of the wave. The *amplitude transmission* coefficient is $1 - \beta(b)$. We require $0 \leqslant \beta(b) \leqslant 1$ and for a completely absorbing disk, $\beta(b) = 1$. The plane wave is moving in the $+z$ direction and the disk is in the $z = 0$ plane. Then the Huygens principle tells us that the amplitude at point $(\mathbf{r})$ can be found by regarding each point on the plane $z = 0$ as the source of a spherical wave. The

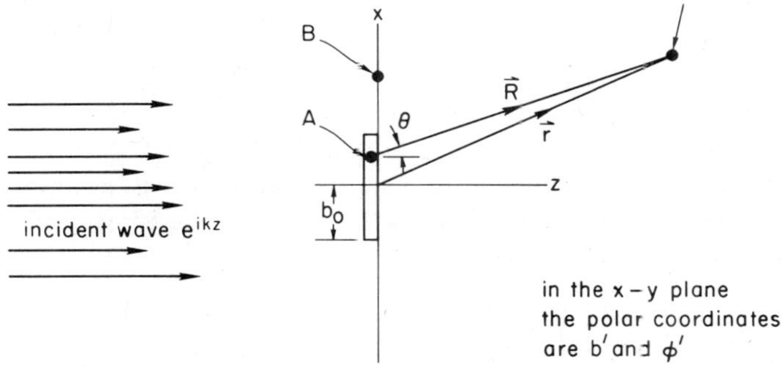

Fig. 2-2 Scattering of a plane wave by an absorbing disk.

amplitude of the spherical wave is proportional to the amplitude of the incident wave at the $z = 0$ plane, taking into account the absorption. The spherical waves from all these sources must be summed together; this includes sources off the disk such as point B in Fig. 2-2 as well as points on the disk such as point A. Equation 2-6 is the quantitative description of this summation process.

$$\psi(\mathbf{r}) = \frac{-ik}{4\pi}\left[\int_0^{2\pi}\int_{b=0}^{b=b_0} d\phi' b'\, db'(1+\cos\theta')(1-\beta(b'))\frac{e^{ikR}}{R} + \int_0^{2\pi}\int_{b=b_0}^{b=0} d\phi' b'\, db'(1+\cos\theta')(1)\frac{e^{ikR}}{R}\right] \tag{2-6}$$

Here θ' is the angle between the normal to the disk and the direction of the vector $\mathbf{R}$; the factor $(1 + \cos\theta')$ sometimes does not appear in this optical formula, $\cos\theta'$ having been set equal to 1 for small angles. Indeed we shall soon do just that. The first integral is over the disk and the second is outside the disk. We note that taking the disk to be at $z = 0$ made the incident wave amplitude at the plane e^{ikz} reduce to just 1. Now if we want the *scattered* wave at $(\mathbf{r})$, we must subtract the incident wave at $(\mathbf{r})$ given by

$$\psi_{\text{inc}}(\mathbf{r}) = \frac{-ik}{4\pi}\int_0^{2\pi}\int_{b=0}^{b=\infty} d\phi' b'\, db'(1+\cos\theta')(1)\frac{e^{ikR}}{R} \tag{2-7}$$

Doing it this way removes the annoying problem of the limit at $b = \infty$. Subtracting Eq. 2-7 from 2-6, we find

$$\psi_{\text{scat}}(\mathbf{r}) = \frac{-ik}{4\pi}\int_0^{2\pi}\int_{b=0}^{b=b_0} d\phi' b'\, db'(1+\cos\theta')(-\beta(b'))\frac{e^{ikR}}{R} \tag{2-8}$$

Now we make the first of many small angle approximations, a convenient but sometimes dangerous habit in high energy diffraction theory. For small angles we set $\cos\theta' = 1$, and Eq. 2-8 becomes

$$\psi_{\text{scat}}(\mathbf{r}) = \frac{-1}{4\pi}\int_0^{2\pi}\int_{b=0}^{b=b_0} d\phi' b' db' \left[-2ik\beta(b')\frac{e^{ikR}}{R}\right] \tag{2-9}$$

Exercise 2-1. Justify the appearance of the factor k as part of the normalization term in the equations in this section.

Equation 2-9 only differs from the first Born approximation, Eq. 2-5, in that the former uses a two-dimensional source (a disk) and the latter uses a three-dimensional source. To make the connection between the two different kinds of sources, let us give the disk a nonzero thickness. Let the *absorption per unit length*, along the z axis, be $\beta_L(\mathbf{r})$ for this thick disc. Then Eq. 2-9 has the form

$$\psi_{\text{scat}}(\mathbf{r}) = \frac{-1}{4\pi} \int d^3r' \left[-2ik\beta_L(\mathbf{r}')\, e^{ikz'} \frac{e^{ikR}}{R} \right] \tag{2-10}$$

Exercise 2-2. Justify Eq. 2-10.

Comparing Eq. 2-10 with Eq. 2-5, we see that for a quantum mechanical potential to act in the scattering process in a manner analogous to the action of an absorbing disk in optical scattering, $U(\mathbf{r})$ must be of the form

$$U_{\text{abs}}(\mathbf{r}) = -2ik\beta_L(\mathbf{r}) \tag{2-11}$$

where $\beta_L(\mathbf{r})$ is real and greater than zero. This idea—that a negative, purely imaginary potential acts as an absorptive potential—is used extensively in our discussion of elastic scattering.

Of course, Eqs. 2-10 and 2-5 are so similar because we have been using the same wave physics. But the two different points of view teach us two useful things. First, our solution is more relativistic than we might have thought at first, because light after all is a relativistic wave. Second, we are not restricted to the use of real potentials; we can also use imaginary, absorptive potentials. Thus Eq. 2-5 is our primary equation for the first Born approximation, *and $U(\mathbf{r})$ is* allowed to be complex.

It is conventional to simplify Eq. 2-5. As shown in Fig. 2-1, $\mathbf{R} = \mathbf{r} - \mathbf{r}'$ and for r much greater than r', $R \approx r - \mathbf{r}\cdot\mathbf{r}'/r$. Remembering that $\mathbf{k}$ is the momentum in the $+z$ direction and that $\mathbf{k}'$ is the momentum in the $\mathbf{r}$ direction, we find that the Born approximation Eq. 2-5 becomes

$$\psi_{\text{scat}}(\mathbf{r}) = -\frac{e^{ikr}}{4\pi r} \int d^3r' \, [U(\mathbf{r}')\, e^{i(\mathbf{k}-\mathbf{k}')\cdot\mathbf{r}'}] \tag{2-12}$$

Comparing with Eq. 2-3, we find the well-known result

$$f(\theta) = -\frac{1}{4\pi} \int d^3r' [U(\mathbf{r}')\, e^{i(\mathbf{k}-\mathbf{k}')\cdot\mathbf{r}'}] \tag{2-13}$$

This is the basic equation for $f(\theta)$ in the first Born approximation to the optical model.

2-4 Fourier Transform Interpretation of Elastic Scattering. To develop the Fourier transform interpretation of small angle elastic scattering, we first denote by $\mathbf{q}$ the three-momentum transferred from one particle to the other particle in the scattering process. (We define only for this section $q = |\mathbf{q}|$.) In the center of mass, we have

$$k = |\mathbf{k}| = |\mathbf{k}'|, \qquad \mathbf{q} = \mathbf{k} - \mathbf{k}', \qquad q^2 = |\mathbf{q}|^2 = 2k^2(1 - \cos\theta)$$

$$= 4k^2 \sin^2\left(\frac{\theta}{2}\right) \tag{2-14a}$$

For small angles, this reduces to the simple form

$$q \approx k\theta \tag{2-14b}$$

We can also think of $f(\theta)$ as a function of q; thus Eq. 2-13 becomes

$$f(\mathbf{q}) = -\frac{1}{4\pi}\int d^3r' \, [U(\mathbf{r}')\, e^{i\mathbf{q}\cdot\mathbf{r}'}] \tag{2-15}$$

We have written the last equation in a form to emphasize that $f(\mathbf{q})$ is the Fourier transform of $U(\mathbf{r})$. Thus, as first mentioned in Ch. 1, the elastic scattering amplitude provides a way to examine the structure of the interaction. As a simple example, we supposed that $U(\mathbf{r})$ consist only of a thin, spherically symmetric shell of the form

$$U(\mathbf{r}) = U_0\delta(r - r_0) \tag{2-16}$$

That is, we are considering the scattering from only the surface of a sphere. Then

$$f_{\text{shell}}(\mathbf{q}) = -\frac{r_0 U_0}{q}\left[\sin qr_0\right] \tag{2-17}$$

Thus the scattering from a shell leads to a wave with a period in q of $2\pi/r_0$ or a period in θ of $\approx 2\pi/(kr_0)$. The q in the denominator damps out the scattering at large angles.

Exercise 2-3. Consider the slightly more complicated case of the scattering from a sphere of uniform density, that is

$$U(\mathbf{r}) = U_0', \qquad r \leqslant r_0$$

$$U(\mathbf{r}) = 0, \qquad r > r_0 \tag{2-18}$$

Show that

$$f_{\text{sphere}}(\mathbf{q}) = -\frac{U_0'}{q^3}\left[\sin qr_0 - qr_0 \cos qr_0\right] \tag{2-19}$$

Finally, verify the expansions in Eqs. 2-20 and 2-21.

For small angles we can expand Eq. 2-17, and we find

$$f_{\text{shell}}(\mathbf{q}) = -r_0^2 U_0 \left[1 - \frac{(kr_0\theta)^2}{6} + \cdots\right] \tag{2-20}$$

Similarly, expanding Eq. 2-19, we obtain

$$f_{\text{sphere}}(\mathbf{q}) = \frac{-r_0^3 U_0'}{3}\left[1 - \frac{(kr_0\theta)^2}{10} + \cdots\right] \tag{2-21}$$

We see that the small angle behavior of the scattering from a spherical shell is not very different from the small angle scattering from a solid sphere. This then is the explanation for the statement made in Ch.1 that small angle scattering tells us little about the internal structure of an interaction. Even structures as different as a spherical shell and a solid sphere give rather similar small angle behavior. This is a basic property of the three-dimensional Fourier transform and is not therefore peculiar to the approximations we are using in this chapter.

2-5 The Relativistic Problem. There is no way to rigorously and directly extend the methods of the last three sections into relativistic quantum mechanics. The fundamental problem is that we have no way to include the production or annihilation of particles in the optical model; yet such processes must occur when the kinetic or potential energies are larger than the particle masses. Of course the correct framework to use is quantum field theory. But field theory has not yet led to a really satisfactory theory of elastic scattering; if it had, we would not still be using the optical model! Therefore, we shall proceed relying on analogy and physical insight.

As an example of how one can treat the relativistic problem in a semiphenomenological way, we first briefly describe a method developed by Serber (2SE) and others. They adopt the intuitive approach of assuming a four-potential (V_0, $\mathbf{V}$) which in the center of mass system has $\mathbf{V} = 0$. If both particles have the same mass, such as in proton–proton elastic scattering, both particles have the same center of mass energy E. Then in analogy to the Klein–Gordon equation, we write

$$[\mathbf{p}^2 + m^2 - [E - V_0(\mathbf{r})]^2]\psi(\mathbf{r}) = 0 \tag{2-22}$$

where $\mathbf{p}$ is the three-momentum operator and m is the mass of either particle. Eq. 2-22 can then be rewritten in the form

$$[\mathbf{p}^2 + m^2 - E^2]\,\psi(\mathbf{r}) = [-2EV_0(\mathbf{r}) + V_0(\mathbf{r})^2]\,\psi(\mathbf{r})$$

Replacing $\mathbf{p}$ by $-i\nabla$ and using $\mathbf{k}$ for the c-number three-momentum of either particle in the center of mass, we obtain the wave equation

$$[\nabla^2 + k^2]\,\psi(\mathbf{r}) = [2EV_0(\mathbf{r}) - V_0(\mathbf{r})^2]\,\psi(\mathbf{r}) \tag{2-23}$$

If we assume $E \gg V_0(\mathbf{r})$ for all $(\mathbf{r})$, Eq. 2-23 reduces to

$$[\nabla^2 + k^2]\,\psi(\mathbf{r}) = 2EV_0(\mathbf{r})\,\psi(\mathbf{r}) \tag{2-24}$$

which is a relativistic analog to Eq. 2-1.

Yet another way to treat the relativistic problem is to develop the absorption concept from the optical analogy. This development is probably the most satisfying physically because we can ignore the problem of making Lorentz-covariant potentials. We suppose that each particle i consists of a distribution of matter whose distribution amplitude is given by a function $\rho_i(\mathbf{r})$ in its rest system. When the particles pass one another there is a mutual absorption that depends on the $\rho_i(\mathbf{r})$ of each particle and on the distance $\mathbf{r}$ between the particles. We consider the particles in the center of mass system, and take them to be initially moving in the $+z$ and $-z$ directions. As the scattering occurs, the particles will move in arcs and will no longer move exactly in the $+z$ and $-z$ directions, but we now ignore this condition. That is, we assume for the purposes of calculating the mutual absorption that the particles continue to move exactly in the $+z$ and $-z$ directions. This approximation is obviously good only for small angles; thus this is a small angle theory. Finally, for fixed $\mathbf{r}$, we integrate over the spatial extensions of the distributions to obtain a $\beta_L(\mathbf{r})$ analogous to that used in Sec. 2-3. Of course if the particles are relativistic, the Lorentz contraction must be applied to their individual distributions before the integration. This is easily done, but for simplicity we shall not do this in this section. We may also allow the presence of a real potential $V(\mathbf{r})$.

We now make the physically reasonable assumption that we simply use Eq. 2-13 as it stands, except that the particles are now relativistic. We give the equation a new number because it is now relativistic and

$$U(\mathbf{r}) = 2kV(\mathbf{r}) - 2ik\beta_L(\mathbf{r})$$

$$V(\mathbf{r}) \text{ real,} \qquad \beta_L(\mathbf{r}) \qquad \text{real} \qquad \text{and} \qquad \geqslant 0 \tag{2-25}$$

The imaginary term is justified by the optical analogy of Sec. 2-3. The presence of $2k$ in the real term arises by analogy to Eq. 2-24 or by the argument that in going from a nonrelativistic to a relativistic situation, one replaces the mass m by the total particle energy E. Either paths leads to $2EV(\mathbf{r})$, but in the high energy region we can replace E by the momentum k. Such changes are not important because in a sense $V(\mathbf{r})$, as well as $\beta_L(\mathbf{r})$, are defined by the application of Eq. 2-25.

2-6 Elastic Scattering Models Using the First Born Approximation. A simple model for elastic scattering consists of replacing each particle by a disk with a constant absorption coefficient. Thus let particle a be represented by a disk of radius b_a and let particle b be replaced by a disk of radius b_b. Now when the distance between the centers called b is greater than $b_a + b_b$, there will be no absorption. When b is less than $|b_a - b_b|$, there will be maximum absorption. For intermediate values of b there will be intermediate amounts of absorption. Thus even this simple model gives a slightly complicated $U(b)$. It is pedagogically desirable that we start with even a simpler model. Therefore we make the unrealistic assumption (for hadrons) that one of the particles, say particle a, has zero radius. Then we have

$$U(b, z) = -ik2\beta\delta(z) \qquad b \leqslant R$$

$$U(b, z) = 0, \qquad b > R \tag{2-26}$$

where R is the fixed radius of the disk and $0 \leqslant \beta \leqslant 1$.

Using these conditions and carrying out the z' integration, noting that $\mathbf{k} \cdot \mathbf{r}' = kz'$, and defining the two-dimensional vector $\mathbf{b}$ in the xy plane:

$$f(\theta) = \frac{ik\beta}{2\pi} \int_0^{2\pi} d\phi' \int_0^R b'\, db'\, e^{-i\mathbf{k}' \cdot \mathbf{b}'} \tag{2-27}$$

Without any loss of generality, we can put $\mathbf{k}'$ in the xz plane. Then $\mathbf{k}' \cdot \mathbf{b}' = kb' \sin\theta \cos\phi'$, where ϕ' is the angle between the x axis and the vector $\mathbf{b}'$. The θ is the angle between $\mathbf{k}'$ and the z axis, thus is the scattering angle. Using the integrals of App. B, we write

$$f(\theta) = ik\beta \int_0^R b'db'\, J_0[b'k \sin\theta] = ik\beta R^2 \left[\frac{J_1(Rk \sin\theta)}{Rk \sin\theta}\right] \tag{2-28}$$

where J_0 and J_1 are the Bessel functions of the first kind. Now $(\mathbf{k} - \mathbf{k}')$ is the three-momentum transferred in the collision. And since this is the center of mass system, we have the exact relationship $t = -(\mathbf{k} - \mathbf{k}')^2$. Also for small angles, we have

$$\sqrt{-t} \approx k \sin\theta \tag{2-29}$$

Exercise 2-4. The presence of the Bessel function in Eq. 2-28 is caused by the disk configuration. Compare Eq. 2-28 with Eq. 2-19 at small angles and show that the small angle elastic amplitude is similar. (Equation 2-19 was derived for the case of a uniformly absorbing sphere.)

Figure 2-3 is a plot of $f(\theta)$ versus $Rk \sin\theta$ using Eq. 2-28. We see a strong forward peak much like that which occurs in high energy elastic scattering. We also have oscillations or dips at the larger angles.

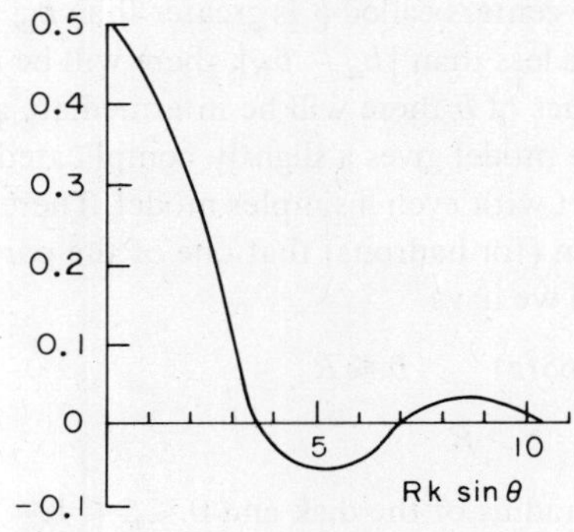

Fig. 2-3 The elastic scattering amplitude for an absorbing disk.

But let us be more quantitative. Since $J_1(z) = (z/2) - (z/2)^3/2! + \cdots$ (App. B), for sufficiently small angles ($\sin\theta \approx \theta$), we have

$$f(\theta) = \frac{ik\beta R^2}{2}\left[1 - \frac{(Rk\theta)^2}{8}\right] \tag{2-30}$$

As discussed in Ch. 1, we know that at small angles we represent the differential cross section by $d\sigma/dt = A\, e^{-b|t|}$. Then

$$A = \frac{\pi}{k^2}|f(0)|^2 \tag{2-31a}$$

$$f(\theta) = f(0)\exp\left[-bk^2(1-\cos\theta)\right] \approx f(0)\left[1 - \frac{bk^2\theta^2}{2}\right] \tag{2-31b}$$

Comparing Eqs. 2-30 and 2-31 we see that

1. $f(\theta)$ will be purely imaginary (2-31c)
2. $f(0) = ik\beta R^2/2$ (2-31d)

3. $A = \pi\beta^2 R^4/4$ and is independent of the energy (2-31e)
4. $R^2 = 4b$ (2-31f)

Let us see how good this model is. From experimental data, $b = 8\ (\text{GeV}/c)^{-2}$. Therefore, $R = \sqrt{32}\,\hbar c = 1.1 \times 10^{-13}$ cm. From nuclear physics we know that this is a very reasonable number for the radius of nucleon–nucleon interactions. We can also see that the angular behavior is correct up to terms in θ^2 because Eqs. 2-30 and 2-31b have the same functional form.

Exercise 2-5. Compare the amplitudes given by Eq. 2-28 and the exact form of Eq. 2-31b over a range of values extending a factor of 10 down from their 0° values.

From Eq. 2-28 we have

$$\frac{d\sigma}{dt} = \pi\beta^2 R^4 \left[\frac{J_1(Rk\sin\theta)}{Rk\sin\theta}\right]^2 \approx \pi\beta^2 R^4 \left[\frac{J_1(R\sqrt{|t|})}{R\sqrt{|t|}}\right]^2 \tag{2-32}$$

The total *elastic* cross section σ_{el} is obtained by the integration over all angles of this differential cross section, which is carried out approximately in Ex. 2-6. We find

$$\sigma_{el} \approx \pi\beta^2 R^2 \tag{2-33}$$

Exercise 2-6. Using a numerical or approximate integral method, derive Eq. 2-33.

The total elastic scattering cross section σ_{el} is caused by the *diffraction* of the incident wave around the absorbing disk; thus σ_{el} represents diffractive elastic scattering, commonly called diffractive scattering. The diffractive scattering amplitude is always purely imaginary, as pointed out previously. We may of course also have nondiffractive elastic scattering coming from the equivalent of a real potential—the diffractive scattering coming from the equivalent of an imaginary potential.

What happens to the part of the incident wave that is absorbed by the disk? In particle scattering the flux represents energy, momentum, and other quantities (e.g., charge and baryon number), which must be conserved. Therefore, the absorbed flux goes into inelastic reactions with a cross section σ_{inel}.

Exercise 2-7. Show that $\sigma_{inel} \approx (2\beta - \beta^2)\pi R^2$ by directly calculating the absorbed flux.

Using

$$\sigma_{\text{inel}} \approx (2\beta - \beta^2)\pi R^2 \tag{2-34}$$

and noting that the total cross section σ_{tot} is given by $\sigma_{\text{tot}} = \sigma_{\text{el}} + \sigma_{\text{inel}}$ we find

$$\sigma_{\text{tot}} \approx 2\beta\pi R^2 \tag{2-35}$$

If the absorption is complete, $\beta = 1$ and we have $\sigma_{\text{tot}} \approx 2\pi R^2$. This famous result says that for a completely absorbing disk, the total cross section is twice the geometric cross section; half the total cross section comes from the absorption and half from the diffractive elastic scattering.

We have calculated σ_{tot} here by a roundabout but physically obvious method. To shorten the work in the future, however, we will use the optical theorem $\text{Im}\, f(0) = (k/4\pi)\sigma_{\text{tot}}$, as discussed in Sec. 1-7 and Ch. 6. From Eq. 2-30, $\text{Im}\, f(0) = k\beta R^2/2$. Inserting this in the optical theorem immediately yields Eq. 2-35 in its exact form, $\sigma_{\text{tot}} = 2\beta\pi R^2$.

In high energy scattering theory it is customary to use the quantity $1 - \alpha = \beta$. We then have the perhaps more familiar equations (α is the *transmission* coefficient)

$$\sigma_{\text{el}} \approx (1-\alpha)^2\pi R^2, \qquad \sigma_{\text{inel}} \approx (1-\alpha^2)\pi R^2, \qquad \sigma_{\text{tot}} = 2(1-\alpha)\pi R^2 \tag{2-36}$$

We found from the exponential slope of the differential cross section $R \approx 1.1 \times 10^{-13}$ cm in this model. Then the total cross section would be 76β mb. For pion on protons at high energy $\sigma_{\text{tot}} = 25$ mb. Therefore $\beta = 1 - \alpha = 0.33$. But we have one additional experimental number to check– σ_{el}, which is about 4 mb. From Eq. 2-33 we find $\pi\beta^2R^2 = 4.2$ mb. Thus using just the parameters β and R we have fitted the small angle elastic behavior, σ_{tot} and σ_{el}; and the values of the parameters used are quite reasonable.

Exercise 2-8. Carrying out the same type of analysis for pp and $\bar{p}p$ elastic scattering, determine the parameters and interpret the difference between the different systems. Use the data in Ch. 1.

Thus we see that the large forward peak in high energy elastic scattering can be said to be due primarily to diffractive scattering from an absorptive body, the absorption in turn representing the inelastic and nondiffractive elastic scattering. Considering the assumptions we have made and the crudeness of the model, this is a very satisfactory result.

Next we are going to improve the model by eliminating the unphysical concept of a sharp-edged disk of zero thickness. To do this we must specify the amplitudes for the matter distribution of each of the particles. A distribution that is both mathematically convenient and physically reasonable is the Gaussian function. For particle a we write

$$\rho_a(\mathbf{r}_a) = g_a\, e^{-r_a^2/R_a^2} \tag{2-37}$$

and we write a similar function for b, where g_b and R_b can differ from g_a and R_a. If we think of these two distributions passing through each other with a mutual absorption, we can fold these Gaussian distributions together to obtain a Gaussian expression for $U(\mathbf{r})$

$$U(\mathbf{r}) = -2ik\eta\, e^{-r^2/R^2} \tag{2-38}$$

Here we have ignored the Lorentz contraction of the distributions. We have explicitly exhibited the factors $-2ik$. By comparison with Eq. 2-25, we see that $\eta \exp(-r^2/R^2)$ is the general amplitude absorption coefficient $\beta_L(\mathbf{r})$. For a particle passing along the z axis, we define the special absorption coefficient β_0, where

$$\beta_0 = \left[\eta \int_{-\infty}^{+\infty} e^{-r^2/R^2}\, dz\right]_{x=0,\,y=0} = \eta\sqrt{\pi R^2} \tag{2-39}$$

For later use, we note that by definition

$$0 \leqslant \beta_0 \leqslant 1 \tag{2-40}$$

We now rewrite Eq. 2-38 in the form

$$U(\mathbf{r}) = -2ik\, \frac{\beta_0}{\sqrt{\pi R^2}}\, e^{-r^2/R^2} \tag{2-41}$$

Inserting this in Eq. 2-13, we obtain

$$f(\theta) = \frac{ik\beta_0 R^2}{2}\, e^{-(R^2/4)(\mathbf{k}-\mathbf{k}')^2} \tag{2-42}$$

which the reader probably expected because the Fourier transform of a Gaussian funtcion is a Gaussian function. Therefore, we find the very convenient expression

$$\frac{d\sigma}{dt} = \frac{\pi\beta_0^2 R^4}{4}\, e^{-(R^2/2)|t|} \tag{2-43}$$

Thus in the first Born approximation the exponential form for the diffraction peak comes from a Gaussian absorption density. Also, we observe that the differential cross section is free from dips and bumps of the type that occurred when we used a sharp-edged disk.

Exercise 2-9. Fill in the steps leading to Eq. 2-43.

This last observation is important. First it shows that by changing the absorption coefficient $\beta_L(\mathbf{r})$ as a function of r, we can change the larger angle part of the diffraction peak without changing significantly the small angle part of the peak. Therefore, we can fit any reasonable sort of diffraction peak behavior. Second, the observation shows that the shape of the diffraction peak provides information about the assumed distribution of matter in the particles (but see the discussion after Ex. 2-3). A smoothly decreasing diffraction peak comes from a smoothly decreasing matter distribution. A diffraction peak that drops through a series of dips and bumps comes from a matter distribution which decreases suddenly.

Returning to $d\sigma/dt$ in Eq. 2-43 and integrating from $|t| = 0$ to ∞, we find

$$\sigma_{\text{el}} = \frac{\beta_0{}^2\pi R^2}{2} \tag{2-44}$$

Equation 2-42 and the optical theorem Eq. 1-10 yield

$$\sigma_{\text{tot}} = \beta_0 2\pi R^2 \tag{2-45}$$

$$\sigma_{\text{inel}} = \left(2\beta_0 - \frac{\beta_0{}^2}{2}\right)\pi R^2 \tag{2-46}$$

Comparing Eq. 1-9 with Eq. 2-43, we see that

$$R^2 = 2b, \quad \text{Gaussian distribution} \tag{2-47}$$

Exercise 2-10. Compare the predictions of the Gaussian distribution with the predictions of the disk model given earlier in this section. Explain the difference between Eq. 2-31f and Eq. 2-47. What value of β_0 is needed to fit typical high energy elastic and total cross section data?

As we have already mentioned, one problem with this Gaussian model or indeed with any model in which we use a spherically symmetric function for $\beta_L(\mathbf{r})$ is the obligation to allow for the Lorentz contraction of the particle structure along the direction of motion. That is, we should not take the matter distribution to be independent of the particle velocity. The simple disk model was not subject to this problem because its thickness in the direction of motion was always taken to be zero. It is not difficult to allow for the Lorentz contraction in the Gaussian distribution, and this is done quite frequently in the literature. When the particles have very high energies compared with their masses, the Lorentz contraction will flatten the particles until they resemble disks. To simplify the calculation, we assume identical particles moving along the z axis, each with velocity β in the cms. To introduce the Lorentz contraction, we simply replace r^2 in Eq. 2-38 by the expression $x^2 + y^2 + \gamma^2 z^2$, where $\gamma = [1 - \beta^2]^{-1/2}$.

Exercise 2-11. Justify the foregoing replacement and carry out the first Born approximation calculation of the elastic and total cross sections. Discuss the effect of this replacement on the calculated quantities.

2-7 The Optical Model and Photon–Hadron Elastic Scattering. In this brief section we present an interesting application of the optical model to photon–hadron elastic scattering. The photon is not a hadron; therefore one might wonder how to think about small angle, high energy photon–proton elastic scattering. However, using the optical model and a few other fundamental but simple ideas, one can understand high energy photon–proton elastic scattering in terms of three types of elastic processes. First, there is the purely electromagnetic scattering given by the Klein–Nishina formula (2BJ). This formula, representing the lowest order electromagnetic interaction between the photon and the proton (taken as a Dirac particle) is

$$\left(\frac{d\sigma}{d\Omega}\right)_{KN} = \frac{\alpha^2}{2M^2}\left(\frac{k'}{k}\right)^2\left[\frac{k'}{k} + \frac{k}{k'} - \sin^2\theta_L\right] \tag{2-48}$$

All quantities are in the laboratory system; k and k' are the initial and final momenta of the photon, θ_L is laboratory scattering angle of the photon and M is the proton mass. Of course we should add a term for the anomalous magnetic moment of the proton, but we can use Eq. 2-48 to estimate the size of this differential cross section. For $t = 0$, $k' = k$ and $\theta = 0$; therefore, $d\sigma/d\Omega \leqslant \alpha^2/M^2$, or

$$\left(\frac{d\sigma}{dt}\right)_{KN} \leqslant \frac{0.074}{k^2}\frac{\mu b}{(\mathrm{GeV/c})^2} \tag{2-49}$$

where k is in GeV/c.

The second type of elastic process we must consider is the diffractive elastic scattering caused by the photon–proton total hadronic cross section. As shown in Fig. 1-1, this total cross section is about 0.11 mb at high energies. Applying the optical view, we argue that this total hadronic cross section is caused by an absorptive interaction. Now we do not precisely know what to take as the range $R(\gamma p)$ of this absorptive interaction. We know it should be of roughly the same range as the corresponding πp or pp absorptive interaction. But we do not know, for example, if $R(\gamma p)$ should equal $R(\pi p)$ or should differ from $R(\pi p)$. Thus we take $b(\gamma p)$ (using $R^2 = 4b$) as having a value between 0.5 $b(\pi p)$ and 1.5 $b(\pi p)$

[i.e., 4 to 12 $(\text{GeV}/c)^{-2}$]. Finally, neglecting any real part to the elastic amplitude, we write

$$\left(\frac{d\sigma}{dt}\right)_{\text{hadronic}} = \frac{\sigma_{\text{tot}}^2}{16\pi} e^{-b|t|} = 0.6\, e^{-b|t|} \frac{\mu\text{b}}{(\text{GeV}/c)^2} \tag{2-50}$$

Comparing Eq. 2-50 with Eq. 2-49 we see that at high momenta and for $|t|$ values up to several $(\text{GeV}/c)^2$ the hadronic scattering is several powers of 10 larger than the scattering given by the Klein–Nishina formula; hence the latter can be ignored.

The third type of elastic process to be considered is somewhat subtler. In the interactions of a photon with any charged particle, purely electromagnetic inelastic

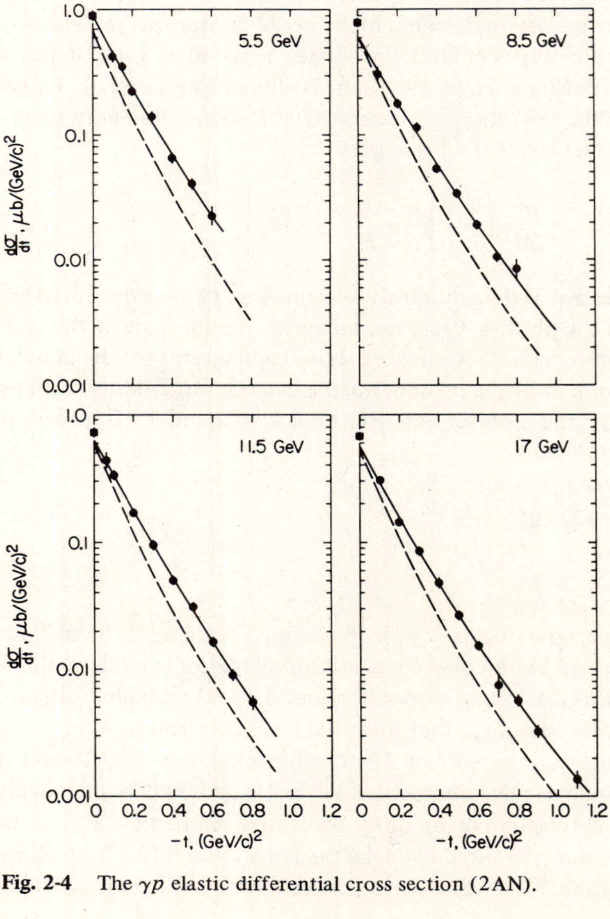

Fig. 2-4 The γp elastic differential cross section (2AN).

processes occur. These are processes like the production of electron–positron pairs. By the optical theorem these purely electromagnetic inelastic processes are also related to the imaginary part of the forward scattering amplitude, and by the optical model these processes lead to diffractive elastic scattering. Furthermore, these processes have total cross sections in the tens of millibarns. But these processes have nothing to do with the strong interactions. Therefore, the width of the diffraction peak from these processes has nothing to do with the range of the strong interactions or the size of the proton. In fact, these purely electromagnetic processes all produce particles in the very forward direction, the average transverse momentum being given by mc, where m is the mass of the electron and c is the velocity of light. The diffraction peak from these purely electromagnetic processes will also have a width given by $t = -m^2c^2$. This peak is extremely narrow compared with the hadronic diffraction and will be negligibly small at the t values at which we consider the latter peak.

Therefore, the high energy, photon–proton diffraction peak must be represented by Eq. 2-50, except at very small angles. Among the measurements in Fig. 2-4, we see a diffraction peak very similar to that found in hadron–hadron elastic scattering. Only the magnitude is much smaller, in agreement with Eq. 2-50. Using the simple exponential form for the peak and using $|t|$ values out to 0.6 $(\text{GeV}/c)^2$, b is found to be about 6 $(\text{GeV}/c)^{-2}$ for all the curves in Fig. 2-4.

2-8 The Glauber Derivation of the Eikonal Approximation. One of the principal difficulties with the first Born approximation is its failure to account for the possibility of several scatterings on the same potential. To put this difficulty another way, the incident wave is assumed to scatter at a single point in the potential, we then integrate that interaction point over the entire potential. But certainly the incident wave is affected on its passage through the potential toward that interaction point, and the scattered wave is affected by that potential after it leaves the interaction point. One obvious effect is simply the change in the momentum of the incident or scattered wave inside the potential as the potential increases in strength toward the center. The other main difficulty with the first Born approximation—that we never obtained a truly relativistic treatment—will remain with our derivation of the eikonal model, since we are not doing field theory or S-matrix theory.

We are following the derivation given by Glauber (2GL) in a series of lectures. For a very full explanation of this method and its limitations, those lectures, which are relatively easy to read, should be consulted. Glauber carries out the derivation for the Schrödinger equation, but we can follow through the same derivation using the relativistic wave equation

$$[\nabla^2 + k^2]\,\psi(\mathbf{r}) = U(\mathbf{r})\psi(\mathbf{r}) \tag{2-51}$$

where $U(\mathbf{r})$ has the meaning given in Eq. 2-25. We use the words "follow through the derivation" purposely, because as we have explained, we cannot treat the relativistic problem rigorously by this method.

Looking at Fig. 2-1, we take the incident wave to be moving in the $+z$ direction. But we assume that this incident wave $\psi_{\text{inc}}(\mathbf{r})$ once inside the potential region is not just a plane wave but can be written in the form

$$\psi_{\text{inc}}(\mathbf{r}) = e^{ikz}\phi(\mathbf{r}) \tag{2-52}$$

where $\phi(\mathbf{r}) = 1$ at $z = -\infty$. Inserting $\psi_{\text{inc}}(\mathbf{r})$ into Eq. 2-51, we find

$$\left[2ik\frac{d}{dz} + \nabla^2\right]\phi(\mathbf{r}) = U(\mathbf{r})\phi(\mathbf{r}) \tag{2-53}$$

We now assume that the potential varies so slowly with $\mathbf{r}$ that we can neglect the second derivative in Eq. 2-53. That is, we assume

$$|\nabla^2\phi(r)| \ll \left|2ik\frac{d\phi(\mathbf{r})}{dz}\right| \tag{2-54}$$

Then Eq. 2-53 becomes

$$2ik\frac{d\phi(\mathbf{r})}{dz} = U(\mathbf{r})\phi(\mathbf{r})$$

and

$$\phi(\mathbf{r}) = \exp\left[\frac{-i}{2k}\int_{-\infty}^{z} U(x,y,z')\,dz'\right] \tag{2-55}$$

We then insert this expression for $\phi(\mathbf{r})$ into $\psi_{\text{inc}}(\mathbf{r})$, and

$$\psi_{\text{inc}}(\mathbf{r}) = e^{ikz}\exp\left[-\frac{i}{2k}\int_{-\infty}^{z} U(x,y,z')\,dz'\right] \tag{2-56}$$

Finally we return to Eq. 2-4, which yields

$$\psi_{\text{scat}} = -\frac{1}{4\pi}\int d^3r\left[U(\mathbf{r})\,e^{ikz}\exp\left(-\frac{i}{2k}\int_{-\infty}^{z} U(x,y,z')\,dz'\right)\frac{e^{ikR}}{R}\right]$$

We can simplify this equation just as Eq. 2-13 was deduced from Eq. 2-5. We obtain

$$f(\theta) = -\frac{1}{4\pi}\int d^3r \left[e^{i(\mathbf{k}-\mathbf{k}')\cdot\mathbf{r}} U(\mathbf{r}) \exp\left(-\frac{i}{2k}\int_{-\infty}^{z} U(x,y,z')\,dz'\right)\right] \qquad (2\text{-}57)$$

This is called the *eikonal approximation*.

The eikonal approximation, a concept taken out of classical wave optics, involves two restrictions. To understand these restrictions it is best to start with a simple example. Consider the real, nonrelativistic potential $U(\mathbf{r})$ shown in

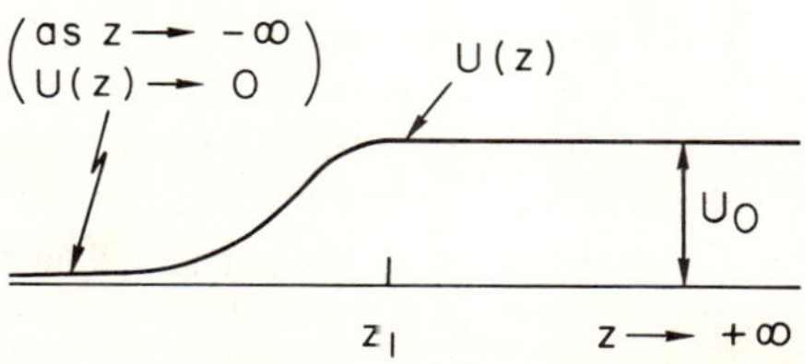

Fig. 2-5 A potential to illustrate the meaning of the eikonal approximation.

Fig. 2.5. The value of $U(\mathbf{r})$ is zero at $z = -\infty$, rising to a constant value (for $z > z_1$) of U_0, and

$$\omega_1 = \frac{1}{2k}\int_{-\infty}^{z} U(z')\,dz' \qquad (2\text{-}58)$$

Equation 2-56 yields

$$\psi_{\text{inc}}(\mathbf{r}) = e^{ikz - i\omega}$$

$$\omega = \frac{U_0}{2k}(z - z_1) + \omega_1$$

Thus the eikonal approximation changes the incident wave by a z-dependent phase. In this simple example, the phase increases linearly with z for $z > z_1$. Let us suppose next that at z_1, $U(\mathbf{r})$ acquires an absorptive component such that

$$z > z_1, \qquad U = U_0 - 2ik\beta$$

where β is real and positive. Now

$$\psi_{\text{inc}}(\mathbf{r}) = e^{ikz - i\omega - \beta(z - z_1)}$$

and we see that in addition to acquiring a z-dependent phase, the incident wave is also being absorbed as it passes through the potential.

Exercise 2-12. When $U_0/2k$ is purely real and much less than the kinetic energy, show that the change of phase can be interpreted as due to the change in momentum of the particle in the potential.

When $U(\mathbf{r})$ is a function of x, y, and z, the same interpretation holds that was used for the constant $U(\mathbf{r})$ in the foregoing example. The real parts of U will lead to z-dependent changes of phase of the plane wave; the imaginary parts of U will lead to absorption of the wave if negative, and if positive, will lead to creation of a larger plane wave! The wave acquires these changes as it moves in the z direction; that is the significance of the integration being over z' only. However, these changes of phase and intensity do not change the direction of propagation of the wave. The plane wave in the z direction remains a plane wave in the z direction. Since the direction of the plane wave represents the direction of the particle, the term $\phi(\mathbf{r})$ does not change the direction of the particle. The direction of the particle only changes at the interaction point $\mathbf{r}$ when the particle scatters on the potential $U(\mathbf{r})$. To repeat, this form of the eikonal model does *not* consist of continuous scattering or bending of the particles as they pass each other. Rather, as the particles move on straight-line paths parallel to their incident directions, they acquire phase changes and intensity changes. But they change direction only once at the interaction point. This is one restriction on the application of the model that can cause errors at large angles, depending on the form of $U(\mathbf{r})$ (2SU, 2SA).

To see the other restriction on the eikonal approximation, insert the solution given in Eq. 2-55 in the inequality of Eq. 2-54. We find

$$|U(\mathbf{r})| \gg \frac{1}{4k^2}|U(\mathbf{r})|^2, \qquad |U(\mathbf{r})| \gg \frac{1}{2k}\left|\frac{dU(\mathbf{r})}{dz}\right| \tag{2-59}$$

Thus there are two conditions. In the nonrelativistic case $U(\mathbf{r}) = 2mV(\mathbf{r})$ these become

$$\frac{k^2}{2m} \gg \frac{|V(\mathbf{r})|}{4}, \qquad |V(\mathbf{r})| \gg \frac{\lambda}{4\pi}\left|\frac{dV(\mathbf{r})}{dz}\right| \tag{2-60}$$

The first condition says that the kinetic energy must be greater than the potential energy; we dare not believe the numerical factor $\frac{1}{4}$ too specifically. The second condition is that the amount that the potential changes in a wavelength $\lambda = 2\pi/k$ must be much smaller than the potential itself in that region.

We define

$$\chi_i(\mathbf{r}) = -\frac{1}{2k}\int_{-\infty}^{z} U(x, y, z')\,dz' \tag{2-61}$$

which is a conventional notation. Thus $\chi_i(\mathbf{r})$ represents the phase and intensity change of the incident wave from $z = -\infty$ to the point of interaction at $\mathbf{r}$. Equation 2-57 then becomes

$$f(\theta) = -\frac{1}{4\pi} \int d^3r \, [e^{i(\mathbf{k}-\mathbf{k}')\cdot\mathbf{r}} \, U(\mathbf{r}) \, e^{i\chi_i(\mathbf{r})}] \tag{2-62}$$

We can now easily determine the relation of the eikonal model to the first Born approximation. We carry out the expansion

$$e^{i\chi_i(\mathbf{r})} = 1 + i\chi_i(\mathbf{r}) + \cdots \tag{2-63}$$

If we insert this expansion into Eq. 2-62 and take the first term alone, we obtain

$$f(\theta) = -\frac{1}{4\pi} \int d^3r \, [e^{i(\mathbf{k}-\mathbf{k}')\cdot\mathbf{r}} \, U(\mathbf{r})] \tag{2-64}$$

which is identical to Eq. 2-13—the first Born approximation. Thus the first Born approximation is the approximation to the eikonal model that is obtained if the series of Eq. 2-63 is truncated after its first term. We also have now a simple explanation for the condition, $Va/\hbar v \ll 1$, for the validity of the first Born approximation (see Sec. 2-2). Here Va represents the product of the strength of the potential and the spatial extent (along one dimension) of the potential, and v is the velocity. In the nonrelativistic case we estimate the integral in Eq. 2-61 by the quantity $Ua = 2mVa$. We find

$$|\chi_i| \approx \frac{mVa}{k} = \frac{Va}{\hbar v}$$

where we have inserted $\hbar$ to make χ_i dimensionless. Thus the condition $Va/\hbar v \ll 1$ is just the condition that the higher terms in the series of Eq. 2-63 can be neglected.

Equation 2-62, which gives the eikonal approximation for $f(\theta)$, can be solved directly only by numerical methods. Although this can be done on a computer, it has become conventional to make one more approximation that simplifies the equation and in particular clarifies the physical meaning of the approximation. We define the vector $\mathbf{b}$ in the xy plane and call the unit vector in the z direction $\hat{z}$. We have already defined $\mathbf{q} = \mathbf{k} - \mathbf{k}'$ as the three-momentum transferred in the center of mass. Then we have

$$(\mathbf{k} - \mathbf{k}') \cdot \mathbf{r} = \mathbf{q} \cdot \mathbf{r} = q_\perp b + q_\parallel z \tag{2-65}$$

Here we have defined $q_\perp$ as the component of $\mathbf{q}$ that is perpendicular to $\mathbf{k}$, that is, perpendicular to the z axis. The $q_\parallel$ is the component of $\mathbf{q}$ along $\mathbf{k}$. The integral in Eq. 2-62 is over that region of $\mathbf{r}$ where $U(\mathbf{r})$ and $\chi_i(\mathbf{r})$ are nonzero. In physically realizable situations this region is roughly of the same extent in b and in z. For the small angle scattering with which we are concerned, however, we can neglect $q_{\|}$ with respect to $q_{\perp}$. Therefore, we write

$$(\mathbf{k}-\mathbf{k}')\cdot\mathbf{r}=\mathbf{q}\cdot\mathbf{r}\approx\mathbf{q}\cdot\mathbf{b} \tag{2-66a}$$

We also note the exact relationships

$$-\mathbf{k}'\cdot\mathbf{b}=\mathbf{q}\cdot\mathbf{b}=kb\sin\theta\cos\phi \tag{2-66b}$$

Exercise 2-13. For $b = z$, work out the limits on t and on θ if the approximation in Eq. 2-66a is to be in error by less than 10%. Using the approximation of Eq. 2-66a, Eq. 2-62 becomes

$$f(\theta)=\frac{k}{2\pi i}\int d^2b\, e^{i\mathbf{q}\cdot\mathbf{b}}\left\{\exp\left[\frac{-i}{2k}\int_{-\infty}^{+\infty}U(\mathbf{b},z)\,dz\right]-1\right\} \tag{2-67}$$

Exercise 2-14. Derive Eq. 2-67.

Next define

$$\chi(\mathbf{b})=-\frac{1}{2k}\int_{-\infty}^{+\infty}U(\mathbf{b},z)\,dz \tag{2-68}$$

in analogy with Eq. 2-61. Thus

$$\chi(\mathbf{b})=\chi_i(x,y,z=\infty)$$

Eq. 2-67 becomes

$$f(\theta)=\frac{k}{2\pi i}\int d^2b\, e^{i\mathbf{q}\cdot\mathbf{b}}[e^{i\chi(\mathbf{b})}-1] \tag{2-69}$$

which is the most common form of the eikonal model.

If $U(\mathbf{r})$ has cylindrical symmetry about the z axis, which is usually assumed in applications of the eikonal model to spinless particles, a further simplification can be made. We define the angle ϕ as the angle between the vector $\mathbf{b}$ and the x axis. Then we can integrate over ϕ (App. B) and obtain

$$f(\theta)=\frac{k}{i}\int_0^{\infty} bdb\;\{[J_0(kb\sin\theta)](e^{i\chi(b)}-1)\}. \tag{2-70}$$

Here we have used Eq. 2-66b, and we can write $\chi(\mathbf{b})$ simply as $\chi(b)$. This formula is often used as the starting point for eikonal model calculations. We observe that the small angle approximation has been used to obtain both this equation and Eq. 2-69.

In Eq. 2-70, the only explicit parameter is the variable b. This variable is called the *impact parameter*. In nonrelativistic scattering from a fixed scattering center, it represents the perpendicular distance between the incident particle trajectory and the scattering center, if the particle has a wavelength sufficiently small to permit it to be localized in the region of the potential (Fig. 2-6). In the

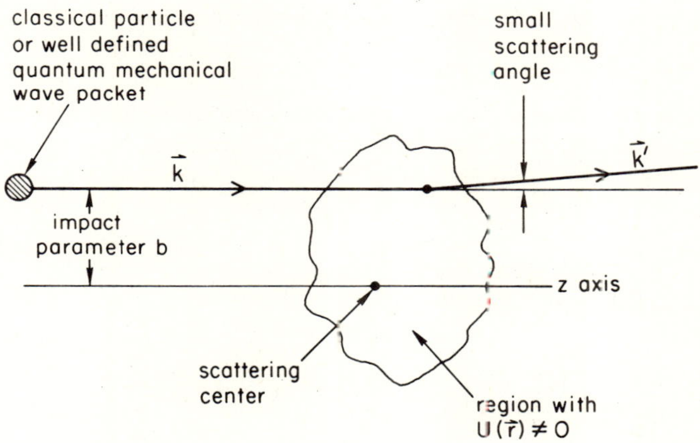

Fig. 2-6 Significance of the impact parameter b in small angle, high energy elastic scattering.

relativistic scattering of the two particles on each other, b again represents that perpendicular distance, if the particles can be localized. The interpretation is only useful for us if the scattering angle is small.

The applications of the eikonal approximation reduce to the problem of finding or calculating $\chi(b)$. We have traveled far from the original inspiration for the optical model, which was the use of an absorptive potential taken from the optical analogy. But we can make four general remarks about how $\chi(b)$ must behave. Certainly $\chi(b)$ must be zero when b is much larger than the range of the strong interactions. Also, $\chi(b)$ can not vary greatly within a wavelength. If such variations exist, we can not use the eikonal approximation. Next, $\chi(b)$ must be closely related to the models we used in the Born approximation to the optical model, since those models were quite good representations of the diffraction peak in elastic scattering. Finally, for an absorptive interaction $U(\mathbf{r})$ is a negative imaginary quantity. From Eq. 2-68 it then follows that for an absorptive interaction, $\chi(b)$ must be a positive imaginary quantity.

Unfortunately, except for a disk with a uniform absorption coefficient, there are no easily worked examples of the use of the eikonal model. There are no simple functions that can be used for $U(\mathbf{r})$ if we demand that these functions be physically reasonable and that the required integrations can be easily carried out analytically. The classic work is that of Serber (2SE), who used a Yukawa function at small impact parameters and a Gaussian function at large impact parameters to fit *pp* elastic scattering. More recent applications involve choosing $\chi(\mathbf{r})$ on the basis of theoretical considerations such as Regge trajectory phenomenology (Ch. 17).

However, we can learn a good deal about the difference between the eikonal approximation and the Born approximation in the following way. We first set

$$\chi(b) = i\lambda\, e^{-b^2/R^2}, \qquad \lambda \text{ real} \tag{2-71}$$

This is a physically realistic choice of $\chi(b)$ for an absorptive potential that satisfies the criterion put forth in the next to last paragraph. Next we expand Eq. 2-69, writing

$$f(\theta) = \frac{k}{2\pi i} \int d^2b\, e^{i\mathbf{q}\cdot\mathbf{b}} \sum_{n=1}^{\infty} \frac{[-\lambda\, e^{-b^2/R^2}]^n}{n!}$$

$$= \frac{kR^2}{2i} \sum_{n=1}^{\infty} \left[\left(\frac{(-\lambda)^n}{n(n!)} \right) \exp \frac{-q_\perp{}^2 R^2}{4n} \right] \tag{2-72}$$

The first term is

$$f_1(\theta) = \frac{ikR^2\lambda}{2} \exp \frac{-q_\perp{}^2 R^2}{4} \tag{2-73}$$

which is the first Born approximation and has the same form as Eq. 2-42. Recalling that for small angles $t = -q_\perp{}^2$, we see that $f(\theta)$ is an infinite sum of exponentials in t. Explicitly, we have

$$f(\theta) = \sum_{n=1}^{\infty} f_n(\theta)$$

$$f_n(\theta) = \frac{kR^2(-\lambda)^n}{2in(n!)}\, e^{b_n t/2}, \qquad b_n = \frac{R^2}{2n} \tag{2-74}$$

We observe that as n increases, the exponential slope b_n decreases. Thus as n increases, each succeeding $f_n(\theta)$ contributes a less and less steep term to the elastic scattering amplitude. If we assume that λ is less than 1, then at $t = 0$ the lowest n terms will be most important. We have illustrated this in Fig. 2-7, where the absolute value of each $f_n(\theta)$ appears on a semilogarithmic scale. It is clear that the higher order terms in n produce a long tail on the diffraction peak. At very large t, in fact, the major contribution to the differential cross section will be from the higher order terms. Of course the full amplitude $f(\theta)$ is obtained by adding the $f_n(\theta)$ terms coherently, taking account of the alternation of sign (2FR).

We have used the $\chi(b)$ defined in Eq. 2-71 because it is physically reasonable and because the term by term integration was easily done. However, the type of contribution made by the higher order terms to the differential cross section is

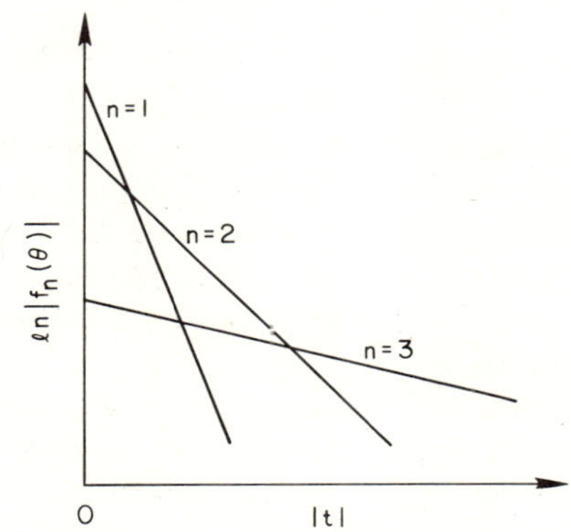

Fig. 2-7 Semilogarithmic plot of $|f_n(\theta)|$ in Eq. 2-74.

certainly very general. The higher order terms in the expansion of the eikonal contribute mainly to the larger t differential cross section and may even dominate that part of the cross section.

Exercise 2-15. Make a more quantitative study of Eq. 2-72, by first connecting λ with σ_{tot} through the optical theorem. Then choose a physical value for R and calculate the coherent contribution of the first two terms in the series to $d\sigma/dt$.

As discussed in the last section, the eikonal model may be described in the following way. The particles move along straight-line paths, acquiring complex phase changes until they reach the interaction point. At that point they change their directions all at once and leave the interaction region without further phase change. This description applies to Eq. 2-62. Using the more approximate form of Eq. 2-69, we can give a more symmetric description, that is, more symmetric with respect to initial and final states. We say that the particles acquire a total complex phase change $\chi(b)$ and that the angular dependence only occurs in the factor $e^{i\mathbf{q}\cdot\mathbf{b}}$, the phase change being independent of scattering angle θ.

Now we can go backward from Eq. 2-69 to obtain yet another form of the eikonal model:

$$f(\theta) = -\frac{1}{4\pi} \int d^3r \left[e^{i(\mathbf{k}-\mathbf{k}')\cdot\mathbf{r}} U(\mathbf{r})\, e^{i\chi_f(\mathbf{r})} \right] \tag{2-75a}$$

$$\chi_f(\mathbf{r}) = -\frac{1}{2k} \int_z^\infty U(x, y, z')\, dz' \tag{2-75b}$$

Exercise 2-16. Derive Eqs. 2-75 and compare them with Eq. 2-62.

Equations 2-75 can be interpreted to mean that first the particles scatter through the angle θ at the interaction point, and only then do they begin to acquire complex phase. They continue to acquire that complex phase until they are widely separated in the final state. Thus we can look on the particles as first acquiring phase and then scattering, or the converse. But we cannot look on the particles as acquiring phase, scattering at the interaction point, and then acquiring more phase, because if this were the case, the effect of the interaction would have to be counted twice.

As we have already mentioned, the derivation of the eikonal approximation, which is used here, is incomplete. First, terms are missing even in the nonrelativistic case. These missing terms arise when we allow several small angular, changes $\theta_1, \theta_2, \ldots, \theta_n$ to occur at different points in the interaction region, the total scattering angle being $\theta = \theta_1 + \theta_2 + \cdots + \theta_n$. Extensive discussions of the missing nonrelativistic terms are given by Refs. 2SU and 2SA.

Second, as we have already remarked, our derivation is not based on relativistic quantum field theory; therefore, it is not a fundamental derivation. Indeed, there is no rigorous derivation of the eikonal approximation based on field theory. There are, however, some semiphenomenological derivations based on summations over an infinite number of Feynman diagrams. Such derivations are closely connected with the theoretical study of high energy inelastic interactions in which many particles are produced—high multiplicity reactions. For the optical theorem shows that there must be an intimate connection between small angle elastic scattering and the total cross section. And at high energies, the total cross section consists mainly of such high multiplicity reactions. Hence we postpone further discussion of the derivation of the eikonal approximation from field theory until Ch. 8, when we consider high multiplicity reactions.

Exercise 2-17. Use the optical theorem, Eq. 1-10, and Eq. 2-69 to obtain the tidy result

$$\sigma_{\text{tot}} = 2 \int d^2b \left[1 - e^{i\chi(\mathbf{b})} \right] \tag{2-76}$$

Discuss the limitations that Eq. 2-76 imposes on $\chi(\mathbf{b})$.

Exercise 2-18. Consider the disk model defined by Eq. 2-26. Apply the eikonal approximation to this model. Also use Eq. 2-76 to obtain the eikonal extension of Eq. 2-35, namely

$$\sigma_{\text{tot}} = 2\pi R^2 [1 - e^{-\beta}] \tag{2-77}$$

REFERENCES

2AN R. L. Anderson et al., *Phys. Rev. Lett.* **25**, 1218 (1970).

2BJ J. D. Bjorken and S. D. Drell, *Relativistic Quantum Mechanics* (McGraw-Hill, New York, 1964).

2FR S. Frautschi and B. Margolis, *Nuovo Cimento* **56**, 1155 (1968).

2GL R. J. Glauber, in *Lectures in Theoretical Physics,* W. E. Brittin and L. G. Dunham, Eds. (Interscience, New York, 1959).

2LA L. D. Landau and E. M. Lifshitz, *Quantum Mechanics, Non-Relativistic Theory* (Addision-Wesley, Reading, Mass., 1958).

2ME A. Messiah, *Quantum Mechanics* (North-Holland, Amsterdam, 1961).

2RO L. S. Rodberg and R. M. Thaler, *Introduction to the Quantum Theory of Scattering* (Academic Press, New York, 1967).

2SA D. S. Saxon and L. I. Schiff, *Nuovo Cimento* **6**, 614 (1957).

2SC L. I. Schiff, *Quantum Mechanics* (McGraw-Hill, New York, 1968).

2SE R. Serber, *Rev. Mod. Phys.* **36**, 649 (1964).

2SU R. L. Sugar and R. Blankenbecler, *Phys. Rev.* **183**, 1387 (1957).

CHAPTER

3

The Partial Wave Analysis and the Spin-Orbit Formalism

3-1 Introduction. In this chapter we begin to apply to elementary particle processes two of the basic principles of quantum mechanics—the quantization of angular momentum and the conservation of angular momentum. Our applications are specifically to the partial wave analysis, to the partial wave form of the eikonal approximation, and to the spin-orbit formalism. Many readers, already well acquainted with this material, may wish to read only Secs. 3-3 and 3-4, which concern the application of the partial wave analysis to the optical model.

In the last two chapters we developed one of the basic concepts of strong interaction physics—the optical model. In this chapter we develop another basic concept—the partial wave analysis. This concept, based as it is on such fundamental ideas as the conservation of particle flux and the conservation of total angular momentum, has multiple uses in the study of strong and electromagnetic interactions. In fact, we use it in this chapter to give another derivation of the eikonal approximation.

The chapter begins with a review of the partial wave analysis for spinless particles. This analysis is then used to derive the eikonal model, Sec. 3-4. Section 3-5 contains a brief discussion of nonrelativistic spin states followed by a review of the spin-orbit formalism. The use of the spin-orbit formalism is illustrated by the derivation of the equations of meson-nucleon elastic scattering. The partial wave forms of the S-matrix and T-matrix are introduced in Secs. 3-3 and 3-7.

3-2 Review of the Partial Wave Analysis for Spinless Particles. First we review the partial wave analysis for spinless particles (3RO, 3SC, 3LA, 3ME), giving the major equations but no proofs. In the cms, the total wave function for elastic scattering can be written (Sec. 2-2) in the approximate form

$$\psi(\mathbf{r}) = e^{ikz} + f(\theta)\frac{e^{ikr}}{r} = \psi_{\text{inc}}(\mathbf{r}) + \psi_{\text{scat}}(\mathbf{r}) \tag{3-1}$$

This form for the approximate wave functions holds even when the particles are relativistic in the cms. The interaction is restricted only in that it must be sufficiently localized to cause the scattered wave to take the asymptotic form given in Eq. 3-1.

The plane wave e^{ikz} can be expanded (App. B) in the form

$$e^{ikz} = \sum_{l=0}^{\infty} (i^l)(2l+1)P_l(\cos\theta)j_l(kr) \tag{3-2}$$

Here $P_l(\cos\theta)$ is a Legendre polynomial normalized to 1 at $\cos\theta = 1$, and $j_l(kr)$ is a spherical Bessel function. The properties of j_l are given in App. B; θ is the scattering angle in the cms, **k** is the momentum, and the incident particles move along the z axis.

As kr goes to infinity, $j_l(kr)$ goes to $(kr)^{-1}\sin(kr - l\pi/2)$; thus Eq. 3-2 becomes

$$e^{ikz} = \frac{1}{2ikr}\sum_{l=0}^{\infty} (2l+1)P_l(\cos\theta)[e^{ikr} - (-1)^l e^{-ikr}] \tag{3-3}$$

and the plane wave is resolved into outgoing and incoming spherical waves. The scattered wave can also be written in a form that resembles Eq. 3-3 because $f(\theta)$ can be expanded in a series of Legendre polynomials. We write

$$\psi_{\text{scat}}(\mathbf{r}) = \frac{e^{ikr}}{2ikr}\sum_{l=0}^{\infty} (2l+1)(a_l)P_l(\cos\theta) \tag{3-4}$$

and the total wave has the form

$$\psi(\mathbf{r}) = \frac{1}{2ikr} \sum_{l=0}^{\infty} (2l+1)P_l(\cos\theta)[(1+a_l)\,e^{ikr} - (-1)^l\,e^{-ikr}] \tag{3-5}$$

Thus the scattering consists of a modification of the outgoing spherical wave, but no change in the incoming spherical wave; the problem is to calculate that modification.

Let us now discuss a limitation on the possible modifications of the outgoing spherical wave. Recall that the probability current density is defined by (using $\hbar = 1$)

$$\mathbf{J}(\mathbf{r}) = \frac{1}{2im}\,[\psi^*\nabla\psi - \psi\nabla\psi^*] \tag{3-6}$$

The radial current density of the outgoing spherical wave alone is

$$J_{r,\mathrm{out}} = \frac{1}{4mkr^2}\left[\sum_l (2l+1)P_l(\cos\theta)\right]^2$$

on a sphere with $r \to \infty$. Integrating this equation over the surface of the sphere and using the orthonormal properties of the Legendre polynomials, we find that the total radial outward flux is the sum of the flux in each angular momentum state (F_l).

$$\text{total flux of outgoing spherical wave} = \sum_{l=0}^{\infty} F_l$$
$$F_l = \frac{\pi}{mk}(2l+1) \tag{3-7}$$

The essence of the partial wave analysis is that the conservation of angular momentum requires that the F_l's be individually conserved. In each state the flux removed from F_l must go either into elastic scattering in the angular momentum state l or into inelastic scattering channels in the state l. It has become conventional to write the outgoing elastically scattered wave of Eq. 3-4 in the following form, where a_l has been replaced by $S_l - 1$

$$\psi_{\mathrm{scat}}(\mathbf{r}) = \frac{e^{ikr}}{2ikr}\sum_{l=0}^{\infty}(2l+1)P_l(\cos\theta)[S_l - 1] \tag{3-8}$$

The form $S_l - 1$ emphasizes that we are being careful not to double count the outgoing spherical wave flux in the plane wave and that S_l shall take on a much more general meaning in the form of the S-matrix. The meaning of S_l can also be illustrated by the expression for the *total* outgoing spherical wave denoted by $\psi_{out}(\mathbf{r})$

$$\psi_{out}(\mathbf{r}) = \frac{e^{ikr}}{2ikr} \sum_{l=0}^{\infty} (2l+1)P_l(\cos \theta)S_l \tag{3-9}$$

Finally we obtain the *basic equation*

$$f(\theta) = \frac{1}{2ik} \sum_{l=0}^{\infty} (2l+1)P_l(\cos \theta)[S_l - 1] \tag{3-10}$$

Now suppose that in the angular momentum state l, β_l of the *amplitude* is absorbed into inelastic channels. Then the fraction of the amplitude retained in the unscattered wave plus the elastically scattered wave is $1 - \beta_l$, and the flux retained in the unscattered plus elastically scattered l states is $F_l[1 - \beta_l]^2$. We can also calculate the retained flux by inserting Eq. 3-9 in Eq. 3-6 and going through the calculation that led to Eq. 3-7. We find that the flux retained in the l state is $F_l|S_l|^2$. Equating these two results gives

$$|S_l|^2 = [1 - \beta_l]^2 \tag{3-11}$$

Again it is conventional to define $1 - \beta_l = \alpha_l$ such that

$$|S_l|^2 = \alpha_l^2 \tag{3-12}$$

Since β_l means the fraction of an amplitude absorbed, it is restricted to the values $0 \leqslant \beta_l \leqslant 1$, hence $0 \leqslant \alpha_l \leqslant 1$. We have purposely chosen the nomenclature to emphasize the analogy with the Born approximation treatment of the optical model; there β for a disk represented the fraction of the plane wave amplitude absorbed, and here β_l is the fraction of the amplitude in the angular momentum state l absorbed.

Equation 3-12 and the statement $0 \leqslant \alpha_l \leqslant 1$ are basic limitations on S; they come from our basic ideas in quantum mechanics and occur again and again in elementary particle physics under the general designation of the unitarity condition. Here we see the unitarity condition in the partial wave analysis formalism for spinless particles; we shall see it in more general and more complicated forms. But it always depends on the principal of the conservation of probability flux; the flux removed from the incident beam must go into elastic or inelastic channels–it cannot be lost.

We often write S_l in the form

$$S_l = \alpha_l\, e^{2i\delta_l} \tag{3-13}$$

where δ_l is a real number. If there is no absorption, $\alpha_l = 1$ and δ_l is the *phase shift.* Sometimes in the literature, in analogy to $\chi(\mathbf{b})$ of Sec. 2-8, we find

$$S_l = e^{i\chi_l} \tag{3-14}$$

so that χ_l is complex. Sometimes $\chi_l/2$ is called the phase shift; and with that definition the phase shift will be complex if α_l is not 1.

Using our basic equation, Eq. 3-10, and the optical theorem or the flux concept, the various cross sections in each partial wave can be calculated. Thus we have

$$\sigma_{\text{el},l} = \frac{\pi}{k^2}\,[2l+1]\,|1 - S_l|^2, \qquad \sigma_{\text{inel},l} = \frac{\pi}{k^2}\,[2l+1]\,[1 - |S_l|^2],$$

$$\sigma_{\text{tot},l} = \frac{\pi}{k^2}\,[2l+1]\,[2(1 - \text{Re}\, S_l)] \tag{3-15}$$

and, summing over all the partial waves,

$$\sigma_{\text{el}} = \sum_{l=0}^{\infty} \sigma_{\text{el},l}, \qquad \sigma_{\text{inel}} = \sum_{l=0}^{\infty} \sigma_{\text{inel},l}, \qquad \sigma_{\text{tot}} = \sum_{l=0}^{\infty} \sigma_{\text{tot},l} \tag{3-16}$$

Exercise 3-1. Derive Eq. 3-15.

In this discussion the S_l's are simply a series of complex numbers. But most readers have probably realized that we have here the beginning of the S-matrix idea, which will gradually be developed into a very general concept. In the process, we will find that a related concept–the T-matrix–is also very useful and is in fact more useful in explicit calculations. Returning to Eq. 3-10, we define

$$T_l = \frac{S_l - 1}{2i} \tag{3-17}$$

so that

$$f(\theta) = \frac{1}{k} \sum_{l=0}^{\infty} (2l+1)\, T_l P_l(\cos\theta)$$

Here the factor $2i$ is an arbitrary convention. At this point we must warn the reader that there are two or three different ways of defining the T-matrix through Eq. 3-17. Some authors extract an additional factor of 2 from T. Others have more important differences from our definition. For example, they extract functions of the momenta from T, usually center of mass momenta. All these definitions are advantageous in that they simplify some expressions, but they complicate other expressions. It would be confusing to try to follow through several conventions in this book. We have defined T as it is commonly used for discrete states in low energy meson-baryon reactions (see Ch. 5).

We emphasize again that all these relationships hold for relativistic as well as nonrelativistic particles.

3-3 Partial Wave Analysis at High Energy. In the high energy region where the particle momenta are above several GeV/c, partial waves with l greater than 8 or 10 must be used to fit the elastic differential cross section. To understand why this is true and to demonstrate the partial wave explanation of the diffraction peak, we consider a simple model for S_l. We assume that we can ignore the phase of S_l and set $\delta_l = 0$. Then $S_l = \alpha_l$ (α_l is a real number between 0 and 1). We have thus constructed a purely absorptive model for elastic scattering. Then we have

$$f(\theta) = \frac{1}{2ik} \sum_{l=0}^{\infty} (2l + 1)P_l(\cos \theta)[\alpha_l - 1] \tag{3-18}$$

Now we simplify the model by assuming that all partial waves with $l \leqslant L$ have the $\alpha_l = \alpha$. We further assume that for $l > L$ there is no absorption, thus for $l > L$, $\alpha_l = 1$. Equation 3-18 reduces to

$$f(\theta) = \frac{(\alpha - 1)}{2ik} \sum_{l=0}^{L} (2l + 1)P_l(\cos \theta), \qquad 0 \leqslant \alpha \leqslant 1 \tag{3-19}$$

The Legendre polynomials $P_l(\cos \theta)$ for $l = 0, \ldots, 4$ are shown in Fig. 3-1. We see that as θ approaches 0 all the polynomials approach 1. At small angles, therefore, the polynomials will add together in Eq. 3-19 to give a sum approaching

$$\sum_{l=0}^{L} (2l + 1) = (L + 1)^2$$

But at larger angles the polynomials show all sorts of behavior, and in the summation we expect a substantial cancellation. There will also be a substantial cancellation at 180°. Therefore, $f(\theta)$ will be relatively large only at the small angles. Knowing this, we directly obtain a diffractive like behavior in this model. Furthermore, if we desire a narrower diffraction peak we must combine more

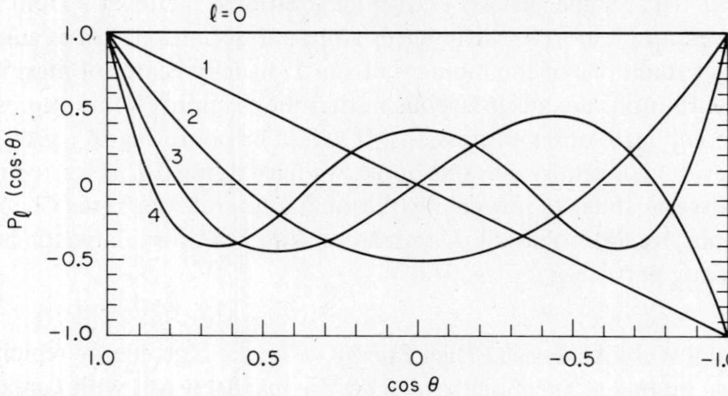

Fig. 3-1 Plot of $P_l(\cos\theta)$ for $l = 0, \ldots, 4$.

and higher order Legendre polynomials. We can in fact carry out the summation in Eq. 3-19 by using the relationships in App. C.

Exercise 3-2. Show that the use of these foregoing relations leads to Eqs. 3-20:

$$f(\theta) = \frac{\alpha - 1}{2ik} [P'_{L+1}(\cos\theta) + P'_L(\cos\theta)] \tag{3-20a}$$

For L large and z close to 1 we can take $P'_{L+1}(z) \approx P'_L(z)$ and obtain

$$f(\theta) = \frac{\alpha - 1}{ik} P'_L(\cos\theta) \tag{3-20b}$$

Exercise 3-3. For L of the order of 8 or 10, plot the Eq. 3-20b and observe its diffraction peak features.

As observed in Ex. 3-3, Eq. 3-20 yields a diffraction peak that drops off through a series of bumps and dips. This is not a necessary feature of this use of partial wave analysis. It is the result of the specific model that set $\alpha_l = \alpha$ for $l \leqslant L$ and $\alpha_l = 1$ for larger l. This sharp change in α_l at $l = L$ produces the oscillations at large angles, just as the sharp-edged disk model produced oscillations in the optical model $f(\theta)$, Sec. 2-6. The relation between these two models is discussed in the next section.

For a more realistic model, α_l must change smoothly with l. We do not discuss such models here, but it is clearly quite easy to set one up. Figure 3-2 illustrates how the α_l's in such a model should vary with l; the coefficients α_l were obtained by fitting the experimental differential cross section for $\pi^- p$ elastic scattering at 5 GeV/c under the assumption that all δ_l were zero. We observe that the absorption is almost complete for small l values, since the α_l are close to zero.

As l approaches 10 the absorption decreases to zero, and α_l rises to 1. This behavior is characteristic of all high energy hadron-hadron elastic scattering A convenient, often used model for α_l is

$$\alpha_l = 1 - e^{-l^2/L^2} \tag{3-21}$$

Here L is a constant indicating roughly where the partial wave amplitude becomes small. The right-hand side of Eq. 3-21 is obviously reminiscent of a Gaussian function, and in Ex. 3-4 we see that this is not a coincidence.

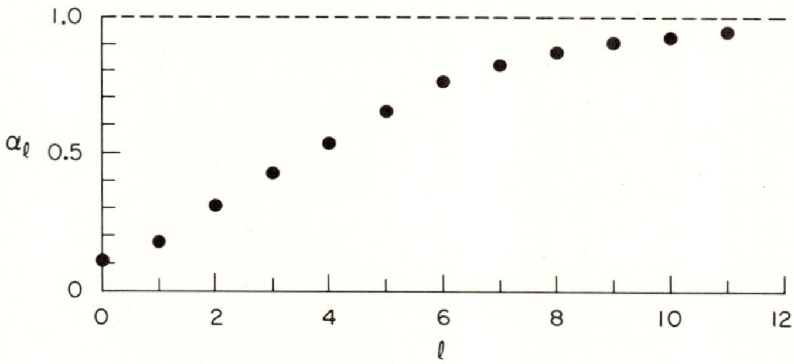

Fig. 3-2 Experimental values of α_l in Eq. 3-18 for 5 GeV/c $\pi^- p$ elastic scattering (3PE).

3-4 Derivation of the Eikonal Approximation from the Partial Wave Analysis. Let us suppose that an interaction requires a very large number of partial waves, allowing us to replace the summation in Eq. 3-10 by an integral. To do this we must replace the Legendre polynomial with a function having l as a continuous parameter and satisfying the angular part of the separated wave equation

$$\left\{(1 - z^2)\frac{d^2}{dz^2} - (2z)\frac{d}{dz} + l(l+1)\right\} P_l(z) = 0, \qquad z = \cos\theta \tag{3-22}$$

Such functions are called Legendre functions, and their properties are summarized in App. C.

Taking l now to be a continuous real positive parameter, $P_l(z)$ is a hypergeometric function where (see App. C)

$$P_l(z) = F\left(-l, l+1; 1; \frac{1-z}{2}\right) \tag{3-23}$$

We can now rewrite Eq. 3-10 in the integral form

$$f(\theta) = \frac{1}{2ik} \int_0^\infty dl[(2l + 1)P_l(\cos\theta)(S(l) - 1)] \tag{3-24}$$

where we have replaced S_l by the continuous function $S(l)$. How much physical meaning have we retained? That depends on S_l. If S_l varies smoothly and slowly with l, we can pick a function $S(l)$ that takes on the values S_l at integral l and, at least for large l, makes the integral a good estimate of the summation. On the other hand, if S_l fluctuates wildly with l, we have lost our physical feeling for $S(l)$ and we should not proceed with this method. If we compare Eq. 3-24 with Eq. 2-70 we see some resemblance.

In App. C we observe that for θ small *and* l large compared with 1 we can approximate

$$P_l(\cos\theta) \approx J_0(l \sin\theta)$$

In Eq. 3-24 we also neglect the 1 in $2l + 1$ for l large. Finally, we rewrite $S(l)$ in the form

$$S(l) = e^{i\chi(l)}$$

where we take $\chi(l)$ to be the continuous extension of the quantity χ_l defined at the end of Sec. 3-2. With these approximations and changes, we write Eq. 3-24 in the form

$$f(\theta) = \frac{1}{ik} \int_0^\infty l\,dl\,[J_0(l \sin\theta)(e^{i\chi(l)} - 1)] \tag{3-25}$$

Up to this point we have made only mathematical changes in Eq. 3-24; we now make a change in the physical meaning. Classically, when a particle with momentum $\mathbf{p}$ has an impact parameter b (see the discussion after Eq. 2-70), the angular momentum of the particle about that center is $|\mathbf{p}|b$. Semiclassically, if we can localize the particle (because of its small wavelength) in a region that is small compared with b, we can equate the particle's quantum mechanical angular momentum $l\hbar$ to $|\mathbf{p}|b$. Then we have

$$l = \frac{|\mathbf{p}|b}{\hbar} = kb \tag{3-26a}$$

Here we are claiming that we can associate the impact parameter b with l in a meaningful way. Using Eq. 3-26a, we write Eq. 3-25 in the final form

$$f(\theta) = \frac{k}{i} \int_0^\infty b\,db\,[J_0(kb \sin\theta)(e^{i\chi(b)} - 1)] \tag{3-27}$$

This equation is identical to Eq. 2-70, thus we have derived the small angle form of the eikonal model from the partial wave analysis. In this last equation we have simply rewritten $\chi(kb)$ as $\chi(b)$

We have several comments on this derivation. First, we note that the derivation connects the behavior of the phase shifts in the partial wave analysis with the $\chi(b)$ of the eikonal model. This is a useful connection because in the partial wave analysis we have been able to avoid complications like the relativistic extension of the potential concept. It is also a useful derivation because it demonstrates from another point of view the reasonableness of the eikonal model.

More specifically, this derivation provides a guide to the form of $\chi(b)$. Suppose a particular inelastic interaction produces absorption mostly in the Lth angular momentum state. Then we expect $\chi(l)$ to be large and absorptive when

$$L - \tfrac{1}{2} \lesssim l \lesssim L + \tfrac{1}{2}$$

Going to the impact parameter representation, we expect $\chi(b)$ to be large and absorptive when

$$\frac{L - \frac{1}{2}}{k} \lesssim b \lesssim \frac{L + \frac{1}{2}}{k} \tag{3-26b}$$

Here we have used Eq. 3-26a. Thus we might construct a $\chi(b)$ of the form

$$\chi(b) = i\beta, \qquad \frac{L - \frac{1}{2}}{k} \lesssim b \lesssim \frac{L + \frac{1}{2}}{k} \tag{3-26c}$$

$$\chi(b) = 0, \qquad \text{otherwise}$$

where β is real and positive.

Suppose on the contrary that absorption takes place in all the lower l states; the degree of absorption smoothly decreasing with increasing l as illustrated in Fig. 3-2. Then there will be absorption at all the corresponding values of b given by Eq. 3-26a, and the greatest absorption will be at $b = 0$. This will lead exactly to the forms of $\chi(b)$ that we used in the last chapter. A simple and much-used example of absorption that decreases smoothly with b is our old friend the Gaussian function of Eq. 2-71.

$$\chi(b) = i\lambda\, e^{-b^2/R^2} \tag{3-26d}$$

Exercise 3-4. Start with this Gaussian absorptive function, setting $\lambda = 1$. Expand $e^{i\chi(b)}$ to lowest order in $\chi(b)$, reason backward from the continuous to the discrete case, and justify Eq. 3-21. Thus show that Eq. 3-21 is the discrete partial wave analogy of the first Born approximation for a Gaussian absorptive potential. Relate L in Eq. 3-21 to R in Eq. 2-42. Finally, demonstrate that if the expansion in $\chi(b)$ is *not* made, we obtain a slightly more complicated version of Eq. 3-21 which is the analogy of the eikonal approximation for a Gaussian absorptive potential.

3-5 Nonrelativistic Spin States and the Spin-Orbit Formalism. In a nonrelativistic situation, the spin state of an elementary particle is easily described. Suppose the particle has intrinsic spin s. Then the spin state is described with reference to an arbitrary axis *fixed in space.* For convenience, we usually call that axis the $+z$ axis. If the component of the spin along that axis is denoted by m, the spin state is denoted by $|s, m\rangle$. Formally, if the angular momentum operator is $\mathbf{J}$

$$\mathbf{J}^2 |s, m\rangle = s(s+1)|s, m\rangle$$
$$J_z |s, m\rangle = m|s, m\rangle \tag{3-28}$$

Now this description and m itself are independent of the velocity of the particle as long as that velocity is nonrelativistic.

This nonrelativistic description is very useful in nuclear physics and low energy elementary particle physics, where the only particles with spin, nuclei, and nucleons are nonrelativistic. But let us see why the nonrelativistic description cannot be used directly for a relativistic particle. Consider first a particle at rest and select a z axis to specify J_z. Now give that particle a velocity $\mathbf{v}$ of relativistic magnitude and arbitrary direction. If we regard the z axis of quantization to be fixed to the particle, that z axis will seldom remain parallel to a z axis fixed in space. In fact, the two z axes will only remain parallel if $\mathbf{v}$ is parallel to, or perpendicular to, the z axis.

Nevertheless, nonrelativistic spin states can be useful in relativistic situations, for we can make a Lorentz transformation to the rest frame of a particle. In that rest frame we can of course use nonrelativistic spin states. In particular, we see in Ch. 11 that such spin states are useful when discussing polarization states. From now on we designate these nonrelativistic spin states by the phrase *rest frame spin states.* We wish to distinguish these states from the *helicity frame spin states* introduced in Ch. 9. Wherever possible, we use the notation of Eq. 3-28 for the rest frame spin states, with the magnetic quantum number a *Latin alphabet* letter. The helicity state notation is described in Ch. 9.

We remind the reader that knowing that parity is conserved in strong and electromagnetic interactions, we can assign an intrinsic parity η to each hadron (3GA, 3RO). The parity is most easily defined using rest frame spin states. If P is the parity operator

$$P|s, m\rangle = \eta|s, m\rangle, \qquad \eta = \pm 1 \tag{3-29}$$

We now briefly consider the spin-orbit formalism applied to a nonrelativistic system consisting of two particles with intrinsic spins s_1 and s_2 and orbital angular momentum l. Books like that of Rose (3RO) treat this extensive subject. We first consider two angular momentum states represented by the eigenfunctions $|s_1 m_1\rangle$ and $|s_2 m_2\rangle$. We can express the direct product of these two states as states with total angular momentum J and total magnetic quantum number M. These total angular momentum states are represented by the symbol $|s_1 s_2 JM\rangle$. The total angular momentum J has all values given by the relation

$$J = |s_1 - s_2|, \qquad |s_1 - s_2| + 1, \ldots, |s_1 + s_2| - 1, \qquad |s_1 + s_2| \tag{3-30}$$

but M has only the value

$$M = m_1 + m_2$$

We usually write the direct product of $|s_1 m_1\rangle$ and $|s_2 m_2\rangle$ in the form $|s_1 s_2 m_1 m_2\rangle$. That is,

$$|s_1 m_1\rangle |s_2 m_2\rangle = |s_1 s_2 m_1 m_2\rangle \tag{3-31}$$

The relations between the direct product states $|s_1 s_2 m_1 m_2\rangle$ and the total angular momentum states $|s_1 s_2 JM\rangle$ are given by the Clebsch–Gordan coefficients. Thus we have

$$|s_1 s_2 m_1 m_2\rangle = \sum_{\substack{|s_1 - s_2| \leqslant J \leqslant |s_1 + s_2| \\ M = m_1 + m_2}} C^{s_1 s_2 J}_{m_1 m_2 M} |s_1 s_2 JM\rangle \tag{3-32}$$

In the future we shall not write the limitations on J and M explicitly, those restrictions being built into the Clebsch–Gordan coefficients themselves. For the values of the coefficients that we use, the angular momentum states are normalized, and

$$\langle s_1 m_1 | s_1 m_1' \rangle = \delta_{m_1 m_1'}$$

The properties of the Clebsch–Gordan coefficients are given in App. C.

In the spin-orbit formalism we first couple the spins of the two particles to give a total spin S whose values range from $|s_1 + s_2|$ to $|s_1 - s_2|$. This total spin S has a total magnetic quantum number $m_s = m_1 + m_2$. The total spin state is represented by the symbol $|s_1 s_2 S m_s\rangle$. We emphasize that unless one of the spins is zero, the coupling of the two spin states leads to a sum of total spin states; that summation is given by Eq. 3-32.

We next couple each of these total spin states $|s_1 s_2 S m_s\rangle$ with the orbital angular momentum state $|l m_l\rangle$, again using Eq. 3-32. This leads to a set of total angular momentum states with total angular momentum J and total magnetic quantum number M. Allowed values of J are given by the relation

$$|l - S| \leqslant J \leqslant |l + S|$$

and M is given by

$$M = m_l + m_s = m_l + m_1 + m_2$$

The total angular momentum state is represented by the symbol $|s_1 s_2 SlJM\rangle$.

As an example of this coupling procedure, we consider neutron–proton elastic scattering. Both particles have spin $\frac{1}{2}$; thus S will be zero or 1. For a given l, J will then be $l+1$, l, or $l-1$. The situation is simpler in the scattering of pions or kaons on nucleons. There is only one nonzero spin, we can ignore the S coupling stage, and J is $l+\frac{1}{2}$ or $l-\frac{1}{2}$.

We conclude this section with a discussion of some parity questions. Physical states made up of two or more particles may or may not be parity eigenstates. For example, two spinless particles in a single orbital angular momentum state $|lm\rangle$ form a parity eigenstate with parity $(-1)^l$. But a plane wave consisting of two spinless particles is not a parity eigenstate because it is a superposition of orbital angular momentum states of both parities (see Eq. 3-2). We are concerned here only with the spin-orbit formalism applied to the simple case of states with just two particles. The particles 1 and 2 have intrinsic parities η_1 and η_2. The parity of the total spin state $|s_1 s_2 Sm_s\rangle$ is given by

$$P|s_1 s_2 Sm_s\rangle = P|s_1 m_1\rangle|s_2 m_2\rangle = \eta_1 \eta_2 |s_1 s_2 Sm_s\rangle$$

and the parity of the direct product state $|s_1 s_2 Sm_s\rangle|lm_l\rangle$ is given by

$$P|s_1 s_2 Sm_s\rangle|lm_l\rangle = \eta_1 \eta_2 (-1)^l |s_1 s_2 Sm_s\rangle|lm_l\rangle$$

It then follows that

$$P|s_1 s_2 SlJM\rangle = \eta_1 \eta_2 (-1)^l |s_1 s_2 SlJM\rangle \tag{3-33}$$

Thus $|s_1 s_2 SlJM\rangle$ is a parity eigenstate.

3-6 Spin Zero Meson-Nucleon Elastic Scattering in the Spin-Orbit Formalism. We now consider the elastic scattering of a spin zero meson on a nucleon. This is the simplest example of the use of the spin-orbit formalism in scattering theory. This example also provides an extension of the concepts introduced in Sec. 3-2 and supplies the basic equations used in Ch. 5. The pion and kaon are pseudoscalar mesons (parity $= -1$), but the same equations hold for scalar particles (parity $= +1$). We follow the method of Sec. 3-2. The calculation is carried out in the center of mass system, the incident meson and nucleon move in the $-z$ and $+z$ directions, respectively, and the direction of quantization is along the $+z$ axis. The incident plane wave

$$\psi_{\text{inc}}(\mathbf{r}) = e^{ikz}|\tfrac{1}{2}m\rangle \tag{3-34}$$

can be resolved as before into outgoing and incoming spherical waves. The outgoing spherical wave is

$$\text{outgoing part of } \psi_{\text{inc}}(\mathbf{r}) = \frac{e^{ikr}}{2ikr} \sum_{l=0}^{\infty} (2l+1) P_l(\cos\theta) \,|\tfrac{1}{2}m_s\rangle$$

The conservation of the probability flux still holds, of course. The conservation of the orbital angular momentum l is now replaced by the conservation of the total angular momentum J. Therefore we must resolve the spherical wave into $|\tfrac{1}{2}0SlJM\rangle$ states. Using the Clebsch–Gordan coefficients described in the last section, the outgoing spherical wave in $\psi_{\text{inc}}(\mathbf{r})$ becomes

$$\frac{e^{ikr}}{2ikr} \sum_{l=0}^{\infty} [4\pi(2l+1)]^{1/2} \sum_J C^{1/2\,l\,J}_{m_s 0 M} |\tfrac{1}{2}lJM\rangle \tag{3-35}$$

Here we abbreviate $|\tfrac{1}{2}0SlJM\rangle$ to $|\tfrac{1}{2}lJM\rangle$.

The interaction modifies this outgoing wave, the modification being different in each different $|\tfrac{1}{2}lJM\rangle$ state. The rotational invariance of the interaction tells us, however, that the modification cannot depend on M. Furthermore, it is only J, not l, which is conserved; in the general case, therefore, l can change. Thus we write in analogy to Eq. 3-9 the *total* outgoing spherical wave in the form

$$\psi_{\text{out}}(\mathbf{r}) = \frac{e^{ikr}}{2ikr} \sum_{l=0}^{\infty} [4\pi(2l+1)]^{1/2} \sum_J C^{1/2\,l\,J}_{m_s 0 M} \sum_{l'} S_{Jll'} |\tfrac{1}{2}l'JM\rangle \tag{3-36}$$

where $S_{Jll'}$ implicitly contains constraints on the possible values of l'.

In the particular case of the scattering of spin $\tfrac{1}{2}$ particles on spin 0 particles the difference between l and l' can only be ± 1. However, parity conservation imposes a further restriction. Since the particles do not change in an elastic scattering, the change in the total parity is just the change in the parity of the orbital angular momentum. The conservation of parity prohibits the change of l by ± 1, hence $S_{Jll'}$ reduces to $S_{Jl}\delta_{ll'}$. Inserting this in Eq. 3-35 and subtracting the outgoing spherical wave part of the incident wave, we find the scattered wave to be

$$\psi_{\text{scat}}(\mathbf{r}) = \frac{e^{ikr}}{2ikr} \sum_{l=0}^{\infty} [4\pi(2l+1)]^{1/2} \sum_J C^{1/2\,l\,J}_{m_s 0 M} [S_{Jl} - 1] \,|\tfrac{1}{2}lJM\rangle \tag{3-37}$$

When the same flux conservation considerations used in Sec. 3-2 are applied here, they give the conditions

$$S_{Jl} = \alpha_{Jl}\, e^{2i\delta_{Jl}}; \qquad \alpha_{Jl}, \delta_{Jl} \text{ real}; \qquad 0 \leqslant \alpha_{Jl} \leqslant 1 \tag{3-38}$$

The generalization of Eq. 3-1 is

$$\psi(\mathbf{r}) = e^{ikr}\,|\tfrac{1}{2}m\rangle + \frac{e^{ikr}}{r}\sum_{m'} f^{(\circ)}_{m'm}(\theta,\phi)\,|\tfrac{1}{2}m'\rangle \tag{3-39}$$

That is the incident wave with spin state $|\frac{1}{2}m\rangle$ can yield a scattered wave with either of two spin states $|\frac{1}{2}m\rangle$ or $|\frac{1}{2}{-m}\rangle$. We use the superscript $^{(\circ)}$ to designate spin-orbit amplitudes (to distinguish them from the helicity amplitudes that are used later). Continuing to follow the method of Sec. 3-2 as carried out in Ex. 3-5, we finally obtain the complicated-looking formula

$$f^{(\circ)}_{m'm}(\theta,\phi) = \frac{1}{2ik}\sum_{l=0}^{\infty}[4\pi(2l+1)]^{1/2}\sum_{J}\{C^{1/2lJ}_{m0M}C^{1/2lJ}_{m'm_lM}[S_{Jl}-1]\,Y_l^{m_l}(\theta,\phi)\} \tag{3-40}$$

in which $Y_l^m(\theta,\phi)$ is a spherical harmonic (App. C).

For this system the Clebsch–Gordan coefficients are rather simple and we can make the last equation more explicit. The coefficients we need are given in App. C.

Exercise 3-5. Derive Eq. 3-40 and 3-43.

In analogy to Eq. 3-17, we define

$$T_{Jl} = \frac{S_{Jl}-1}{2i} \tag{3-41}$$

To simplify the symbolism we write

$$f^{(\circ)}_{+1/2+1/2}(\theta,\phi) = f^{(\circ)}_{++} \qquad \text{and} \qquad f^{(\circ)}_{-1/2+1/2}(\theta,\phi) = f^{(\circ)}_{-+} \tag{3-42}$$

Thus we have

$$f^{(\circ)}_{++} = \frac{1}{k}\sum_l\,[(l+1)T_{l+1/2,l} + lT_{l-1/2,l}]P_l(\cos\theta)$$

$$f^{(\circ)}_{-+} = \frac{1}{k}\sum_l\,[T_{l+1/2,l} - T_{l-1/2,l}]\,e^{+i\phi}\sin\theta\,P_l'(\cos\theta) \tag{3-43}$$

$$f^{(\circ)}_{--} = +f^{(\circ)}_{++}, \qquad f^{(\circ)}_{+-} = -f^{(\circ)}_{-+}$$

The relations $f_{++}^{(\circ)} = f_{--}^{(\circ)}$ and $f_{+-}^{(\circ)} = -f_{-+}^{(\circ)}$ are a result of parity conservation. We accept that result as explicitly derived here, but Ch. 10 presents a thorough discussion of the restrictions on the amplitudes resulting from parity conservation. The amplitude $f_{++}^{(\circ)}$ is called the *nonspin flip* amplitude and $f_{-+}^{(\circ)}$ is called the *spin flip* amplitude. If the target nucleons are not polarized, the differential cross section is obtained by averaging over the initial nucleon polarizations states and summing over the final states, yielding

$$\frac{d\sigma}{d\Omega^*} = |f_{++}^{(\circ)}|^2 + |f_{-+}^{(\circ)}|^2 \tag{3-44}$$

Polarization effects are not discussed here, but we consider polarization and related phenomena when the helicity formalism is introduced and developed.

A number of equations and conventions have been useful in the application of Eq. 3-43. Using Ex. 3-6, we first observe that the total elastic cross section σ_{el} can be expressed as the sum of the total partial wave elastic cross sections $\sigma_{el,Jl}$.

Exercise 3-6. Derive the following equations for an unpolarized target, starting from Eq. 3-44.

$$\sigma_{el} = \sum_{l=0}^{\infty} [\sigma_{el,l+1/2,l} + \sigma_{el,l-1/2,l}] \tag{3-45}$$

$$\sigma_{el,Jl} = \frac{\pi}{2k^2}[2J+1]|S_{Jl} - 1|^2 = \frac{2\pi}{k^2}[2J+1]|T_{Jl}|^2 \tag{3-46}$$

The maximum value of $|S_{Jl} - 1|$ is obtained when $\alpha_{Jl} = 1$ and $2\delta_{Jl} = 180°$, yielding $|T_{Jl}| = 1$. Thus the maximum value of $\sigma_{el,Jl}$ is

$$\sigma_{el,Jl,\max} = \frac{2\pi}{k^2}[2J+1] \tag{3-47}$$

This is yet another consequence of the unitarity condition. The property of σ_{el} that enables us to write it as the sum of the separate $\sigma_{el,Jl}$ elastic cross sections appears to occur rather fortuitously in the derivation of Ex. 3-6. In Ch. 10 we see that this separation property has a natural explanation in the helicity formalism.

We can now use the optical theorem to derive the total and absorption cross sections. In the optical theorem we use that part of the forward scattering amplitude in which there is no change in the spin states of the particles, as discussed in Ch. 6. Thus we use only the nonspin flip amplitude $f_{++}^{(\circ)}$. In scalar meson–nucleon scattering, the spin flip amplitude $f_{-+}^{(\circ)}$ is, for special reasons, zero at zero degrees. This can be seen directly in Eq. 3-43. However, in more complicated spin situations such as in nucleon–nucleon scattering, it is possible to have a nonzero forward scattering amplitude that involves changes in the spin states. This is discussed in Ch. 10. We obtain

$$\sigma_{\text{inel},Jl} = \frac{\pi}{2k^2}[2J+1][1-|S_{Jl}|^2] = \frac{2\pi}{k^2}[2J+1][\text{Im}\,T_{Jl}-|T_{Jl}|^2]$$

$$\sigma_{\text{tot},Jl} = \frac{\pi}{2k^2}[2J+1][2-2\,\text{Re}\,S_{Jl}] = \frac{2\pi}{k^2}[2J+1][\text{Im}\,T_{Jl}] \tag{3-48}$$

3-7 The *S*- and *T*-Matrices in Two-Body Inelastic Interactions. In Secs. 3-2 and 3-6 we developed the *S*- and *T*-matrix concept in elastic scattering. Of course, by allowing $\alpha \neq 1$ in Eq. 3-13 we have implicitly included the effect of inelastic interactions. But in this section we explicitly develop the *S*- and *T*-matrix concept for a particular type of inelastic interaction—the two-body inelastic interaction. Examples of such reactions are $\pi^- + p \rightarrow \pi^0 + n$, $\pi^- + p \rightarrow K^0 + \Lambda^0$, as well as reactions in which one or both of the final state particles is a strong interaction resonance (e.g., $\pi^- + p \rightarrow \rho^0 + n$—sometimes called a *quasi-two-body reaction*; Ch. 7).

Consider then the reaction $a + b \rightarrow c + d$, designate the initial state $(a + b)$ by α and the final state $(c + d)$ by β, for simplicity assume that all the particles have spin zero. In analogy to Eq. 3-9, we write the total outgoing spherical wave

$$\psi_{\text{out},\beta\alpha}(\mathbf{r}) = \frac{e^{ik_\beta r}}{2ik_\alpha r}\sum_{l=0}^{\infty}(2l+1)P_l(\cos\theta)S_{\beta\alpha,l} \tag{3-49}$$

where k_α is the initial state center of mass momentum and k_β is the final state center of mass momentum. The choice of k_α for the denominator of Eq. 3-49 is arbitrary; some authors use k_β. From Eq. 3-49 and the conservation of flux in a partial wave (see Sec. 3-2) we find

$$|S_{\beta\alpha,l}|^2 \leqslant 1 \tag{3-50}$$

Our next tasks are to find $\psi_{\text{scat},\beta\alpha}$, in analogy to Eq. 3-8, and to calculate $f_{\beta\alpha}(\theta)$. This is actually easier to do than in the elastic case, for there is no need to subtract from $\psi_{\text{out},\beta\alpha}$ the outgoing spherical wave component of the incident plane wave. Thus $\psi_{\text{scat},\beta\alpha} = \psi_{\text{out},\beta\alpha}$ and

$$f_{\beta\alpha}(\theta) = \frac{1}{2ik_\alpha}\sum_{l=0}^{\infty}(2l+1)P_l(\cos\theta)S_{\beta\alpha,l} \tag{3-51}$$

Finally, in analogy to Eq. 3-17, we define

$$T_{\beta\alpha,l} = \frac{S_{\beta\alpha,l}}{2i} \tag{3-52}$$

for two-body *inelastic* reactions. Note that we have extended the meaning of the S- and T-matrices. The elements are now designated not only by *angular momentum labels* but also by *particle state labels.* In particular, inelastic reactions are described by *off-diagonal* elements in these matrices.

Exercise 3-7. Explain the absence of the -1 in Eqs. 3-51 and 3-52 compared with Eqs. 3-10 and 3-17, respectively.

Clearly we can consider in a similar manner two-body inelastic reactions with particles of nonzero spin, and the spin-orbit formalism can be used in a straightforward manner. In particular, for a reaction such as $\pi^- + p \to K^0 + \Lambda^0$ the formulas are almost exactly the same as those used in the previous section for meson–nucleon elastic scattering. The only difference that may arise will come from intrinsic parity differences between the initial and final states.

Exercise 3-8. Write the equations analogous to Eq. 3-43 for the inelastic reaction $a + b \to c + d$. Take particles a and c to have spin 0 and the same parities. Take particles b and d to have spin $\frac{1}{2}$ and opposite parities.

Returning to Eq. 3-52, we see that we can combine it with Eq. 3-17 by writing

$$T_{\beta\alpha,l} = \frac{S_{\beta\alpha,l} - \delta_{\beta\alpha}}{2i} \tag{3-53}$$

Then for an elastic or inelastic two-body reaction involving only spin zero particles:

$$f_{\beta\alpha} = \frac{1}{k_\alpha} \sum_{l=0}^{\infty} (2l+1) P_l(\cos\theta) T_{\beta\alpha,l} \tag{3-54}$$

$$\sigma_{\beta\alpha,l} = \frac{4\pi}{k_\alpha^2}(2l+1)\,|T_{\beta\alpha,l}|^2 \tag{3-55}$$

$$\sigma_{\beta\alpha} = \sum_{l=0}^{\infty} \sigma_{\beta\alpha,l} \tag{3-56}$$

When $\beta = \alpha$, we have elastic scattering. Also, $\sigma_{\beta\alpha}$ is the total cross section for the reaction in which the state α goes to the state β, and we refer to $\sigma_{\beta\alpha}$ as the *reaction cross section.* We continue to use the term "reaction cross section" when there are more than two particles in the final state.

REFERENCES

3GA S. Gasiorowicz, *Elementary Particle Physics* (Wiley, New York, 1966).

3LA L. D. Landau and E. M. Lifshitz, *Quantum Mechanics, Non-Relativistic Theory* (Addison-Wesley, Reading, Mass., 1958).

3ME A. Messiah, *Quantum Mechanics* (North-Holland, Amsterdam, 1961).

3PE M. L. Perl and M. C, Corey, *Phys. Rev.* **136**, B787 (1964).

3RO M. E. Rose, *Elementary Theory of Angular Momentum* (Wiley, New York, 1957).

3SC L. I. Schiff, *Quantum Mechanics* (McGraw-Hill, New York, 1968).

CHAPTER

4

Selection Rules in Strong and Electromagnetic Interactions

4-1 Introduction. In this brief chapter we summarize the selection rules imposed on strong and electromagnetic interactions by the conservation of electric charge, baryon number, and strangeness. We also formulate the concept of isotopic spin invariance in strong interactions. We have several reasons for introducing these principles at this point—just prior to the discussion of low energy elastic and inelastic processes in Ch. 5. First, the necessity for these principles was initially recognized and their validity was initially tested in low energy processes. In fact, low energy reactions still yield the preponderance of the evidence for isospin invariance and strangeness conservation in strong interactions.

The second reason for introducing the selection rules here is that they limit the kinds of reaction that can occur. These limitations are particularly important at low energies when particle multiplicities are small. Furthermore, for the strong interaction processes that are allowed, the principle of isotopic spin invariance drastically restricts the number of parameters needed to describe these processes.

Isospin invariance and these selection rules restrict in a fundamental way the theoretical formulations of strong and electromagnetic interactions. We can describe those restrictions in elegant ways using symmetry and group theory principles. But we also seek a more physical explanation of isospin invariance and these selection rules. The most common explanations hypothesize that the elementary particles are composite systems made up of a few basic entities. These basic entities are then taken to carry the quantum numbers that enter into the selection rules. The quark theory, Ch. 14, is such an hypothesis.

Although we assume that the reader is familiar with these selection rules, in this chapter we review these rules and emphasize their experimental origins. In Ch. 14, within the broader context of $SU(3)$, we discuss the unsolved problem of the relation of these rules to hadron dynamics.

For background material on this chapter we refer the reader to more general books on elementary particle physics such as 4PI, 4GA, 4WI, and 4OM.

4-2 Additive Selection Rules. We consider first the electric charge Q or q which is conserved in weak, electromagnetic, and strong interactions. Not only is the electric charge conserved, but elementary particles possess that charge in discrete amounts only. The q of an elementary particle is always equal to the charge on the electron multiplied by a positive integer, a negative integer, or zero. Charge conservation was of course known in classical physics, and physicists were not surprised to find charge conservation also holding true in elementary particle physics. On the other hand, the discreteness of Q is purely an elementary particle phenomenon.

There is another elementary particle phenomenon associated with electric charge—the existence of particle–antiparticle pairs. Consider a particle X with mass m, spin s, and electric charge q. Given a sufficiently sensitive search, we find that there also exists a particle with mass m, spin s, and electric charge $-q$. This is the antiparticle $\overline{X}$, discussed further in Sec. 4.4.

If we simply use the conservation of four-momentum, of angular momentum, of parity, and of charge, we can write processes that should occur but do not occur. For example, just those conservation laws would allow a reaction that has never been detected, namely, $p + p \rightarrow \pi^+ + \pi^+$. To explain the absence of such reactions, the concept of baryon number B was constructed. All elementary particles have a baryon number $+1$, 0, or -1. The baryon number is initially defined by assigning $B = +1$ to the proton and $B = 0$ to the pions and by defining the total baryon number B of a state as the sum of the B's of the individual particles in the state. One then finds that the other elementary particles can be given a fixed baryon number, ± 1 or 0, such that the total baryon number is conserved in all elementary particle processes: weak, electromagnetic, strong. For example, $B = -1$ is assigned to the antiproton because of the observed reaction $p + \bar{p} \rightarrow \pi^+ + \pi^-$.

There is a most remarkable connection between baryon number and spin. All hadrons assigned baryon number 0 have zero or integral spin; these particles are of course the mesons. Among the hadrons, all particles with baryon number +1 or −1 have half-integral spin; these particles are the baryons. The only particles with baryon number 0 and half-integral spin are muons, electrons, and neutrinos—all nonhadronic particles. Finally we note that the photon, like the meson, has baryon number 0 and integral spin. Thus excluding the leptons we find

1. If $B = 0$ the spin is 0, 1, 2, ...
2. If $B = \pm 1$ the spin is $\frac{1}{2}, \frac{3}{2}, \ldots$.

We now consider a third conserved quantity—strangeness, denoted by S and conserved only in strong or electromagnetic processes. Like the baryon number, strangeness was introduced to explain why the cross sections of some hadronic reactions are not of the magnitude expected in strong interaction or electromagnetic processes. It is easiest to explain by means of an example. The hadronic reaction $\pi^- + p \rightarrow K^0 + n$ is not prevented by any of the conservation laws we have yet discussed. When the pion momentum is a few GeV/c we might therefore expect the cross section for this reaction to be about the same size as the reaction $\pi^- + p \rightarrow \pi^0 + n$; that is, about a millibarn. Yet the reaction has never been detected, its cross section is many powers of 10 smaller. To explain the absence of this cross section and many other cross sections at the strong or electromagnetic interaction level, the conservation of strangeness was introduced. Particles are assigned a strangeness number S equal to +1, −1, or 0, as listed in App. D. The total strangeness of a state (S) is obtained by adding the strangeness numbers of each particle. This total strangeness cannot change in a strong or electromagnetic process.

The strangeness concept is no more strange than the concepts of baryon number or electric charge. Its name only comes from the historical order of discoveries in elementary particle physics. The need for strangeness conservation was discovered when the physicists had already ceased to wonder at the strangeness of baryon and charge conservation. Of course strangeness conservation is not as universal as baryon or charge conservation; it does not hold in the weak interactions. The relation of strangeness to the weak interactions is itself a vast subject beyond the scope of this book.

There are two more conserved quantities that concern the leptonic particles—the electron, the muon, and the neutrino. There is a conserved quantity associated with the electron called the electron lepton number n_e. The electron has $n_e = +1$. The only other particle having $n_e = +1$ is the neutrino associated with the electron (called the electron neutrino). The only two particles having $n_e = -1$ are the positron and the electron antineutrino. All other particles have $n_e = 0$. A different but analogous quantity, the muon lepton number n_μ, is associated with the muon. The negative muon and the muon associated neutrino are assigned $n_\mu = +1$, and their antiparticles are assigned $n_\mu = -1$. The total n_e and the total n_μ in a state are obtained by adding the respective lepton numbers of the individual particles. For all processes n_e and n_μ are *separately* conserved. Thus processes like $e^- + p \to \mu^- + p$ or $\mu^+ \to e^+ + \gamma$ cannot occur.

The quantities Q, B, S, n_e, and n_μ have the following properties in common:

1. They are all scalar quantities.
2. Their total value for a state is obtained by algebraically adding the quantities assigned to all particles in the state.

Because of these two properties, the selection rules arising from these quantities are called *additive* selection rules.

4-3 Isotopic Spin. In the last section we discussed additive selection rules. We now turn to a nonadditive selection rule—the conservation of isotopic spin, and we need a more complicated mathematical formalism because of this nonadditive property. Historically, the isotopic spin invariance of the strong interactions was discovered in studies of nucleon–nucleon and pion–nucleon systems. Many treatments of isotopic spin therefore start by the consideration of these systems. But as far as we can tell experimentally, all hadrons obey isotopic spin invariance in their strong interactions. Therefore, such obedience is a property of all the hadrons, and we prefer to start with a series of general experimental observations.

1. The hadrons can be arranged in sets. All members of the set possess the same baryon number, the same strangeness, the same spin, the same intrinsic parity, almost the same mass, and in general behave similarly in strong interactions.

2. The members of a set differ in their charges, but a set is found to possess all possible charges within limits q_{max} and q_{min}. That is, we find the complete series $q = q_{\text{max}}, q_{\text{max}} - 1, \ldots, q_{\text{min}}$. No two members of the set possess the same charge.

3. We assign to each set a half-integer or integer i where the number of members in the set is $2i + 1$. We define the set of numbers $i_z = i, i-1, i-2, \ldots, -i$. Observation 2 now allows us to assign an integer or half-integer q_c to each set such that the charge of any member of the set is given by $q = i_z + q_c$. Then we can specify each member of the set by its i_z value. We call a set with $2i + 1$ members an ith multiplet, and we symbolize each member of the multiplet by $|i, i_z\rangle$.

4. We define the *isotopic spin* vector operator **i** to have the same commutation properties as the angular momentum vector operator **J**, namely,

$$[i_x, i_y] = \sqrt{-1}\, i_z, \qquad xyz \text{ cyclic} \tag{4-1}$$

The isotopic spin eigenfunctions $|i, i_z\rangle$ have the same structure as the spin eigenfunctions $|sm\rangle$. This is in agreement with experimental observation 2.

5. Using observation 4, we now consider a physical state made up of two or more particles with isotopic spin states $|i, i_z\rangle$, $|i', i'_z\rangle$, and so forth. Then because **i** and **s** have the same group properties, we can combine the isotopic spin states of the particles using the Clebsch–Gordan coefficients. (Isotopic spin is analogous only to intrinsic physical spin; there is no isotopic spin concept analogous to orbital angular momentum). We obtain eigenstates of the total isotopic spin I and the total isotopic spin "z" component I_z.

6. The total I is conserved in all *strong* interaction processes.

7. The total I_z is conserved in all *strong* interaction processes. We see below that this is mostly a restatement of the conservation of charge.

8. The amplitude for a *strong* interaction process is independent of the total I_z.

From remarks 6, 7, and 8 we see that isotopic spin invariance leads to consequences exactly analogous to total angular momentum conservation: I like J and I_z like M are conserved. The amplitude for an interaction is independent of I_z as it is independent of M. Thus the entire formalism of angular momentum can be directly applied to isotopic spin. This is an enormous convenience in calculating. The formal similarity of isospin and physical spin also means that the two concepts have identical group theoretical structure; this is also very convenient in calculations. However, we know of no physical connection between isospin and intrinsic spin. The isotopic spin invariance of strong interactions and the complete formal similarity between isospin and intrinsic spin are two amazing and not understood conditions.

We have emphasized that isotopic spin invariance holds only for the strong interactions. It is easy to see that it does not hold for electromagnetic interactions. Consider small angle elastic scattering through the Coulomb interaction, ignoring elastic scattering through the strong interaction. Isotopic spin invariance leads to the prediction that the π^+p system and the π^-n system would have the same elastic differential cross section. Yet with the Coulomb interaction alone, the latter system with a neutral member will show no elastic cross section. Thus, in this example, the electromagnetic interaction violates isospin invariance.

The electromagnetic interaction in fact always violates isotopic spin invariance; and the photon does not have a definite isotopic spin I. When isotopic spin considerations are being used in photon–hadron interactions, the photon is said to act as an isotopic scalar ($I = 0, I_z = 0$) *and* as the third component ($I = 1, I_z = 0$)

of an isotopic vector. Some general theoretical ideas about currents (4BE) and some experimental evidence argue against the photon having components in higher isotopic spin multiplets, such as $(I = 2, I_z = 0)$.

Isotopic spin is also not conserved in weak interactions. Isotopic spin considerations are important in weak interaction theory, but that subject is beyond the limits of this book.

Returning to the strong interactions, we find from experiment that

$$q = i_z + \frac{S + B}{2} \tag{4-2}$$

The quantity $S + B$ is called the *hypercharge* Y; thus Eq. 4-2, the Gell–Mann–Nishijima formula, is often written in the form

$$q = i_z + \frac{Y}{2} \tag{4-3}$$

Now the total S, the total B, hence the total hypercharge Y, are conserved in strong interactions. Therefore, the conservation of the total charge Q and the conservation of the total I_z are directly related. Each implies the other. Finally, we note that although the total I is not conserved in electromagnetic interactions, the total I_z is conserved. This follows from Eqs. 4-2 and 4-3.

4-4 Charge Conjugation, Antiparticles, and *G*-Conjugation. The formal similarity between isotopic spin and intrinsic spin is commonly used to construct the following picture of isotopic spin. A mathematical isotopic spin space is constructed, with I_x, I_y, and I_z orthogonal axis. The members of the isospin multiplet then correspond to the possible quantizations of I along I_z, the charge of each member being given by $q = I_z + q_c$. The isotopic spin invariance of strong interactions then corresponds to the statement that strong interactions are invariant to rotations in isotopic spin space. This is analogous to their invariance to rotations in real space. But the analogy cannot be pushed too far. Consider an object in real space described by sets of coordinates (x, y, z). If the object is rotated with respect to the coordinate system, those sets of coordinates will change. But either set of coordinates provides a physically meaningful description of the object. This is not the case for isotopic spin space. An arbitrary rotation in isotopic spin space has no direct physical meaning, and the resulting new description cannot be directly interpreted physically. However, the use of rotations in isotopic spin space is a valuable formal technique, and we use it in this section. Specifically, we employ the isotopic spin rotation operator $e^{-i\theta\hat{n}\cdot\mathbf{I}}$. The analogous angular momentum operator $e^{-i\theta\hat{n}\cdot\mathbf{J}}$, is not discussed until Chs. 6 and 9. Therefore, the reader may wish to delay reading the last part of this section.

The *charge conjugation operator* C was used implicitly in Sec. 4-2. It has the following effect on the charge operator Q, the baryon number operator B, the strangeness operator S, and the lepton number operator n_l ($l = \mu$ or e):

$$
\begin{aligned}
CQC^{-1} &= -Q \\
CBC^{-1} &= -B \\
CSC^{-1} &= -S \\
Cn_lC^{-1} &= -n_l
\end{aligned}
\tag{4-4}
$$

Thus it defines the antiparticle state $|\bar{X}\rangle$

$$C|X\rangle = \eta_c|\bar{X}\rangle \tag{4-5}$$

where η_c is an unknown, particle-dependent phase ($|\eta_c|^2 = 1$) that must be defined or determined separately. Equation 4-5 is the basic definition of an antiparticle.

Strong and electromagnetic interactions are invariant under charge conjugation. We restrict the remainder of our discussion in this section to strong and electromagnetic interactions. Particles with

$$Q = B = S = 0 \tag{4-6}$$

are eigenstates of C; that is, $C|X\rangle = \eta_c|X\rangle$. To determine the *charge parity* η_c for these states, one must have recourse to experimental observations on how these particles are related through interactions to themselves or other particles. We leave further discussion to the exercises. The charge parities of the better-known C eigenstates are given in App. D.

Exercise 4-1. Show that the particles defined in Eq. 4-6 are eigenstates of C. Discuss the effect of charge conjugation on two-particle states. In particular, show that a two-particle state consisting of a particle and its antiparticle is an eigenstate of C.

Exercise 4-2. Using the literature (e.g., Ref. 4PI), show how the charge parity of particles like the π^0, η^0, and ρ^0 is determined.

We would like to generalize the concept of charge conjugation eigenstates to particles that do not satisfy Eq. 4-6. We can do this to some extent by defining the *G-conjugation* operator

$$G = Ce^{-i\pi I_y} \tag{4-7}$$

where $e^{-i\pi I_y}$ produces a rotation of 180° about the y axis in isotopic spin space. Then

$$[G, \mathbf{I}] = 0 \tag{4-8}$$

G-conjugation preserves the essential nature of particle–antiparticle conjugation, for

$$GBG^{-1} = -B, \qquad GSG^{-1} = -S \tag{4-9}$$

Hadrons with $B = S = 0$, π, and ρ mesons for example, are eigenstates of G. And for these particles we can define a *G-parity* η_G where

$$G|X\rangle = \eta_G|X\rangle \tag{4-10}$$

The G-parities of the better-known G eigenstates are given in App. D.

Exercise 4-3. Derive Eqs. 4-8 and 4-9.

Exercise 4-4. Refer to the references given at the end of Sec. 4-1 for a discussion of how the G-parities of particular particles are defined or determined.

Exercise 4-5. Prove the following. If all the particles involved in a strong interation vertex or interaction are G-conjugation eigenstates, the product of the G-parities of all the particles going in must equal the product of the G-parities of all the particles going out. Also show that if only pions are involved, the total number of pions–entering plus leaving–must be even.

Although Eq. 4-5 provides the basic definition of an antiparticle, messy phase questions can arise when we consider the isotopic spin multiplet of an antiparticle in relation to that of a particle. One problem is that C and I_z do not commute. G-conjugation can be helpful here, as is discussed in detail in Ref. 4MA.

4-5 Isotope Spin Considerations in Two-Body Reactions. We conclude this chapter with a very brief discussion of the application of isotopic spin to two-body reactions. We wish to emphasize several principles, using as an example the set of reaction in which both the initial and the final states contain a pion and a nucleon: $\pi^+ + p \to \pi^+ + p$, $\pi^+ + n \to \pi^+ + n$, $\pi^+ + n \to \pi^0 + p$, $\pi^0 + p \to \pi^0 + p$, $\pi^0 + n \to \pi^0 + n$, $\pi^- + p \to \pi^- + p$, $\pi^- + p \to \pi^0 + n$. We have not written down the reversed reactions because we can calculate them by using time-reversal invariance (Secs. 10-4 and 10-5). Still, there are seven different reactions. If we knew nothing about isotopic spin invariance, we would have to measure seven different amplitudes. But the pion-nucleon system has only two possible values of total isotopic spin, $I = \frac{1}{2}$ or $I = \frac{3}{2}$. Therefore, using the principle of isotope spin invariance, there are only two amplitudes to be determined. This is a tremendous simplification and demonstrates the power of the isotopic spin invariance principle. Using an obvious notation, we write the relationships between the product spin states and the total isotopic spin states

$$
\begin{aligned}
|\pi^+p\rangle &= |\tfrac{3}{2}, \tfrac{3}{2}\rangle \\
|\pi^-p\rangle &= \sqrt{\tfrac{1}{3}}\,|\tfrac{3}{2}, -\tfrac{1}{2}\rangle - \sqrt{\tfrac{2}{3}}\,|\tfrac{1}{2}, -\tfrac{1}{2}\rangle \\
|\pi^0n\rangle &= \sqrt{\tfrac{2}{3}}\,|\tfrac{3}{2}, -\tfrac{1}{2}\rangle + \sqrt{\tfrac{1}{3}}\,|\tfrac{1}{2}, -\tfrac{1}{2}\rangle
\end{aligned}
\tag{4-11}
$$

We now apply these considerations to the S- and T-matrices in the spin-orbit formalism, as discussed in Ch. 3. For simplicity we ignore the dependence on the total angular momentum J. Then we have

$$
\begin{aligned}
T_{l,\pi^-p,\pi^-p} &= \langle\pi^-p\,|\,T_l\,|\,\pi^-p\rangle = \\
&\tfrac{1}{3}\langle\tfrac{3}{2}, -\tfrac{1}{2}\,|\,T_l\,|\,\tfrac{3}{2}, -\tfrac{1}{2}\rangle - \sqrt{\tfrac{2}{3}}\,\langle\tfrac{3}{2}, -\tfrac{1}{2}\,|\,T_l\,|\,\tfrac{1}{2}, -\tfrac{1}{2}\rangle \\
&- \sqrt{\tfrac{2}{3}}\,\langle\tfrac{1}{2}, -\tfrac{1}{2}\,|\,T_l\,|\,\tfrac{3}{2}, -\tfrac{1}{2}\rangle + \tfrac{2}{3}\langle\tfrac{1}{2}, -\tfrac{1}{2}\,|\,T_l\,|\,\tfrac{1}{2}, -\tfrac{1}{2}\rangle
\end{aligned}
$$

The second and third term in the last expression are zero by isotopic spin conservation. Therefore, we write

$$T_{l,\pi^-p,\pi^-p} = \tfrac{1}{3}T_l(I = \tfrac{3}{2}) + \tfrac{2}{3}T_l(I = \tfrac{1}{2}) \tag{4-12}$$

and similarly

$$T_{l,\pi^+p,\pi^+p} = T_l(I = \tfrac{3}{2}) \tag{4-13}$$

$$T_{l,\pi^0n,\pi^-p} = \sqrt{\tfrac{2}{3}}\,[T_l(I = \tfrac{3}{2}) - T_l(I = \tfrac{1}{2})] \tag{4-14}$$

We note that when two isotopic spin states contribute to a differential or reaction cross section, that cross section will contain an interference term between the two isotopic spin amplitudes. Thus the inelastic reaction $\pi^- + p \rightarrow \pi^0 + n$ has the reaction cross section

$$\sigma_{l,\pi^0n,\pi^-p} = \frac{2\pi}{k^2}\,[2J + 1]\,|\sqrt{\tfrac{2}{3}}\,T_l(I = \tfrac{3}{2}) - \sqrt{\tfrac{2}{3}}\,T_l(I = \tfrac{1}{2})\,|^2 \tag{4-15}$$

On the other hand, the optical theorem (Sec. 6-7) enables us to separate the isotopic spin contributions to the total cross section in the following way:

$$
\begin{aligned}
\sigma_{\text{tot},l}(\pi^+p) &= \sigma_{\text{tot},l}(I = \tfrac{3}{2}) \\
\sigma_{\text{tot},l}(\pi^-p) &= \tfrac{1}{3}\sigma_{\text{tot},l}(I = \tfrac{3}{2}) + \tfrac{2}{3}\sigma_{\text{tot},l}(I = \tfrac{1}{2})
\end{aligned}
\tag{4-16}
$$

This is a very useful result; it means that the experimental total cross sections can be directly combined to yield the isotopic spin total cross sections $\sigma_{\text{tot},I}(I)$.

Exercise 4-6. Fill in the steps leading to Eqs. 4-12 to 4-16.

REFERENCES

4BE J. Bernstein, *Elementary Particles and Their Currents* (Freeman, San Francisco, 1968).

4GA S. Gasiorowicz, *Elementary Particle Physics* (Wiley, New York, 1966).

4MA A. D. Martin and T. D. Spearman, *Elementary Particle Theory* (North-Holland, Amsterdam, 1970).

4OM R. Omnes, *Introduction to Particle Physics* (Wiley, New York, 1970).

4PI H. Pilkuhn, *The Interaction of Hadrons* (North-Holland, Amsterdam, 1967).

4WI W. S. C. Williams, *An Introduction to Elementary Particles* (Academic Press, New York, 1971).

CHAPTER

5

Two-Body Reactions at Low Energy and Strong Interaction Resonances

5-1 Introduction. In Chs. 1 and 2 we introduced and described some striking phenomena of strong interactions. All these phenomena—the diffraction peak, the weak energy dependence of that peak, and the weak energy dependence of the total cross section—were connected with high energy elastic scattering. In this chapter we introduce another striking and basic phenomenon in strong interactions—the existence of resonant hadronic states. These resonant states have strong effects in low energy and inelastic processes.

As explained in the Preface, our emphasis is on strong and electromagnetic interactions at high energy; we do not propose to present a comprehensive treatment of the low energy interactions of hadrons or of the methods for detecting resonances and discovering their properties. To assure that our present views of strong interactions at high energy are understood, however, some exposure of these subjects is necessary.

The chapter begins with a description of the concept of strong interaction resonances. Next the varieties of behavior that can occur in low energy elastic scattering are illustrated by the πp and pp systems (Sec. 5-3). For a more detailed illustration, we discuss in Sec. 5-4 the low energy, pion–nucleon elastic phase shifts. Here we use the spin-orbit formalism (Ch. 3) and the principle of isotopic spin invariance (Ch. 4). But our major purpose is to illustrate the low energy aspects of strong interaction resonances.

We develop some quantitative aspects of the resonance concept in Secs. 5-5 through 5-8. The Breit–Wigner formula is derived using nonrelativistic potential scattering theory, and the formula is extended to inelastic processes. The characteristics of the Breit–Wigner formula are illustrated using pion–nucleon elastic reactions. The chapter concludes with two brief sections on the relativistic Breit–Wigner formula and the threshold effect.

5-2 Strong Interaction Resonances. One can arrange the hadrons in a rough sequence according to their lifetimes. The sequence starts with the stable proton and the relatively long-lived neutron. Going toward shorter lifetimes, the sequence passes through the charged pions, the kaons, and the hyperons, particles that decay mainly through the weak interactions. These are followed by particles (e.g., Σ^0, π^0, and η^0) that decay through the electromagnetic interaction with lifetimes of the order of 10^{-14} to 10^{-20} sec. Finally we reach particles with lifetimes shorter than 10^{-20} sec. These are the particles that decay through the strong interactions, and they are the particles that hold special interest to us.

We call the particles that decay through the strong interactions, *strong interaction resonances* or simply *resonant states* or *resonances.* The better-established resonances (listed in App. D) and the more stable hadrons are part of a presumably infinite set of hadronic states. The existence of these states, and their masses, spins, electric charges, and other properties, are of course intimately connected with the strong interactions. We have no fundamental understanding of the laws determining the existence of these states. But we do know empirical selection rules, rules that can be systematized by the application of group theory. This and related questions are discussed in Chs. 14 and 17.

Aside from their importance as members of the set of hadrons, the strong interaction resonances provide two avenues for investigation of the strong interactions. First, the decay angular distributions and decay rates of these resonances provide tests of strong interaction theories. Second, these resonances are important in two different kinds of hadronic experiments called *production experiments* and *formation experiments.*

In a *production experiment* the resonance X occurs when one of the inelastic channels (coming from the initial state $a + b$) consists of X plus at least one other particle. For example

$$a + b \to X + c + d, \qquad X \to e_1 + e_2 + \cdots + e_n$$

assuming the resonance X decays to n particles. If the four-momentum of the e_i particle is P_i, the total four-momentum $P_X = \Sigma_i P_i$ is the four-momentum of X. Therefore $M_X^2 = P_X^2$, where M_X is the mass of the resonance. Resonances are found in production experiments by measuring the total four-momentum of some set of particles in the final states for large numbers of events. A concentration of events at a particular M_X indicates a resonance X that decays into that set of particles. In Ch. 8 we present a more elaborate discussion of how resonances are found in production experiments.

In a *formation experiment* the resonance X acts as an intermediate state. Thus the reaction $a + b \to c + d$ may take place through the double process

$$a + b \to X \to c + d$$

Usually X has more than one decay channel. Therefore we also find other reactions, such as

$$a + b \to X \to e + f + g \qquad \text{or} \qquad a + b \to X \to a + b$$

The presence of X in a formation experiment is indicated by some special behavior of the reaction parameters when $s = M_X^2$. Examples of such special behavior are a peak in the total cross section, a peak in a reaction cross section, or a phase shift passing through 90° (Sec. 5-4).

Before proceeding, let us make several remarks on resonances. In this period of elementary particle physics we are not sure how to think about strong interaction resonances in a fundamental way. An older view held that the more massive resonances consisted of metastable states of the less massive and more stable hadrons. Thus the $\Delta(1236)$ [also called the $N^*(1236)$] might be considered to be a metastable state of a pion and a nucleon. That view is not popular today. One tends to think of all the hadrons as being intrinsically the same; it is only the accidents of mass differences and selection rules that make some hadrons more stable than other hadrons. This newer view that all hadrons are equally fundamental requires some explanation of the existence and properties of all the types of hadrons. The most popular explanation is that the hadrons are made up of combinations of more elementary and more basic particles, these more basic particles being limited to just a few different kinds. The quark model is of course one example of such an explanation. But again we must postpone consideration of these questions to a later chapter.

5-3 Illustrations of Low-Energy Hadron–Hadron Elastic Scattering. In Ch. 1 we showed that all high energy, hadron–hadron elastic differential cross sections have as their most prominent feature a diffraction peak. In low energy hadron–hadron elastic scattering there is no prominent feature shared by all systems. In Figs. 5-1 and 5-2 we present the low energy elastic differential cross sections for the $\pi^{\pm}p$ and pp systems. The $\pi^{+}p$ differential cross section shows strong angle and energy dependence. At the relatively low energy of 200 MeV, for example, there is strong angular dependence, produced of course by the dominance of the $\Delta(1238)$ nucleon resonance. The pp differential cross section (Fig. 5-2) presents a remarkable contrast. From 100 to 500 MeV, the differential cross section is isotropic except for the forward peak caused by Coulomb scattering. Even at higher energies the major deviation from isotropy is the gradual growth with energy of the diffraction peak. The pp differential cross section is even independent of energy to 20% in the energy range of 100 to 500 MeV. This can be observed directly by looking at the pp total cross section in Fig. 1-3 and noting that the total inelastic cross section is small or negligible below 500 MeV.

To explain the contrast between the uniformly diffractive behavior of hadronic elastic scattering at high energies and the nonuniformity in hadronic elastic scattering at low energies, we turn to partial wave analysis. As we emphasized in the last chapter, at high energies all the hadronic systems have a very large number of partial waves in the elastic scattering amplitude. The small angle behavior of those partial wave amplitudes is largely determined by the size and opaqueness of the hadrons; these factors not varying greatly over the hadrons considered. But at low energies there are only a few partial waves involved, and the amplitudes of these partial waves depends intimately on the interactions of the particular hadrons. Thus we do not have, and should not expect, uniformity of behavior.

5-4 Pion-Nucleon Phase Shift Analysis at Low Energies. We first consider very low energy $\pi^{+}p$ elastic scattering—below 250 MeV. This allows two simplifications. First, this energy region is below the inelastic threshold, 340 MeV. Hence $\alpha_{Jl} = 1$ (see Eq. 3-38). Second, only a few partial waves are needed, in fact, it is sufficient to use only $l = 0$ and $l = 1$. With these restrictions, Eq. 3-43 reduces to

$$f_{++}^{(\mathrm{o})}(\pi^{+}p) = \frac{1}{k}\,[T_{S33} + (2T_{P33} + T_{P31})\cos\theta]$$

$$f_{-+}^{(\mathrm{o})}(\pi^{+}p) = \frac{1}{k}\,[(T_{P33} - T_{P31})\,e^{+i\phi}\sin\theta] \tag{5-1}$$

where we have used the conventional notation $T_{l,2I,2J}$.

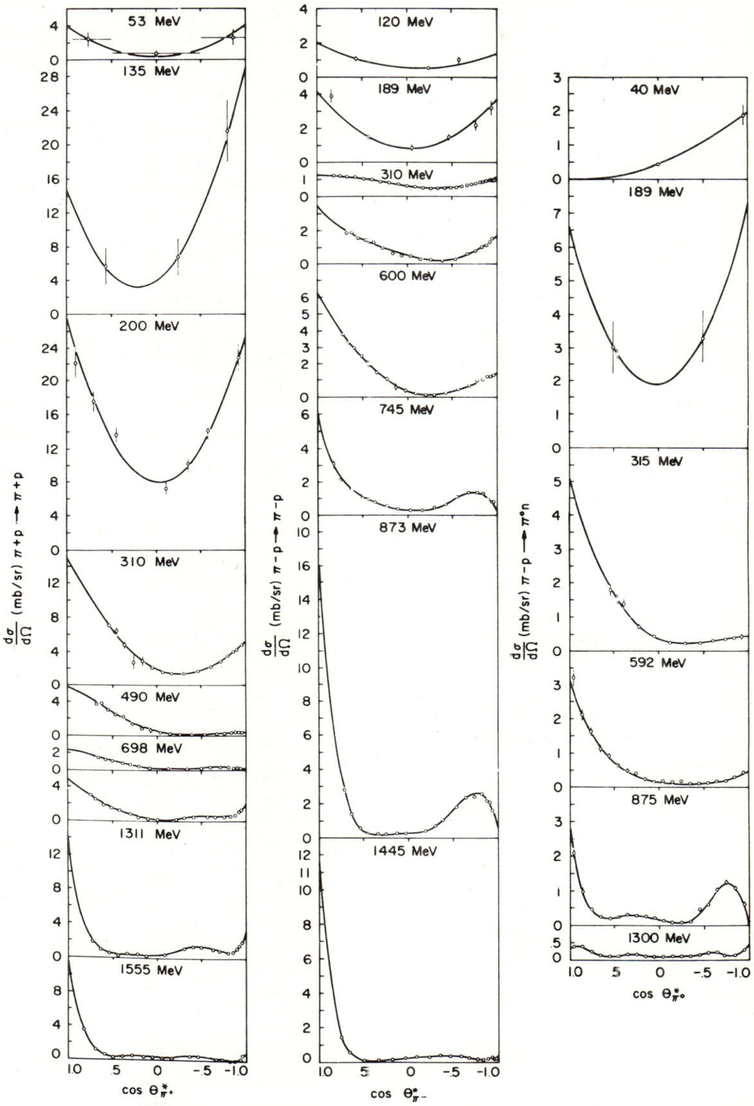

Fig. 5-1 Differential cross sections for low energy π^+p and π^-p elastic scattering and the closely related change exchange reaction $\pi^-p \rightarrow \pi^0 n$ (SCE). The laboratory kinetic energy of the pion is indicated.

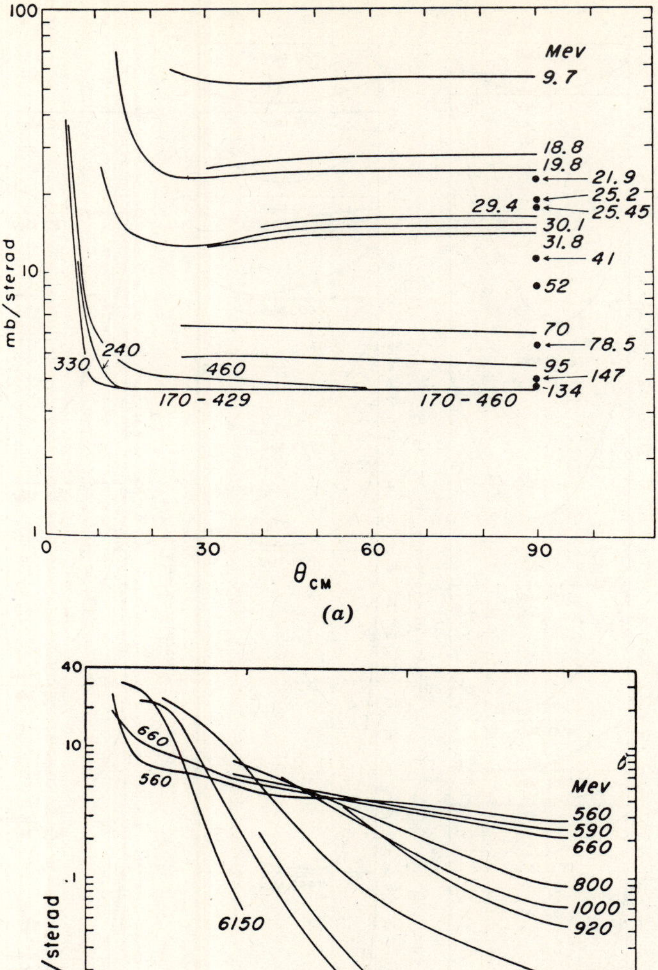

Fig. 5-2 A classic illustration of the differential cross section for low energy *pp* elastic scattering (5HE).

For the π^+p system the differential cross section is then

$$\frac{d\sigma}{d\Omega^*} = \frac{1}{k^2}\,[A + B\cos\theta + C\cos^2\theta] \tag{5-2}$$

where the A, B, C coefficients are complex quadratic functions of the T's.

Exercise 5-1. From Eq. 5-1, evaluate explicitly the A, B, C coefficients.

At low energies it is sometimes convenient to use the phase shift description of the elastic scattering parameters. From Ch. 3, with $\alpha_{Jl} = 1$, we have

$$\sigma_{\text{el},l,2I,2J} = \sigma_{\text{tot},l,2I,2J} = \frac{2\pi}{k^2}\,[2J-1]\,[\sin^2\delta_{l,2I,2J}] \tag{5-3}$$

Exercise 5-2. Complete the steps in the derivation of Eqs. 5-1 and 5-3. Write Eq. 5-2 for π^+p elastic scattering in terms of the phase shifts.

In this energy region, the energy behavior of both the π^+p and π^-p total elastic cross sections is dominated by a large peak at just under 200 MeV (see Figs. 1-2*a* and 1-2*b*). To examine that peak, one must carry out a phase shift analysis of the π^+p differential cross section, Eq. 5-2, and of the π^-p differential cross section represented by a similar equation. One must also use polarization data whose formulation we have not discussed here. When this is done one obtains the phase shifts shown in Fig. 5-3. The carrying out of a phase shift analysis is not easy (5CE, 5JO) especially at higher energies. First, there are statistical errors which mean that one looks not for a unique fit, but only for a best fit. Second, ambiguities arise because the T parameters appear only as squares or products in the differential cross section and polarization formula.

The quantities presented in Fig. 5-3 are the phase shifts $\delta_{l,2I,2J}$, in degrees. The partial wave itself is generally denoted by l, $2I$, $2J$ with $S, P, D, \ldots$ used for $l = 0, 1, 2, \ldots$. Five of the phase shifts are quite small when the energy is less than 300 MeV. The only large phase shift is of course δ_{P33}. (This means $l = 1$, $I = \frac{3}{2}$ and $J = \frac{3}{2}$.) The δ_{P33} passes through 90° at about 180 MeV, and this passage through 90° causes the large peaks in the energy dependence of the π^+p and π^-p total elastic cross sections.

The P_{33} phase shift is the classic example of an elementary particle resonance, the $\Delta(1236)$, and we devote a paragraph and an exercise to examining it. It is such a beautiful example because in discussing it we can neglect the effects of the other phase shifts in the treatment of the π^+p system. Also, as mentioned at the beginning of this section, this is a purely elastic region, thus we have here a purely elastic phase shift. Considering first the π^+p total elastic cross section, we have from Eq. 5-3

$$\sigma_{el}(\pi^+p) = \frac{2\pi}{k^2}\,[4]\,\sin^2(\delta_{P33})$$

Thus the peak in the energy behavior of the total elastic cross section comes simply from the $\sin^2(\delta_{P33})$ factor reaching its maximum value. The differential cross section at the peak has the form

$$\frac{d\sigma_{el}(\pi^+p)}{d\Omega^*} = \frac{1}{k^2}\,[1 + 3\cos^2\theta]\,|T_{P33}|^2 \tag{5-4}$$

which can be obtained directly from Ex. 5-2 or indirectly from Eq. 5-1. The 200-MeV curve in Fig. 5-1 indeed shows this symmetric form.

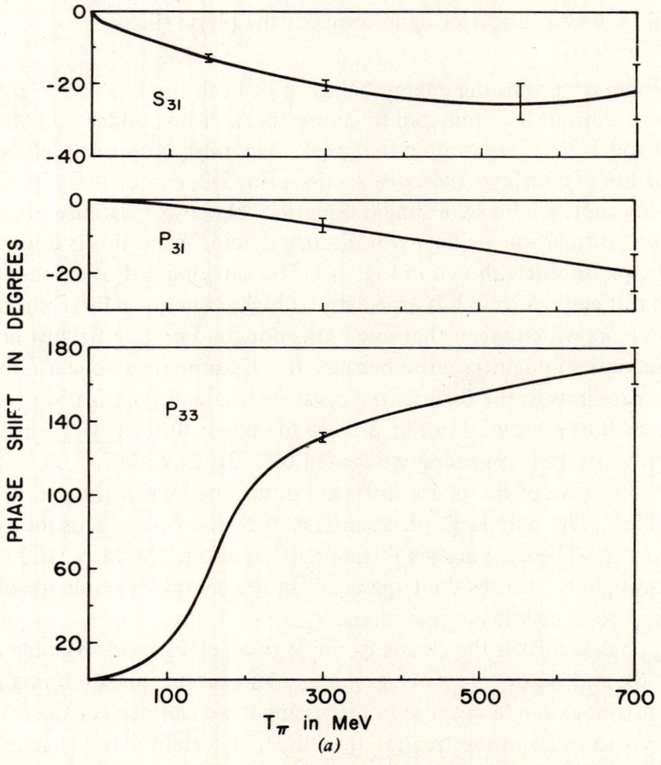

(a)

Fig. 5-3 The S and P wave phase shifts for πp elastic scattering (5CE).

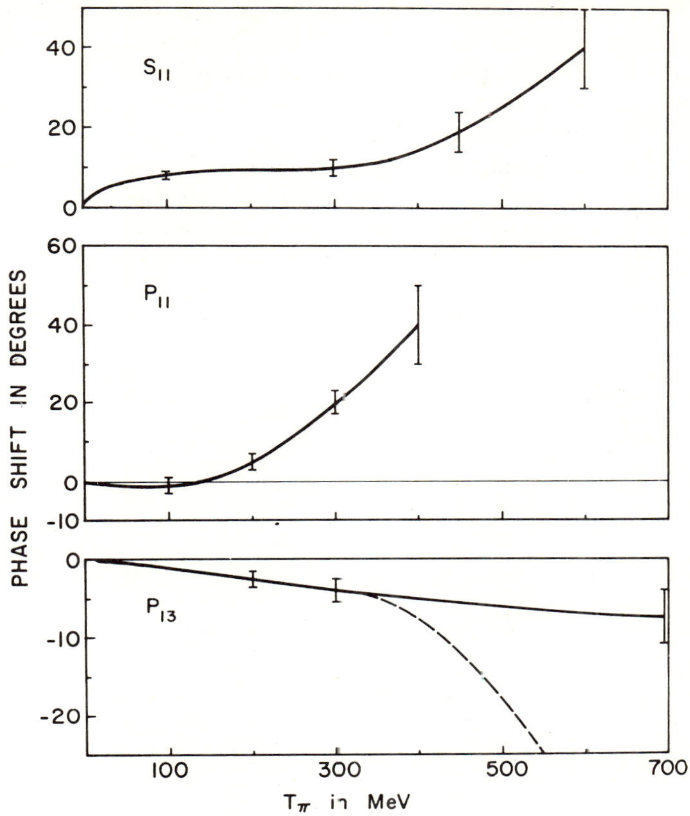

(b)

Fig. 5-3 *(continued)*.

Exercise 5-3. Using Eq. 5-4 carry out numerical fits to the 135- and 200-MeV curves in Fig. 5-1, finding the phase shift and checking it against Fig. 5-3.

Once the threshold for inelastic channels is crossed and α becomes less than 1, the variation of $S_{l,2I,2J}$ with energy becomes more complicated. To represent the variations it is useful to employ the Argand diagram (Fig. 5-4a and 5-4*b*), which consists of a unit circle set in the xy plane. The x coordinate is Im S_n and the y coordinate is $1 - \text{Re}\, S_n$; n represents l, $2I$, $2J$. At zero momentum we expect $\alpha_n = 1$, $\delta_n = 0$ so that all paths of S_n start from the point P. There is no general law to predict the path of a particular S_n; the dashed and dotted lines indicate possible paths. We do know, however, that $\alpha_n = 1$ below the inelastic threshold; hence the paths for low energies must be along the circumference of the unit circle. From Eq. 5-3 it is clear that when α_n is 1, the elastic cross section and the total cross section reach a maximum at $\delta_n = 90°$. This maximum indicates the center of resonance. Thus when $\alpha_n = 1$, a resonance occurs when the path on the Argand diagram passes through the top of the unit circle. The heavy solid line in Fig. 5-4*a* illustrates this behavior. If the phase shift is mostly elastic but not completely elastic, the behavior on the Argand diagram will be rather similar, as illustrated by the dashed line.

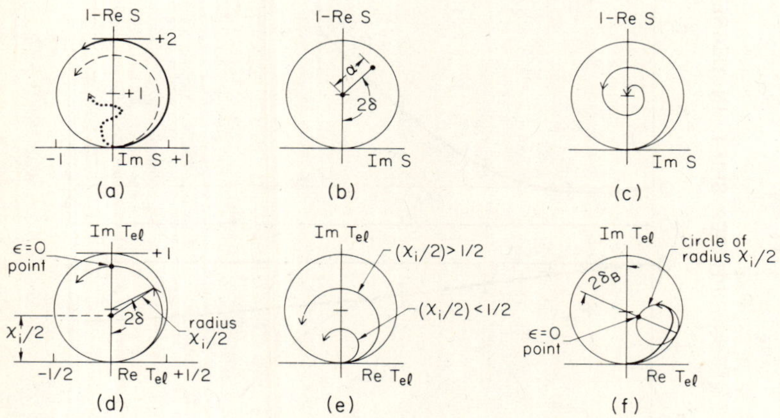

Fig. 5-4 Examples of Argand diagrams as discussed in text.

We next observe that the maximum elastic cross section occurs at 90° regardless of whether the phase shift is increasing or decreasing as it passes through 90°. However, there is a theorem in potential scattering theory, called the Wigner condition, that the derivative of the phase shift with energy is positive (see Sec. 5-5). Thus when potential theory applies as an approximation, or even as a concept, we expect that the phase shift will approach 90° from below. This in fact is what occurs experimentally at low energies when the phase shifts are mostly elastic. To summarize, a resonance is said to occur in a *mostly* elastic phase shift when the phase shift passes through 90°, the phase shift increasing with increasing energy.

The Argand diagram has several properties that are worth noting. Twice the phase shift (i.e., $2\delta_n$) is directly given by the angle shown in Fig. 5-4*b*. The distance from the center of the unit circle to the path gives α_n. We also note the relationships

$$2 \operatorname{Im} T_n = 1 - \operatorname{Re} S_n \qquad \text{and} \qquad 2 \operatorname{Re} T_n = \operatorname{Im} S_n$$

Thus the Argand diagram also represents a plot of 2 Im T_n against 2 Re T_n. Sometimes just Im T_n is plotted versus Re T_n; but then the circle has a radius of $\frac{1}{2}$ rather than 1.

To illustrate the varieties of phase shift behavior that actually occur, we present the experimentally determined Argand diagrams for pion–nucleon elastic scatterings (Fig. 5-5). The classic behavior of the P_{33} phase shift is evident. But other phase shifts have very complicated behavior.

What does one expect these phase shifts to do as the energy increases to the many-GeV region? As discussed in Ch. 3, at very high energies α goes to zero for the low partial waves. Thus we expect the paths of the phase shifts on the Argand diagram to gradually move to the center of the diagram, as in Fig. 5-4c. As the path approaches the center of the diagram, it becomes more difficult to find the value of δ; small errors in the fitting can produce large errors in δ.

5-5 Review of Phase Shift Behavior at Low Energies. In this section we relate S to the scattering potential. To simplify the equations we base the discussion on spinless particles. The introduction of spin complicates the problem if the potential has spin-dependent terms. We start from the nonrelativistic Schrödinger equation, with a spherically symmetric potential $U(r)$ of finite range less than R_0. We seek a solution to the radial wave equations for the lth partial wave:

$$\left[\frac{d^2}{dr^2} + \frac{2}{r}\frac{d}{dr} - \frac{l(l+1)}{r^2} + k^2\right] u_l(k, r) = U(r) u_l(k, r), \qquad r < R_0 \tag{5-6a}$$

$$\left[\frac{d^2}{dr^2} + \frac{2}{r}\frac{d}{dr} - \frac{l(l+1)}{r^2} + k^2\right] v_l(kr) = 0, \qquad r > R_0 \tag{5-6b}$$

This solution in the *far field* must correspond to the radial part of Eq. 3-5, namely,

$$v_l(kr) \underset{r\to\infty}{=} \left[\frac{1}{2ikr}\right] [e^{2i\delta_l} e^{ikr} - (-1)^l e^{-ikr}] \tag{5-7}$$

Therefore, we have

$$v_l(kr) = \frac{i^l}{2} [e^{2i\delta_l} h_l(kr) + h_l^*(kr)] \tag{5-8}$$

where $h_l(z)$ is a Bessel function (App. B).

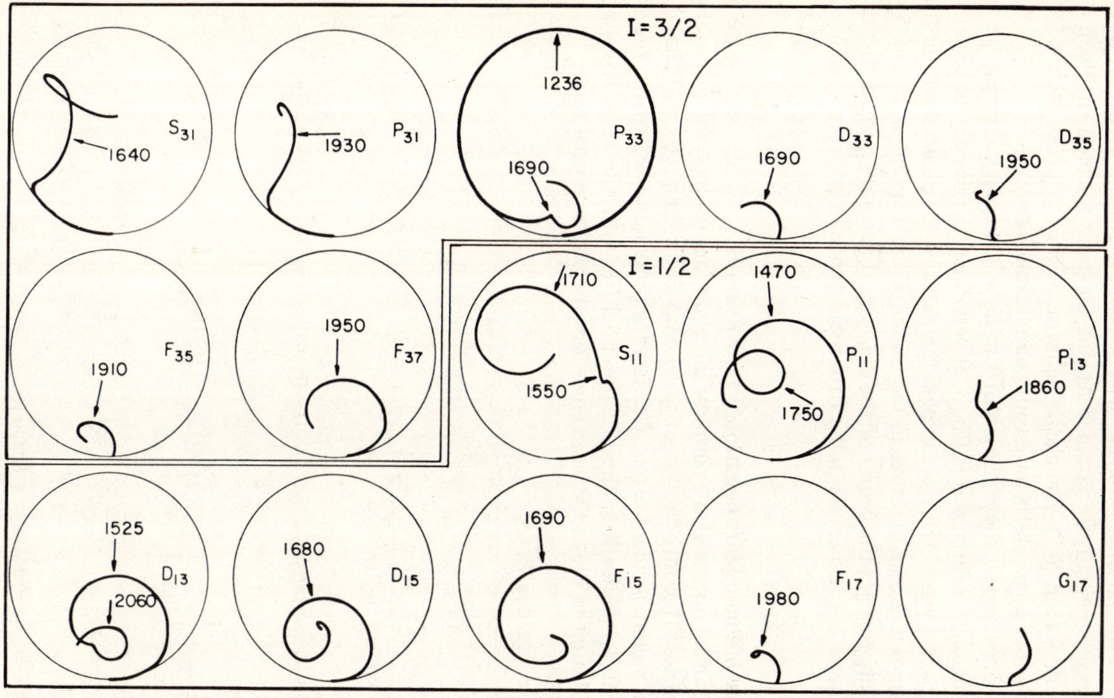

Fig. 5-5 Examples of fitted πp elastic scattering amplitudes on an Argand diagram (5BAR). The parameters of the indicated πp resonances as given in Table 5-1. This is not the only nor the most recent fit to these amplitudes, and some of the resonances indicated here are not established (see table). For the most recent data, use Ref. 5PDG.

We now proceed in the conventional way (5BA, 5SC, 5LA). Define $G_l(k, r)$ to be a logarithmic derivative of $u_l(k, r)$, that is,

$$G_l(k, r) = \frac{\frac{d}{dr} u_l(k, r)}{u_l(k, r)} \tag{5-9}$$

Finally, equating the logarithmic derivatives at $r = R_0$, we obtain

$$\cot \delta_l = \left[\frac{\left[\frac{dn_l(kr)}{dr} - n_l(kr) G_l(k, r) \right]}{\left[\frac{dj_l(kr)}{dr} - j_l(kr) G_l(k, r) \right]} \right]_{r=R_0} \tag{5-10}$$

where n_l and j_l are Bessel functions (App. B).

Exercise 5-4. Fill in the algebraic steps leading to Eq. 5-10. Derive Eq. 5-11.

All the foregoing is low energy theory only in the sense that inelastic channels are not permitted; we have made no specific low energy approximation within our nonrelativistic potential framework. Before making a low energy approximation, we give a general property of $G_l(k, R_0)$ that holds as long as $V(r)$ is independent of energy, or as usually stated, as long as $V(r)$ is velocity independent. The value of $G_l(k, R_0)$ decreases as E_{kin} increases so that

$$\frac{\partial G_l(k, R_0)}{\partial E_{\text{kin}}} \leqslant 0 \tag{5-11}$$

where E_{kin} is the total kinetic energy in the cms. This is the *Wigner condition.*

For sufficiently low energies we can use the relationships of App. B to simplify Eq. 5-10, yielding

$$\cot \delta_l = \frac{1 \cdot 3 \cdot 5 \cdots (2l-1) \cdot 1 \cdot 3 \cdot 5(2l+1)}{(kR_0)^{2l+1}} \left[\frac{l + 1 + R_0 G_l(k, R_0)}{l - R_0 G_l(k, R_0)} \right] \tag{5-12}$$

Here we have discarded the next higher order term, whose magnitude compared with the lowest order term differs by a factor of $(kR_0/2)^2$. Thus if we require $kR_0 < 1$, say, we make an error of about 25%. We certainly dare not put a higher upper limit on kR_0. Unfortunately for the application of this derivation to elementary particle physics, this is a very low limit. Thus in pion–proton scattering, if we take $R_0 = 2 \times 10^{-13}$ cm (to get slightly outside the major portion of the nucleon potential) we find that k, the momentum in the center of mass, must be less than 0.1 GeV/c. This corresponds to a pion laboratory momentum of 0.12 GeV/c, a very low energy situation. Thus we must emphasize again that our considerations here are more illustrative of than directly applicable to, elementary particle physics.

Returning to Eq. 5-12, for a potential weak compared with E_{kin}, the energy dependence of cot δ_l will lie in the term in front of the square bracket and in the denominator in the square bracket.

Exercise 5-5. Work out these last considerations quantitatively for a spherical square well.

For a potential that is not too weak compared with E_{kin} we may find that the *numerator* in the square bracket in Eq. 5-12 is zero. This will give $\cot \delta_l = 0$, $\delta_l = 90°$; thus it is a *resonance condition.* We note that this requires $l + 1 = -R_0 G_l(k, R_0)$, thus ensuring that the denominator, $l - R_0 G_l(k, R_0) = 2l + 1$, is not zero. We can generalize the resonance condition by using Eq. 5-10. In that equation if the numerator in the square bracket is zero and the denominator is not zero, we will have $\delta_l = 90°$.

5-6 The Breit-Wigner Formula in Purely Elastic Scattering. This section is concerned only with elastic scattering, but to prepare for inelastic reactions we begin to use the *total center of mass energy E.* Now returning to Eq. 5-12, at the resonant condition where $l + 1 = -R_0 G_l(k, R_0)$, let E and k be called E_{0l} and k_{0l}. Then for E close to E_{0l} we can express the numerator by the linear term of a Taylor expansion in $(E - E_{0l})$

$$l + 1 + R_0 G_l(k, R_0) = 0 + R_0 [E - E_{0l}] \left[\frac{\partial G_l(k, R_0)}{\partial E}\right]_{E=E_{0l}}$$

Then defining

$$\Gamma_l(E) = -\frac{2k^{2l+1} R_0^{\,2l}}{[1 \cdot 3 \cdot 5 \cdots (2l-1)]^2 \, [\partial G_l(k, R_0)/\partial E]_{E=E_{0l}}} \tag{5-13}$$

Eq. 5-12 reduces to

$$\cot \delta_l = -\frac{2[E - E_{0l}]}{\Gamma_l(E)} \tag{5-14}$$

which is the basic result.

It is convenient to rewrite Eq. 5-14 in terms of the S- and T-matrices. Before doing so, we consider the more general situation in which the interacting particles may have nonzero spins. Then a particular value of l may occur for several different values of the total angular momentum J. But we find experimentally that a single resonance has a *unique* J as well as a *unique* l. Therefore, without taking the space to extend Eqs. 5-6 through 5-14 to particles with spin, we replace the subscript l by the subscript Jl. Returning to Eq. 5-14, a little manipulation leads us to the *Breit–Wigner resonance formulas* (5BL).

$$S_{Jl}(E) - 1 = -\frac{i\Gamma_{Jl}(E)}{(E - E_{0Jl}) + i\Gamma_{Jl}(E)/2} \tag{5-15}$$

$$T_{Jl}(E) = -\frac{\Gamma_{Jl}(E)}{2(E - E_{0Jl}) + i\Gamma_{Jl}(E)} \tag{5-16}$$

We would like to test the Breit–Wigner formula against a strong interaction resonance in a formation experiment. We would also like a minimum amount of interference from other amplitudes. The $\Delta(1236)$ resonance formed in π^+p elastic scattering provides such a case. Dropping the Jl subscript for simplicity, we write

$$\text{Im}\, T = \frac{\Gamma^2(E)}{4(E - E_0)^2 + \Gamma^2(E)} \tag{5-17a}$$

$$\text{Re}\, T = \frac{2(E_0 - E)\Gamma(E)}{4(E - E_0)^2 + \Gamma^2(E)} \tag{5-17b}$$

Figure 5-6 (5BAR) shows the Im T and Re T for the $\Delta(1236)$. The Im T is not symmetric about E_0, as it would be if Γ were a constant. Rather, we see that the high energy side of the resonance is broader and falls off more slowly with $|E - E_0|$. This effect is clearly seen by comparing the experimental data with curve 1 given by the Breit–Wigner formula with Γ constant at 126 MeV. But Eq. 5-13 shows that Γ is a function of E through k^{2l+1} in the numerator and perhaps some k dependence in the differential in the denominator. If we ignore the latter k dependence, we can write

$$\Gamma(E) = \Gamma_0 \left(\frac{k}{k_0}\right)^{2l+1} \tag{5-18}$$

Equation 5-18 in Eq. 5-17a will lead to an asymmetric curve for Im T as shown by curve 2 in Fig. 5-6. Here $\Gamma_0 = 130$ MeV and $l = 1$. Thus the k^{2l+1} term in Eq. 5-18 provides most of the required asymmetry. As discussed by Refs. 5BAR, 5JA, and 5PI, other corrections can be applied to $\Gamma(E)$. The factor $(k/k_0)^{2l}$ or

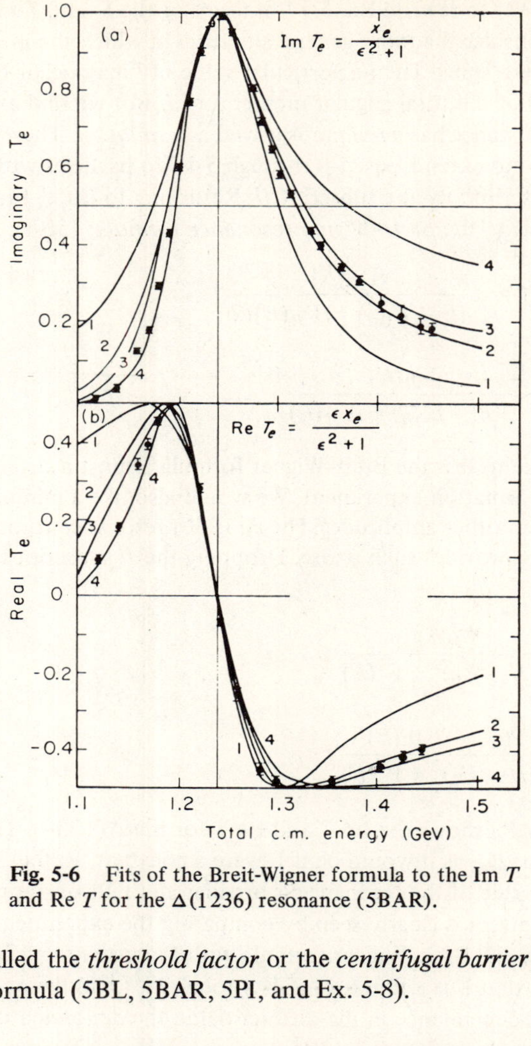

Fig. 5-6 Fits of the Breit-Wigner formula to the Im T and Re T for the $\Delta(1236)$ resonance (5BAR).

$(k/k_0)^{2l+1}$ is called the *threshold factor* or the *centrifugal barrier factor* in the Breit–Wigner formula (5BL, 5BAR, 5PI, and Ex. 5-8).

5-7 The Breit–Wigner Formula, Inelastic Reactions, and *s*-Channel Resonances. Our next task is to extend the Breit–Wigner formula to make it usable in the presence of inelastic as well as elastic two-body reactions. Thus we seek a multichannel Breit–Wigner formula. There is no complete theory of strong interactions; therefore, no general and rigorous derivation of the multichannel formula can be given. Of the various derivations (5DA), we choose a simple, limited derivation, but one we believe illustrates well the physical picture behind the multichannel formula.

In this section we follow a philosophy very similar to that in the compound nucleus resonance theory of nuclear reactions (5BL). Thus we are using very general concepts. Actually we have no choice, for we have no rigorous theory of resonances in strong interactions. We consider a set of two-body reactions in which the initial state $|i\rangle$ can go to different final states $|f\rangle$. Thus we have $a + b \to c + d, a + b \to e + f, \ldots$, as well as the elastic reaction $a + b \to a + b$. For simplicity, all particles are taken to have zero intrinsic spin. The initial and final states are resolved into eigenstates of orbital angular momentum and isotopic spin. The symbols i and f denote the type of particles, l, I, and I_z. We use the unitary property (Ch. 6) and the symmetric property (Sec. 10-4) of the S-matrix.

Since S is a unitary symmetric matrix, it can be diagonalized by a unitary real matrix. If the diagonalizing matrix is C and the diagonal form of S is called S', then $S' = C^\dagger SC$ and $S = CS'C^\dagger$. Therefore

$$S_{fi} = (CS'C^\dagger)_{fi} = \sum_\alpha C_{f\alpha}S'_{\alpha\alpha}C^\dagger_{\alpha i} = \sum_\alpha C_{f\alpha}C_{i\alpha}S'_{\alpha\alpha} \tag{5-19}$$

Inserting S_{fi} from Eq. 5-19 into Eq. 3-17, we obtain

$$T_{fi} = \frac{1}{2i}\left[\sum_\alpha C_{f\alpha}C_{i\alpha}S'_{\alpha\alpha} - \delta_{fi}\right] \tag{5-20}$$

Now $\Sigma_\alpha C_{f\alpha}C_{i\alpha} = \delta_{fi}$ because C is a real unitary matrix. Therefore, Eq. 5-20 can be written

$$T_{fi} = \frac{1}{2i}\left[\sum_\alpha C_{f\alpha}C_{i\alpha}(S'_{\alpha\alpha} - 1)\right] \tag{5-21}$$

Equation 5-21 is simply a mathematical result based on the fact that S' is a unitary symmetric matrix. Let us consider the physical interpretation: $S'_{\alpha\alpha}$ is the S-matrix element for a transition from an initial state $|\alpha\rangle$ to a final state $|\alpha\rangle$; the states α are just related to the physical states $|i\rangle$ by $|i\rangle = \Sigma_\alpha C_{i\alpha}|\alpha\rangle$. Since S' is diagonal, there are only elastic processes between the $|\alpha\rangle$ states. Thus we have replaced the matrix S, which allows elastic and inelastic processes between the states $|i\rangle$, with the matrix S', which allows only elastic processes between a different set of states. If we must know most of the $S'_{\alpha\alpha}$ elements to calculate a particular S_{fi} element, we have gained nothing by these manipulations. But let us make two basic physical assumptions.

1. Only one of the $S'_{\alpha\alpha}$ elements, say $S'_{\gamma\gamma}$, has $S_{\alpha\alpha} - 1 \neq 0$.
2. The elastic scattering from the initial $|\gamma\rangle$ state to the final state $|\gamma\rangle$ is resonant. Specifically, using Eq. 5-15, we have

$$S'_{\gamma\gamma} - 1 = \frac{-i\Gamma_\gamma}{(E - E_{0\gamma}) + i\Gamma_\gamma/2} \tag{5-22}$$

With these two assumptions, Eq. 5-21 reduces to

$$T_{fi} = \frac{-C_{f\gamma}C_{i\gamma}\Gamma_\gamma}{2(E - E_{0\gamma}) + i\Gamma_\gamma} \tag{5-23}$$

Before using Eq. 5-23, let us interpret these assumptions. Suppose that all the members of the set of reactions $|i\rangle \rightarrow |f\rangle$ go through one and only one intermediate state γ, and suppose further that this intermediate state is a strong interaction resonance. Then Eq. 5-23 would be the result. For $C_{i\gamma} = \langle i|\gamma\rangle$ corresponds to the projection of the state $|i\rangle$ onto the state $|\gamma\rangle$. That is, $C_{i\gamma}$ is the overlap of the state $|i\rangle$ with the state $|\gamma\rangle$. Similarly, $C_{f\gamma}$ is the projection of the final state $|f\rangle$ on the state $|\gamma\rangle$.

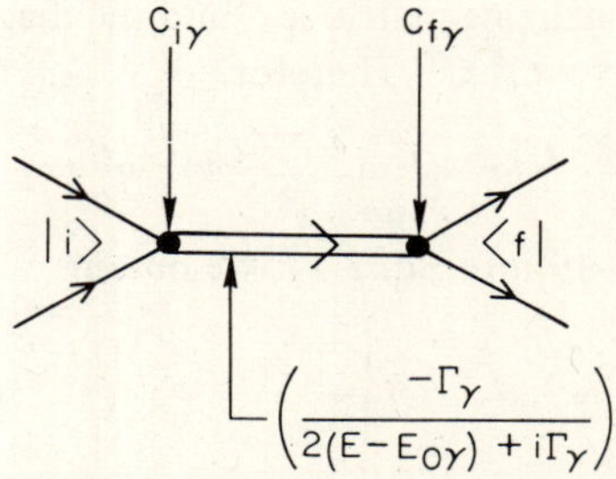

Fig. 5-7 Schematic representation of an s-channel resonance. The double line represents the intermediate but real resonant state $|\gamma\rangle$. This is not a Feynman diagram.

We often think of the state $|\gamma\rangle$ as having a transient but real existence visualized through Fig. 5-7. At the left vertex the state $|i\rangle$ converts into the state $|\gamma\rangle$. The intermediate double line represents the resonant scattering from the initial state $|\gamma\rangle$ to the final state $\langle\gamma|$. Thus the double line is a strong interaction resonance of mass $E_{0\gamma}$. The right vertex represents the conversion of $\langle\gamma|$ into $\langle f|$. Figure 5-7 looks like but is *not* a Feynman diagram. For if we reduce a propagator to its nonrelativistic form, we obtain $(E - E_0)^{-1}$; there is no imaginary part to the denominator. Using the s-channel terminology of Secs. 7-5 and 15-3, we usually refer to the entire phenomenon as an *s-channel resonance.* This resonance has the l, I, and I_z quantum numbers of the states $|i\rangle$, $|\gamma\rangle$, and $|f\rangle$, and it has mass $E_{0\gamma}$. In the terminology of Sec. 5-2, this resonance is the intermediate resonant state X in a formation experiment.

Returning to the discussion preceding Eq. 5-22, the states $|\alpha\rangle$ are often called *eigenstates* (5DA) of the S-matrix. We also write for use in the next section

$$S'_{\alpha\alpha} = e^{2i\delta_\alpha} \tag{5-24}$$

and the δ_α's are called the *eigenphases.* This form is required because $|S'_{\alpha\alpha}|^2 = 1$ for the purely elastic scattering from the state $|\alpha\rangle$ to the state $|\alpha\rangle$. From Eq. 5-22 we note that

$$S'_{\gamma\gamma} = e^{2i\delta_\gamma} = \frac{(E - E_{0\gamma}) - i\Gamma_\gamma/2}{(E - E_{0\gamma}) + i\Gamma_\gamma/2} \tag{5-25}$$

We return to Eq. 5-23 and simplify the notation by defining

$$\Gamma_i = C_{i\gamma}{}^2\Gamma_\gamma, \qquad \Gamma_f = C_{f\gamma}{}^2\Gamma_\gamma \tag{5-26}$$

and by dropping the γ; there is only one intermediate state. Then we write

$$T_{fi} = \frac{-\sqrt{\Gamma_i\Gamma_f}}{2(E - E_0) + i\Gamma} \tag{5-27}$$

Finally, defining

$$\Gamma_{\text{el}} = \Gamma_i, \qquad \Gamma_{\text{inel}} = \Gamma - \Gamma_{\text{el}} \tag{5-28}$$

we have the cross-section formulas for spin zero particles

$$\sigma_{\text{el}} = \frac{\pi}{k^2}\,[2l+1]\,\frac{\Gamma_{\text{el}}{}^2}{(E - E_0)^2 + \Gamma^2/4} \tag{5-29a}$$

$$\sigma_{fi} = \frac{\pi}{k^2}\,[2l+1]\,\frac{\Gamma_{\text{el}}\Gamma_f}{(E - E_0)^2 + \Gamma^2/4} \tag{5-29b}$$

$$\sigma_{\text{inel}} = \sum_{f\neq i} \sigma_{fi} = \frac{\pi}{k^2}\,[2l+1]\,\frac{\Gamma_{\text{el}}\Gamma_{\text{inel}}}{(E - E_0)^2 + \Gamma^2/4} \tag{5-29c}$$

$$\sigma_{\text{tot}} = \frac{\pi}{k^2}\,[2l+1]\,\frac{\Gamma_{\text{el}}\Gamma}{(E - E_0)^2 + \Gamma^2/4} \tag{5-29d}$$

Exercise 5-6. Fill in the steps leading to Eq. 5-29.

We observe that the denominators of all these cross sections are the same. Therefore, all these resonant peaks have the same full width at half maximum, namely, Γ. Thus in principle we can find the parameters of a resonance by observing a peak in the elastic cross section, in the total cross section, or in any inelastic cross section. However, this can rarely be done in practice because most resonant states are overlapped by neighboring resonant states.

It is useful to consider the locus of T_{el} on an Argand diagram when T_{fi} obeys the simple, multichannel Breit–Wigner formula (Eq. 5-27). To do this we first make some simplifying changes of variable. We define

$$\chi_f = \frac{\Gamma_f}{\Gamma} = C_{f\gamma}{}^2 \tag{5-30}$$

In particular, we note that for χ_i, called the *elasticity*, we have

$$\chi_i = \frac{\Gamma_{\text{el}}}{\Gamma} = \frac{\sigma_{\text{el}}}{\sigma_{\text{tot}}} \tag{5-31}$$

We also define the reduced energy difference

$$\epsilon = \frac{2(E_0 - E)}{\Gamma} \tag{5-32}$$

Then

$$T_{\text{el}} = \frac{\chi_i}{\epsilon - i} = \frac{i\chi_i}{2} + \frac{i\chi_i}{2}\left[\frac{i+\epsilon}{i-\epsilon}\right] \tag{5-33}$$

We observe that

$$\left|\frac{i+\epsilon}{i-\epsilon}\right| = 1, \qquad |\chi_i| \leqslant 1 \tag{5-34}$$

Therefore, the locus of T_{el} is a circle of radius $|\chi_i/2|$ centered at $i\chi_i/2$. This is shown in Fig. 5-4*d*.

Figure 5-4*e* presents the locus of T_{el} for two cases: $\chi_i > \frac{1}{2}$ and $\chi_i < \frac{1}{2}$. When χ_i is greater than $\frac{1}{2}$, the top of the circle corresponds to $2\delta = 180°$ or $\delta_i = 90°$. But when χ_i is less than $\frac{1}{2}$, the top of the circle corresponds to $\delta_i = 0°$. Thus depending on the size of the elastic cross section compared with the total cross section (Eq. 5-21), the center of a resonance occurs when the phase shift δ_i is 90° or 0°.

5-8 *s*-Channel Resonances and Nonresonant Backgrounds. Next we must face the physical fact that almost all resonances in formation experiments occur in the presence of nonresonant backgrounds and tails from neighboring resonances. Therefore, we must revise the two simple assumptions used in the derivation of the basic equation, Eq. 5-23. But we cannot do this in a general way unless we also revise the derivation itself. As reviewed by Dalitz and Moorhouse (5DA), when several of the matrix elements $S'_{\alpha\alpha}$ (Eq. 5-24) are nonzero, there is a connection between them such that if one eigenphase goes through a resonance, others will do the same. The connection between the matrix elements $S'_{\alpha\alpha}$ comes about because the $C_{i\alpha}$'s are themselves energy dependent. As the energy changes, the $C_{i\alpha}$'s change, the diagonalization of the S-matrix changes, and the diagonal elements become intermixed. We do not have the space to develop these ideas further, but fortunately one simple case illustrates the effect of backgrounds without nullifying the derivation. We assume that in addition to the isolated resonance in the $|\gamma\rangle$ eigenstate, there is an energy-independent background of the same magnitude and phase in each diagonal eigenstate. Thus we set

$$\begin{aligned} S'_{\alpha\alpha} &= e^{2i\delta_B}, \qquad \alpha \neq \gamma \\ S'_{\gamma\gamma} &= e^{2i(\delta_\gamma+\delta_B)} = \frac{(E - E_{0\gamma}) - i\Gamma_\gamma/2}{(E - E_{0\gamma}) + i\Gamma_\gamma/2}\, e^{2i\delta_B} \end{aligned} \tag{5-35}$$

where δ_B is any real number. This form is required by the condition on the diagonalized S'-matrix that $|S'_{\alpha\alpha}|^2 = 1$. Then Eq. 5-21 leads to

$$T_{fi} = \frac{-\sqrt{\Gamma_f\Gamma_i}\, e^{2i\delta_B}}{2(E - E_0) + i\Gamma} + \frac{e^{2i\delta_B} - 1}{2i}\,\delta_{fi} \tag{5-36}$$

Comparing Eqs. 5-36 and 5-27 we see that we have obtained a more general form for the behavior of an s-channel resonance. We usually make Eq. 5-36 even more general by going beyond the derivation and allowing δ_B to be a *slowly* varying function of the energy.

Exercise 5-7. Derive Eq. 5-36 and prove the statements made next about the behavior of T_{el}.

To illustrate the effect of the background, we note the following properties of the elastic amplitude T_{el}.

1. T_{el} has the form of a circle of radius $\chi_i/2$, as before.
2. But the center of the circle is now displaced and is usually off the imaginary axis. The displacement is given by $[\exp(2i\delta_B) - 1]/2i$.
3. The resonance is rotated about its center by the angle $2\delta_B$. Therefore, the resonance no longer occurs at the top of the circle.

These three points are illustrated in Fig. 5-4*f*. The displacement of the center of the circle adds some confusion to the task of finding the center of the resonance. But the greatest confusion is caused by the rotation by $2\delta_B$. For example, if $\delta_B = 90°$, the resonance will occur at the bottom of the circle. Furthermore, often δ_B itself has some energy dependence, thus adding to the uncertainty. In Fig. 5-4*f* the heavy line shows a possible path of T_{el}.

We can see these effects by looking again at the actual values of T_{el} in the pion–nucleon system in Fig. 5-5. We have indicated on the Argand diagrams the positions of the confirmed resonances according to Ref. 5PDG. These resonances are also listed in Table 5-1 and in App. D. As an example, consider the S_{11}

Table 5-1 Parameters of the πp resonances in Fig. 5-5. The resonances marked with "?" are not established at present. Also, established πp resonances not in Fig. 5-5 are not listed. See App. D and Ref. 5PDG.

Resonance Name	Phase Shift Name	Mass (MeV)	Γ_{total} (MeV)	$\Gamma_{el}/\Gamma_{tot} = \chi_i$
$\Delta(1650)$	S_{31}	1615–1695	130–200	0.29
$\Delta(1910)$	P_{31}	1780–1935	230–420	0.25
$\Delta(1236)$	P_{33}	1230–1236	110–122	1.00
?$\Delta(1690)$	P_{33}			0.10
$\Delta(1670)$	D_{33}	1650–1720	175–300	0.15
?$\Delta(1960)$	D_{35}			0.14
$\Delta(1890)$	F_{35}	1840–1920	135–350	0.17
$\Delta(1950)$	F_{37}	1930–1980	140–220	0.44
$N(1535)$	S_{11}	1500–1600	50–160	0.38
$N(1710)$	S_{11}	1665–1765	100–400	0.65
$N(1470)$	P_{11}	1435–1505	165–400	0.61
$N(1780)$	P_{11}	1650–1860	50–450	0.29
$N(1860)$	P_{13}	1770–1900	180–330	0.25
$N(1520)$	D_{13}	1510–1540	105–150	0.54
?$N(2040)$	D_{13}			0.20
$N(1670)$	D_{15}	1655–1680	105–175	0.42
$N(1688)$	F_{15}	1680–1692	105–180	0.62
?$N(1990)$	F_{17}			0.12

amplitudes. The $N(1710)$ occurs at the top of a clear resonant circle and has a large elasticity (0.65), thus is well established. On the other hand, it is more difficult to be sure about the $N(1535)$. The elasticity is smaller (0.38), leading to a smaller and less evident resonant circle. But the most confusing feature is the low energy tail of the $N(1710)$, which begins to dominate the amplitude just as the $N(1535)$ seems to be reaching its resonant center. Thus the fitting of phase shifts, and the finding of resonances in formation experiments is difficult (5CE, 5JO, 5BAR).

5-9 The Relativistic Breit–Wigner Formula. Although all the work in this chapter has been nonrelativistic, we conclude by giving a relativistic form for the Breit–Wigner formula. First, we rewrite E_0 as m_0 to emphasize that E_0 may be interpreted as the mass m_0 of a resonance. Remember that we have defined E to be the total center of mass energy. Now if $|E - m_0| \ll m_0$, we can write Eq. 5-27 in the form

$$T_{fi} = \frac{-m_0\sqrt{\Gamma_i\Gamma_f}}{2m_0(E - m_0) + im_0\Gamma} \approx \frac{-m_0\sqrt{\Gamma_i\Gamma_f}}{(E^2 - m_0^{\,2}) + im_0\Gamma} \tag{5-37}$$

Finally, setting $E^2 = s$, where s is the usual kinematic variable denoting the square of the total energy in the center of mass, we obtain the *relativistic Breit–Wigner formula*

$$T_{fi} = -\frac{-m_0\sqrt{\Gamma_i\Gamma_f}}{(s - m_0^2) + im_0\Gamma} \tag{5-38}$$

In Sec. 15-11 we give a better justification of this formula. To conclude this section, we observe that the main effect of this on the cross-section formulas of Eq. 5-29 is to change the denominators to $(s - m_0^2)^2 + m_0^2\Gamma^2$.

5-10 Threshold Effects. If we assume in Eq. 5-12 that the dominant energy dependence is outside the square bracket, we can obtain some useful information about the threshold behavior of T_l. We are of course then assuming that the resonance energy in this channel is far from threshold, if the channel has a resonance. Taking the quantity inside the square bracket to be independent of energy, we find

$$\cot \delta_l = \frac{1}{C_l k^{2l+1}} \tag{5-39}$$

where C_l is a constant. Then we have

$$T_l = \frac{C_l k^{2l+1}}{1 - iC_l k^{2l+1}} \tag{5-40}$$

Although this derivation is very crude, the $(k)^{2l+1}$ has some basic significance because it comes directly from the low energy behavior of the Bessel functions and is somewhat independent of the interaction. One problem is that this derivation makes it difficult in strong interactions to determine C_l, hence to estimate the significance of the denominator in Eq. 5-40. But if we assume that near *threshold* (where the momentum k is close to 0) the denominator varies slowly compared to the numerator, we obtain the *threshold effect*

$$T \sim \text{constant} \times k^{2l+1} \tag{5-41}$$

We have already noted in connection with Eq. 5-18 that this *threshold factor* appears in the width Γ of a resonance. But the threshold effect is much more general and is frequently used in nuclear and particle physics.

Exercise 5-8. Show that the threshold factor comes directly from the $z \to 0$ behavior of the Bessel function combination $n_l(z)/j_l(z)$. Hence the right-hand side of Eq. 5-41 is usually called the *centrifugal barrier penetration factor.* Show that the 1 in the exponent may be attributed to a *phase space factor* (Sec. 6-6). Hence the factor in Eq. 5-41 is really the product of the centrifugal barrier penetration factor and the phase space factor.

5-11 An Overview of Resonances. Let us review what we have learned in this chapter. Most important, we have found that we can explain resonant effects in low energy elastic scattering if we take the strong interaction resonances to be real but short-lived particles that can act as real intermediate states. Although we have not discussed it here, the same explanation holds for resonant effects in K^-p elastic scattering and for resonant effects in low energy, inelastic reactions in the πp, K^-p, and γp system.

We have also shown that we can fit the experimental observations on these reactions if we use the Breit–Wigner formula and add various background and correction terms. The Breit–Wigner formula works quite well, even though we derived it from low energy, nonrelativistic, potential scattering theory. In Ch. 15 we see that the Breit–Wigner formula can also be derived from the more general, relativistic concept of poles in the S-matrix. This will explain why the formula is so widely applicable in elementary particle physics.

Finally, we have learned a little bit about the properties of the strong interaction resonances as illustrated by those in the πp system.

Now let us see what we have not learned. Most important, we have not learned anything about the fundamental dynamic nature of the strong interactions resonances. That is, we have not learned why strong interaction resonances exist or why they have their specific properties. In the author's view, the strong interaction dynamics leading to resonances is not understood at present. Therefore, in this book we make no attempt to discuss that subject.

But in terms of the constituent models of hadrons (Ch. 14 and Sec. 14-7) we have an explantion—indeed, a quite detailed explanation—of the properties of the strong interaction resonances. In the author's view such models are not fundamental dynamic theories because of the dynamic uncertainties and paradoxes that now exist in such theories. Nevertheless, the constituent models work very well. In fact, they provide the only explanation of some very general rules that appear to govern which strong interaction resonances actually occur. For example, as discussed in Ex. 14-7, strong interaction resonances are found in the $\pi p, K^-p$, and γp systems. But no strong interaction resonances have been found and confirmed in the K^+p system or in *any baryon–baryon system.* Thus there is a general rule: the resonances do not occur in $B = +1, S = +1$ systems or in $|B| > 1$ systems.

REFERENCES

5BA G. Baym, *Lectures on Quantum Mechanics* (Benjamin, New York, 1969).

5BAR A. Barbaro-Galtieri, *Advances in Particle Physics*, Vol. 2, R. L. Cool and R. E. Marshak, Eds. (Interscience, New York, 1968).

5BL J. W. Blatt and V. F. Weisskopf, *Theoretical Nuclear Physics* (Wiley, New York, 1952); this is the classic book on the Breit–Wigner formula and related topics.

5CE R. J. Cence, *Pion–Nucleon Scattering* (Copyright © by Princeton University Press, Princeton, N.J., 1969). Figures 5-1 and 5-3 reprinted by permission of Princeton University Press.

5DA R. H. Dalitz and R. G. Moorhouse, *Proc. Royal Soc.* (*London*) **318**, 279 (1970).

5HE W. N. Hess, *Rev. Mod. Phys.* **30**, 68 (1958).

5JA J. D. Jackson, *Nuovo Cimento* **34**, 1644 (1964).

5JO C. Johnson and H. Steiner, *Pion-Nucleon Scattering,* G. L. Shaw and D. Y. Wong, Eds. (Wiley-Interscience, New York, 1969).

5LA L. D. Landau and E. M. Lifshitz, *Quantum Mechanics, Non-Relativistic Theory* (Addison-Wesley, Reading, Mass., 1958).

5PDG See the yearly *Review of Particle Properties* by the Particle Data Group; for example, P. Söding et al., *Phys. Letters* **39B**, 1 (1972).

5PI H. Pilkuhn, *The Interaction of Hadrons* (North-Holland, Amsterdam, 1967).

5SC L. I. Schiff, *Quantum Mechanics* (McGraw-Hill, New York, 1968).

CHAPTER

6

The *S*-Matrix, Cross Sections, and Coordinate Transformations

6-1 Introduction. In a rigorous development of the *S*-matrix, two related topics require lengthy discussion. One involves the definition of free particle states in the presence of an interaction; the other relates to the construction and manipulation of wave packets. We cannot devote sufficient space to these topics here; thus we must be content with nonrigorous derivations and definitions. There are a number of thorough books to which the reader can refer if he wishes to study this subject (6GA, 6GO, 6MA, 6RO, 6WE).

The chapter begins with the development of the *S*-matrix concept for states described only by discrete quantum numbers. This simplifies the treatment by temporarily avoiding the normalization problem. The normalization problem for plane wave states is discussed next, however, followed by the development of the S-matrix for the general case. The T-matrix is then defined and the general unitarity relation is derived.

The chapter continues with Sec. 6-6, a summary of the relationships between the T-matrix and experimentally determined quantities—cross sections and decay rates. This is not a profound section, but the formulas it contains are used very frequently in the book. Section 6-7 contains a formal definition and derivation of the optical theorem, a relationship first introduced at the end of Ch. 1.

The remaining third of the chapter is devoted to coordinate transformations—spatial translations, spatial rotations, Lorentz velocity transformations, space inversion, and time reversal. The structure and predictions of strong interaction theory must be invariant under such transformations. If we possessed a complete, fundamental theory of strong interactions, we would not bother much in this book with invariance under coordinate transformations; such invariance would be built into the theory. But we do not possess a complete fundamental theory. Therefore, at many different levels in the discussion of strong interaction dynamics we must use invariance under coordinate transformations to restrict the possible forms of the theory, to make predictions, and to guide the discussion. Our purpose is to develop the mechanisms and notations of coordinate transformations for just such uses. Our approach is practical rather than formal, and we do not discuss the general concept of symmetry transformations, the relation between such transformations and conserved quantities (6ME, 6MA, 6GA), or the Lie theory of continuous transformation groups (6HA).

6-2 The S-Matrix for Discrete States. We suppose that we have a set of free particle states $|\alpha\rangle$ specified by a set of discrete quantum numbers represented by α. The number of such states may be finite or denumerable. One example would be two noninteracting particles whose state is specified by the total angular momentum state $|s_1s_2SLJM\rangle$ (see Ch. 3). Another example is a set of noninteracting particles, each described by a plane wave state, the plane waves being quantized in a box of finite dimensions. We also suppose that the states are *normalized* and *orthogonal,* with

$$\langle\alpha'|\alpha\rangle = \delta_{\alpha'\alpha} \tag{6-1}$$

Here if α represents the parameters $l, m, n, \ldots, \delta_{\alpha'\alpha} = \delta_{l'l}\delta_{m'm}\delta_{n'n}, \ldots$. The set of states is also *complete*, which means that any state $|\phi\rangle$ can be written

$$|\phi\rangle = \sum_\alpha C_\alpha|\alpha\rangle \qquad \text{or} \qquad |\phi\rangle = \sum_\alpha |\alpha\rangle\langle\alpha|\phi\rangle \tag{6-2}$$

We define these states in the Heisenberg picture. Therefore, these states do not change in time. For example, a state $|\alpha\rangle$ consisting of a plane wave of pions incident on a plane wave of protons remains always in that state. If the interaction of the pion and proton results in the production of a kaon and a hyperon, the kaon and the hyperon are represented by a different state $|\beta\rangle$, consisting of a kaon plane wave and a hyperon plane wave. To describe the interaction in the Heisenberg picture, we say that before the interaction, the state of the system coincided with the state $|\alpha\rangle$; and after the interaction the state of the system coincided with the state $|\beta\rangle$.

We now suppose that we have a system that interacts for a finite period of time between time t_1 and time t_2. Before the time t_1 the initial state of the system coincides with the state $|i\rangle$; after the time t_2 the final state of the system coincides with the state $|f\rangle$. The *scattering operator S* is defined as the operator that gives the *transition probability amplitude* S_{fi} for the initial state $|i\rangle$ to go to the final state $|f\rangle$. The symbol S is also used to designate the matrix whose elements give that transition probability amplitude. Thus we have

$$S_{fi} = \langle f|S|i\rangle \tag{6-3}$$

The probability of going from the state $|i\rangle$ to the state $|f\rangle$ is the absolute square of the probability amplitude S_{fi}. We call this probability P_{fi}. Then we have the relations

$$\begin{aligned} P_{fi} &= \langle f|S|i\rangle^*\langle f|S|i\rangle = S_{fi}^*S_{fi} \\ &= S_{if}^\dagger S_{fi} \end{aligned} \tag{6-4}$$

The total probability of going to any state $|f\rangle$ from the initial state $|i\rangle$ is obtained by summing P_{fi} over all the final states $|f\rangle$. It is very important to recognize that we include in this summation the state $|i\rangle$ as one of the final states. Thus we write

$$P_{\text{total}} = \sum_f P_{fi} = \sum_f S_{if}^\dagger S_{fi}$$

But the probability of anything happening plus the probability of nothing happening is 1. Therefore, we obtain

$$\sum_f S_{if}^\dagger S_{fi} = \langle i|S^\dagger S|i\rangle = 1 \tag{6-5}$$

We can derive an additional property of S, however. Consider a state $|j\rangle$ which is a superposition of two *different* orthonormal states $|i\rangle$ and $|i'\rangle$. Define the normalized superposition state

$$|j\rangle = \frac{a|i\rangle + a'|i'\rangle}{(|a|^2 + |a'|^2)^{1/2}}$$

where a and a' are independent arbitrary complex numbers. Replacing the state $|i\rangle$ by this state $|j\rangle$ in Eq. 6-5, we find

$$0 = a^*a'\langle i|S^\dagger S|i'\rangle + aa'^*\langle i'|S^\dagger S|i\rangle$$

This can only be satisfied if

$$\langle i|S^\dagger S|i'\rangle = \sum_f S^\dagger_{if}S_{fi'} = 0 \qquad \text{for} \qquad i \neq i' \tag{6-6}$$

Exercise 6-1. Fill in the steps leading to Eq. 6-6.

Equations 6-5 and 6-6 can be combined

$$\sum_f S^\dagger_{jf}S_{fi} = \delta_{ji} \tag{6-7a}$$

and written in the matrix form

$$S^\dagger S = 1 \tag{6-7b}$$

where I is the unit matrix. Therefore, S is a *unitary* matrix, and Eqs. 6-7 express the *unitarity* of or *unitarity condition* on the S-matrix.

Finally, we note that S includes the probability that there is no interaction. To eliminate this aspect of S we define another matrix, the T-matrix, by

$$S = I + 2iT, \qquad T = \frac{S - I}{2i} \tag{6-8}$$

And this is a more general definition, for discrete states, of the T-matrix introduced in Ch. 3.

If we insert Eq. 6-8 into Eq. 6-7b, we find

$$I = S^\dagger S = I + 2i(T - T^\dagger) + 4T^\dagger T$$

Thus

$$i[T^\dagger - T] = 2T^\dagger T \tag{6-9}$$

which is the *unitarity relation* for discrete states.

The set of two-body reactions discussed in Sec. 3-7 provides an example of the use of the discrete S- and T-matrices. Of course we are dealing with a more general situation here. In particular, the states may consist of more than two particles.

6-3 Normalization of Continuous Plane Wave States. We now begin to extend our considerations to a complete set of plane wave states in which we must use a continuous momentum variable p for the specification. We start with states consisting of a single particle. These states are next specified by the three-momentum $\mathbf{p}$, and α–a discrete parameter representing the spin state of the particle. The orthonormal property is then

$$\langle \mathbf{p}'\alpha' \mid \mathbf{p}\alpha \rangle = N(p)\delta^3(\mathbf{p}' - \mathbf{p})\delta_{\alpha'\alpha} \tag{6-10}$$

where $N(p)$ is a *normalization factor* yet to be determined. As indicated by the notation, it is allowed to be a function of p. The Dirac δ function that appears here is the continuous extension of the Kronecker δ in Eq. 6-1.

The states $|\mathbf{p}\alpha\rangle$ are defined using the Heisenberg picture; thus they do not change in time, and in the coordinate representation for *spinless* particles these states $|\mathbf{p}, \alpha\rangle$ have the form

$$\langle \mathbf{r} \mid \mathbf{p} \rangle = \left(\frac{N(p)}{(2\pi)^3}\right)^{1/2} [e^{-i\mathbf{p}\cdot\mathbf{r}}] \tag{6-11}$$

which is a plane wave of infinite extent. With such a plane wave the particle has an equal probability of being anywhere, and it is not localized at all. Therefore, we cannot speak about particles that are well separated from one another and are hence free particle states. This leads to the problems we mentioned in the introduction to this chapter. We do not discuss them here, remarking that the formation of wave packets provides the way out.

Returning to Eq. 6-10, our first task is to study the normalization factor. Since the states $|\mathbf{p}\alpha\rangle$ form a complete set, we can expand any state $|\phi\rangle$ as a summation over those base states. This is in analogy with Eq. 6-2 for purely discrete states. Now we must replace the summation by an integral. To do this we have to use the *density of states* $W(p)$, which lets us know the number of continuous states between $\mathbf{p}$ and $\mathbf{p} + d\mathbf{p}$. Then in analogy to Eq. 6-2 we have

$$|\phi\rangle = \sum_{\alpha'} \int d^3p' [W(p') \,|\mathbf{p}'\alpha'\rangle\langle \mathbf{p}'\alpha' \mid \phi\rangle] \tag{6-12}$$

Now let $|\phi\rangle$ be the state $|\mathbf{p}\alpha\rangle$ and take the scalar product of both sides of Eq. 6-12 with yet another state $|\mathbf{p}'', \alpha''\rangle$. We obtain

$$\langle \mathbf{p}''\alpha'' \mid \mathbf{p}\alpha\rangle = \sum_{\alpha'} \int d^3p' [W(p')\langle \mathbf{p}''\alpha'' \mid \mathbf{p}'\alpha'\rangle\langle \mathbf{p}'\alpha' \mid \mathbf{p}\alpha\rangle]$$

which, on using Eq. 6-10 and simplifying, becomes

$$N(p)\delta(\mathbf{p}'' - \mathbf{p})\delta_{\alpha''\alpha} = W(p)N(p)N(p)\delta^3(\mathbf{p}'' - \mathbf{p})\delta_{\alpha''\alpha}$$

Canceling the Dirac δ functions (really not a rigorous operation), we find

$$N(p) = \frac{1}{W(p)} \tag{6-13}$$

Thus the normalization factor and the density of states factor are related, as we might have expected. We can choose any pair of functions consistent with Eq. 6-13; there are many choices. Let us choose

$$\begin{aligned} N(p) &= (2\pi)^3 2E \\ W(p) &= [(2\pi)^3 2E]^{-1} \\ E &= +[\mathbf{p}^2 + m^2]^{1/2} \end{aligned} \tag{6-14}$$

The factor $2(2\pi)^3$ is quite arbitrary. The factor E is convenient because the expression

$$\frac{d^3p}{(2\pi)^3 2E}$$

and the normalization

$$\langle \mathbf{p}'\alpha' \mid \mathbf{p}\alpha \rangle = (2\pi)^3 2E\delta^3(\mathbf{p}' - \mathbf{p})\delta_{\alpha'\alpha} \tag{6-15}$$

are Lorentz invariant.

Exercise 6-2. Show that the last two expressions are Lorentz invariant.

With these definitions fixed, we can calculate the number of particles per unit volume. In this calculation we are again manipulating δ functions and infinities in a nonrigorous manner. The total number of particles in a particular $|\mathbf{p}\alpha\rangle$ state is given by Eq. 6-15 for $\mathbf{p}' = \mathbf{p}$. We define the δ function by

$$\delta^3(\mathbf{p}) = \lim_{L\to\infty} \left[\frac{1}{(2\pi)^3} \int\!\!\int\!\!\int_{-L/2}^{+L/2} dx\, dy\, dz\, e^{-i\mathbf{p}\cdot\mathbf{r}} \right]$$

Then $\delta(0)$ has the meaning $V/(2\pi)^3$ where $V = L^3$ is the infinite volume in which the interaction takes place. If we then divide the expression

$$\langle \mathbf{p}\alpha \mid \mathbf{p}\alpha \rangle = (2\pi)^3 2E\delta^3(0) = 2EV$$

by V we obtain the number of *particles per unit volume* $\rho(\mathbf{p}\alpha)$ where

$$\rho(\mathbf{p}\alpha) = 2E \tag{6-16}$$

Until now we have been discussing a single particle, but we can immediately generalize again in the Heisenberg picture to the direct product of two or more free particle, plane wave states

$$|\mathbf{p}_1\alpha_1\rangle|\mathbf{p}_2\alpha_2\rangle\cdots = |\mathbf{p}_1\alpha, \mathbf{p}_2\alpha_2, \ldots\rangle \tag{6-17}$$

The orthonormal property is

$$\langle \mathbf{p}_1'\alpha_1', \mathbf{p}_2'\alpha_2', \ldots | \mathbf{p}_1\alpha_1, \mathbf{p}_2\alpha_2, \ldots\rangle = (2\pi)^3 2E_1 \delta^3(\mathbf{p}_1' - \mathbf{p}_1)(2\pi)^3 2E_2 \times \delta^3(\mathbf{p}_2' - \mathbf{p}_2)\cdots \delta_{\alpha_1'\alpha_1}\delta_{\alpha_2'\alpha_2}\cdots \tag{6-18}$$

and the *density of states* is given by

$$\frac{1}{(2\pi)^3 2E_1}\frac{1}{(2\pi)^3 2E_2}\cdots \tag{6-19}$$

6-4 The *S*-Matrix for Continuous States. With the normalization business concluded, albeit in a nonrigorous way, we can proceed to develop the *S*-matrix in complete analogy with the treatment in Sec. 6-2. Now $|i\rangle$ and $|f\rangle$ represent states of the form expressed in Eq. 6-17. As before, we define the scattering operator S where $S_{fi} = \langle f|S|i\rangle$. To construct the generalization of Eqs. 6-7, we must extend the meaning of the summation over all final states of the quantity $(S_{if}^\dagger S_{fi})$. We mean by the summation the quantity

$$\sum_{f\alpha}\int\left[\prod_{n=1}^{N_f}\frac{d^3p_n}{(2\pi)^3 2E_n}(S_{if}^\dagger S_{fi})\right]$$

Here we have explicitly indicated that the summation consists of both the integration over the momentum variables of every particle and the summation over all the spin states α of the final state. We also have noted explicitly the density of states factor for each final state particle. There are N_f particles in the final state f. Finally, there is a summation over all the different final states. Thus this is a very complicated object. (Also see Eq. F-1b.)

Our next task is to evaluate the analog of Eqs. 6-7, which give the unitarity property of the *S*-matrix. We expect that the Kronecker δ in Eq. 6-7a must in part be replaced by a Dirac δ function for the momentum variables. The relation in fact is

$$\sum_{f\alpha}\int\left[\prod_{n=1}^{N_f}\frac{d^3p_n}{(2\pi)^3 2E_n}S_{jf}^\dagger S_{fi}\right] = \delta_{ji}\delta_{\alpha_j\alpha_i}\prod_{n=1}^{N_i}[(2\pi)^3 2E_n\delta^3(\mathbf{p}_{j,n} - \mathbf{p}_{i,n})] \tag{6-20a}$$

The somewhat tedious derivation is relegated to Ex. 6-3. In Eq. 6-20a the two Kronecker δ functions indicate symbolically that the expression is zero unless $|j\rangle$ and $|i\rangle$ have the same particles with the same internal quantum number. This ensures that for each particle in the state $|i\rangle$ there is an identical particle in the state $|j\rangle$. Therefore, the product of the Dirac δ functions makes sense, each δ function referring to the pair of identical particles in the states $|i\rangle$ and $|j\rangle$. In Eq. 6-20a our use of infinite plane wave states produces an infinite value for the total probability P_{total}. Thus we have

$$P_{\text{total}} = \sum_{f\alpha} \int \left[\prod_{n=1}^{N_f} \frac{d^3p_n}{(2\pi)^3 2E_n} S_{if}^\dagger S_{fi} \right] = \prod_{n=1}^{N_i} [(2\pi)^3 2E_n \delta^3(0)] \tag{6-21}$$

However, we can continue formal manipulations with these quantities. We define

$$\hat{I}_{ji} = \delta_{ji}\delta_{\alpha_j\alpha_i} \prod_{n=1}^{N_i} [(2\pi)^3 2E_n \delta^3(\mathbf{p}_{j,n} - \mathbf{p}_{i,n})] \tag{6-22}$$

We also note that

$$\sum_{f\alpha} \int \left[\prod_{n=1}^{N_f} \frac{d^3p_n}{(2\pi)^3 2E_n} \hat{I}_{jf}\hat{I}_{fi} \right] = \hat{I}_{ji} \tag{6-23}$$

Exercise 6-3. Derive Eq. 6-20a and prove the relationship in Eq. 6-23. Our treatment, which follows Ref. 6BU, is somewhat unorthodox. We are caught between a desire to give some physical feeling for Eq. 6-20b and the problems posed by the infinities introduced by using plane waves.

Thus the conventional form of the unitarity property of the S-matrix

$$S^\dagger S = \hat{I} \tag{6-20b}$$

is rather symbolic. In particular, $\hat{I}$ is a complicated operator. To emphasize this, we have used the symbol $\hat{I}$.

We call S_{fi} the transition probability amplitude for a transition from the state i to the state f, emphasizing that it is the probability amplitude after the process has been completed; that is, after a very long time has been allowed to elapse. (It is *not* the probability amplitude per unit time.) All our experience in macroscopic and microscopic physics tells us that such a quantity cannot depend on how we define the coordinate system in which the process takes place. Thus S_{fi} must be unchanged by spatial translations or rotations, since these are transformations that only change the definition of the coordinate system. It is less evident, but also true according to the principles of special relativity, that S_{fi}

is not changed by Lorentz velocity transformations. Thus S_{fi} is a *Lorentz scalar.* This is a very important property of the S-matrix.

6-5 The Unitarity Relation. In this section we construct the T-matrix for the plane wave states. The interesting part of the S-matrix is the part that leads to a final state different from the initial state. We isolate this part by defining the T-matrix through the equation

$$S_{fi} = \hat{I}_{fi} + i(2\pi)^4 \delta^4(P_f - P_i) T_{fi} \tag{6-24}$$

where P_f and P_i are the total four-momenta of the final state and the initial state, respectively. This equation is the analog of the discrete state definition for the T-matrix, Eq. 6-8; the four-dimensional Dirac delta function expressing the conservation of four-momentum has been explicitly exhibited. The quantity $\hat{I}_{fi}$ has already been discussed.

As with the discrete T-matrix, there are several different definitions for the T-matrix between continuous states. Some authors omit $(2\pi)^4$ or use a $-i$. Others insert kinematic quantities, dividing T by functions of the energy. All these definitions have advantages and disadvantages. When using a book or paper with which one is not acquainted, it is not always straightforward to find the definition of T and other related normalization conventions. One method of answering these questions is to look at the final expressions for differential cross sections and compare them with the same expressions in a formalism known to the reader.

We observe that in our definition of the T-matrix (Eq. 6-24) the T-matrix elements T_{fi} are Lorentz scalars. This follows because S_{fi} is a Lorentz scalar, $\hat{I}_{fi}$ is a Lorentz scalar as shown explicitly in Eq. 6-22, and $\delta^4(P_f - P_i)$ is a Lorentz scalar.

Our next task is to derive the general unitarity relation for continuous states, extending the unitarity relation for discrete states (Eq. 6-9). To derive the unitarity relation, we replace the S-matrix in Eq. 6-20b—or better, in the more explicit Eq. 6-20a—by the right-hand side of Eq. 6-24. Canceling the $\hat{I}_{ji}$ on both sides of the equation, we find

$$0 = \sum_{f\alpha} \int \left\{ \prod_{n=1}^{N_f} \frac{d^3p_n}{(2\pi)^3 2E_n} \left[i\hat{I}_{jf} T_{fi} \delta^4(P_f - P_i) - i T^\dagger_{fj} \hat{I}_{fi} \delta^4(P_f - P_j) \right] \right.$$
$$\left. + \left[T^\dagger_{jf} T_{fi} \right] (2\pi)^4 \left[\delta^4(P_f - P_i) \delta^4(P_f - P_j) \right] \right\} \tag{6-25}$$

After the cancellation of δ function, Eq. 6-25 reduces to

$$i[T^{\dagger}_{ji} - T_{ji}] = (2\pi)^4 \sum_{f\alpha} \int \left[\prod_{n=1}^{N_f} \frac{d^3p_n}{(2\pi)^3 2E_n} T^{\dagger}_{jf} T_{fi} \delta^4(P_f - P_i) \right] \tag{6-26}$$

This is the *general unitarity relation.* We finally have left the ambiguities behind us, arriving at an equation whose mathematical and physical meaning can be made clear. Here T_{fi} is the transition amplitude between infinite plane wave, free particle states normalized as indicated in Eq. 6-18, $T, (d^3p/2E)$, and $\delta^4(P)$ are all Lorentz scalars—thus both sides of the equation are Lorentz scalars.

To demonstrate the physical meaning of Eq. 6-26, let us take the $\pi^- p$ system and suppose that there are just three open channels: $\pi^- p, K^+\Sigma^-$, and $\pi^0\pi^- p$. We designate these channels by the letters a, b, and c; and we designate the transition amplitude T to go from, say, channel a to channel c by $T_{ca}[P_{c_1}P_{c_2}P_{c_3}, P_{a_1}P_{a_2}]$. The subscript numerals on the momenta indicate the particles in the channel. For simplicity, we assume that the particles have spin zero, which is of course not true for the baryons. We wish to write down the unitarity relation for the reaction

$$\pi^- + p \to K^+ + \Sigma^-$$
$$P_{a_1} + P_{a_2} \to P_{b_1} + P_{b_2}$$

Under each particle we have indicated its momentum. With these assumptions and this notation, Eq. 6-26 becomes

$$i[T^{\dagger}_{ba}[P_{b_1}P_{b_2}, P_{a_1}P_{a_2}] - T_{ba}[P_{b_1}P_{b_2}, P_{a_1}P_{a_2}]] =$$

$$(2\pi)^4 \int \frac{d^3P'_{a_1}\, d^3P'_{a_2}}{4(2\pi)^6 E'_{a_1}E'_{a_2}} \delta^4(P'_a - P_a) T^{\dagger}_{ba}[P_{b_1}P_{b_2}, P'_{a_1}P'_{a_2}]$$

$$\times\, T_{aa}[P'_{a_1}P'_{a_2}, P_{a_1}P_{a_2}] + (2\pi)^4 \int \frac{d^3P'_{b_1}\, d^3P'_{b_2}}{4(2\pi)^6 E'_{b_1}E'_{b_2}}$$

$$\times\, \delta^4(P'_b - P_a) T^{\dagger}_{bb}[P_{b_1}P_{b_2}, P'_{b_1}P'_{b_2}] T_{ba}[P'_{b_1}P'_{b_2}, P_{a_1}P_{a_2}]$$

$$+\, (2\pi)^4 \int \frac{d^3P'_{c_1}\, d^3P'_{c_2}\, d^3P'_{c_3}}{8(2\pi)^9 E'_{c_1}E'_{c_2}E'_{c_3}} \delta^4(P'_c - P_a) T^{\dagger}_{bc}[P_{b_1}P_{b_2}, P'_{c_1}P'_{c_2}P'_{c_3}]$$

$$\times\, T_{ca}[P'_{c_1}P'_{c_2}P'_{c_3}, P_{a_1}P_{a_2}] \tag{6-27}$$

Even in this very simple case we see that for exact calculations we need a great amount of data. Thus to calculate T_{ba} for a particular set of initial and final momenta we need not only to know T_{ba} for all sets of momenta but also everything about the other four transition amplitudes T_{aa}, T_{ba}, T_{bc}, and T_{ca}.

The uses of the unitarity equation in explicit calculations are restricted to special circumstances in which the amount of information to be inserted is small. The best-known application is the optical theorem. Other applications require an approximation in which only a few known amplitudes are inserted on the right-hand side to find some unknown amplitude on the left-hand side. We supply examples of these uses later in the book, but one is mentioned now. If the energy is below threshold for inelastic reactions, or if the inelastic reactions are small compared to the elastic scattering process, Eq. 6-26 reduces to

$$i[T_{\text{el}}^{\dagger} - T_{\text{el}}] = (2\pi)^4 \sum_{\alpha} \int \left[\frac{d^3P_1 d^3P_2}{(2\pi)^6 4E_1E_2} T_{\text{el}}^{\dagger} T_{\text{el}} \delta^4(P_f - P_i) \right] \tag{6-28}$$

where T_{el} is the elastic scattering amplitude. This is called the *elastic unitarity* equation.

The theoretical uses of the unitarity relations of Eqs. 6-26 and 6-28 are enormous. Such relationships are very important in dispersion relations and in the general treatment of the analytic properties of the S- and T-matrices. The unitarity relations are so generally useful in elementary particle theory because their derivation is so general. No matter how little we know about the details of a particular theory, as long as it is based on the conversation of probability, we can use the unitarity relations in that theory.

6-6 Cross Sections and Decay Rates. Our next task is to connect the T-matrix with the differential and total cross sections—the quantities that are experimentally determined. The relationship we derive is rigorous, but the derivation is not rigorous because of our use of infinite plane wave, free particle states. It is somewhat simpler to start with the derivation of the decay rate of a single unstable particle in the initial state $|i\rangle$. The probability of the state $|i\rangle$ decaying to all final states $|f\rangle$ different from the state $|i\rangle$ is

$$\sum_f P_{fi} = \sum_{f\alpha} \int \left[\prod_{n=1}^{N_f} \frac{d^3p_n}{(2\pi)^3 2E_n} (S_{if}^{\dagger} S_{fi}) \right] \qquad f \neq i$$

which reduces to

$$\sum_f P_{fi} = \sum_{f\alpha} \int \prod_{n=1}^{N_f} \frac{d^3p_n}{(2\pi)^3 2E_n} [|T_{fi}|^2 (2\pi)^4 \delta^4(P_f - P_i)(2\pi)^4 \delta^4(P_f - P_i)] \tag{6-29}$$

The δ^2 function leads to infinities in Eq. 6-29. We get around this problem in the conventional way of interpreting $\delta^3(0)$ as equal to $V/(2\pi)^3$, as we did in the derivation of Eq. 6-16. Similarly, we interpret

$$\delta(P_f^0 - P_i^0) \underset{P_f^0 \to P_i^0}{=} \delta(0) = \frac{T}{2\pi}$$

where T is the total time (and also is infinite). But the probability per unit volume per unit time is a finite quantity. Extracting VT from the right-hand side of Eq. 6-29 and dividing both sides by VT, we obtain

$$\sum_f \text{(Prob. per unit volume per unit time)}_{fi} =$$

$$(2\pi)^4 \sum_{f\alpha} \int \left[\prod_{n=1}^{N_f} \frac{d^3p_n}{(2\pi)^3 2E_n} \, |T_{fi}|^2 \delta^4(P_f - P_i) \right]$$

If we divide both sides of this equation by the particle density, we obtain the probability of the decay per unit time of one particle. We call this the decay rate and use the symbol Λ. The particle density is given as $2E_i$ in Eq. 6-16. Thus we have

$$\Lambda = \frac{(2\pi)^4}{2E_i} \sum_{f\alpha} \int \left[\prod_{n=1}^{N_f} \frac{d^3p_n}{(2\pi)^3 2E_n} \, |T_{fi}|^2 \delta^4(P_f - P_i) \right] \tag{6-30}$$

The lifetime of the particle τ is given by

$$\tau = \frac{1}{\Lambda} \tag{6-31}$$

We next turn to the calculation of cross sections, starting with the total cross section. There are now two particles in the initial state, designated by the subscripts $i1$ and $i2$ with the individual particle densities $\rho(i1) = 2E_{i1}$, $\rho(i2) = 2E_{i2}$. Using Eq. 6-29 for a two-particle initial state, replacing one of the four-dimensional δ functions by VT, and dividing both sides of Eq. 6-29 by VT yields

$$\sum_f \text{(Prob. per unit volume per unit time)}_{fi}$$

$$= \sum_{f\alpha} \int \left[\prod_{n=1}^{N_f} \frac{d^3p_n}{(2\pi)^3 2E_n} \, |T_{fi}|^2 \delta^4(P_f - P_i)(2\pi)^4 \right]$$

At this point it is simplest to work in the laboratory system where particle 2 is the target particle. Dividing both sides of the last expression by $2E_{i2}$, we obtain

$$\sum_f (\text{Prob. per target particle per unit time})_{fi} =$$

$$\frac{(2\pi)^4}{2E_{i2}} \sum_{f\alpha} \int \left[\prod_{n=1}^{N_f} \frac{d^3p_n}{(2\pi)^3 2E_n} |T_{fi}|^2 \delta^4(P_f - P_i) \right]$$

To remove the time factor and obtain the total cross section, we divide the last expression by the flux of incident particles, which in the laboratory system is

$$(\text{density of particle 1})(\text{velocity of particle 1 in lab.}) = 2E_{i1} v_{i1,\text{lab}}$$

This final division yields

$$\sigma_{\text{tot}} = \frac{(2\pi)^4}{4E_{i1}E_{i2}v_{i1,\text{lab}}} \sum_{f\alpha} \int \left[\prod_{n=1}^{N_f} \frac{d^3p_n}{(2\pi)^3 2E_n} |T_{fi}|^2 \delta^4(P_f - P_i) \right] \tag{6-32}$$

Now the total cross section is a Lorentz scalar. This statement may seem a little strange because the total cross section has the units of area. However, if we make a Lorentz transformation using a velocity not parallel to the direction of relative motion, we must redefine the transverse plane used in the definition of the total cross section. This keeps the total cross section a constant quantity. Since σ_{tot} is a Lorentz scalar and that part of the right-hand side of Eq. 6-32 after the summation sign is a Lorentz scalar, we must be able to write the remaining part of the right side of Eq. 6-32 as a Lorentz scalar.

Defining

$$FD_{\text{lab}} = 4E_{i1}E_{i2}v_{i1\text{lab}} = 4|\mathbf{p}_{i1\text{ lab}}|m_{i2} \tag{6-33}$$

where FD is the flux of particle 1 multiplied by the density of particle 2, we can write FD in the invariant form

$$FD = 4[(P_{i1}\cdot P_{i2})^2 - m_{i1}{}^2 m_{i2}{}^2]^{1/2} \tag{6-34}$$

Thus Eq. 6-32 can be written in the invariant form

$$\sigma_{\text{tot}} = \frac{(2\pi)^4}{4[(P_{i1}\cdot P_{i2})^2 - m_{i1}{}^2 m_{i2}{}^2]^{1/2}} \sum_{f\alpha} \int \left[\prod_{n=1}^{N_f} \frac{d^3p_n}{(2\pi)^3 2E_n} |T_{fi}|^2 \delta^4(P_f - P_i) \right] \tag{6-35}$$

In any Lorentz frame in which the momenta of the particles are parallel to each other, it is convenient to use the form

$$\sigma_{\text{tot}} = \frac{(2\pi)^4}{4E_{i1}E_{i2}|\mathbf{v}_{i1} - \mathbf{v}_{i2}|} \sum_{f\alpha} \int \left[\prod_{n=1}^{N_f} \frac{d^3p_n}{(2\pi)^3 2E_n} |T_{fi}|^2 \delta^4(P_f - P_i) \right] \tag{6-36}$$

where v_{i1} and v_{i2} are the velocities of the particles in the initial state. We note that in the center of mass

$$FD = 4p_i^* \sqrt{s} \tag{6-37}$$

where p_i^* is the magnitude of the three-momentum of either initial particle in the cms.

Finally we define the differential cross section into a final state f by carrying out a partial differentiation. Note that there must be an integration over at least four variables to remove the δ function. If there is no integration over the first N particles in the final state, we have

$$\frac{d\sigma_{fi}}{d^3p_1 \cdots d^3p_N} = \frac{(2\pi)^4}{4E_{i1}E_{i2}|\mathbf{v}_{i1} - \mathbf{v}_{i2}|} \left[\prod_{n=1}^{N} \frac{1}{(2\pi)^3 2E_n} \right]$$

$$\times \int \left[\prod_{n=N+1}^{N_f} \frac{d^3p_n}{(2\pi)^3 2E_n} |T_{fi}|^2 \delta^4(P_f - P_i) \right] \tag{6-38}$$

This differential cross section is not a Lorentz scalar because of the first N energy factors in the denominator of the right-hand side.

Two-body reactions are so commonly studied in elementary particle physics that it is worthwhile to work out Eq. 6-38 specifically for the two-body case. We use the notation $1 + 2 \to 3 + 4$.

Particle j has mass m_j, spin s_j, spin state α_j, four-momentum P_j, center of mass three-momentum $\mathbf{p}_j$, and center of mass energy E_j. In the center of mass Eq. 6-38 yields

$$\left(\frac{d\sigma}{d\Omega^*}\right)(s\theta\phi\alpha_3\alpha_4\alpha_1\alpha_2) = \frac{1}{64\pi^2 s}\left(\frac{p_f^*}{p_i^*}\right)|T(s\theta\phi\alpha_3\alpha_4\alpha_1\alpha_2)|^2 \tag{6-39}$$

where $p_i^* = |\mathbf{p}_1| = |\mathbf{p}_2|$, $p_f^* = |\mathbf{p}_3| = |\mathbf{p}_4|$.

We have explicitly indicated the variables on which T and $d\sigma/d\Omega^*$ depend. The amplitude $f(s\theta\phi\alpha_3\alpha_4\alpha_1\alpha_2)$ is defined in the center of mass by

$$\frac{d\sigma}{d\Omega^*} = |f|^2 \tag{6-40}$$

Comparing these two equations, we have

$$f(s\theta\phi\alpha_3\alpha_4\alpha_1\alpha_2) = \frac{1}{8\pi\sqrt{s}}\left(\frac{p_f^*}{p_i^*}\right)^{1/2} T(s\theta\phi\alpha_3\alpha_4\alpha_1\alpha_2) \tag{6-41}$$

This is a very useful relation, for example, in determining how T has been defined. For elastic scattering, we write

$$f_{\text{el}} = \frac{1}{8\pi\sqrt{s}} T_{\text{el}} \tag{6-42}$$

Finally, we consider a convenient invariant form of the differential cross section when the particles are spinless or when we have summed over all the final spin states and averaged over all the initial spin states. We define

$$|T|^2_{\text{av}}(s\theta\phi) = \frac{1}{(2s_1+1)(2s_2+1)} \sum_{\alpha_1\alpha_2\alpha_3\alpha_4} |T(s\theta\phi\alpha_3\alpha_4\alpha_1\alpha_2)|^2$$

and then

$$\frac{d\sigma}{d\Omega^*}(s\theta\phi) = \frac{1}{64\pi^2 s}\left(\frac{p_f^*}{p_i^*}\right)|T|^2_{\text{av}}(s\theta\phi) \tag{6-43}$$

But in the last equation there cannot be any dependence of $|T|^2_{\text{av}}$ on ϕ. This is because there is no quantity in the initial state that can be used to construct an $\boldsymbol{x}$ axis to use as a reference axis for ϕ. Using $d\sigma/dt = (1/(2p_i^*p_f^*))(d\sigma/d\cos\theta)$, we have

$$\frac{d\sigma}{dt} = \frac{|T|^2_{\text{av}}}{64\pi s(p_i^*)^2}$$

where $d\sigma/dt$ is a Lorentz scalar. At very high energies we obtain the very simple form

$$\frac{d\sigma}{dt} = \frac{|T|^2_{\text{av}}}{16\pi s^2} \tag{6-44}$$

6-7 The Optical Theorem. The optical theorem is derived directly from the general unitarity relation (Eq. 6-26). We consider the special case in which the final state j is identical to the initial state i, the initial state i consisting of just two particles. We require:

1. That j has the same two particles as i.
2. That neither particle change momentum in going from i to j.
3. That neither particle change its spin state in going from i to j.

These specifications define elastic scattering at zero degrees with no change of spin state of either particle and are summarized in the phrase "forward, no spin flip, elastic scattering." We have inserted the words "no spin flip" to emphasize that neither particle can change spin state if the state j is to be identical to the state i. Usually those words are omitted and the shorter phrase "forward elastic scattering" is used. But as Ch. 10 reveals, it is possible for both particles to change their spin states and still give a nonzero forward scattering amplitude.

Since the state j is the same as the state i, we put

$$i[T_{ji}^{\dagger} - T_{ji}] = i[T_{ii}^{*} - T_{ii}] = 2\,\mathrm{Im}\,T_{ii}$$

Then, under the conditions given in the last paragraph, Eq. 6-26 becomes

$$2\,\mathrm{Im}\,T_{\mathrm{el},i}(0) = (2\pi)^4 \sum_{f\alpha} \int \left[\prod_{n=1}^{N_f} \frac{d^3p_n}{(2\pi)^3 2E_n} \,|\,T_{fi}\,|^2 \delta^4(P_f - P_i) \right] \tag{6-45}$$

where we have inserted the subscript el, i to indicate that this is the elastic scattering amplitude from the state i. Now using Eq. 6-35, we can replace the right-hand side of Eq. 6-45 by

$$\sigma_{\mathrm{tot},i} 4\,[(P_{i1}\cdot P_{i2})^2 - m_{i1}{}^2 m_{i2}{}^2]^{1/2}$$

to obtain

$$\mathrm{Im}\,T_{\mathrm{el},i}(0) = 2\,[(P_{i1}\cdot P_{i2})^2 - m_{i1}{}^2 m_{i2}{}^2]^{1/2}\,\sigma_{\mathrm{tot},i} \tag{6-46a}$$

The notation (0) symbolizes specifications 2 and 3 of the first paragraph of this section. Equation 6-46a is the *optical theorem* in covariant form. In the laboratory system

$$\mathrm{Im}\,T_{\mathrm{el},i}(0) = 2\,|\,\mathbf{p}_1|_{\mathrm{lab}}\, m_2 \sigma_{\mathrm{tot},i} \tag{6-46b}$$

and in the center of mass system

$$\mathrm{Im}\,T_{\mathrm{el},i}(0) = 2p^* \sqrt{s}\,\sigma_{\mathrm{tot},i} \tag{6-46c}$$

where p^* is the magnitude of the center of mass three-momentum of the incident particle. Using the center of mass amplitude $f(s00\alpha_{-}\alpha_2\alpha_1\alpha_2)$ defined in Eqs. 6-40 and 6-41, we write

$$\mathrm{Im}\,f_{\mathrm{el},i}(0) = \frac{p^*}{4\pi}\,\sigma_{\mathrm{tot},i} \tag{6-46d}$$

Finally, we note that in Eqs. 6-46, the only variable is implicitly s.

Before using the optical theorem generally, we must develop the concept of two-particle helicity states and apply that concept to elastic scattering. This will allow us to determine the minimum number of independent elastic scattering amplitudes, thus to use the optical theorem in the most economical way. This is done in Chs. 9 and 10.

6-8 An Overview of Coordinate Transformations. In this section we draw a general picture of coordinate transformations using spatial translations as an example. We consider a physical system described by $f(\mathbf{r})$ where $\mathbf{r}$ is a position vector in three-dimensional space. Classically, $f(\mathbf{r})$ might represent a matter density distribution. Quantum mechanically, $f(\mathbf{r})$ is a wave function of the state $|f\rangle$ in the coordinate representation; namely, $f(\mathbf{r}) = \langle \mathbf{r}|f\rangle$.

Let us carry out a translation through the vector distance $\mathbf{a}$. There are two ways of thinking about this translation.

1. To think of the physical system being shifted by the distance $\mathbf{a}$ is the *active* view of the transformation. That is, we hold the coordinate system fixed and change the position of the physical system.
2. On the other hand, we may hold the physical system fixed and shift the coordinate system by the vector distance $-\mathbf{a}$. This is the *passive* view of the transformation.

These are equivalent ways of describing a transformation, but in thinking about a transformation it is best to be clear about which view is being adopted. Otherwise, sign, phase, and related errors can occur.

We suppose that a physical system, described originally by $f(\mathbf{r})$, is translated by a distance $\mathbf{a}$ and is now described by $f'(\mathbf{r})$. The point of the physical system that is now at $\mathbf{r}_1$ was formerly at $\mathbf{r}_1 - \mathbf{a}$; therefore, we have

$$f'(\mathbf{r}) = f(\mathbf{r} - \mathbf{a}) \tag{6-47}$$

We note that we obtain the same result from the passive view. To be sure that the meaning of Eq. 6-47 is understood, we put it into words. The function $f'(\mathbf{r})$, which describes the physical system after the translation, has a value at the point $\mathbf{r}$ given by the value of the original function $f(\mathbf{r})$ at the point $\mathbf{r} - \mathbf{a}$. The original function $f(\mathbf{r})$ described the system before the translation.

Exercise 6-4. Work out Eq. 6-47 explicitly for a plane wave $\exp i\mathbf{p}\cdot\mathbf{r}$ and for a Gaussian in x, $\exp -x^2$.

For further development we must define an *operator* $U(a)$ which connects the original state $|f\rangle$ to the translated state $|f'\rangle$. We write formally

$$|f\rangle \rightarrow |f'\rangle = U(\mathbf{a})|f\rangle \tag{6-48}$$

For this operator to be useful, it must have the same meaning for some set of states $|f\rangle, |g\rangle, |h\rangle, \ldots$. We can in fact find such an operator using the coordinate representation. The Taylor expansion of Eq. 6-47 yields

$$f'(\mathbf{r}) = f(\mathbf{r}) - \mathbf{a}\cdot\nabla f(\mathbf{r}) + \cdots = \sum_{n=0}^{\infty} (-\mathbf{a}\cdot\nabla)^n \frac{f(\mathbf{r})}{n!}$$

which can be written as

$$f'(\mathbf{r}) = [e^{-\mathbf{a}\cdot\nabla}]f(\mathbf{r}) = U(\mathbf{a})f(\mathbf{r}) \qquad \text{or} \qquad U(\mathbf{a}) = e^{-\mathbf{a}\cdot\nabla} \tag{6-49}$$

Of course functions of differential operators, such as the exponential in the last equation, are to be interpreted by expanding the function in a power series.

Exercise 6-5. Apply the last equation to the first two examples in Ex. 6-4. Show that the same result is obtained as in Ex. 6-4.

In the coordinate representation, the three-dimensional momentum operator $\mathbf{p}$ has the form

$$\mathbf{p} = -i\nabla \tag{6-50}$$

and we now have

$$U(\mathbf{a}) = e^{-i\mathbf{a}\cdot\mathbf{p}} \tag{6-51}$$

Thus we have an explicit expression for $U(\mathbf{a})$ which is independent of the representation.

The spatial translation is defined by the rule that the physical point that was at the coordinate point $\mathbf{r}$ before the translation is now at the point $\mathbf{r}'$, where $\mathbf{r}' = \mathbf{r} + \mathbf{a}$. We write this as

$$\mathbf{r} \rightarrow \mathbf{r}' = \mathbf{r} + \mathbf{a} \tag{6-52}$$

Frequently we do not wish to write the rule explicitly. We then define an operator $T(\mathbf{a})$ which gives the coordinates of the point of a physical system *after* the translation in terms of the coordinates of the point *before* the translation. We take the active view here, which means that both the new and old coordinates refer to the original, unchanged coordinate system. Thus we write Eq. 6-52 in the symbolic form

$$\mathbf{r} \rightarrow \mathbf{r}' = T(\mathbf{a})\mathbf{r} \tag{6-53}$$

To relate $T(\mathbf{a})$ to $U(\mathbf{a})$ we rewrite Eq. 6-48 in the form

$$f(\mathbf{r}) \rightarrow f'(\mathbf{r}) = U(\mathbf{a})f(\mathbf{r}) \tag{6-54}$$

where $f(\mathbf{r})$ describes a classical or quantum mechanical system. Then, always taking the active view, we can also write

$$f'(\mathbf{r}') \equiv f(\mathbf{r}) \tag{6-55}$$

$$f'(T(\mathbf{a})r) \equiv f(\mathbf{r}) \tag{6-56a}$$

$$f'(\mathbf{r}') \equiv f(T(\mathbf{a})^{-1}\mathbf{r}') \tag{6-56b}$$

where the $\equiv$ sign means that f' written in terms of $\mathbf{r}'$ has identically the same functional form as f written in terms of $\mathbf{r}$. This identity—or better, tautology—enables us to relate $U(\mathbf{a})$ to $T(\mathbf{a})$. Reference 6GO discusses this in detail. Also, there is

$$T^{-1}(\mathbf{a})T(\mathbf{a}) = I \tag{6-57}$$

where I is a unit operator.

Equations 6-53 through 6-57 have been derived for spatial translations. But they are more general; and they are equally true and equally useful for spatial rotations and Lorentz velocity transformations.

6-9 The Inhomogeneous Lorentz Group. The coordinate transformations we are about to discuss belong to a set of mathematical groups that together form the *inhomogeneous Lorentz transformation group*. This group consists of those transformations which do not change the scalar product

$$(x^\mu - y^\mu)(x_\mu - y_\mu) \tag{6-58}$$

where x and y are any two space-time points. We shall not prove or discuss the group nature of this set of transformations (6GA, 6MA, 6WE, 6HA). But we do make implicit use of the group properties; for example, the product of two spatial rotations is a spatial rotation. We must also remember that the group is not commutative. In general, for two transformation operators, $U_1U_2 \neq U_2U_1$.

The inhomogeneous Lorentz group consists of the following groups.

1. *The Group of Space and Time Translations.* The physical point at space-time point x is translated by the four-vector displacement $a = (a^0, a^1, a^2, a^3)$. The rule (taking the active view) is

$$x^\mu \to x'^\mu = x^\mu + a^\mu \tag{6-59}$$

2. *The Group of Spatial Rotations.* The rotations in three-dimensional space are discussed in detail in Ch. 9. We only note here that the rotation is specified by the real, orthogonal matrix A with the determinant of A equal to $+1$. The rule for the transformation of the coordinates is

$$r_i \to r_i' = \sum_{j=1}^{3} A_{ij}r_j \qquad t \to t' = t \tag{6-60}$$

Here we have again taken the active view that A describes the rotation in which a physical point particle goes from $\mathbf{r}$ to $\mathbf{r}'$. Since the rotation preserves $\Sigma_{i=1}^{3}(x^i - y^i)(x_i - y_i)$ and does not change $(x^0 - y^0)(x_0 - y_0)$, it does not change the quantity in Eq. 6-58.

3. *The Group of Lorentz Velocity Transformations.* Here it is convenient to take the passive view. Consider a physical particle described in a coordinate system. In that coordinate system the particle's position is described by the unprimed space-time point x. Do not change the physical system at all, but give the coordinate system a velocity $-\mathbf{v}$. Now the physical particle's position is described in the transformed coordinate system by the primed quantity x'. We say that the coordinate system has received a *boost* of velocity $-\mathbf{v}$. But we may also take the active view and say that the physical particle has received a *boost* by velocity $\mathbf{v}$. For a boost of the physical particle along only the x_j axis, $j = 1, 2$, or 3, the Lorentz velocity transformation is

$$\begin{aligned} x_j' &= \gamma(x_j + v_j t) \\ t' &= \gamma(t + v_j x_j) \\ x_k' &= x_k \qquad \text{for} \quad k \neq j \end{aligned} \tag{6-61}$$

where

$$\gamma = (1 - v_j^2)^{-1/2} \tag{6-62}$$

The principles of special relativity say that space and time are such that the quantity described in Eq. 6-58 is not changed under the transformation described in Eq. 6-61.

4. *The Restricted Lorentz Group.* The boosts described by Eq. 6-61 and the rotations described by Eq. 6-60 can be described by a single 4 x 4 matrix, often called Λ. The matrix Λ is real because A is real and the coefficients in Eq. 6-61 are real. For a spatial rotation, we write

$$\Lambda = \begin{bmatrix} 1 & 0 & 0 & 0 \\ 0 & & & \\ 0 & & \mathrm{A} & \\ 0 & & & \end{bmatrix} \tag{6-63}$$

For a boost with $j = 1$ in Eq. 6-61

$$\Lambda = \begin{bmatrix} \gamma & \gamma v_1 & 0 & 0 \\ \gamma v_1 & \gamma & 0 & 0 \\ 0 & 0 & 0 & 0 \\ 0 & 0 & 0 & 0 \end{bmatrix} \tag{6-64}$$

We write for the combined group

$$x^\mu \to x'^\mu = \Lambda_\nu{}^\mu x^\nu \tag{6-65}$$

The combined group of spatial rotations and boosts is called the *restricted homogeneous Lorentz group* or just the *restricted Lorentz group.*

5. *The Restricted Inhomogeneous Lorentz Group.* Combining the restricted Lorentz group with the group of space and time translations, we obtain the *restricted inhomogeneous Lorentz group.* The transformations are now defined by the rule that a physical point originally at x is transferred to the point x' where

$$x^\mu \to x'^\mu = a^\mu + \Lambda_\nu{}^\mu x^\nu \tag{6-66}$$

6. *The inhomogeneous Lorentz or Poincaré Group.* In the active view, under space inversion a point particle at the point (x^0, x^1, x^2, x^3) goes to the point $(x^0, -x^1, -x^2, -x^3)$. Thus space inversion is defined by

$$\Lambda_\pi = \begin{bmatrix} 1 & 0 & 0 & 0 \\ 0 & -1 & 0 & 0 \\ 0 & 0 & -1 & 0 \\ 0 & 0 & 0 & -1 \end{bmatrix} \tag{6-67}$$

We use the subscript π (for parity) to denote a space inversion.

Similarly for a time inversion transformation

$$\Lambda_T = \begin{bmatrix} -1 & 0 & 0 & 0 \\ 0 & 1 & 0 & 0 \\ 0 & 0 & 1 & 0 \\ 0 & 0 & 0 & 1 \end{bmatrix} \tag{6-68}$$

When space and time inversions are combined with the restricted inhomogeneous Lorentz group, we obtain the *inhomogeneous Lorentz group*, also called the *Poincaré group.*

6-10 Infinitesimal Coordinate Transformations. The operator U, which represents the effect of a coordinate transformation on a quantum mechanical system, is in general a rather intractable object. The explicit form of U depends on the system and the representation. We would like to avoid the consideration of explicit forms in studying the invariant properties of the system under the transformations. And indeed, we can often do just that by using infinitesimal coordinate transformation. In the next section we apply the concept to the groups comprising the restricted, inhomogeneous Lorentz group.

Consider a coordinate transformation operator $U(\alpha)$ which depends on a single continuous parameter α. An example is the operator $U_z(\theta)$, which represents a rotation by an angle θ about the z axis. Suppose $U(\alpha)$ has the following properties:

1. $U(\alpha_2)U(\alpha_1) = U(\alpha_2 + \alpha_1)$. (6-69)
2. $U(\alpha)$ is a continuous function of α. That is, as we vary α continuously, $U(\alpha)$ also varies continuously. Or in other words, $U(\alpha + \delta\alpha)$ can be made arbitrarily close to $U(\alpha)$ by making $\delta\alpha$ sufficiently small.
3. $U(0) = I$, the identity operator.
4. $U(\alpha)$ can be obtained from I by a continuous change of α.

These four postulates, which are the conditions for a single-parameter, connected Lie group (6HA), are obeyed by all the restricted, inhomogeneous Lorentz transformations. The theory of such Lie groups says that we can write (to first order in $\delta\alpha$)

$$U(\alpha + \delta\alpha) - U(\alpha) = i\delta\alpha L \tag{6-70}$$

where L, which is *independent* of the parameter α, is called the *generator* of the group. The i is an arbitrary convention.

An *infinitesimal coordinate transformation* is a transformation in which the parameter α is only an infinitesimal amount $\delta\alpha$ different from zero. Designating the related transformation operator by $U(\delta\alpha)$, Eq. 6-70 yields

$$U(\delta\alpha) = I + i\delta\alpha L \tag{6-71}$$

Thus the properties of an infinitesimal transformation operator are fixed by the properties of the generator L.

Exercise 6-6. Use the example given previously of $U_z(\theta)$ to illustrate the discussion up to Eq. 6-70.

To make full use of the infinitesimal transformation operators we do two things:

1. We connect the generators, when possible, with quantum mechanical operators whose properties are well known. Indeed, in the next section we show that within a phase, the generator of the group of space and time translations is the linear momentum operator; and the generator of the group of spatial rotations is the angular momentum operator.
2. We learn the commutation properties of the generators of the various groups, for that enables us to interchange the order of the transformation operators.

These concepts are clarified by the specific examples of the next section.

Returning to Eq. 6-71, and using Eq. 6-69, we observe that we can directly integrate this equation to obtain the well-known and beautiful result

$$U(\alpha) = e^{i\alpha L} \tag{6-72}$$

Finally, we make the important observation that since U is a unitary operator, the generator L is a *hermitian operator.*

6-11 Infinitesimal Operators of the Restricted Inhomogeneous Lorentz Group

1. *The Group of Space and Time Translations.* This is a particularly simple case. Consider a wave function $f(\mathbf{r})$ in the coordinate representation. Take the active view and suppose that the physical system is translated by the infinitesimal vector displacement $\delta\mathbf{a}$. The new wave function is then given by the equation following Eq. 6-48 (to first order in $\delta\mathbf{a}$). Explicitly, we have

$$\begin{aligned} f(\mathbf{r}) \to f'(\mathbf{r}) &= U(\delta\mathbf{a})f(\mathbf{r}) = f(\mathbf{r}) - \delta\mathbf{a}\cdot\nabla f(\mathbf{r}) \\ &= (I - \delta\mathbf{a}\cdot\nabla)f(\mathbf{r}) \end{aligned} \tag{6-73}$$

Therefore,

$$\mathbf{L} = i\nabla = -\mathbf{p}$$

Exercise 6-7. Prove the generator of the time translation is $L^0 = i(\partial/\partial t) = P^0$. Derive Eqs. 6-74 and 6-75, which are written in four-vector notation. In deriving Eq. 6-75, show that Eq. 6-69 is obeyed by the translation transformation.

$$U(\delta a) = I + i(\delta a)_\mu P^\mu \tag{6-74}$$

$$U(a) = e^{ia_\mu P^\mu} \tag{6-75}$$

Equations 6-74 and 6-75 were derived for a very simple situation in the coordinate representation. Although we do not prove it here, this result is very general. And it is very useful, for we have connected the generator of space and time translations with the linear momentum operator—an operator whose properties are very well known. For example, the commutation properties of P^μ are

$$[P^\mu, P^\nu] = 0 \qquad \text{for} \quad \text{all } \mu, \nu \tag{6-76}$$

2. *The Group of Spatial Rotations.* We consider a single, spinless particle, following the same path we have just taken: constructing the coordinate transformation, considering the infinitesimal transformations, and identifying the

generators. We know from classical mechanics that any rotation in three-space can be produced by a sequence of rotations about the x, y, and z axis. Thus we only need to consider rotations about these axes and use operator multiplication.

In particular, we consider a rotation about the z axis denoted by the coordinate transformation operator $T(\theta_z)$; we have symbolically $\mathbf{r} \to \mathbf{r}' = T(\theta_z)\mathbf{r}$.

And representing the rotation by the matrix A (see Sec. 6-9), we have

$$A = \begin{bmatrix} \cos\theta_z & -\sin\theta_z & 0 \\ \sin\theta_z & \cos\theta_z & 0 \\ 0 & 0 & 1 \end{bmatrix} \tag{6-77}$$

We define θ_z as being positive for a right-hand rotation about the z axis, noting again that we are taking the active view. For an infinitesimal transformation $T(\delta\theta_z)$ we write

$$\begin{aligned} x' &= x - y\,\delta\theta_z \\ y' &= x\,\delta\theta_z + y \\ z' &= z \end{aligned} \tag{6-78}$$

Using Eq. 6-56b gives

$$f'(\mathbf{r}') \equiv f(T(\delta\theta_z)^{-1}\mathbf{r}') \tag{6-79}$$

$$f'(x', y', z') \equiv f(x' + y'\,\delta\theta_z, -x'\,\delta\theta_z + y', z')$$

$$f'(x', y', z') = \left[1 + \delta\theta_z\left(y'\frac{\partial}{\partial x'} - x'\frac{\partial}{\partial y'}\right)\right] f(x', y', z')$$

$$= [1 + i\delta\theta_z L_z] f(x', y', z')$$

Dropping the primes, we have

$$L_z = i\left(x\frac{\partial}{\partial y} - y\frac{\partial}{\partial x}\right) = -(\mathbf{r} \times \mathbf{p})_z \tag{6-80}$$

Therefore, the generator L_z is the negative of the quantum mechanical operator $(\mathbf{r} \times \mathbf{p})_z$, which is of course the z component of the orbital angular momentum operator.

Using $\mathbf{J}$ to denote the orbital angular momentum operator and generalizing from the coordinate representation, we have

$$U(\delta\theta_z) = I - i\delta\theta_z J_z \tag{6-81}$$

Next we observe that

$$U(\theta_z')U(\theta_z) = U(\theta_z' + \theta_z)$$

Therefore, the condition of Eq. 6-69 is fulfilled and

$$U(\theta_z) = e^{-i\theta_z J_z} \tag{6-82}$$

The same arguments hold for the other axis, and for $j = 1, 2, 3$, we have

$$U(\delta\theta_j) = I - i\delta\theta_j J_j \tag{6-83}$$

$$U(\theta_j) = e^{-i\theta_j J_j} \tag{6-84}$$

Equation 6-84 is even more general than we have yet indicated. It is not proved here, but it is the general rotation operator for systems with half-integral as well as integral spin; **J** is simply the corresponding angular momentum operator. Finally, we may consider a rotation by an angle θ about an arbitrary axis whose direction is given by the unit vector $\hat{n}$. Then we have

$$U(\theta, \hat{n}) = e^{-i\theta\hat{n}\cdot J} \tag{6-85}$$

The commutation relations for J and P are

$$[J_i, J_j] = i\epsilon_{ijk}J_k, \qquad [p_i, J_j] = i\epsilon_{ijk}p_k, \qquad [P^0, J_i] = 0 \tag{6-86}$$

Exercise 6-8. Verify the commutation relations in Eq. 6-86 using the coordinate representation of the operators for a single spinless particle. For more general proofs see Refs. 6GA and 6MA. Appendix A defines ϵ_{ijk}.

3. *The Boost Group.* In Sec. 6-9 the Lorentz velocity transformation or boost transformation was labeled by the velocity v_j of the boost. (The axis along which the boost takes place is indicated by the index j.) If a boost of velocity v_j' follows a boost of velocity v_j, the equivalent boost is given by the velocity

$$v_j'' = \frac{v_j' + v_j}{1 + v_j' v_j} \tag{6-87}$$

Thus the velocities v_j do not meet the additive requirement of Eq. 6-69. But if we define the quantity ξ_j by

$$v_j = \tanh \xi_j \tag{6-88}$$

we obtain

$$\xi_j'' = \text{arctanh } v_j'' = \xi_j' + \xi_j \tag{6-89}$$

and the quantities ξ_j have the desired additive property. Equation 6-61 becomes

$$\begin{aligned} x_j' &= x_j \cosh \xi_j + t \sinh \xi_i \\ t' &= x_j \sinh \xi_j + t \cosh \xi_j \\ x_k' &= x_k, \qquad k \neq j \end{aligned} \tag{6-90}$$

The parameter ξ is sometimes called the *boost parameter* or the *velocity parameter* (Sec. 9-4). For the infinitesimal transformation $U(\delta\xi_j)$ we have, in the coordinate representation,

$$U(\delta\xi_j) = 1 - \delta\xi_j \left(t \frac{\partial}{\partial x_j} + x_j \frac{\partial}{\partial t} \right) \tag{6-91}$$

Exercise 6-9. Derive Eq. 6-91 using a method analogous to Eq. 6-79.

It is conventional to define, in the coordinate representation (using non-relativistic notation)

$$K_j = -i \left(t \frac{\partial}{\partial x_j} + x_j \frac{\partial}{\partial t} \right) \tag{6-92}$$

such that

$$U(\delta\xi_j) = 1 - i\delta\xi_j K_j$$

The additive property of the ξ_j parameter then yields

$$U(\xi_j) = e^{-i\xi_j K_j} \tag{6-93}$$

Thus for the boosts, as for the spatial rotations and space-time translations, we have obtained a simple exponential form. As before, we can move out of the coordinate representation and give Eq. 6-93 a more general meaning. In Sec. 9-4 we work out some useful, explicit examples of the operator $U(\xi_j)$. We do not work out or use explicit forms of the generator K_j. Our main purpose in introducing the generator K_j is to emphasize the analogy between Lorentz velocity transformations and spatial rotations.

We conclude this section by exhibiting the commutation relations between K_j and the generators of the restricted, inhomogeneous Lorentz group. These relations are true for the general case, including spin. But we only derive them, in the next exercise, for spinless particles in the coordinate representation.

$$\begin{aligned}
[K_i, P^0] &= -ip_i \\
[K_i, p_j] &= -i\delta_{ij}P^0 \\
[J_i, K_j] &= i\epsilon_{ijk}K_k \\
[K_i, K_j] &= -i\epsilon_{ijk}J_k
\end{aligned} \tag{6-94}$$

Exercise 6-10. Use the coordinate representations of P, K, and J to calculate the commutation relations in Eq. 6-94.

6-12 Space Inversion and Time Reversal. To the best of our knowledge the strong interactions are invariant under space inversion and time reversal. We show in Ch. 10 that this invariance is very useful in the reduction of the number of independent amplitudes in two-body reactions. And this is our principal use of the properties of the strong interactions under these transformations. Therefore, we make no use of the rather elaborate formalism that has been developed for these transformations, particularly in quantum field theory (see, e.g., Refs. 6GA, 6RO, and 6ME). We usually depend on the simple effects of space inversion and time reversal on displacement four-vectors, energy-momentum four-vectors, and angular momentum three-vectors.

We conclude this section with a few remarks. We use $\mathscr{P}$ to denote the space inversion operator and T to denote the time reversal operator. Like the other coordinate transformation operators previously discussed, $\mathscr{P}$ is unitary and linear. But T is antiunitary and antilinear (6ME).

Finally we note that if an interaction is invariant under *time reversal*, the transition probability for the initial state $|i\rangle$ to go to the final state $|f\rangle$ is equal to the transition probability for the initial state $\mathrm{T}|f\rangle$ to go the final state $\mathrm{T}|i\rangle$. Thus time reversal invariance involves two kinds of changes.

1. $|i\rangle \rightarrow |i'\rangle = \mathrm{T}|i\rangle, |f\rangle \rightarrow |f'\rangle = \mathrm{T}|f\rangle$

By $U_T|\phi\rangle$ we mean that in a coordinate representation; t in $\phi(x)$ would be changed to $-t$; in a momentum representation $\mathbf{p}$ would be changed to $-\mathbf{p}$. The s_z (of particles with spin) would be changed to $-s_z$.

2. The initial and final states $|i'\rangle$ and $|f'\rangle$ are interchanged. For example, suppose the state i consists of spinless particles with momenta $\mathbf{p}_1, \mathbf{p}_2 \ldots$; and the final state f consists of spinless particles with momenta $\mathbf{p}_1', \mathbf{p}_2', \ldots$. If the S-matrix is invariant under time reversal, we will have

$$\langle -\mathbf{p}_1, -\mathbf{p}_2, \ldots, |S| - \mathbf{p}_1', -\mathbf{p}_2', \ldots\rangle = \langle \mathbf{p}_1', \mathbf{p}_2', \ldots, |S|\mathbf{p}_1, \mathbf{p}_2 \ldots\rangle \tag{6-95}$$

Nonzero spins complicate the situation, and we refer the reader to Sec. 10-4.

REFERENCES

6BU H. Burkhardt, *Dispersion Relation Dynamics* (North-Holland, Amsterdam, 1969).

6GA S. Gasiorowicz, *Elementary Particle Physics* (Wiley, New York, 1966).

6GO M. L. Goldberger and K. M. Watson, *Collision Theory* (Wiley, New York, 1964).

6HA M. Hamermesh, *Group Theory* (Addison-Wesley, Reading, Mass., 1962).

6MA A. D. Martin and T. D. Spearman, *Elementary Particle Physics* (North-Holland, Amsterdam, 1970).

6ME A. Messiah, *Quantum Mechanics* (North-Holland, Amsterdam, 1961).

6NE R. G. Newton, *Scattering Theory of Waves and Particles* (North-Holland, Amsterdam, 1961).

6RO L. S. Rodberg and R. M. Thaler, *Introduction to the Quantum Theory of Scattering* (Academic Press, New York, 1967).

6ROM P. Roman, *Theory of Elementary Particles* (North-Holland, Amsterdam, 1961).

6WE J. Werle, *Relativistic Theory of Reactions* (North-Holland, Amsterdam, 1966).

CHAPTER

7

The General Description of Inelastic Interactions—I

7-1 Introduction. Until this point, the book has been predominantly about elastic scattering: how the data look at high and low energies, the optical model of high energy elastic scattering, the importance of resonances in low energy elastic scattering. We shall have more to say about elastic scattering. However, the early chapters served to present the data on elastic scattering and to develop some basic theoretical ideas. In this chapter and the next, we do the same for inelastic interactions. Of course in two chapters it is not possible to present many of the existing inelastic data or even to describe all the different patterns and regularities found in inelastic phenomena. It would not even be wise to describe these patterns and regularities in a series of neighboring chapters; such a presentation would be indigestible. Therefore, our plan is to present only the general aspects of inelastic interactions in this and the next chapter. Then as additional theory is developed in later chapters, other aspects of the inelastic interaction are introduced and discussed as illustrations of the theory.

At high energies, 80 to 90% of the total cross section goes into a great variety of *inelastic* channels. And in most of these inelastic channels many particles are produced in the final state. These are the *many-body* or *multiparticle* inelastic reactions. Our first task in this chapter is to sketch a broad picture of high energy inelastic reactions—a picture that emphasizes the many-body reactions. Fortunately, very high energy data on many-body reactions now exist, the highest energy data coming from the CERN 30-GeV proton–proton colliding beams. We begin by describing in Sec. 7-2 the topological composition of the total inelastic cross section. The particle multiplicities as a function of energy are then discussed. There are a few simple observations that summarize the angle and momentum spectra of the individual particles in many-body reactions, and these observations are summarized in Sec. 7-3. Having sketched the broad picture, we devote the remainder of the chapter to a description of the high energy behavior of *two-body* and *quasi-two-body* reactions. We return to the *many-body* reactions in Ch. 8.

A two-body inelastic reaction is a reaction $a + b \to c + d$ in which at least one of the final state particles is different from one of the initial state particles. Examples are $\pi^- + p \to \pi^0 + n$ and $\pi^- + p \to K^+ + \Sigma^-$. But there is a larger class of inelastic reactions which are almost two-body or, put more concisely, are *quasi-two-body*. These are reactions in which one or both of the final state particles are strong interaction resonances. Examples are $\pi^- + p \to \rho^- + p$ and $\pi^+ + p \to \pi^+ + N^{*+}$. From one point of view there is no intrinsic difference between two-body and quasi-two-body inelastic reactions; one or both of the final state particles in the latter simply have very short lifetimes. But in analyzing experimental data, it is often difficult to separate the quasi-two-body reaction from other reactions that have the same long-lived particles in the final state. We do not discuss this problem in this chapter; nor do we discuss the related question of the transition region where the lifetime of a final state particle becomes so short that its identification as a unique particle is in doubt. The full treatment of the behavior of two-body and quasi-two-body reactions requires the presentation of both differential cross-section and polarization information. In this chapter we are only concerned with the differential cross section; the description of typical polarization behavior is postponed to Chs. 11 and 12. To avoid having to write repeatedly "two-body and quasi-two-body reactions," we define the phrase "two-body reactions," unless further qualified, to include both two-body and quasi-two-body reactions.

We begin our discussion of two-body processes in Sec. 7-4 with a description of the forward and backward peaks that dominate the differential cross sections of high energy, two-body inelastic reactions. The explanation of these peaks through one-particle exchange is then given. Although this explanation is simple and beautiful from the qualitative viewpoint, great quantitative problems are associated with it. The attempts to resolve these quantitative problems are discussed later in the book. Next we introduce the subject of backward peaks in elastic reactions, since the theory of backward elastic peaks is identical to the theory of forward and backward peaks in two-body inelastic reactions.

In Sec. 7-7 we take up the energy dependence of two-body inelastic reactions and provide some illustrations of the simple power law behavior of the cross sections. This discussion of energy dependence leads us naturally to the consideration of diffractive inelastic processes—inelastic processes whose energy dependencies resemble elastic scattering rather than inelastic scattering.

In Sec. 7-9 we survey the entire field of high energy strong interactions to emphasize what is usually called the overall *peripheralism* of these reactions. We discuss this term and introduce what we hope is a less ambiguous term, *forward dominance*.

7-2 Topological Cross Sections and Particle Multiplicity. In a bubble chamber, spark chamber, or scintillation counter system, it is easy to observe the number of *charged* particles coming out of an interaction. The number of charged prongs is called the charged particle multiplicity and is denoted by N_{ch}. The cross section for the production of final states with N_{ch} charged prongs is called the *topological cross section* $\sigma_{top}(N_{ch}, s)$. The energy dependence of the topological cross sections is shown in Fig. 7-1 for the pp system. As the energy increases, the topological cross sections for the larger values of N_{ch} rise comparatively quickly and flatten out. At the highest energy in this figure, the $N_{ch} < 8$ topological cross sections are decreasing, while higher multiplicity topological cross sections are small but increasing. We expect that at still higher energies these higher multiplicity cross sections will in turn and flatten out and then perhaps decrease. Thus the high energy inelastic cross section is made up of a greater number of topological cross sections, all varying differently with energy. Yet as pointed out in Ch. 1, the total inelastic cross section at high energies is independent of energy or is a slowly varying function of energy. We do not understand this phenomenon in detail, but the optical model of Ch. 2 provides a semiclassical explanation.

Exercise 7-1. Amplify the last statement.

Thus as the energy increases, both the spread in N_{ch} and the average value of N_{ch} (called $\langle N_{ch} \rangle$) increase. This is to be expected—more particles are produced because more energy is available for conversion into mass. We do not know the precise relationship between $\langle N_{ch} \rangle$ and the total center of mass energy, $E_{cms} = \sqrt{s}$. Usually the relationship is represented by a *power law*

$$\langle N_{ch} \rangle = cs^n \tag{7-1a}$$

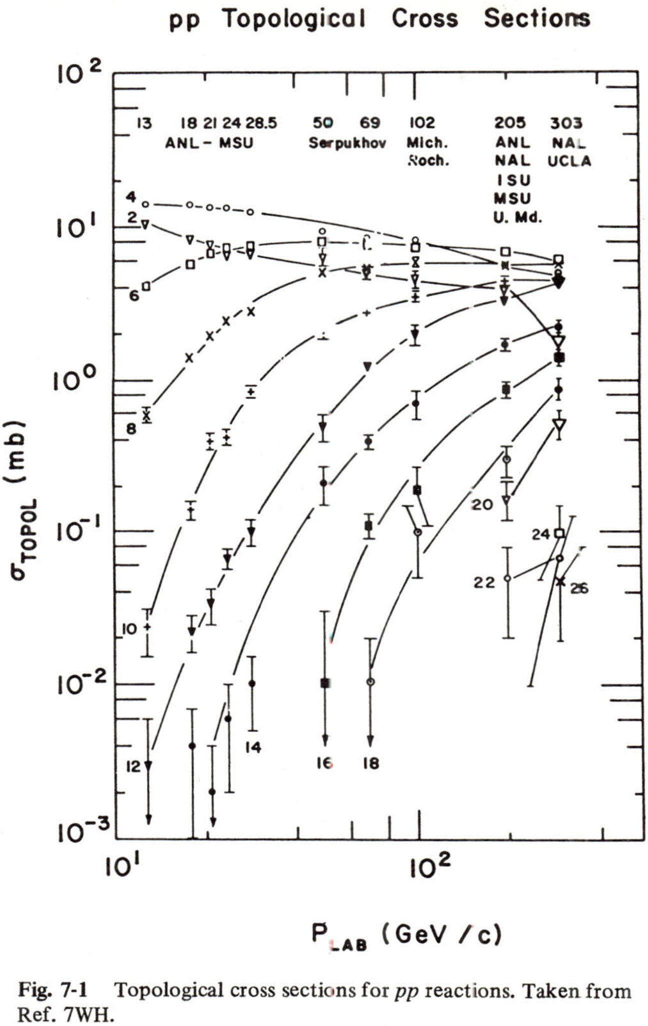

Fig. 7-1 Topological cross sections for *pp* reactions. Taken from Ref. 7WH.

or by *logarithmic law*

$$\langle N_{ch} \rangle = a + b \ln s \tag{7-1b}$$

Figure 7-2 plots $\langle N_{ch} \rangle$ for *pp* collisions semilogarithmically; the high energy approximation $s \approx 2mp_{lab}$ is used. Both the power law, with $n \sim 0.25$ to 0.35, and the logarithmic law provide rough but useful fits to the data. Of course there is no reason to expect a perfect fit by either of the simple functions in Eq. 7-1.

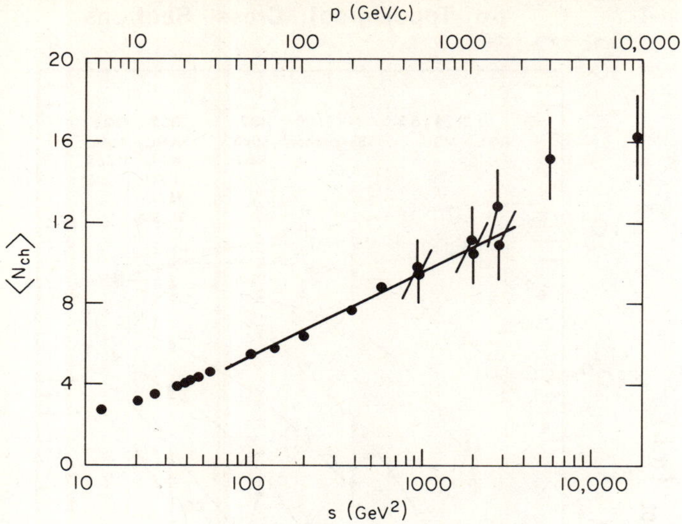

Fig. 7-2 The average charged particle multiplicity $\langle N_{ch} \rangle$, in *pp* reactions versus *s*. The line represents the logarithmic law (Eq. 7-lb), with $b = 1.8$. The two highest points are from emulsion measurements on cosmic rays and can have large systematic errors. The next six points are early measurements using *pp* colliding beams. Their errors should decrease with time.

But we like to think in terms of such simple functions to allow simple comparisons with theoretical models (see Ch. 8). Actually the logarithmic law with a bit of power law

$$N_{\text{ch}} = a + b \ln s + cs^n \tag{7-1c}$$

is the best fit.

Exercise 7-2. Existing data show that $\langle N_{\text{ch}} \rangle$ in πp collisions is quite similar in magnitude and energy behavior to $\langle N_{\text{ch}} \rangle$ in *pp* collisions. Consult the literature for further information. What are the differences?

Exercise 7-3. As we discuss next, the particles contributing to $\langle N_{\text{ch}} \rangle$ are mostly pions. Show that if all the available energy went into particle production we would obtain a power law with $n = \frac{1}{2}$, which is very different from what actually occurs.

The quantity $\langle N_{\text{ch}} \rangle$ is just one of the parameters describing particle multiplicity distributions. But before we can discuss these distributions in more detail we must consider what types of particles are produced in inelastic collisions.

Thus the next observation of interest is the relative abundance of the different types of particles produced in inelastic processes. The general rule is very simple. At all high energies that can be attained at accelerators, if we sum over *all* inelastic channels, we find that most of the produced particles are

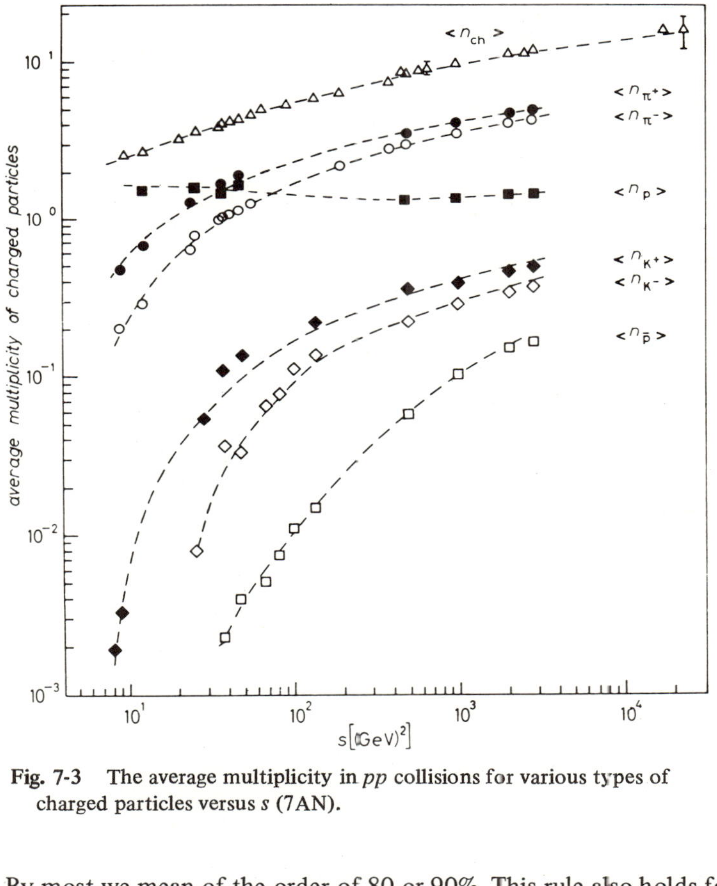

Fig. 7-3 The average multiplicity in *pp* collisions for various types of charged particles versus *s* (7AN).

pions. By most we mean of the order of 80 or 90%. This rule also holds for all the combinations of initial state particles that have been studied: nucleon–nucleon, pion–nucleon, kaon–nucleon, and photon–nucleon. After the pions, the particles are produced in the following order of decreasing copiousness: nucleons; K^+ mesons; and taken together, K^- mesons, hyperons, and antibaryons (Fig. 7-3). This rule is for a *summation* of the particle production rates over *all momenta and angles* of the produced particles. The relative production rates can vary substantially in different kinematic regions (see Ch. 8).

Exercise 7-4. Consult the literature (7IC) for details on the relative particle production rates. In particular, note that antiproton production approaches proton production at very high energies in the kinematic region of very slow particles in the center of mass (Fig. 8-4).

At lower energies the reasons for the preponderance of pion production are straightforward. The strange particle production is inhibited relative to pion production at these energies because of the requirement of strangeness conservation and the relatively higher mass of the strange particles. Antibaryon production is inhibited by the requirement of baryon conservation and by the larger mass of the antibaryons. As the energy increases, the relative production of strange particle and antibaryons should increase; and it does increase. But the production of these particles never equals pion production except in special kinematic regions (Ex. 7-4). Pion production still predominates. This puzzle is related to the subject of Ex. 7-3 and we return to it in Ch. 8.

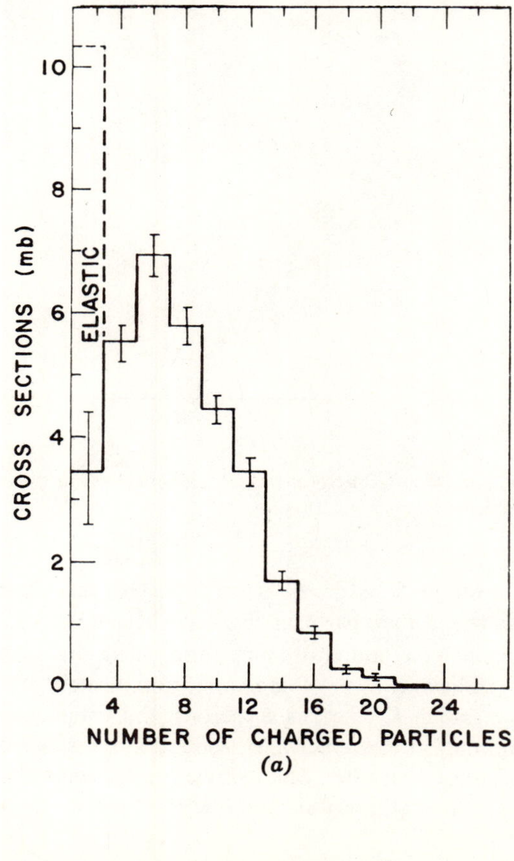

(a)

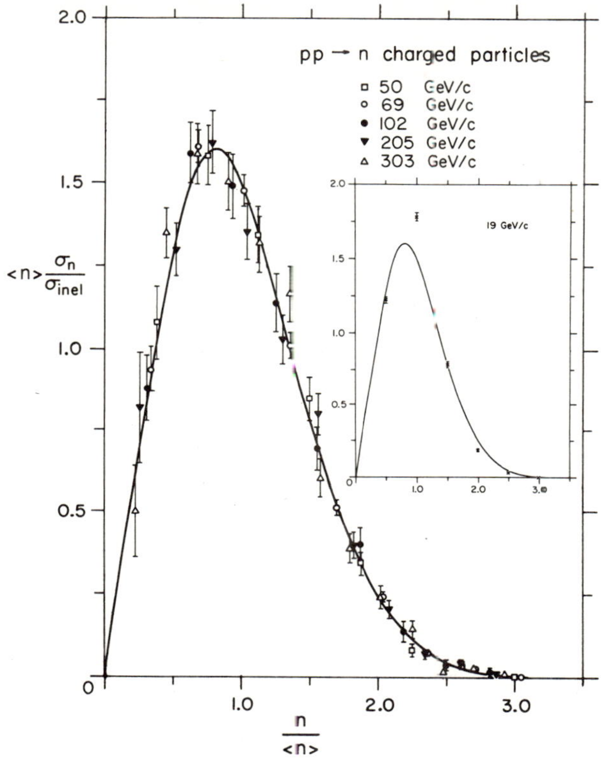

(b)

Fig. 7-4 The charge particle multiplicity distribution for *pp* reactions (*a*) $\sigma_{tot}(N_{ch})$ versus N_{ch} at 200 GeV (7CH), (*b*) a composite for many different energies (7SL). The scaling law for the multiplicity distribution used here does not work at lower energies, as shown for 19 GeV/c in the inset (Fig. 7-4*b*). The variables are explained in the text. The curve in (*b*) is *not* a Poisson distribution.

One tends to emphasize the charged hadrons because they are most easily studied, and until this point we have ignored the neutral hadrons. But recently data have been acquired on the multiplicities of the π^0 mesons. It is found that the π^0 is also produced copiously at high energies; and we have the simple rule

$$\langle N_{\pi^+}\rangle \approx \langle N_{\pi^-}\rangle \approx \langle N_{\pi^0}\rangle \tag{7-2}$$

at the same *s* is we sum over all inelastic channels. Therefore, although we continue to emphasize the charged hadrons, we keep in mind that to the best of our knowledge the π^+, π^-, and π^0 behave very similarly in many-body reactions.

Having established that at the energies of existing accelerators, the pion is the most abundantly produced particle in inelastic collisions, we return to the consideration of multiplicity distributions. If we plot $\sigma_{top}(N_{ch}, s)$ versus N_{ch} at fixed s, we obtain a multiplicity distribution. Figure 7-4*a* shows that distribution for an initial *pp* system at high energy. This distribution is typical of other energies and other initial systems, such as πp and γp. It is convenient to define

$$P(N_{ch}, s) = \frac{\sigma_{top}(N_{ch}, s)}{\sigma_{inel}(s)} \tag{7-3}$$

Thus one often sees plots of $P(N_{ch}, s)$ versus N_{ch}. The statement that the distribution in Fig. 7-4*a* is typical of other energies is reinforced by considering $P(N_{ch}, s)$ as a function of $N_{ch}/\langle N_{ch}\rangle$. As shown in Fig. 7-4*b*, we obtain an energy-independent relation

$$P(N_{ch}, s) = \frac{1}{\langle N_{ch}\rangle} f\left(\frac{N_{ch}}{\langle N_{ch}\rangle}\right) \tag{7-4}$$

This equation is a convenient basis for theoretical speculation and data summation.

We have only discussed the multiplicity distributions for N_{ch}; we have not tried to separate out the pion, nucleon, and K meson contributions. As we noted earlier, pions make up most of the charged particles; hence we are actually studying charged pion multiplicity distributions. We return to the subject of multiplicity distributions in Ch. 8.

7-3 Single-Particle Angle and Momentum Distributions. In this section we continue to describe the general behavior of the many-body inelastic reactions. To fully describe the angle and momentum distributions of the particles produced in a high multiplicity inelastic reaction is an impossible task. In studying inelastic channels in which four or more particles are produced, therefore, we have to find ways to summarize or condense these distributions. The simplest approach is to select one particle—a π^-, K^+, or p, for example—whose angle and momentum distribution is to be examined, and to integrate over the kinematic parameters of all the other particles. These are the *single-particle angle* and *momentum distributions* discussed in this section. Although it is an old idea to look at such distributions, they have recently been given a new name. They are a type of *inclusive distribution*; and reactions so treated are called *inclusive reactions.* The word "inclusive" refers to the inclusion together of all final states that produce the specific particle, or set of specific particles, being studied. We introduce the single-particle distributions using a rather old-fashioned method —the Peyrou plot.

The Peyrou plot provides a simple and instructive way to study single-particle distributions. As illustrated in Fig. 7-5, the Peyrou plot shows the distribution of the center of mass, three-momentum vectors of the particles for a single inelastic channel. The ordinate is $p_\perp$, the momentum component of the secondary particle perpendicular to the direction of the incident particle. This component is of course invariant to a Lorentz transformation from the laboratory system to the cms. The abscissa $p_\parallel^*$ (also called p_l^*) is the momentum component in the center of mass along the direction of the incident particle. We note that the allowed area for the vectors is a semicircle, as in Fig. 7-5. The radius of the semicircle depends on the total energy, the number and masses of the particles produced in the collision, and the mass of the particle whose distribution is being presented.

Exercise 7-5. Check by calculation the radii given in some of the Peyrou plots in this section. Show that at very high energy and not too high multiplicity, the radii depend only on the total energy.

Figure 7-5 is the Peyrou plot for the reaction $K^-p \to K^-p\pi^+\pi^-$. The plots for some 2-, 4-, 6-, and 8-prong interactions of the π^+p system are presented in Figs. 7-6*a* and 7-6*b*. These typical Peyrou plots lead to the following observations.

1. In most cases the points are not uniformly distributed over the area of the semicircle. Even along arcs of fixed radius, the point densities are not uniform. With some exceptions, the distributions extend further along the $\pm p_\parallel^*$ axis than they do in the $p_\perp$ direction. Thus in general the particle momenta are not isotropic in the center of mass system.
2. But at higher multiplicities there is some approach to isotropy in the center of mass system, particularly for the pions.
3. There is a strong tendency, particularly at low multiplicities, for the incident particles to preserve their initial momentum in direction and magnitude. This is most evident for the proton in all the figures. The same tendency is also evident for the K^- in Fig. 7-5 and for the π^+ in Fig. 7-6. Of course when we say that the incident particle tends to preserve its momentum, we do not mean that the particle passes through the reaction unchanged. That statement has no meaning in strong interaction physics. We only mean that there is a particle in the final state, identical to a particle in the initial state, and that final state particle has very roughly the same momentum as the initial state particle. This is called the *leading particle effect.*
4. When the proton changes to a neutron, the neutron distributions, like the proton distributions, are concentrated in the backward hemisphere. Thus the particle carrying the baryon number of the incident proton tends to preserve the momentum of the incident proton.

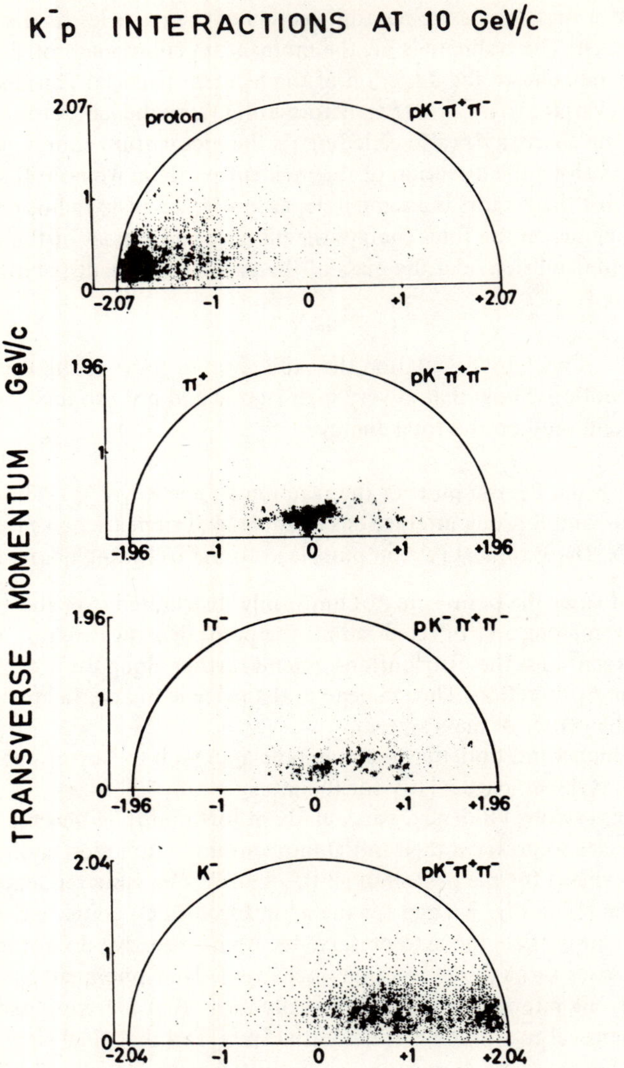

Fig. 7-5 Peyrou plot for the reaction $K^-p \to K^-p\pi^+\pi^-$ at 10 GeV/c. Taken from Ref. 7CZ.

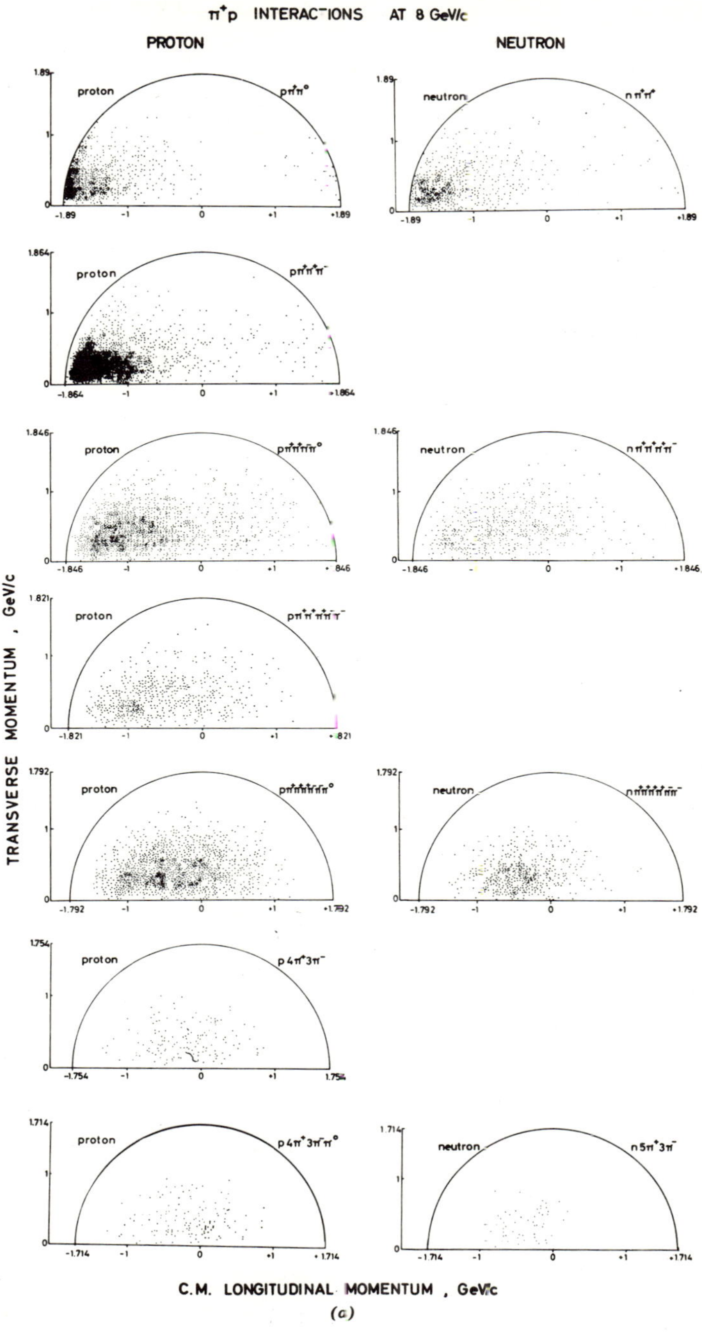

π⁺p INTERACTIONS AT 8 GeV/c
PROTON
NEUTRON
TRANSVERSE MOMENTUM , GeV/c
C.M. LONGITUDINAL MOMENTUM , GeV/c
proton
neutron

(a)

Fig. 7-6.

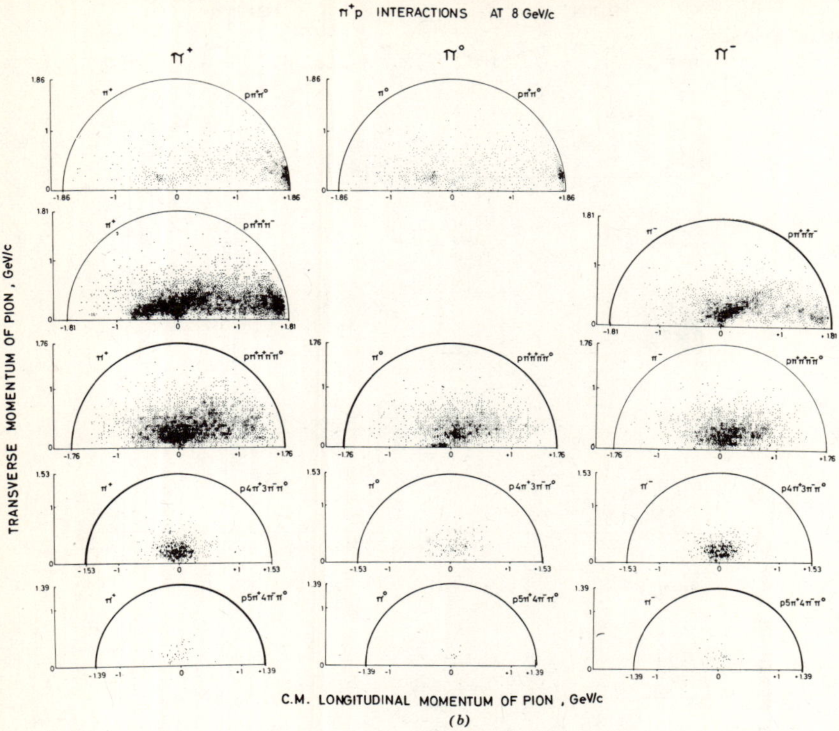

Fig. 7-6 Peyrou plots for π^+p reactions at 8 GeV/c. Taken from Ref. 7CZY.

5. Observations 1, 3, and 4 are sometimes summarized by the statement that the lower multiplicity inelastic interactions are *peripheral* or show *peripheralism.* This concept is defined and developed in Sec. 7-9.

Perhaps the most remarkable feature of many-body reactions is the rapid decrease, illustrated by these Peyrou plots, in the transverse momentum distribution as the transverse momentum increases. These are Peyrou plots for intermediate energies; but very high energy reactions show the same effect, as discussed in Ch. 8. The rapid decrease in the transverse momentum distribution is emphasized by the *experimental* observation that the average transverse momentum $\langle p_\perp \rangle$ is

$$\langle p_\perp \rangle \sim 0.3 \text{ to } 0.4 \text{ GeV/c} \tag{7-5a}$$

This observation hold even at very high energies where the average energy per produced particle is 5 or 10 GeV!

The transverse momentum distributions for a particle in a many-body reaction i can be roughly described by the functions

$$\rho_i(p_\perp^2) = \frac{d\sigma_i}{d(p_\perp^2)} = A_i\, e^{-a_i p_\perp} \tag{7-6a}$$

or

$$\rho_i(p_\perp^2) = \frac{d\sigma_i}{d(p_\perp^2)} = B_i\, e^{-b_i p_\perp^2} \tag{7-6b}$$

where σ_i is the reaction cross section for a fixed interval in $p_\parallel^*$ or the entire reaction cross section obtained by integrating over $p_\parallel^*$. These forms fit roughly all the many-body reaction channels that make up an inelastic cross section; and the parameters a_i or b_i depend very little on the particular channel. Therefore one finds experimentally that the transverse momentum distribution of a particle from the *entire* inelastic cross section can also be fit by

$$\rho(p_\perp^2) = \frac{d\sigma_{\text{inel}}}{d(p_\perp^2)} = A\, e^{-ap_\perp} \tag{7-7a}$$

or

$$\rho(p_\perp^2) = \frac{d\sigma_{\text{inel}}}{d(p_\perp^2)} = B\, e^{-bp_\perp^2} \tag{7-7b}$$

As an aside, we note that we prefer $d\sigma/d(p_\perp^2)$ to $d\sigma/dp_\perp$ because, given that Eqs. 7-6 and 7-7 are reasonable fits, the $d\sigma/dp_\perp$ distribution would have a zero at $p_\perp = 0$.

Restricting our considerations to the small $p_\perp^2$ region, $p_\perp^2 \lesssim 1$ (GeV/c)^2, the choice of function is a matter of taste and the particular reaction. Neither function is perfect, and sometimes one fits the data better than the other. We take up first Eq. 7-7a. Independent of energy and the nature of the initial hadronic system, a usually falls in the range of 4 to 8 (GeV/c)$^{-1}$. Now considering Eq. 7-7b, we observe that the function on the right-hand side is the same as the exponential function of t used in elastic scattering, Eq. 1-9. Therefore, as we discuss in Ch. 8, the second form emphasizes the connection between diffractive elastic scattering and the transverse momentum distribution in many-body reactions. Comparing with the experimental observation, we put

$$\text{for } a = 6\ (\text{GeV/c})^{-1} \qquad \langle p_\perp \rangle = 0.33\ \text{GeV/c} \tag{7-5b}$$

To conclude our discussion, the transverse momentum distributions of the particles are

1. Only weakly dependent on the total energy, on the multiplicity, or on the type of particle.

2. Up to equivalent laboratory energies of 2000 GeV, $\langle p_\perp \rangle$ is in the range of 0.35 ± 0.05 GeV/c.

This is certainly a remarkable property of inelastic, strong interaction reactions and should be understandable on rather general principles. We attempt to make it so in the next chapter and in Ch. 22, but for the remainder of this chapter we turn our attention to two-body interactions.

7-4 Forward and Backward Peaks in Two-Body Inelastic Reactions. At high energies almost all two-body inelastic reactions show a striking behavior in their differential cross sections—the differential cross sections are dominated by strong peaks in the forward or backward direction. This behavior is fully as striking as the diffraction peak in elastic scattering. To describe this phenomenon more precisely, we consider the reaction

$$a + b \to c + d$$

and define θ_{ac} as the angle between $\mathbf{p}_a$ and $\mathbf{p}_c$ in the center of mass and $\mathbf{p}_i$ the three-momentum of particle i. We also define the squares of the four-momentum transfers

$$t_{ac} = (P_a - P_c)^2, \qquad t_{ad} = (P_a - P_d)^2 \tag{7-8}$$

where P_i is the four-momentum of particle i. When $s \gg m_i^2$ for all i, $|t_{ac}|$ is small when θ_{ac} is close to 0° and $|t_{ad}|$ is small when θ_{ac} is close to 180°. We can then rephrase our initial statement as follows. At high energies almost all two-body inelastic differential cross sections show strong peaks when $|t_{ac}|$ or $|t_{ad}|$ is small.

Let us turn to some illustrations. Figure 7-7, a classic example, shows five different quasi-two-body final states all produced from the initial state π^+p. In all these cases we observe a strong peak when $|t_{ac}|$ is small. A much weaker peak occurs when $|t_{ad}|$ is small. Note which particle is called c and which is called d. It is cumbersome to keep defining which particles are a and c and to have to specify whether we mean t_{ac} or t_{ad} when we refer to the four-momentum transfer. Therefore, the following very convenient system has been developed for reactions in which one of the incident particles (namely, a) is a meson or photon, and the other particle (namely, b) is a baryon. We define particle c to be that final state particle which has baryon number 0. Then we call θ_{ac} simply θ, we call t_{ac} simply t, and we denote t_{ad} by u. Also, the small t region is called the *forward* inelastic scattering region and the small u region is called the *backward* inelastic scattering region. This nomenclature is in analogy to the designation of the small t region in elastic scattering as the forward elastic scattering region. If both particles in the initial state are baryons, the nomenclature just given does not apply; hence t and u must be explicitly defined for each case.

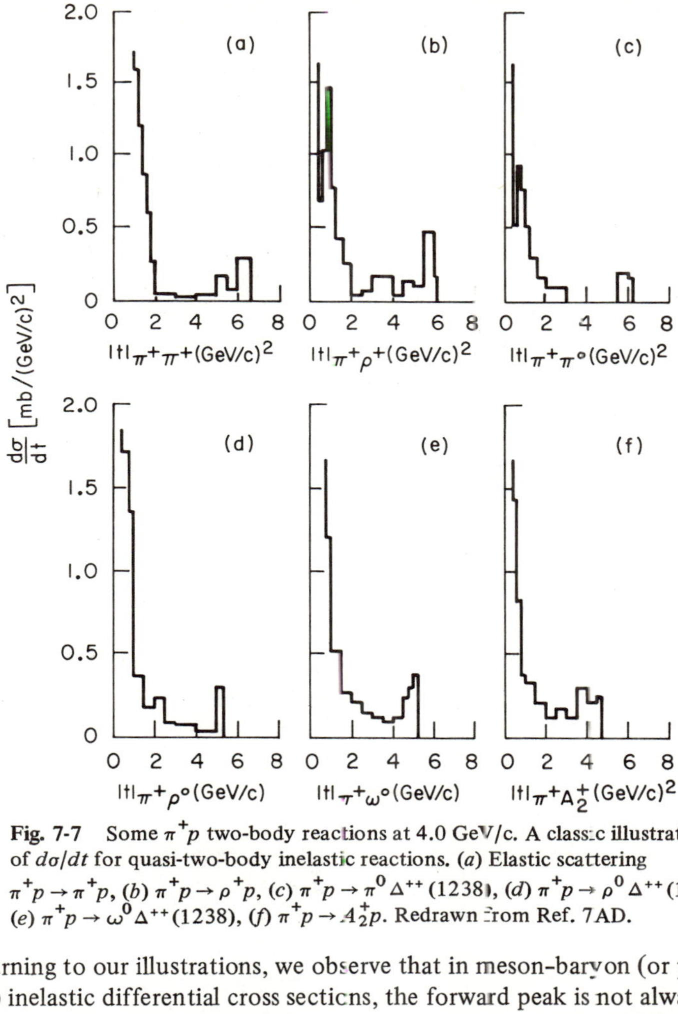

Fig. 7-7 Some π^+p two-body reactions at 4.0 GeV/c. A classic illustration of $d\sigma/dt$ for quasi-two-body inelastic reactions. (*a*) Elastic scattering $\pi^+p \to \pi^+p$, (*b*) $\pi^+p \to \rho^+p$, (*c*) $\pi^+p \to \pi^0\Delta^{++}(1238)$, (*d*) $\pi^+p \to \rho^0\Delta^{++}(1238)$, (*e*) $\pi^+p \to \omega^0\Delta^{++}(1238)$, (*f*) $\pi^+p \to A_2^+p$. Redrawn from Ref. 7AD.

Returning to our illustrations, we observe that in meson-baryon (or photon-baryon) inelastic differential cross sections, the forward peak is not always stronger then the backward peak, as is the case in the reactions of Fig. 7-7. To see this, we look at the associated production differential cross sections in Fig. 7-8. We remind the reader that in accordance with our nomenclature, θ is defined between the incident pion and the final *kaon*. The reactions $\pi^- + p \to K^0 + \Lambda^0$ and $\pi^- + p \to K^0 + \Sigma^0$ are dominated by forward peaks, *but* $\pi^- + p \to K^+ + \Sigma^-$ is dominated by a backward peak. Another reaction dominated by a backward peak is $K^- + p \to \pi^+ + \Sigma^-$. However, for reasons given in the next section, most high energy, two-body inelastic reactions have differential cross sections dominated by forward peaks.

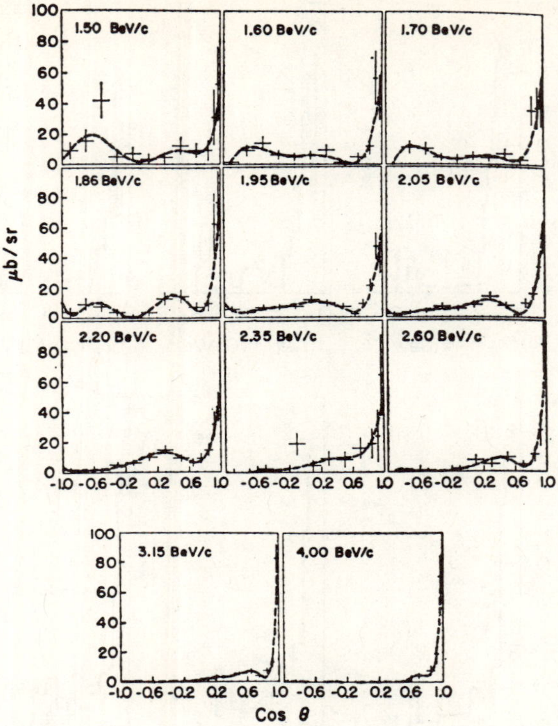

(a)

Fig. 7-8.

Thus there is at least a superficial similarity between the shape of the differential cross section of high energy elastic scattering and the shape of the differential cross section in high energy, two-body inelastic scattering. But whereas the forward peak in elastic scattering always has the approximate form $e^{-b|t|}$ at small t, the forward peak in inelastic scattering shows much greater variety in the t dependence. Two examples of different types of t behavior of the forward peak are provided by the photoproduction reactions $\gamma + p \to \pi^+ + n$ (Fig. 7-9) and $\gamma + p \to K^+ + \Lambda^0$ (Fig. 7-10). The former exhibits relatively slow t dependence for $|t| > 0.01$ (GeV/c)^2. But when $|t| < 0.01$ (GeV/c)^2, there is a very sharp peak. This very forward peak, if fit by $e^{-b|t|}$, yields $b \gtrsim 50$ (GeV/c)$^{-2}$! Conversely, the latter reaction shows a dip at $t \sim 0$!

Although $e^{-b|t|}$ does not represent the detailed behavior of most forward or backward peaks, it is instructive to use it as a rough measure of the steepness of a peak. In Table 7-1 we present b values for forward peaks in a number of pion-proton and kaon-proton interactions. We see that the values of b vary from 2 to

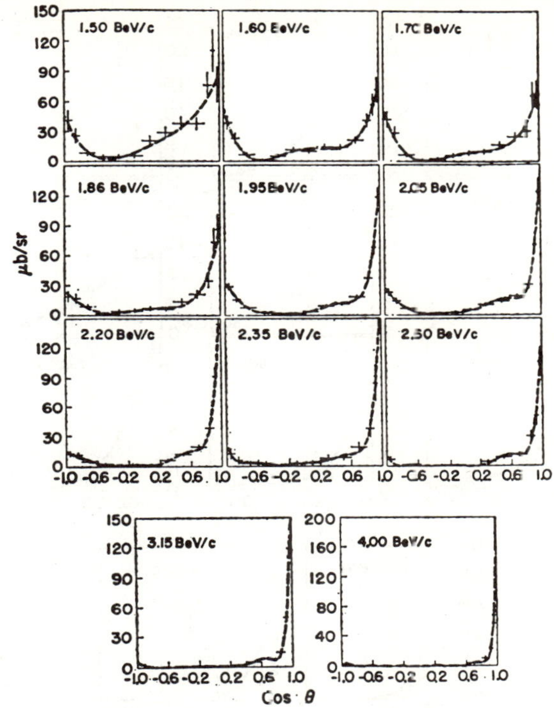

(*b*)

Fig. 7-8.

17 $(\mathrm{GeV/c})^{-2}$. Thus the inelastic scattering b parameters are distributed over a large range of numerical values compared with the elastic scattering b parameters.

As the energy increases into the tens-of-GeV range, the forward and backward peaks become the only easily measurable parts of the differential cross sections. That is, the remainder of the angular region simply has such a small differential cross section that it is difficult to collect events from that region. Therefore, in the tens-of-GeV range the only part of a differential cross section for a two-body inelastic process that is usually displayed is the small t or small u region. The photoproduction processes in Figs. 7-9 and 7-10 are illustrative.

Before going on to the next section we must work out some messy but necessary kinematics. In elastic scattering, $t = 0$ at $\theta = 0$, but the situation is more complicated in inelastic scattering. Consider the reaction $a + b \to c + d$ where all the masses may be different and t_{ac} is given by

$$t_{ac} = m_a^2 + m_c^2 - 2E_aE_c + 2p_ap_c \cos \theta_{ac} \tag{7-9}$$

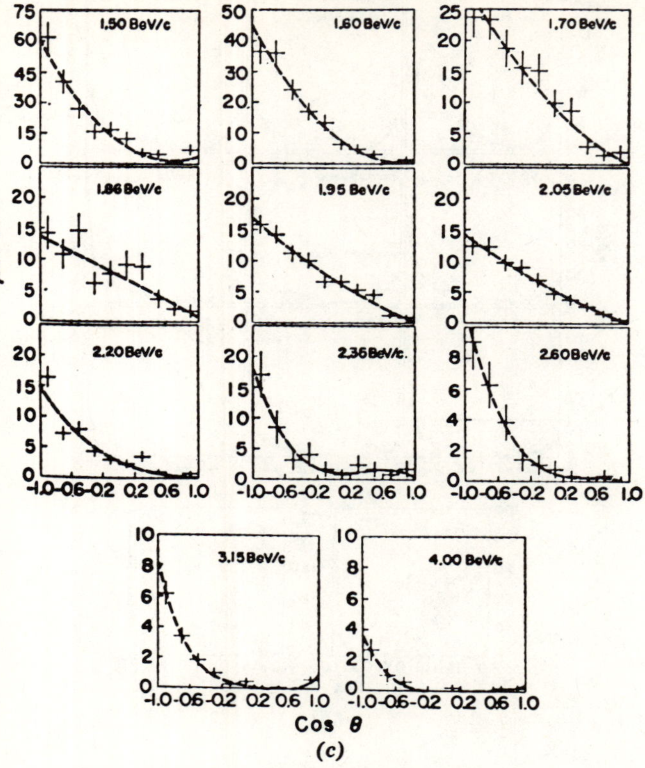

Fig. 7-8 Differential cross sections for the associated production reactions (*a*) $\pi^-p \to K^0\Lambda^0$, (*b*) $\pi^-p \to K^0\Sigma^0$, (*c*) $\pi^-p \to K^+\Sigma^-$(7DA).

where the energies E_i, magnitude of three-momenta p_i, and the angle θ_{ac} are all in the barycentric system. The value of t at $\theta_{ac} = 0$ is

$$t_{ac}(\theta_{ac} = 0) = m_a{}^2 + m_c{}^2 - 2E_aE_c + 2p_ap_c = t_{ac,\,\min} \tag{7-10}$$

It is conventional to use the notation $t_{ac,\,\min}$ for $t(\theta = 0)$, although it is seldom correct because t_{ac} is usually negative. Thus we write

$$t_{ac} = t_{ac,\,\min} - 2p_ap_c(1 - \cos\theta_{ac}) \tag{7-11}$$

Exercise 7-6. Derive Eq. 7-9 and the equations for $t_{\min}$ given in App. A.

7-5 Forward Peaks, Backward Peaks, and One-Particle Exchange. The forward and backward peaks that dominate high energy, two-body inelastic

Table 7-1 Properties of typical two-body inelastic reactions.[a]

Reaction	*t*-Channel Exchange: Type	Dominant Trajectories	Momentum Range (GeV/c)	n	Average slope $(\mathrm{GeV}/c)^{-2}$
$\pi^- p \to K^0 \Lambda$	$S_t = 1$	$K^*(890), K^*(1420)$	3–16	1.6	7, break at $\lvert t \rvert \cong 0.4$
$K^- p \to \pi^0 \Lambda$		$K^*(890), K^*(1420)$	3–14	1.7	6, break at $\lvert t \rvert \cong 0.4$
$\bar{K}^0 p \to \pi^+ \Lambda$		$K^*(890), K^*(1420)$	2–75	1.5	7, break at $\lvert t \rvert \cong 0.4$
$\bar{K} p \to \pi \Sigma$		$K^*(890), K^*(1420)$	2–14	1.5	8, break at $\lvert t \rvert \cong 0.4$
$\pi p \to K \Sigma$		$K^*(890), K^*(1420)$	3–14	1.5	9, break at $\lvert t \rvert \cong 0.4$
$\pi^- p \to \pi^0 n$	$S_t = 0, I_t = 1$	ρ	5–50	1.09 ± 0.03	9 + forward dip
$K^- p \to \bar{K}^0 n$		ρ, A_2	4–35	1.5 ± 0.1	5
$K^+ n \to K^0 p$		ρ, A_2		1.3	5
$\pi^- p \to \eta^0 n$		A_2	10–45	1.38 ± 0.07	
$\pi^- p \to X^0 n$		A_2	3.6–50	1.2 ± 0.1	
$\pi^+ p \to \pi^0 \Delta^{++}$		ρ	2–13	1.6 ± 0.1	9 + dip
$K^+ p \to K^0 \Delta^{++}$		ρ	3–12.7	1.9 ± 0.1	5 + forward dip
$\bar{K}^0 p \to K^- \Delta^{++}$		ρ		1.9	5
$\pi^- p \to \rho^0 n$		$\pi,$	4–20	2.07 ± 0.04	10
$K^+ n \to K^*(890) p$		π, ρ, A_2	2.3–12.7	2.3	9 + break
$K^- p \to K^*(890) p$		π, ρ, A_2	3.9–14.3	2.19 ± 0.06	6 + break
$K^- p \to \bar{K}^*(1420) p$		π, ρ, A_2	3.9–14.3	2.10 ± 0.13	6.5 + break
$K^+ p \to K^*(890) p$		f_0, ρ, π	3–12.7	1.78 ± 0.10	4
$\pi^+ p \to \rho^0 \Delta^{++}$		π	4–22	1.56 ± 0.07	$b_1 \simeq 17, b_2 \simeq 3$
$\pi^+ p \to f^0 \Delta^{++}$		π	5–22	1.16 ± 0.09	$b_1 \simeq 11, b_2 \simeq 2$
$\pi^- p \to g(1710) p$			8–18	1.4 ± 0.4	$b_1 \simeq 10, b_2 \simeq 2$
$\pi^+ p \to \rho^+ p$	$S_t = I_t = 0, G_t = -1$	ω, π	3–16	1.80 ± 0.08	8 + break
$\pi^- p \to \rho^- p$		ω, π	3–16	1.80 ± 0.08	8 + break
$K_L^0 P \to K_S^0 p$		ω	1.5–8	2.1	9 + break

[a] The columns are in order: the reaction, the S_t (strangeness) and I_t (isotopic) of the *t*-channel exchange, the dominant exchange particle or Regge trajectory (Ch. 17), the range of incident momenta for the indicated n and b values, n in $\sigma \sim s^{-n}$, b in $d\sigma/dt \sim A\, e^{-b|t|}$, with notes on deviations from the exponential behavior; b_1, b_2 refers to a peak with an initial exponential slope b_1 and a larger $\lvert t \rvert$ exponential slope b_2. Taken from Ref. 7GI.

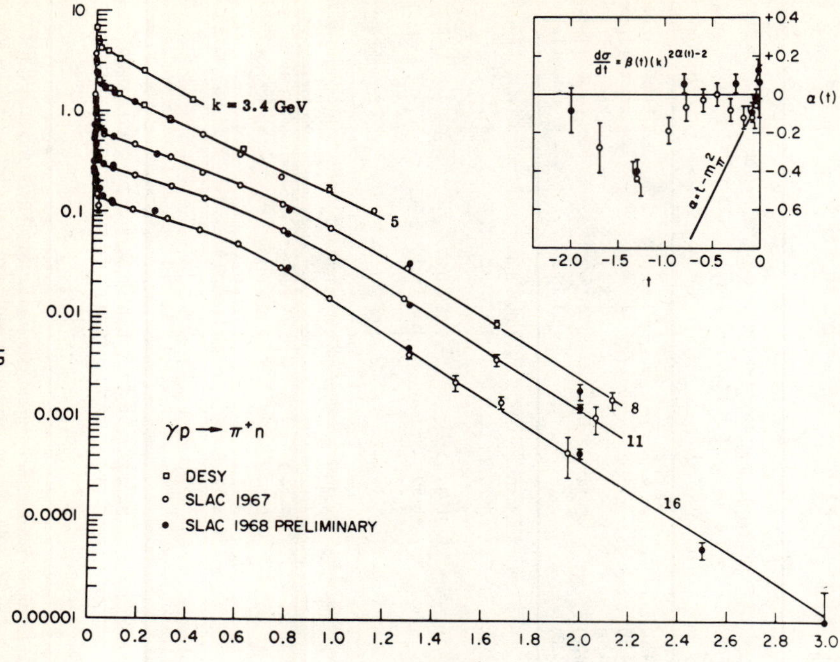

Fig. 7-9 The $d\sigma/dt$ for $\gamma + p \rightarrow \pi^+ + n$. Note the very sharp forward peak that occurs in the $|t| < 0.02$ (GeV/c)^2 interval (7DI).

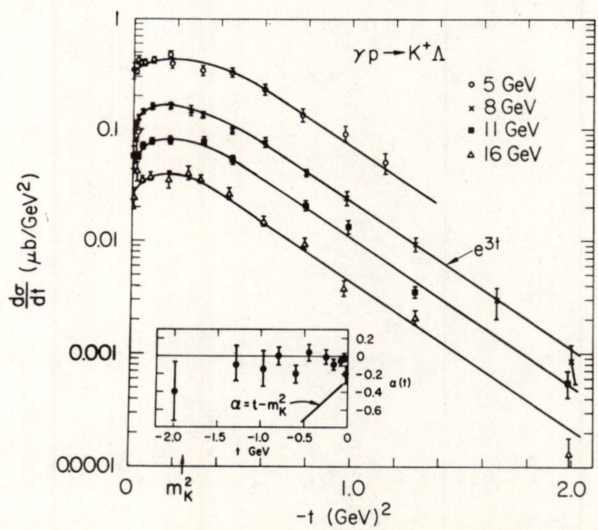

Fig. 7-10 The $d\sigma/dt$ for $\gamma + p \rightarrow K^+ + \Lambda^0$. Note the dip as $t \rightarrow 0$ (7DI).

reactions have a beautiful and simple qualitative explanation—one-particle exchange. This explanation forms the subject matter of this section. But we must warn the reader that the explanation is not simple from the quantitative viewpoint. Indeed, we do not have a complete quantitative understanding of one-particle exchange. However, many of the later chapters of this book are devoted to developing our partial understanding through several-particle exchange, absorption corrections, Regge theory, and duality theory.

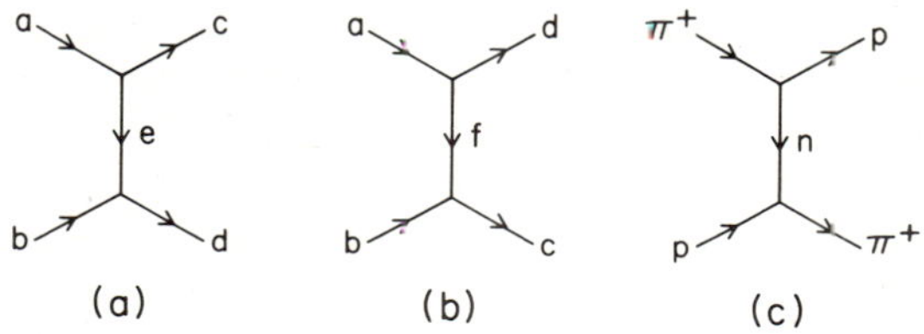

Fig. 7-11 Feynman diagrams for one-particle exchange explanation of (*a*) forward peak at small t_{ac}, (*b*) backward peak at small t_{ad}, (*c*) backward peak in π^+p elastic scattering.

At this point we are content with the simplest model of one-particle exchange. As shown in Fig. 7-11*a* we assume that the inelastic interaction $a + b \to c + d$ occurs through the exchange of *one virtual* particle e. We avoid all spin and charge complications by taking all of the particles, including e, to have spin zero and charge zero. The mass of the ith particle is m_i. The differential cross section from Ch. 12 is

$$\frac{d\sigma}{dt} = \frac{|g_{aec}g_{bed}|^2}{64\pi s(p_i^*)^2} \frac{1}{[|t_{ac}| + m_e{}^2]^2} \tag{7-12}$$

where t_{ac} is defined in Eq. 7-9 and g_{aec} and g_{bed} are the coupling parameters at the indicated vertices. In the simplest theory we take the g's to be constants.

For $|t_{ac}| \gg m_e^2$, $d\sigma/dt$ will decrease as $1/|t_{ac}|^2$, producing a prominent peak at small $|t_{ac}|$. Thus the exchange of a single virtual particle leads to a peak at small $|t_{ac}|$ values, provided $m_e{}^2$ is small compared with the t_{ac} range under consideration. Thus we have a qualitative explanation of peaks at small t values in two-body inelastic reactions. Obviously by considering the one-particle exchange diagram of Fig. 7-11*b* we can explain a peak in the differential cross section at small values of t_{ad}.

But the one-particle exchange model does much more than explain how a peak can occur at small t values; it predicts whether a peak will occur at small t_{ac} or small t_{ad} in any particular reaction. This predictive power is based on three rules.

1. The exchanged particle e is a known hadron. It may be a relatively stable hadron such as a pion or a relatively unstable hadron such as a ρ^0 meson.

2. At each vertex the following quantities, and perhaps others, are conserved: charge, total isotopic spin, baryon number, strangeness, parity.

3. Most g-coupling parameters are roughly of the same size. Then in a particular reaction, the forward or backward peak dominates according to which peak is associated with the smallest mass exchanged particle.

For a general application of these rules, consider the large class of reactions discussed in Sec. 7-4, in which a meson or photon (particle a) interacts with a nucleon (particle b). We take particle c to be the final state particle with baryon number 0. Then there will be a peak at small t_{ac}–namely, a forward peak–if there is a particle e that can couple at both vertices of Fig. 7-11a according to rule 2. Usually there is such a particle: a pion, ρ^0 meson, or kaon, for example. Thus there is usually a forward peak in such reactions. For such reactions to show a backward peak–namely, a peak at small t_{ad}–there must exist a baryon or baryon resonance to serve as the exchanged particle in Fig. 7-11b. But by rule 3, even if there is such a backward peak, the forward peak will dominate. Thus for the class of reactions under discussion, we refer to the forward peaks as being caused by meson exchange and the backward peaks as being caused by baryon exchange. We do not include in this class the diffractive inelastic processes discussed in Sec. 7-8.

An exception to the dominance of forward peaks in this set of reactions is provided by the reaction $\pi^- p \to K^+ \Sigma^-$ illustrated in Fig. 7-8c. Again the three rules explain this condition. There is no known meson of charge +2 and strangeness +1 to couple to the $\pi^- K^+$ vertex. Thus there is no forward peak in this reaction.

There are no well-established forward or backward peaks whose existence and relative magnitudes cannot be explained by the three rules. Of course this statement is a bit circular. One way to find a hadron, for use in rule 1, is to find a forward or backward peak in a differential cross section which requires such a hadron for its explanation. But we can say that the existence of all hadrons used in rule 1 has always been confirmed by other means. General recipes for determining which hadrons may exist have been derived from quark and $SU(3)$ models (Ch. 14).

Thus we see that the exchange of a single virtual particle provides a beautiful qualitative explanation for the forward and backward peak phenomena in two-body inelastic interactions. But very serious discrepancies occur when one tries to calculate the precise shape and energy dependence of these peaks. We discuss those discrepancies in Chs. 12 and 13.

We conclude with an explanation of some conventional terminology. In the reaction $a + b \to c + d$, let $s = (P_a + P_b)^2 = (P_c + P_d)^2$, where the P_i's are four-momenta. We usually call $a + b \to c + d$ the *s-channel.* Hence the term *s-channel resonances* in Sec. 5-8. And referring to Fig. 7-11a, we call the reaction

$a + \bar{c} \to \bar{b} + d$, or its inverse, the *t-channel reaction* ($\bar{b}$ and $\bar{c}$ being antiparticles of b and c, respectively). This is discussed in detail in Sec. 15-3. We note here only that the one-particle exchange diagram (Fig. 7-11*a*) is usually said to represent *t-channel exchange.*

7-6 Forward and Backward Peaks in Elastic Scattering. While reading the last two sections, the reader may have wondered why we do not use the concept of one-particle exchange to explain the forward peak—that is, the diffraction peak—in elastic interactions. There are several interlocking answers to the question represented. Before giving those answers, we note that one-particle exchange is believed to contribute to the forward peak in elastic scattering. The exact amount of the contribution of course varies with the elastic scattering system and with the theory used to extract the contribution. We say more about this in the chapters on Regge theory, but we do know that the dominant process in the forward elastic peak is diffraction, not one-particle exchange. (In making this statement I have obviously excluded the pomeron from the category of known hadrons to be used according to rule 1 of Sec. 7-5.)

We have the following reasons for not taking one-particle exchange to be the main cause of the forward peak in elastic scattering:

1. At high energies, all hadron–hadron systems show very similar forward peaks in elastic scattering. But one-particle exchange would contribute quite differently to the different systems. For example, one-pion exchange can occur in proton–proton elastic scattering. By the rules of Sec. 7-5 this would be the dominant one-particle contribution. But one-pion exchange cannot contribute to pion–nucleon elastic scattering because a three-pion vertex is not allowed in strong interactions (see Ex. 4-5). The lowest mass hadron which could be exchanged is the ρ meson. Therefore pion–nucleon and proton–proton elastic scattering should exhibit quite different forward peaks. But they exhibit quite similar peaks. Therefore, some other process—namely, diffraction—must dominate one-particle exchange.

2. Through the optical theorem, the forward elastic amplitude is directly related to the total cross section. The total cross section is a summation over all types of interactions. There could not possibly be a simple relationship between that summation and the forward amplitude of a particular one-particle exchange process. (I must note that we are not considering the pomeron to be a particle here.)

3. The optical model explanation of the forward elastic peak is so natural and fits the data so well that we cannot ignore it. But the optical model is not a semiclassical description of one-particle exchange. Therefore, having accepted

the optical model of high energy elastic scattering, we must accept the one-particle exchange model as an *additional* contribution to the elastic scattering, *not* as an alternative explanation.

We now turn our attention to backward peaks in elastic scattering. As illustrated in Fig. 7-11*c* these can be produced by the exchange of a baryon. Using the rules of Sec. 7-5, we expect backward peaks in π^+p, π^-p, K^+p, and np elastic scattering. But we do not expect backward peaks in K^-p, pp, or $\bar{p}p$ elastic scattering. Or more precisely, we expect smaller backward peaks in the latter systems compared with the backward peaks in the former systems.

Exercise 7-7. Justify the last two statements.

The predictions cited are confirmed by experiment. The backward peaks in pion–nucleon elastic scattering appear in Fig. 7-12. The variable used in this figure is u, the square of the four-momentum transfer between the incident meson and the scattered nucleon. We observe that the π^+p backward peak exhibits a rather sharp peak, followed by a dip and then a rounder second peak—quite a complicated structure. If the reader looks back to Fig. 1-7*b* he will see how this backward peak fits in with the full differential cross section of the π^+p system. Next we look at Fig. 22-1, a medium high energy measurement of the K^+p and K^-p differential cross section. We observe that the K^+p system shows a relatively strong backward peak. But the K^-p system has a relatively small backward peak.

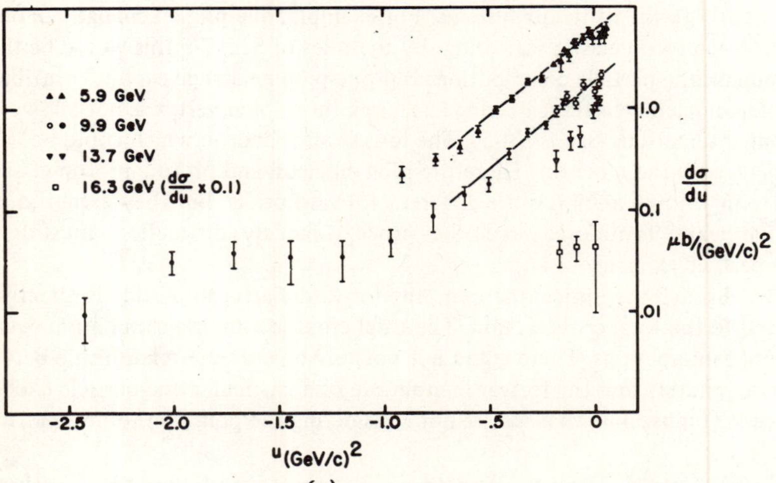

(a)

Fig. 7-12.

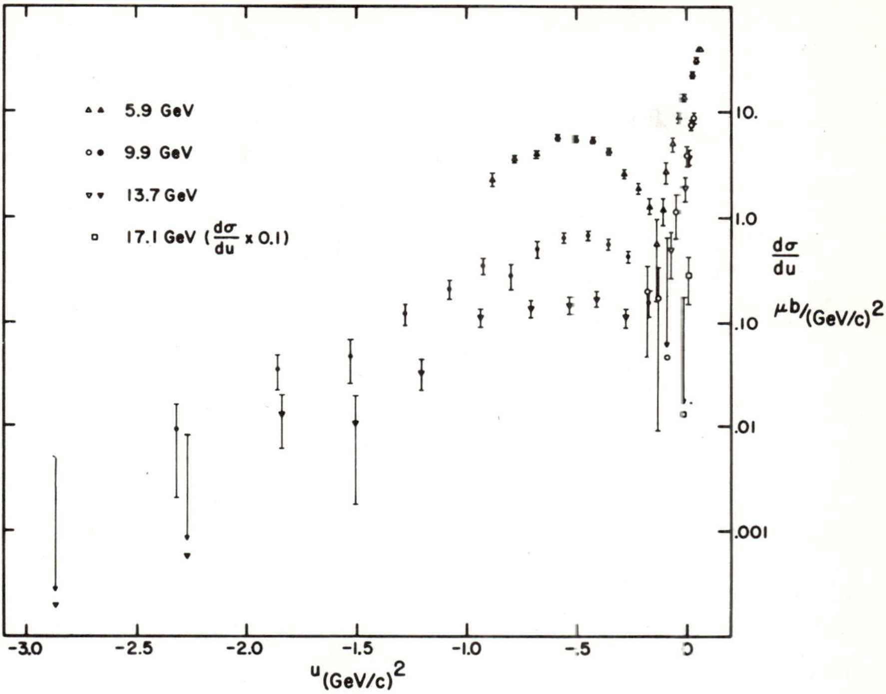

(*b*)

Fig. 7-12 The $d\sigma/du$ for backward elastic scattering for (*a*) $\pi^- p$, (*b*) $\pi^+ p$ (7OR).

Exercise 7-8. Use the equations in App. A to explain the values of u in Fig. 7-12.

7-7 Energy Dependence of Two-Body Inelastic Reactions. Once the multi-GeV energy range is attained, the total elastic cross section $\sigma_{el}(s)$ varies very slowly with energy (Fig. 1-11). The cross sections for most two-body inelastic reactions varies much more rapidly with energy. Typical examples are given in Fig. 7-13. Denoting the reaction cross section for the reaction $a + b \to c + d$ by $\sigma_{ab\to cd}$, it has been found that a good fit to the data can be obtained by Eq. 7-13, once the energy is *well above threshold.*

$$\sigma_{ab\to cd} \sim p_{\text{lab}}^{-n} \sim s^{-n} \tag{7-13}$$

Typical values for n are given in Table 7-1. Since n in the table exceeds 1, $\sigma_{ab\to cd}$ decreases very rapidly with increasing energy. There is a small class of reactions

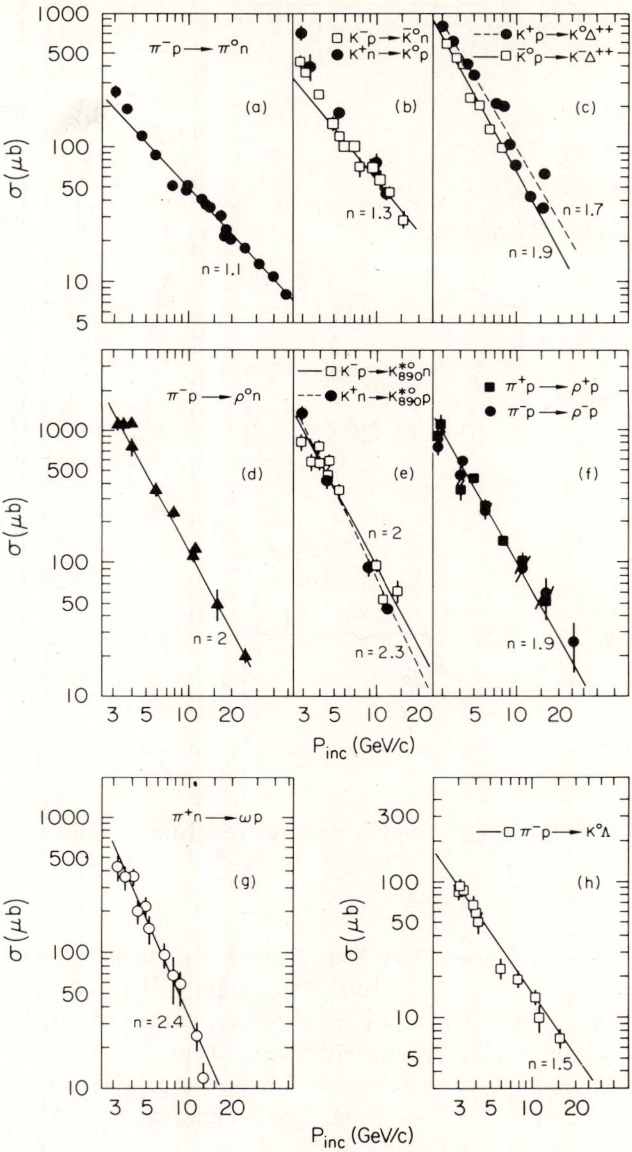

Fig. 7-13 The momentum dependence of various two-body inelastic cross sections; n is defined in Eq. 7-13. Taken from Ref. 7BA.

in which n is smaller than 1—the so-called *diffractive inelastic* reactions. We discuss these reactions in the next section. For the present, we consider only the much larger class of reactions in which n exceeds 1.

As discussed in Sec. 7-4, the high energy, two-body inelastic reactions are dominated by forward or backward peaks. Therefore it is the energy dependence of these peaks that is being described by Eq. 7-13. Once the several-GeV region is attained, there are no drastic changes in these peaks with respect to their domination over the differential cross section. That is, if a forward peak, or a backward peak, is dominant in the several-GeV region, it remains dominant as the energy increases. There are, of course, energy dependent changes in the shape of the peaks. In particular, secondary peaks, dips, and breaks in the larger $|t|$ or larger $|u|$ region of these peaks can be very energy dependent. These effects are treated later in the book.

The reactions with dominant forward peaks have n values in the range of roughly 1.0 to 3.0. The reactions with dominant backward peaks have larger values of n. Thus the smaller values of n are associated with meson exchange and the larger values of n are associated with baryon exchange.

In this section we have been able to easily codify the energy dependence of large classes of inelastic two-body reactions. Unfortunately, the explanation of this behavior is still in doubt. We must return several times to the problem of how to understand this behavior, and in the end we fail to develop a complete understanding or rigorous calculational method.

7-8 Diffractive Inelastic Processes in Two-Body and Few-Body Reactions.

As we mentioned in the last section, there is a small class of two-body inelastic reactions for which the cross sections do not decrease rapidly with increasing energy. The most beautiful example is the reaction $\gamma + p \to \rho^0 + p$, whose cross section versus the incident photon momentum is shown in Fig. 7-14. The cross

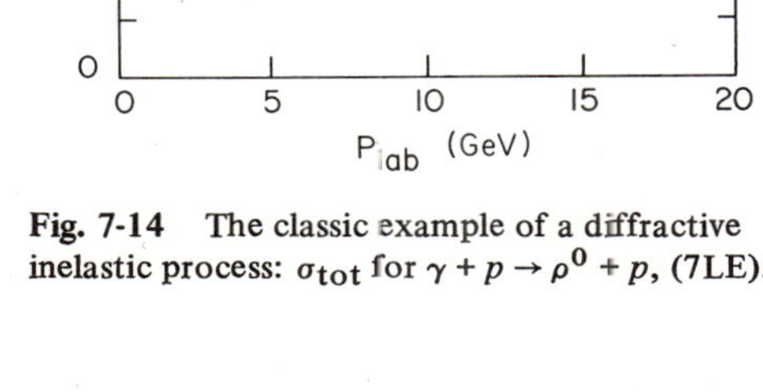

Fig. 7-14 The classic example of a diffractive inelastic process: σ_{tot} for $\gamma + p \to \rho^0 + p$, (7LE).

section falls off very slowly with energy—at about the same rate as the elastic cross sections. Other examples are provided by the reactions $p + p \to N^*(1.40) + p$, $p + p \to N^*(1.52) + p$, and $p + p \to N^*(1.69) + p$ (7ED). Using Eq. 7-13 to fit the energy dependence of the cross sections for these reactions, we obtain the n values given in Table 7-2 (7LE, 7ED).

Table 7-2 n in $\sigma \sim s^{-n}$ for various diffractive inelastic reactions and elastic reactions (momentum range; 5 to 20 or 30 GeV/c). Taken from Refs. 7ED and 7LE.

Reactions	n
$\gamma p \to \rho^0 p$	~0.2
$pp \to N^*(1.40)p$	0.5 ± 0.3
$pp \to N^*(1.52)p$	0.34 ± 0.06
$pp \to N^*(1.69)p$	0.56 ± 0.06
$\pi^- p \to A_1^- p$	0.41 ± 0.11
$K^0 p \to Q^0 p$	0.59 ± 0.16
$\pi^- p \to \pi^- p$	~0.2
$pp \to pp$	0.2 ± 0.05
$K^+ p \to K^+ p$	0.09 ± 0.03
$K^- p \to K^- p$	0.39 ± 0.04

Such processes are called *diffractive inelastic* reactions because in their energy dependence and in the shape of their differential cross section *forward peak* they closely resemble *diffractive elastic* scattering. That is:

1. $\sigma(ab \to cd)$ varies with energy as $\sigma(ab \to ab)$.
2. The forward peak that dominates $d\sigma/dt$ $(ab \to cd)$ has the form $e^{-b|t|}$, with b in the range of values given by elastic scattering.

Table 7-2 illustrates these observations. These diffractive inelastic processes seem to be limited to reactions in which, when going from particle a to c and from particle b to d, there is *no* change of:

1. Baryon number.
2. Charge.
3. Strangeness.
4. *G*-parity.
5. Isotopic spin.

We note that conditions 1 through 5 are also obeyed by elastic scattering. Thus, in addition to the similarity in energy dependence, there are further likenesses between these diffractive inelastic processes and elastic scattering. But the similarity is not complete because in these diffractive inelastic processes there can be a change of intrinsic spin and intrinsic parity in going from particle a to c or in going from particle b to d. Nevertheless the similarities have led to the hypothesis that diffractive inelastic scattering can be explained by the same infinite set of Feynman diagrams that one hopes can explain diffractive elastic scattering. Thus if Fig. 7-15*a* is the diagram for diffractive elastic diagram, Fig. 7-15*b* is the diagram for diffractive inelastic scattering. The cross-hatched areas in these diagrams represent a summation over the diagrams of Fig. 8-10. Of course, the equivalence cannot be exact because changes of spin and parity are allowed in Fig. 7-15*b*. The theoretical understanding of diffractive processes is still rudimentary, and other explanations have been given (7SA, 7LE, 7CO).

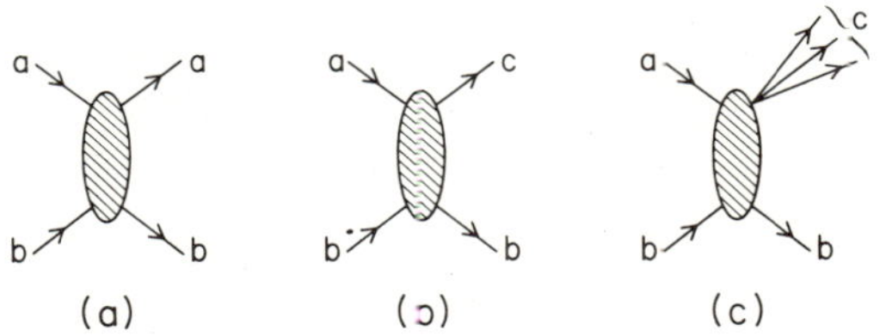

Fig. 7-15 All the cross-hatched areas in (*a*), (*b*), and (*c*) represent roughly the same infinite set of Feynman diagrams: (*a*) corresponds to elastic scattering, (*b*) and (*c*) correspond to diffractive inelastic reactions. Other proposed mechanisms for few-body diffractive processes, such as the Deck effect (7LE, 7SA), are not shown here.

The subject of diffractive inelastic processes is both enriched and confused because these processes also occur in few-body interactions. Consider the reaction

$$\pi^- + p \rightarrow \pi^- + \pi^+ + \pi^- + p \tag{7-14}$$

whose three-pion invariant mass distribution is shown in Fig. 7-16. This mass plot, which is typical for this reaction at high energies, has a huge peak in the low mass region. This broad low mass peak is called the A peak, the A bump, or the A enhancement. We must be blunt and state that at present there is no agreement on just what constitutes this enhancement. There are certainly one or more resonant particles in the enhancement. In particular, the upper mass region of the A enhancement contains the A_2 resonance—a resonance with a mass of about 1300 MeV. There is no consensus about whether the lower mass region of this enhancement also contains a resonance—the A_1 resonance—or whether this lower mass region is a nonresonant kinematic or dynamic effect. But there is good evidence that diffractive inelastic processes are involved in the production of all or part of the A_1 region in the A enhancement. The A_1 region is defined to extend between the three-pion invariant mass limits of 1000 and 1200 MeV. We shall summarize this and related observations.

1. The total cross section for the production of the A_1 region is a slowly decreasing function of energy. Specifically $\sigma(\pi^- p \to A_1^- p) \sim p_{\text{lab}}^{-0.4}$.

2. Enhancements in the low mass region are found in the three-meson invariant mass in the reaction

$$K^\pm + p \to K^\pm + \pi^+ + \pi^- + p \tag{7-15}$$

The low mass region in the $K^\pm p$ reactions is called the Q bump, Q peak, or Q enhancement—the variety of names demonstrating our uncertain understanding.

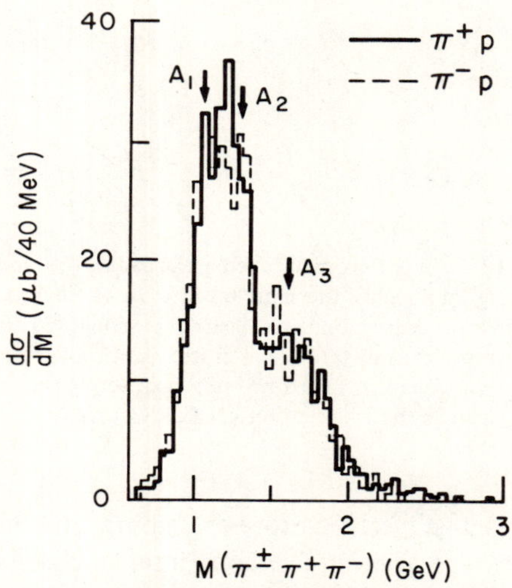

Fig. 7-16 The A enhancement in the reactions $\pi^\pm p \to \pi^\pm p \pi^+ \pi^-$ (7LE).

The total cross section for the production of the Q enhancement also decreases relatively slowly with energy—the energy dependence is similar to that of the A_1 (Table 7-2).

3. Enhancements like the A_1 and the Q are produced coherently on nuclei. Since, in the interests of clarity we have restricted this chapter to inelastic reactions on free nucleons, we refer the reader to the literature (7SA, 7FI).

The subject of diffractive inelastic reactions is particularly interesting because it involves *coherence*, a concept that appears in many branches of physics. Unfortunately, we cannot take the space to introduce the coherence concept here (7FI) or to discuss further our present rather tangled theoretical understanding of diffractive inelastic processes.

7-9 Forward Dominance in High Energy Interactions. Let us consider the following broad observations on the high energy interactions of incident particles $a + b$. We describe these observations in the *center of mass* Lorentz frame, but they are easily translated to the laboratory Lorentz frame.

1. High energy elastic scattering $a + b \to a + b$ is dominated by a forward peak (Chs. 1 and 2). In that peak: final particle a has about the same three-momentum as initial particle a, final particle b has about the same three-momentum as initial particle a, and the transverse momenta of the particles in the final state are relatively small.
2. High energy, two-body inelastic reactions $a + b \to c + d$ are dominated by forward or backward peaks (Sec. 7-4). Particle c has about the same three-momentum as particle a, particle d has about the same three-momentum as particle b, and the transverse momenta in the final state are relatively small.
3. Some two-body and few-body reactions show diffractive inelastic scattering whose properties are roughly described (Sec. 7-8) by observation 1.
4. If the multiplicity is not too high, most many body reactions show the following features: the transverse momenta are relatively small in the final state; much of the three-momentum of each incident particle is carried off by one particle or by a group of particles making small angles with the incident particle (Sec. 7-3).
5. In observations 1 through 4 the particle or group of particles that seem to carry off the three-momenta of the incident particles have the following characteristics. First, they have the same baryon number as the incident particle. Second, they have the same strangeness as the incident particles. Or if that is not possible, the change of strangeness from the incident particles to the final state particles is minimal.
6. The average particle multiplicity at high energies is much less than the multiplicity that is energetically possible (Sec. 7-2).

These six observations are major examples of what are called *peripheral* phenomena in high energy physics. Used in this way, the word "peripheral" has a very broad meaning, but it also has a more limited meaning in particle physics in connection with the *peripheral model* of two-body inelastic reactions. We describe this model in Ch. 18. Although there is some relation between the two uses of the word, we find it best to restrict its use in this book to the peripheral model.

Therefore, we have to construct a new phrase to describe all the processes contained in observations 1 through 6. The phrase—*forward dominance*—indicates that in the center of mass the differential cross sections or angular distributions are dominated by particles going *forward* along the directions of motion of *both incident particles.* It also says that transverse momenta are relatively small. Finally, it suggests that drastic changes in the strangeness, or changes in the baryon number, have relatively low probability, these changes being defined from each incident particle to the particle, or group of particles, which is carrying the incident particle's three-momentum. We include *backward* peaks in this concept of forward dominance, for by the term *forward* we mean that in the center of mass the particles tend to go forward along the directions of motion of either incident particle.

REFERENCES

7AD M. Aderholz et al., *Phys. Rev.* **138B**, 897 (1965).

7AN M. Antinucci et al., *Let. Nuovo Cimento*, **6**, 121 (1973).

7BA R. Barloutaud, in *Proceedings of the Fourth International Conference on High Energy Collisions, 1972* (Rutherford High Energy Laboratory, Chilton, England, 1972).

7CH G. Y. Charleton et al., *Phys. Rev. Lett.* **29**, 515 (1972).

7CO G. Cohen-Tannoudji et al., *Nucl. Phys.* **B37**, 77 (1972).

7CZ O. Czyzewski, in *Proceedings of the Fourteenth International Conference on High Energy Physics, 1968* (CERN, Geneva, 1968).

7CZY O. Czyzewski, in *Proceedings of the Topical Conference on High Energy Collisions of Hadrons, 1968*, CERN Report 68-7 (CERN, Geneva, 1968).

7DA O. I. Dahl et al., *Phys. Rev.* **163**, 1430 (1967).

7DI R. Diebold, in *Proceedings of the Boulder Conference on High Energy Physics*, K. T. Mahanthappa, W. D. Walker, and W. E. Brittin, Eds. (Colorado Associated University Press, Boulder, 1970).

7ED R. M. Edelstein et al., *Phys. Rev.* **D5**, 1073 (1972).

7FI C. M. Fisher, in *Methods in Subnuclear Physics*, Vol. I (Gordon and Breach, New York, 1968). A thorough introduction to the relation of coherence to diffractive inelastic interactions.

7GI G. Giacomelli, in *Proceedings of the Sixteenth International Conference on High Energy Physics, 1972* (National Accelerator Laboratory, Batavia, Ill., 1973).

7IC A good source for data on many-body reactions and references to particular experiments are the review articles in the recent proceedings of the International Conferences on High Energy Physics. These conferences are held every two years.

7JA M. Jacob, in *Proceedings of the Sixteenth International Conference on High Energy Physics, 1972* (National Accelerator Laboratory, Batavia, Ill., 1973).

7LE D. W. G. S. Leith, in *Proceedings of the Sixteenth International Conference on High Energy Physics, 1972* (National Accelerator Laboratory, Batavia, Ill., 1973).

7OR J. Orear et al., *Phys. Rev. Lett.* **21**, 389 (1968).

7SA H. Satz and K. Schilling, in *Proceedings of the Colloquium on Multiparticle Dynamics, 1971* (Research Institute for Theoretical Physics, Helsinki, Finland, 1971).

7SL P. Slattery, *Phys. Rev. Lett.*, **29**, 1624 (1972).

7WH J. Whitmore, in *Proceedings of the Sixteenth International Conference on High Energy Physics, 1972* (National Accelerator Laboratory, Batavia, Ill., 1973).

CHAPTER

8

The General Description of Inelastic Interactions—II

8-1 Introduction. In this, our second chapter on inelastic reactions, we discuss many-body reactions at high energies—the experimental data and the theoretical models. This subject was introduced in Secs. 7-2 and 7-3, but here we go into much more detail. We begin with the general expression for the differential cross section for a many-body inelastic reaction, separating that expression into a matrix element factor and a phase space factor. This leads to the simple phase space model of many-body inelastic reactions. The model is of great use to experimentalists in the search for resonances and the determination of their properties, but in Secs. 8-2 and 8-3 we devote relatively little time to those uses of phase space. Our major interest is in the mechanism of many-body processes, and in Sec. 8-4 we begin to study these processes with the Fermi statistical model. We know this model and the simple thermodynamic model described in Sec. 8-5 to be wrong. But they are worth discussing because they provide a great deal of insight into how many-body inelastic reactions can be understood using statistical ideas.

In Sec. 8-6 we develop the modern language for discussing many-body reactions; and we describe the major experimental observations in this field. In particular we introduce the scaling law for one-particle momentum distributions. To begin to understand the mechanisms behind these experimental observations, we construct in Sec. 8-7 a special phase space model incorporating a sharp cutoff in the transverse momentum. Although this model is rather unphysical it provides a good foundation for discussing more physical models.

In Secs. 8-8 through 8-10 we discuss these more physical models of many-body reactions. Here we had to try to describe the many models that have been developed during the last few years, when the subject of many-body reactions returned to fashion. Lack of space prevents comprehensive treatment of these models, and we have discussed a few of what we hope are the major aspects of the more prominent models. Section 8-9 also contains a general discussion of the relation between many-body reactions and elastic diffractive scattering.

Because the field of inelastic interactions is so large, even two chapters cannot summarize the entire field. Therefore we have omitted a number of subjects. For example, the discussions of electron–nucleon and muon–nucleon inelastic scattering and hadron production in electron–positron collisions are postponed until a later chapter.

8-2 Phase Space and Many-Body Reactions. To begin we consider the general formula, Eq. 6-38, for the differential cross section for an inelastic interaction in which N particles are produced:

$$\frac{d\sigma_N}{d^3p_1 \cdots d^3p_{N'}} = \frac{(2\pi)^4}{4E_{i1}E_{i2}|\mathbf{v}_{i1} - \mathbf{v}_{i2}|} \prod_{n=1}^{N'} \frac{1}{(2\pi)^3 2E_n}$$

$$\times \int \prod_{n=N'+1}^{N} \frac{d^3p_n}{(2\pi)^3 2E_n} |T_N|^2 \delta^4(P_f - P_i) \qquad (8\text{-}1)$$

Here only the first N' particles in the final state of N particles are detected; we integrate over the momenta of the remaining $N - N'$ particles. An old model for inelastic interactions, first proposed by Fermi (8FE), consists of the simple assumption that $|T_N|^2$ is *independent* of the individual momenta of the particles in the final state. The matrix element $|T_N|^2$ may be dependent on the total energy ($\sqrt{s}$), on the number and types of particles in the final state, and on the spin states of the particles. The crucial assumption in this model that $|T_N|^2$ is independent of $p_i (i = 1, 2, \ldots, N)$ allows $|T_N|^2$ to be removed from the integral in Eq. 8-1.

Then we define a factor $D_N(s)$

$$D_N(s) = \frac{(2\pi)^4 |T_N|^2}{4E_{i1}E_{i2}|\mathbf{v}_{i1} - \mathbf{v}_{i2}|(2\pi)^{3N}} \tag{8-2}$$

where $D_N(s)$ depends only on the number, type, and spin states of the final state particles. We also define the *phase space factor* $R_N(P_i^0, \mathbf{P}_i)$

$$R_N(P_i^0, \mathbf{P}_i) = \int \left[\prod_{n=1}^{N} \frac{d^3p_n}{2E_n} \delta^4(P_f - P_i) \right] \tag{8-3}$$

Here the four-momentum is denoted by $P = (P^0, \mathbf{P})$. In the cms the phase space factor is $R_N(\sqrt{s}, 0)$. Then Eq. 8-1 becomes in the center of mass

$$\frac{d\sigma_N}{d^3p \cdots d^3p_{N'}} = D_N(s) \frac{dR_N(P_i^0, \mathbf{P}_i)}{d^3p \cdots d^3p_{N'}} \tag{8-4}$$

Equations 8-2 through 8-4 are the basic equations for a *simple phase space model* of many-body inelastic reactions.

In the cms $dR_N(\sqrt{s}, 0)/d^3p \cdots d^3p_{N'})$ has no preferred direction in space, for unpolarized incident particle $D_N(s)$ also has no preferred direction in space. Therefore, in these models all momentum distributions would be isotropic in the center of mass system. But since we already know that the center of mass momentum distributions are in general not isotropic, it follows that the simple phase space model is wrong.

However, we can use the simple phase space model to separate the behavior enforced by the phase space from the dynamic behavior of the matrix element T_N. For example, we know that for low multiplicities the incident particles tend to preserve their directions of motion. This behavior must come from T_N because in the cms R_N does not contain any information on the directions of motion of the incident particles. T_N, not R_N, carries all the information on the spin, parity, and isotropic spin behavior of the interaction. Finally, the additive conservation rules for strangeness, baryon number, and lepton number are carried by T_N; R_N is concerned only with the conservation of four-momentum. From the discussion in Sec. 6-4, R_N is an invariant under Lorentz transformations. Therefore R_N is a Lorentz invariant phase space factor. The Lorentz invariance is produced by the presence of the E_n's in the denominator of Eq. 8-3. If these E_n factors are removed from R_N and placed in $D_N(s)$, we obtain a phase space factor $R_N^{(\text{nonrel})}$ that is not Lorentz invariant. There is no reason for selecting one phase space factor in preference to the other, unless it is known that the pure s dependence of $D_N(s)$ can be obtained by selecting one of the forms. At present it is customary to use the Lorentz invariant phase space factor of Eq. 8-3. Until

1960 it was customary to use the non-Lorentz invariant form (8MIL, 8KR). We emphasize that neither form is correct; that is, it is not possible in either form to cancel out the dependence of T_N on individual particle momenta.

The analytic calculation of $R_N(\sqrt{s}, 0)$ and its derivatives is very difficult if N is greater than 3 or 4. Therefore, our plan is to discuss the useful $N = 3$ case to illustrate some of the concepts. We then present some examples for larger N. However, let us begin with the simplest case, $N = 2$. For two particles, 1 and 2, with masses m_1 and m_2, we write

$$R_2(\sqrt{s}, 0) = \int \frac{d^3p_1 d^3p_2}{4E_1E_2} \delta^4(P_i - P_f) = \frac{\pi p}{\sqrt{s}} \tag{8-5}$$

where p is the magnitude of the three-momentum of either particle in the center of mass, and from App. A we have

$$p^2 = \frac{(s - m_1{}^2 - m_2{}^2)^2 - 4m_1{}^2m_2{}^2}{4s} \tag{8-6}$$

When s is very large compared with the masses, $p = \sqrt{s}/2$ and

$$R_2(\sqrt{s}, 0) = \frac{\pi}{2}, \qquad s \to \infty \tag{8-7}$$

The only interesting differential is

$$\frac{dR_2(\sqrt{s}, 0)}{d\Omega_{\text{cms}}} = \frac{p}{4\sqrt{s}} \tag{8-8}$$

Exercise 8-1. Derive Eqs. 8-5 through 8-8.

For three particles with masses m_1, m_2, and m_3, we have

$$R_3(\sqrt{s}, 0) = \int \frac{d^3p_1 d^3p_2 d^3p_3}{8E_1E_2E_3} \delta^4(P_i - P_f) \tag{8-9}$$

To evaluate this we first derive an interesting recursion relation between R_N and R_{N-1}. Now

$$R_N(\sqrt{s}, 0) = \int \frac{d^3p_N}{2E_N} \left(\prod_{n=1}^{N-1} \frac{d^3p_n}{2E_n} \right) \delta^4\left(P_N + \sum_{n=1}^{N-1} P_n - P_i\right) \tag{8-10}$$

We observe that

$$\int \left(\prod_{n=1}^{N-1} \frac{d^3p_n}{2E_n} \right) \delta^4 \left(\sum_{n=1}^{N-1} P_n - (P_i - P_N) \right)$$

is $R_{N-1}(P_i^0 - P_N^0, -\mathbf{P}_N)$. Since R_N is Lorentz invariant, we can write

$$R_{N-1}(P_i^0 - P_N^0, -\mathbf{P}_N) = R_{N-1}(\sqrt{s_{N-1}}, 0)$$

where

$$s_{N-1} = (P_i^0 - P_N^0)^2 - (\mathbf{P}_N)^2 \tag{8-11}$$

Therefore, we have the recursion relation

$$R_N(\sqrt{s}, 0) = \int \frac{d^3p_N}{2E_N} \left[R_{N-1}(\sqrt{s_{N-1}}, 0) \right] \tag{8-12}$$

where s_{N-1} is given by Eq. 8-11. As an example, in the three-particle case, using Eqs. 8-5, 8-6, and 8-12, we have

$$R_3(\sqrt{s}, 0) = \frac{\pi}{2} \int \frac{d^3p_3}{2E_3} \left\{ \frac{[(s_2 - m_1^2 - m_2^2)^2 - 4m_1^2 m_2^2]^{1/2}}{s_2} \right\} \tag{8-13}$$

where

$$s_2 = (\sqrt{s} - E_3)^2 - p_3^2 = s + m_3^2 - 2E_3\sqrt{s} \tag{8-14}$$

Even this is a messy integral. Therefore although the recursion formula is simple it quickly leads to complicated integrals.

Exercise 8-2. Evaluate the integral in Eq. 8-13 for the ultrarelativistic case by setting all the masses equal to zero. Evaluate the integral when all the particles are nonrelativistic pions (8NY).

We began this section by observing that the simple phase space model is in general *incorrect* because it predicts that the center of mass momentum distributions will be isotopric, and experiment shows that for the lower multiplicities the center of mass momentum distributions are decidedly nonisotropic. Therefore, applications of the simple phase space model must be restricted to the higher multiplicities. But we still must ascertain the predictive power of the model even at these higher multiplicities. To do so, we turn to the simple phase space model calculation of single-particle momentum distributions.

The single-particle distributions follow directly from Eq. (8-12), thus we have

$$\frac{dR_N(\sqrt{s}, 0)}{dp_N} = \frac{2\pi p_N^2}{E_N} R_{N-1}(\sqrt{s_{N-1}}, 0) \tag{8-15}$$

Again the only simple example is for $N = 3$ and

$$\frac{dR_3(\sqrt{s}, 0)}{dp_3} = \frac{\pi^2 p_3^2 [(s_2 - m_1^2 - m_2^2)^2 - 4m_1^2 m_2^2]^{1/2}}{E_3 s_2} \tag{8-16}$$

where p_3 is the magnitude of the three-momentum of particle 3 in the center of mass, dR_3/dp_3 is zero when $p_3 = 0$ (its minimum value), and dR_3/dp_3 is also zero when p_3 assumes its maximum value.

Exercise 8-3. Show that p_3 is a maximum when $s_2 = (m_1 + m_2)^2$ to prove the last statement.

Exercise 8-4. Draw the curve represented by Eq. 8-16 when $m_1 = m_2 = m_3 = m$ and $s = 20m^2$.

To compare the *simple phase space model* with experiment, consider an inelastic channel with N particles in the final state and reaction cross section σ_N. In terms of the transverse momentum ($p_\perp$) and the longitudinal center of mass momentum ($p_\parallel^*$) defined in Sec. 7-3, the single-particle distributions are

$$\frac{d\sigma_N}{dp_\perp} = D_N(s) \frac{dR_N(\sqrt{s}, 0)}{dp_\perp} \tag{8-17a}$$

$$\frac{d\sigma_N}{dp_\parallel^*} = D_N(s) \frac{dR_N(\sqrt{s}, 0)}{dp_\parallel^*} \tag{8-17b}$$

Typical $d\sigma/dp_\perp$ and $d\sigma/dp_\parallel^*$ distributions are shown in Fig. 8-1, which compares simple phase space model prediction with the data in 8 GeV/c π^+p interactions. These figures illustrate the following important observations. For lower multiplicities the phase space prediction is wrong. But when the multiplicity is 8, the phase space prediction is correct. Thus the phase space kinematics have an increasingly important effect as the multiplicity increases. At the highest multiplicities, the momentum distributions are entirely controlled by the phase space factor. The matrix element T_N has little obvious effect. There is an important limitation on the concept that at high multiplicities the kinematic variable distributions are given by the simple phase space model. The requirement of high multiplicity is a *relative* requirement. At incident laboratory momenta of 10 GeV/c, eight or ten charged prongs is a high multiplicity. But at 1000 GeV/c, the average number of charged prongs $\langle N_{ch} \rangle$ is about ten, as in

Fig. 7-2. At 1000 GeV/c, therefore, eight or ten charged prongs is a moderate, not high, multiplicity. At 100,000 GeV/c, eight or ten charged prongs may represent a low multiplicity. As we discuss in Ex. 8-16, it is only for the highest multiplicity at any energy that we expect the simple phase space model to make correct predictions about distributions.

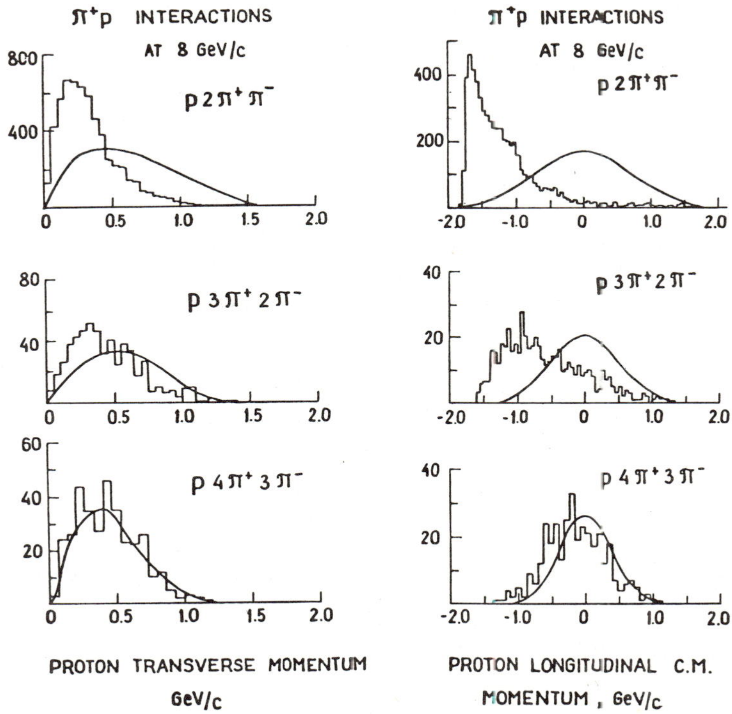

Fig. 8-1 Transverse and longitudinal momentum distributions for protons in some π^+p reactions at 8 GeV/c (8CZ).

8-3 Phase Space, Two-Particle Distributions, and Dalitz Plots. In the last section we discussed single-particle distributions. This is a large topic in itself and is changing rapidly with respect to both the underlying concepts and the methods of data presentation. Therefore we restrict our discussion to (1) simple three-particle systems and (2) invariant mass distributions in multiparticle systems.

The application of phase space considerations to three-particle systems was first emphasized by Dalitz, and the concept of the Dalitz plot is probably familiar. Here we provide only a summary of the Dalitz plot concept with the intent of illustrating the use of phase space considerations and the formalism developed in the previous section. We also review how the Dalitz plot is used to find strong interaction resonances. We do not discuss the extensive use of the Dalitz plot to find the spin and parity of resonances, having previously explained this is outside our scope (8NY, 8KA).

Start with $R_3(\sqrt{s}, 0)$ as given in Eq. 8-9 and carry out the integration over $\mathbf{p}_3$. Then write

$$R_3(\sqrt{s}, 0) = \int \frac{d^3p_1 d^3p_2}{8E_1E_2E_3} \delta(P_f^0 - P_i^0) \tag{8-18}$$

where

$$E_3^2 = m_3^2 + (\mathbf{p}_1 + \mathbf{p}_2)^2 = m_3^2 + 2p_1p_2 \cos\theta + \mathbf{p}_1^2 + \mathbf{p}_2^2 \tag{8-19}$$

recalling that all these variables are in the cms. Take $\mathbf{p}_1$ to have spherical angles θ_1, ϕ_1 with respect to an x, y, z coordinate system fixed in space, and take $\mathbf{p}_2$ to have spherical angles θ, ϕ with respect to a coordinate system fixed in $\mathbf{p}_1$. Then write

$$R_3(E_{\text{cms}}, 0) = \pi^2 \int dE_1\, dE_2; \qquad \frac{dR_3(E_{\text{cms}}, 0)}{dE_1\, dE_2} = \pi^2 \tag{8-20}$$

Exercise 8-5. Complete the steps leading to Eq. 8-20.

Thus $dR_3/dE_1\, dE_2$ is independent of E_1 and E_2. In other words, the phase space factor $R_3(\sqrt{s}, 0)$ as a function of E_1 and E_2 is a constant. Hence we have

$$\frac{d\sigma}{dE_1\, dE_2} = D_{fi}\pi^2 \tag{8-21a}$$

$$\frac{d\mathcal{N}}{dE_1\, dE_2} = CD_{fi}\pi^2 \tag{8-21b}$$

for the simple phase space model, $\mathcal{N}$ being the number of events.

A graphical examination or a graphical portrayal of the behavior of $d\sigma/dE_1\, dE_2$ or $d\mathcal{N}/dE_1\, dE_2$ can be obtained by means of the *Dalitz plot* illustrated in Fig. 8-2*a*. The dotted area inside the closed boundary is the kinematic region fixed by the conservation of four-momentum. If T_{fi} is independent of E_1 and E_2, then $d\sigma/dE_1\, dE_2$ and $d\mathcal{N}/dE_1\, dE_2$ will be constants over that region. In the analysis of experiments it is common to indicate an event with energies E_1, E_2 by a dot at the point E_1, E_2 on the Dalitz plot. Then a uniform distribution of dots indicates a constant $d\mathcal{N}/dE_1\, dE_2$, hence a T_{fi} independent of E_1 and E_2. On the other hand, a nonuniform distribution of dots (Fig. 8-2*c*) indicates the influence of T_{fi}. The extent of the kinematic region is given by rewriting Eq. 8-19 in the form (where $E = E_{\text{cms}}$)

$$E^2 - 2EE_1 - 2EE_2 + 2E_1E_2 = m_3^2 - m_1^2 - m_2^2 + 2(E_1^2 - m_1^2)^{1/2} \times (E_2^2 - m_2^2)^{1/2} \cos\theta \tag{8-22}$$

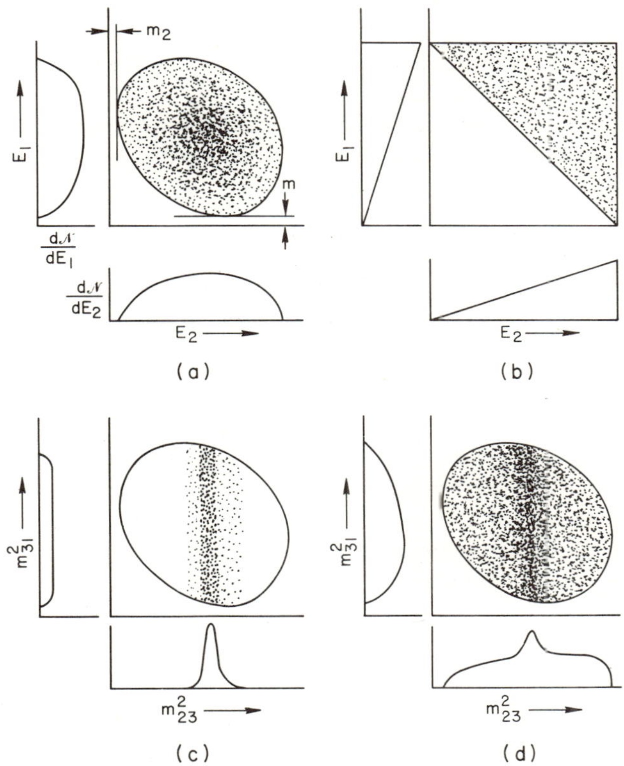

Fig. 8-2 Illustrative plots for a three-particle system: (*a*) a uniform distribution in phase space, (*b*) the boundary becomes triangular at very high energies (*c*) the reaction $a + b \to X + 1$, $X \to 2 + 3$ without background, (*d*) the same reaction with background.

and allowing $\cos\theta$ to vary from $+1$ to -1. The boundary of the region is obtained by setting $\cos\theta = \pm 1$, yielding a second-order curve. The one-dimensional distributions $d\mathcal{N}/dE_1$ and $d\mathcal{N}/dE_2$ also appear in Fig. 8-2*a*.

Exercise 8-6. Show that the boundary of the Dalitz plot is obtained by setting $\cos\theta = \pm 1$. Also show that the boundary is a second-order curve but that it is not a single ellipse. Show that the one-dimensional distributions $d\mathcal{N}/dE_1$ and

$d\mathcal{N}/dE_2$ are projections of the event distribution onto the E_1 and E_2 axis. The explicit construction of the boundary of the Dalitz plot is tedious, and the details can be found in the references given at the beginning of this section. At very high energies, however, when the masses of the particles can be neglected, the boundaries become straight lines forming a right isosceles triangle (Fig. 8-2*b*).

There is no direct way to extend the E_1, E_2 Dalitz plots to higher multiplicities. For more than three particles in the final state. there are too many independent kinematic variables.

In Sec. 5-2 we first discussed the production of strong interaction resonances in *production* experiments. Explicitly, the resonant particle X is produced in the inelastic reaction

$$a + b \rightarrow X + e + f \tag{8-23}$$

The resonant particle X may decay into a two-particle final state, $X \rightarrow c + d$, in which case the experimenter actually observes the reaction

$$a + b \rightarrow c + d + e + f \tag{8-24}$$

Or X may decay into more than two particles, in which case the experimenter observes five or more particles in the final state. Our object is to show how the experimenter studies the kinematic variable distributions in the final state to fino proof of the existence of X.

For simplicity we start with the three-particle final state $a + b \rightarrow 1 + 2 + 3$, and we define the invariant masses

$$\begin{aligned} m_{12}^2 &= (P_1 + P_2)^2 \\ m_{23}^2 &= (P_2 + P_3)^2 \\ m_{31}^2 &= (P_3 + P_1)^2 \end{aligned} \tag{8-25}$$

where the P_i's are four-momentum vectors, and $P_{\text{total}} = P_1 + P_2 + P_3$. In the center of mass

$$m_{ij}{}^2 = (P_{\text{total}} - P_k)^2 = s - 2\sqrt{s}E_k + m_k{}^2 \tag{8-26}$$

Therefore, we have

$$dE_1\, dE_2 = \frac{1}{4s}\, dm_{23}^2\, dm_{31}^2, \qquad \frac{dR_3(\sqrt{s}, 0)}{dm_{23}^2\, d_{31}^2} = \frac{\pi^2}{4s} \tag{8-27}$$

If T_{fi} is a constant, $d\mathcal{N}/dm_{23}^2\, dm_{31}^2$ is constant, and the Dalitz plot in m_{23}^2 and m_{31}^2 is uniformly populated.

Suppose that the reaction $a + b \to 1 + 2 + 3$ takes place only through the two-step process involving the production of a resonant particle X with mass M_x; that is

$$a + b \to 1 + X, \qquad X \to 2 + 3 \tag{8-28}$$

Then the Dalitz plot will only show events in a narrow band centered on the line $m_{23}^2 = M_x^2$ (Fig. 8-2*c*). The width of the band depends on the experimental mass resolution and on the lifetime of resonant state X. Usually the reaction $a + b \to 1 + 2 + 3$ occurs partly through the two-step process of Eq. 8-28 and partly through a more general process that populates the entire Dalitz plot. The result appears in Fig. 8-2*d*. Figure 8-3 is a beautiful example in which two resonant states, the $K^*(890)$ and $\Delta(1236)$, are produced in the reaction $K^+ + p \to K^0 + p + \pi^+$ at GeV/c.

For more than three particles in the final state, we search for two-particle resonances in the single variable distribution $d\mathcal{N}/dm_{ij}^2$. This is a direct extension of the previous discussion. Of course if we have no idea where a resonance might appear, all possible particle combinations must be tried. For N particles in the final state, $(N)(N-1)/2$ plots of $d\mathcal{N}/dm_{ij}^2$ are necessary. The search can be further

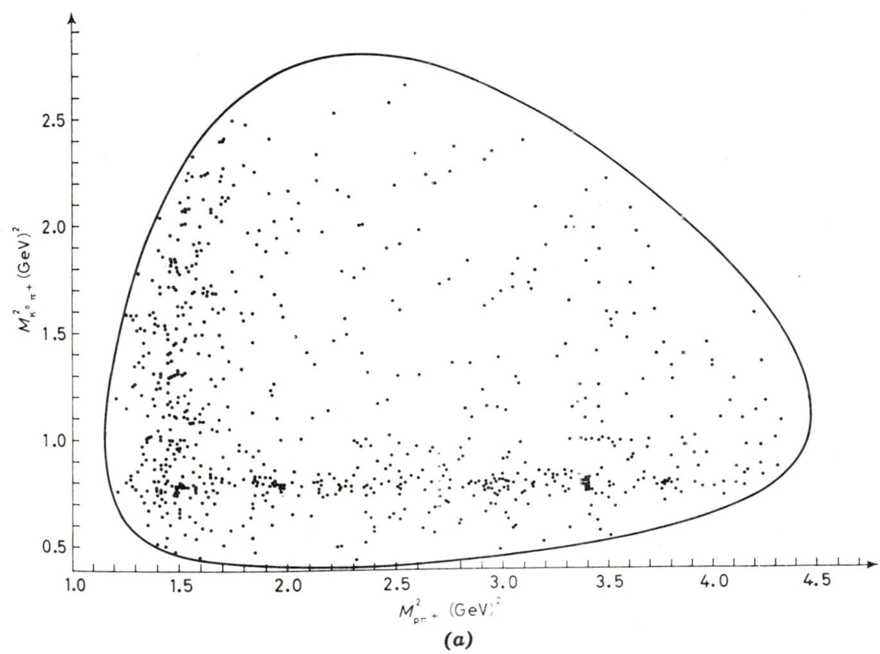

(*a*)

Fig. 8-3.

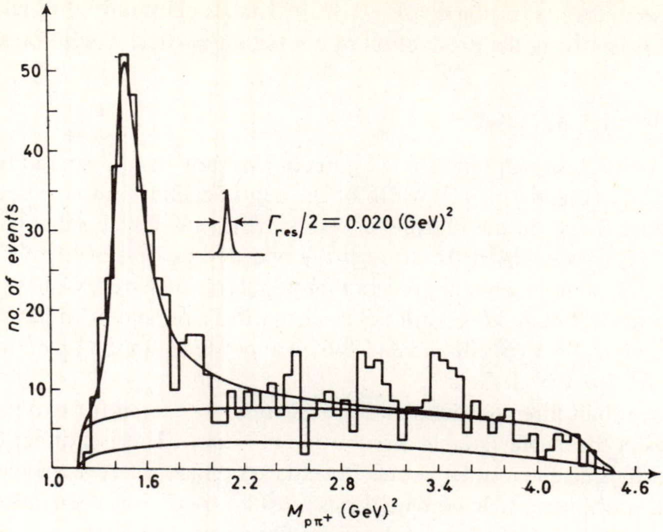

(b)

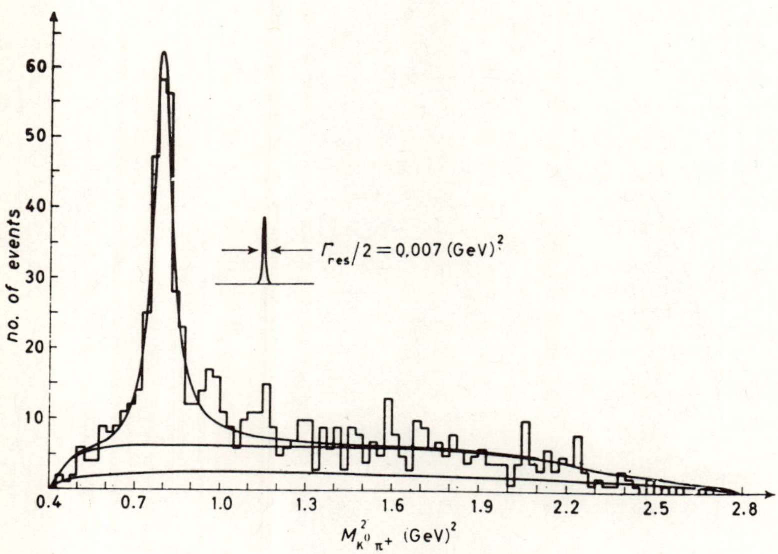

(c)

Fig. 8-3 A beautiful Dalitz plot with two resonances, the $K^*(890)$ and $\Delta(1236)$, produced in the reaction $K^+ + p \to K^0 + p + \pi^+$ at 3 GeV/c: (a) the Dalitz plot, (b) the projection on the $m^2_{p\pi}$ axis, (c) the projection on the $M^2_{K\pi}$ axis (8FER).

extended if resonances with three or more particle decay modes are considered. For example, to search for a resonance with a three-particle decay mode in a final state with $N > 3$ particles, form the invariant mass

$$m_{ijk}^2 = (P_i + P_j + P_k)^2$$

Here the P_i's are four-momentum vectors. Then a plot of the distribution $d\mathcal{N}/dm_{ijk}^2$ will show a peak centered on a mass M_x if the resonance X has the decay mode

$$X \to i + j + k$$

It is not always straightforward to determine whether a peak in a $d\mathcal{N}/dm_{ij}^2$ plot in a production experiment is caused by the production of a resonance. The following comments are applicable to this problem.

1. The width of the peak depends on the lifetime of resonance and on the experimental mass resolution. If the resonant peak has a Breit–Wigner shape (see Secs. 5-6 through 5-9) then $\Delta M_x^2 \sim \Gamma_x^2/4 + \Delta m_{ij}^2$, where Δm_{ij} is the experimental resolution on m_{ij}.
2. For the peak to be easily detected, the width just discussed must be small compared to the range in m_{ij} given by $m_{ij,\max} - m_{ij,\min}$. Otherwise it is difficult to distinguish a resonant peak from a matrix element dependent on individual particle momenta.
3. The peak must be statistically significant. In deciding what is statistically significant, remember that if hundreds of mass plots are studied, random fluctuations will be observed whose probabilities are about 1 in 100. Usually we demand not only high statistical significance for a peak but also that the peak appear at the same M_x value at different incident energies and in different experiments.
4. Even if there are no resonant peaks, the experimental distributions seldom follow the simple phase space model predictions; we have already observed that the simple phase space model prediction is only followed by distributions from high multiplicity events. When examining the statistical significance of a possible resonant peak, therefore the background distributions used must usually be more complicated than those given by the simple phase space model.
5. The matrix element T_{fi} can produce peaks in mass plots which stand out above the phase space distribution and yet are not resonances. Such peaks, called *enhancements*, are discussed in Sec. 7-8.

8-4 The Fermi Statistical Model. In this section we describe the *Fermi statistical model* (8FE), a simple phase space model in which the concept of statistical equilibrium is used to justify the model and to derive a formula for

$|T_N|^2$. This model is beautifully explained in a paper by Fermi, and we cannot improve on the author's explanation. Therefore we quote directly from the paper. Fermi speaks of the interaction of "two nucleons [which] collide with very great energy." It is only necessary to replace the phrase "two nucleons" by the phrase "two hadrons" to make the model applicable to any high energy hadron–hadron collision.

> When two nucleons collide with very great energy in their center of mass system this energy will be suddenly released in a small volume surrounding the two nucleons. We may think pictorially of the event as of a collision in which the nucleons with their surrounding retinue of pions hit against each other so that all the portion of space occupied by the nucleons and by their surrounding pion field will be suddenly loaded with a very great amount of energy. Since the interactions of the pion field are strong we may expect that rapidly this energy will be distributed among the various degrees of freedom present in this volume according to statistical laws. One can then compute statistically the probability that in this tiny volume a certain number of pions will be created with a given energy distribution. It is then assumed that the concentration of energy will rapidly dissolve and that the particles into which the energy has been converted will fly out in all directions.
>
> It is realized that this description of the phenomenon is probably as extreme, although in the opposite direction, as is the perturbation theory approach. On the other hand, it might be helpful to explore a theory that deviates from the unknown truth in the opposite direction from that of the conventional theory. It may then be possible to bracket the correct state of fact in between the two theories. One might also make a case that a theory of the kind here proposed may perhaps be a fairly good approximation to actual events at very high energy, since then the number of possible states of the given energy is large and the probability of establishing a state to its average statistical strength will be increased by the very many ways to arrive at the state in question.
>
> The statement that we expect some sort of statistical equilibrium should be qualified as follows. First of all there are conservation laws of charge and of momentum that evidently must be fulfilled. One might expect further that only those states that are easily reachable from the initial state may actually attain statistical equilibrium. So, for example, radiative phenomena in which photons could be created will certainly not have time to develop. The only type of transitions that are believed to be fast enough are the transitions of the Yukawa theory. A succession of such transitions starting with two colliding nucleons may lead only to the formations of a number of charged or neutral pions and also presumably of nucleon–anti-nucleon pairs. The discussion shall be limited, therefore, to these particles only. Notice the additional conservation law for the difference of the numbers of the nucleons and the anti-nucleons.

The crucial parameter in the theory is the volume Ω into which the energy of the two colliding hadrons, to use Fermi's phrase, "is dumped." This volume Ω is fixed by the range of the strong interactions; thus roughly

$$\Omega \approx \left(\frac{\hbar}{m_\pi c}\right)^3 \tag{8-29}$$

where m_π is the pion mass. This is only an approximation, and we may take Ω to be somewhat energy dependent. For example, Fermi takes Ω to be Lorentz contracted from an initial volume Ω, because of the motion of the incident particles. But we shall not specify the energy dependence beyond supposing that Ω may be a function of s.

Fermi's statistical assumption is that in the *center of mass* $|T_N|^2$ for N particles in the final state is proportional to the probability that all N particles are contained in the volume Ω. As discussed in Sec. 6-3, the normalization convention of this book is that particle n with energy E_n has a particle density of $2E_n$ per unit volume. Therefore, for N particles in the final state

$$|T_N|^2 = A(s) \prod_{n=1}^{N} (2E_n\Omega) \tag{8-30}$$

where $A(s)$ is a proportionality constant that may be s dependent. Inserting this in Eq. 6-36 yields the partial total cross section $\sigma_N(s)$ for the production of N particles. The summation over spin states in Eq. 6-36 is replaced by the multiplicative factor

$$\sum_\alpha \rightarrow \prod_{n=1}^{N} (2S_n + 1) \tag{8-31}$$

where S_n is the spin of the nth particle. Finally, since we are using statistical ideas, we should correct for the presence of identical particles in the final state. If there are n_a particles of type a, we should divide the phase space by the term $n_a!$ Fermi discusses this term but does not use it. Putting all this together and remembering that it is in the center of mass, we have

$$\sigma_N(s) = \frac{(2\pi)^4 A(s)\left[\prod_{n=1}^{N} (2S_n + 1)\right]}{E_{i1}E_{i2}|\mathbf{v}_{i1} - \mathbf{v}_{i2}|[n_a!n_b!\ldots]} \int \left[\prod_{n=1}^{N} \frac{(2E_n\Omega)\, d^3p_n}{(2\pi)^3 2E_n}\, \delta^4(P_f - P_i)\right] \tag{8-32}$$

This is the basic formula for the Fermi statistical model. There have been many elaborations of the model (8LA). It is possible to impose conservation of total isotopic spin and to crudely impose conservation of total angular momentum. One can also use different Ω's for different kinds of particles.

Slightly different statistical assumptions can be used. Finally, $SU(3)$ invariance can also be imposed. We shall not follow these elaborations. As we have already observed, the Fermi statistical model is in general wrong because it conflicts with the nonisotropy of the center of mass momentum distributions. We note, however, that elaborations of the Fermi statistical model have been successful (8LA) in fitting intermediate and low energy antiproton-proton annihilations. This can be understood by assuming that the large amount of mass energy released in the annihilation dominates the kinetic energy brought into the reaction.

To develop some feeling for the workings of statistical models in many-body reactions, we calculate the multiplicity-energy relationship predicted by the Fermi model. To our surprise we get a reasonable answer that agrees with the power law relationship in Eq. 7-1a. For simplicity, we assume that the final state is made up only of pions—π^+, π^-, and π^0 in equal numbers (namely, there are $N/3$ of each). Combining some s-dependent factors, Eq. 8-32 reduces to

$$\sigma_N(s) = \frac{B(s)}{((N/3)!)^3} \int \left[\prod_{n=1}^{N} \frac{(2E_n\Omega)\, d^3p_n}{(2\pi)^3 2E_n} \delta^4(P_f - P_i) \right] \tag{8-33}$$

There are now two ways to proceed. For the first way we cancel the $2E_n$ terms in the integral and obtain

$$\sigma_N(s) = \frac{B(s)}{((N/3)!)^3} \left(\frac{\Omega}{(2\pi)^3} \right)^N R_N^{(\text{nonrel})}(\sqrt{s}, 0) \tag{8-34}$$

where

$$R_N^{(\text{nonrel})}(\sqrt{s}, 0) = \int \left[\prod_{n=1}^{N} d^3p_n\, \delta^4(P_f - P_i) \right] \tag{8-35}$$

and $R_N^{(\text{nonrel})}$ is the non-Lorentz invariant phase space factor mentioned in Sec. 8-2. This is even more difficult to calculate than the Lorentz invariant phase space factor R_N, although an explicit expression can be derived if all the particles have zero mass (8MIL, 8KR).

The second and more modern way to proceed from Eq. 8-33 is to extract the $(2E_n\Omega)$ terms from the integral. We do this by replacing E_n in the numerator of the integral by some average value $\bar{E}$. Then

$$\sigma_N(s) = \frac{B(s)}{((N/3)!)^3} \left[\frac{2\bar{E}\Omega}{(2\pi)^3} \right]^N R_N(\sqrt{s}, 0) \tag{8-36}$$

In the general case, $R_N(\sqrt{s}, 0)$ cannot be calculated explicitly, but as shown in Ex. 8-7, we can give an explicit formula if all the particles have zero mass.

Exercise 8-7. Using the recursion relation of Eq. 8-12, show that for massless particles $R_N(\sqrt{s}, 0)$ is given by Eq. 8-37. This derivation is not long but it can be difficult (8MIL, 8BY).

$$R_N^{(m=0)}(E_{\text{cms}}, 0) = \frac{\pi^{N-1} E_{\text{cms}}^{2N-4}}{2^{N-1}(N-1)!(N-2)!} \tag{8-37}$$

To get some feeling for the predictions of the model we assume that all the particles are massless. Thus we are considering the extreme relativistic case, and

$$\sigma_N(s) = C(s)\left(\frac{\bar{E}\Omega}{8\pi^2}\right)^N \frac{E_{\text{cms}}^{2N}}{(N-1)!(N-2)!((N/3)!)^3} \tag{8-38}$$

Here we have combined with $B(s)$ some additional terms that do not depend on N. For multiplicity calculations, the crucial parameter is $(\bar{E}\Omega/8\pi^2)$, which has the units of $(\text{energy})^{-2}$.

It is convenient to define

$$E_{\text{stat}}^2 = \left(\frac{\bar{E}\Omega}{8\pi^2}\right)^{-1} \tag{8-39}$$

(We note that E_{stat} is not standard nomenclature.)

The final result is

$$\sigma_N(s) = C(s)\frac{(E_{\text{cms}}/E_{\text{stat}})2N}{(N-1)!(N-2)!((N/3)!)^3} \tag{8-40}$$

At a fixed total energy E_{cms}, the multiplicity distribution depends only on E_{cms} and E_{stat}. If $E_{\text{cms}}/E_{\text{stat}} \gg 1$, $\sigma_N(s)$ first increases with N. But at larger N, the factorials in the numerator dominate and $\sigma_N(s)$ decreases again. Thus we obtain, at least qualitatively, the multiplicity distributions exhibited in Fig. 7-4.

Exercise 8-8. Select a value for E_{stat} such that Eq. 8-40 roughly fits the 200-GeV *pp* multiplicity distribution given in Fig. 7-4. More elaborate models use a value of E_{stat} of 1 or 2 GeV. Calculate the multiplicity distribution at 2000 GeV using the same E_{stat} you used at 200 GeV. Finally, calculate the average multiplicity at each energy. In comparing your calculations with Fig. 7-4, remember that Fig. 7-4 shows only the multiplicity of the charged particles. Derive Eq. 8-41.

If E_{stat} is independent of energy, Eq. 8-40 also predicts how the multiplicity distribution changes with energy. Rather than calculating $\langle N \rangle$, it is easier to calculate N_{peak}, the number of particles at the peak of the multiplicity distribution, and to make the reasonable approximation that $\langle N \rangle$ and N_{peak} have the same energy dependence. Using the Stirling factorial approximation, Eq. 8-40 yields, for $N \gg 1$,

$$\frac{d[\ln \sigma_N(s)]}{dN} \approx \text{constant} + \ln \left(\frac{E_{\text{cms}}}{E_{\text{stat}}}\right)^2 - \ln [N(N-1)(N-2)] \tag{8-41}$$

Setting the right-hand side of Eq. 8-41 to zero and taking $N \gg 1$, we obtain

$$\langle N \rangle \sim N_{\text{peak}} \sim c_0 + c_1 \left(\frac{E_{\text{cms}}}{E_{\text{stat}}}\right)^{2/3} \sim c_0 + c_2 s^{1/3} \tag{8-42}$$

where the c are constants. This is a crude power law prediction; it says that $\langle N \rangle$, the average multiplicity, is proportional to $s^{1/3}$. Comparing this with the power law given in Eq. 7-1a, and considering the discussion that follows the equation, we recall that experimentally $n \sim 0.25$ to 0.35. Given the chain of crude approximations that led to this result and the basic incorrectness of the Fermi statistical model, this is a pretty good prediction!

8-5 The Simple Thermodynamic Model and Multiplicities. We continue this chapter with a brief look at a variation of the statistical model—the *thermodynamic model.* We assume that the multiplicities are so high that we can apply thermodynamic considerations to the final state. One of the earliest and clearest treatments of this idea is given by Fermi in the paper to which we have already referred (8FE). For simplicity we assume that the final state consists only of positive, negative, or neutral pions and that each of these charge states has equal statistical weight. We then think of a *gas* of pions confined in the volume Ω with total energy E_{cms} in the center of mass. We also take the pions to be extremely relativistic, thus allowing us to neglect their mass.

The statistical mechanics of massless pions is very similar to the statistical mechanics of photons—they both obey Bose-Einstein statistics. At a temperature T, the most probable number of photons per unit frequency per unit volume is

$$\frac{d\mathcal{N}(\nu)}{d\nu} = \left[\frac{8\pi\nu^2}{c^3}\right]\left[\frac{1}{e^{h\nu/KT} - 1}\right]$$

where K is the Boltzmann constant and ν is the frequency. Since $d\mathcal{N}(\nu)/d\nu$ multiplied by $(h\nu)$ gives the famous black-body distribution law derived by Planck, we are dealing with a beautiful concept announced in 1900! To use this equation for massless pions, we need only observe that the photon of a fixed energy has two available spin states and the pion has three available charge states. Therefore, the most probable number of pions per unit energy per unit volume is

$$\frac{d\mathcal{N}(E)}{dE} = \frac{3}{2}\frac{E^2}{\pi^2\hbar^3c^3}\frac{1}{e^{E/KT}-1} \tag{8-43}$$

where E is the energy of the pion.

We have used the term "most probable number" because these equations give the most probable distribution. If we integrate the right-hand side of Eq. 8-43 for $0 \leqslant E \leqslant \infty$ we will obtain $\mathcal{N}_{\text{most probable}}$, the *total number* of pions per unit volume in the most probable distribution. Other distributions will give a smaller or larger number of total pions. Then the usual thermodynamic concepts allow us to equate $\mathcal{N}_{\text{most probable}}$ to $\langle\mathcal{N}\rangle$, the average total number of pions per unit volume for all the different distributions. Carrying out the integration and setting $\hbar = c = 1$, we write

$$\langle\mathcal{N}\rangle = \frac{3.60}{\pi^2}(KT)^3 \tag{8-44}$$

From Eq. 8-44 we know $\mathcal{N}$ if we know the temperature T. To find T we first calculate U, the total energy per unit volume, given by

$$U = \int_0^\infty \left[\frac{3E^3}{2\pi^2}\right]\left[\frac{1}{e^{E/KT}-1}\right] dE \tag{8-45}$$

Carrying out the integration, we obtain

$$U = \frac{9.74}{\pi^2}(KT)^4 \tag{8-46}$$

But U, the total energy per unit volume is also given by

$$U = \frac{E_{\text{cms}}}{\Omega} \tag{8-47}$$

Combining Eqs. 8-46 and 8-47, we find the simple result

$$(KT) = 1.00 \;\frac{E_{\text{cms}}}{\Omega}^{1/4} \tag{8-48}$$

Rather than evaluate T, we insert KT into Eq. 8-44 and obtain

$$\langle\mathcal{N}\rangle = 0.36 \;\frac{E_{\text{cms}}}{\Omega}^{3/4} \tag{8-49}$$

Taking Ω to be independent of energy in Eq. 8-49 and noting that the average particle multiplicity $\langle N \rangle$ is just $\langle \mathcal{N} \rangle \Omega$, we have

$$\langle N \rangle = [\text{constant}] E_{\text{cms}}^{0.75} \sim cs^{0.375} \tag{8-50}$$

This is not only a simple result, but it is also in crude agreement with experiment. The reader should compare this result with Eq. 8-42 and the discussion following that equation. Having achieved this remarkable result, we must leave this subject. The thermodynamic model, like the statistical model, has been elaborated and extended. We refer to it briefly in the next section, where we give recent references.

Exercise 8-9. Check the numerical constant given in Eq. 8-49. If Ω is given roughly by (mass of the pion)$^{-3}$ what is the corresponding value of the temperature T?

8-6 Scaling in Inclusive Reactions. In Sec. 7-3 we defined *inclusive reactions* and *inclusive distributions.* Let us give a more formal definition. Consider *all* reactions of particles a and b such that

$$a + b \to c + (\text{one or more hadrons of any number or type}) \tag{8-51}$$

Or more succinctly

$$a + b \to c + \text{anything} \tag{8-52}$$

The reaction in Eq. 8-52 is called a *one-particle inclusive reaction* in which particle c is produced. The angle and momentum distributions of particle c are *one-particle inclusive distributions.* A *two-particle inclusive reaction* is

$$a + b \to c + d + \text{anything} \tag{8-53}$$

And the joint angle and momentum distributions of particles c and d are *two-particle inclusive distributions.* For lack of space, all detailed discussion of inclusive reactions in this book is limited to one-particle inclusive reactions.

In Sec. 7-3 we also described some of the general experimental properties of the one-particle inclusive distributions. In particular, we observed that there is a sharp cutoff in the tranverse momentum distribution; and furthermore, at high energies the form of that distribution is independent of the energy. Another general observation can be made about the longitudinal momentum distribution, but let us first describe these distributions more formally. We know that the total cross section $\sigma_{\text{tot}}(s)$ and d^3p/E are Lorentz invariants. Therefore we can define a Lorentz invariant momentum distribution

$$f_c(\mathbf{p}_c, s) = \frac{E_c d\sigma_{\text{tot}}(s)}{d^3 p_c} \tag{8-54}$$

for particle c produced in the interaction of incident particles a and b. We note that although f_c is Lorentz invariant, the quantity $\mathbf{p}_c$ used to specify it is not Lorentz invariant. We confine our considerations to the usual experimental situation in which particles a and b are unpolarized. It is very useful to replace $\mathbf{p}$ by $p_\perp$, $p_\parallel$, and ϕ. We remind the reader that $p_{\perp c}$ is the momentum component transverse to the direction of motion of particles a and b; and $p_{\parallel c}$ is the momentum component along that direction, taken as + when particle c is moving in the same direction as particle a. In addition, $p_{\perp c}$ is invariant to Lorentz velocity transformations along the direction of motion of particles a and b; but of course, $p_{\parallel c}$ and E_c are not invariant under such transformations. Returning to Eq. 8-54, we rewrite it as follows:

$$f_c(p_{\perp c}, p_{\parallel c}, s) = \frac{E_c d\sigma_{\text{tot}}(s)}{\pi dp_{\perp c}^2\, dp_{\parallel c}} \tag{8-55}$$

In the future we omit the subscript c unless it is required for clarity.

To compare $f(p_\perp, p_\parallel, s)$ with experiment we go to a particular Lorentz frame, namely, the center of mass frame of particles a and b. We have already indicated that since the $p_\perp$ distribution has a sharp cutoff as the energy increases, the average value of $p_\parallel^*$ must increase. To at least crudely cancel this effect, we define the variable

$$x_c = \frac{p_{\parallel c}^*}{p_{\max,c}^*} \tag{8-56}$$

As the next paragraph indicates, this is a very happy choice, for it leads to a very simple description of the experimental observations. The * indicates the center of mass of the $a + b$ system, and $p_{\max,c}^*$ is the maximum momentum that particle c can have in that system. Thus the range of x is

$$-1 \leqslant x \leqslant 1 \tag{8-57}$$

Exercise 8-10. Some authors use the definition

$$x_c = \frac{2p_{\parallel c}^*}{\sqrt{s}} \tag{8-58}$$

Derive the relationship between definitions 8-57 and 8-58 for the case of n particles in the final state with masses $m_1, m_2, \ldots, m_n$ and total energy $\sqrt{s}$. Show that at high energies $s \gg m_i^2$ the definitions are equivalent.

Finally we write Eq. 8-55 in the form

$$f(p_\perp, x, s) = \left(\frac{E}{\pi p_{\max}^*}\right) \frac{d\sigma_{\text{tot}}(s)}{dp_\perp^2\, dx} \tag{8-59}$$

We now receive our reward for going through this sequence of manipulations. We find experimentally, with the exceptions noted below, that at very high energies $f(p_\perp, x, s)$ is *independent* of s; that is,

$$f(p_\perp, x, s) = f(p_\perp, x) \tag{8-60}$$

The beautifully simple observation is that at high energies, not only is the form of $f(p_\perp, x, s)$ independent of s but the *magnitude* of $f(p_\perp, x, s)$ is also independent of s. The f is said to obey a *scaling law* in which f depends only on the *scaling variables* $p_\perp$ and x. Put another way, f is said to *scale* with energy according to the scaling variables $p_\perp$ and x. This form of scaling in one-particle inclusive reactions is usually called *Feynman scaling* (8FEY). In Sec. 8-8 we discuss a closely related form of scaling based on the fragmentation model.

Figures 8-4*a* and 8-4*b* illustrate the scaling law behavior for particles produced in pp collisions. We do not have such high energy data available for πp or Kp collisions, but lower energy data for pions (Fig. 8-5) agree with the scaling law. We note the following exceptions to the scaling law:

1. At the energies of existing accelerators, the scaling law should be limited to $p_\perp < 1$ or 2 GeV/c. At higher transverse momentum either scaling has not yet set in at the energies of existing accelerators, or nonscaling production mechanisms may be involved. This is discussed in Ch. 22. We should remember, however, that almost all secondary particles have $p_\perp$ much smaller than 1 or 2 GeV/c. Hence the scaling law applies to all but a small portion of the total cross section.
2. The scaling law for secondary pions holds at incident particle laboratory energies as low as 10 or 20 GeV. For other secondary particles, particularly antiprotons, the scaling behavior begins at higher incident particles energies. Also, over a particular energy range, the deviation from scaling behavior may be dependent on x.

To conclude this section we discuss the form of $f(p_\perp, x)$. As shown in Fig. 8-4*a*, at fixed x, $f(p_\perp, x)$ falls off rapidly with $p_\perp$. This is what we expect from our discussion in Sec. 7-3. Adopting the form first given in Eq. 7-7a, we write

$$f(p_\perp, x) \approx A(x)\, e^{-a(x)p_\perp} \tag{8-61}$$

although we recognize that this is sometimes a poor fit to the data. As we have indicated, $a(x)$ in general may be a function of x. But over limited ranges of x, $a(x)$ varies slowly with x, and

$$f(p_\perp, x) \sim A(x)\, e^{-ap_\perp}, \qquad a \sim \text{constant} \tag{8-62}$$

Exercise 8-11. Consult the recent literature (8IC) for numerical illustrations of the dependence of $a(x)$ on the type of secondary particle and on x.

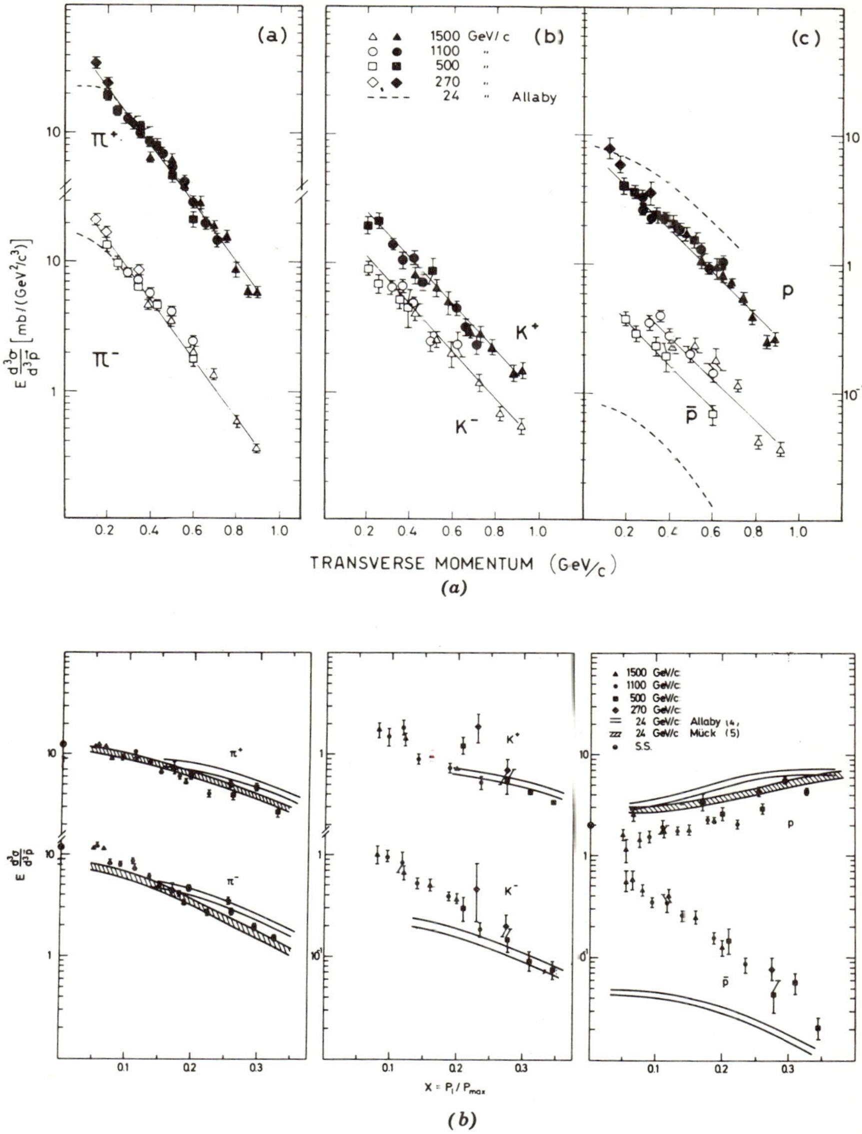

Fig. 8-4 Illustration of Feynman scaling for the invariant inclusive cross section $E(d\sigma/d^3p)$ in pp collisions (*a*) versus $p_\perp$ at $x = 0.16$ (8BE) and (*b*) versus x at fixed $p_\perp$. Taken from Ref. 8JA. The symmetry about $x = 0$ for pp collisions obviates the need to show the $-1 \leqslant x \leqslant 0$ region.

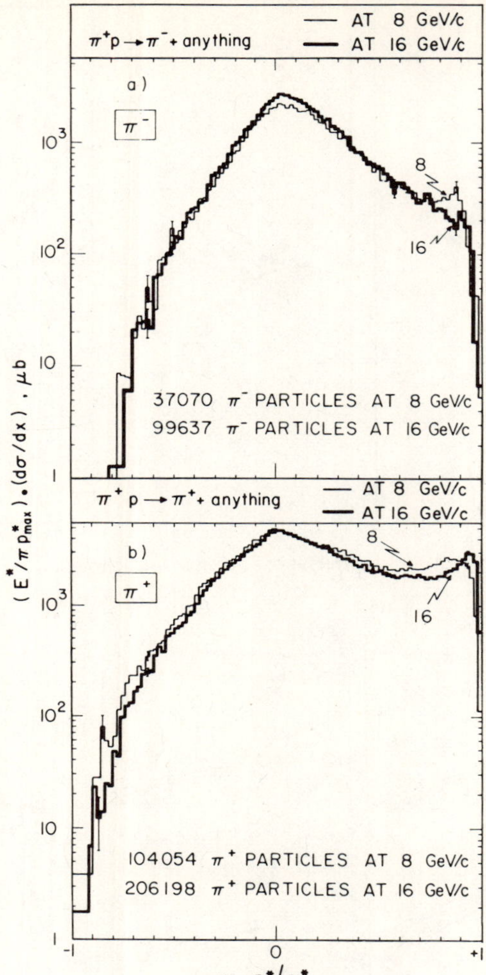

Fig. 8-5 The invariant inclusive cross section $E(d\sigma/d^3p)$ in πp collisions, integrated over $p_\perp$ (8BEA).

We also note that $a(x)$ in Eq. 8-61 depends on the secondary particle: $a(x)$ is largest for pions, smallest for protons and antiprotons, and of intermediate magnitude for K mesons.

As shown in Figs. 8-5 and 8-6 the dependence of $f(p_\perp, x)$ on x at fixed $p_\perp$ is more complicated. For we have the *leading particle effect*, first described in Sec. 7-3 and beautifully illustrated in Fig. 8-6. This means that in πp and Kp

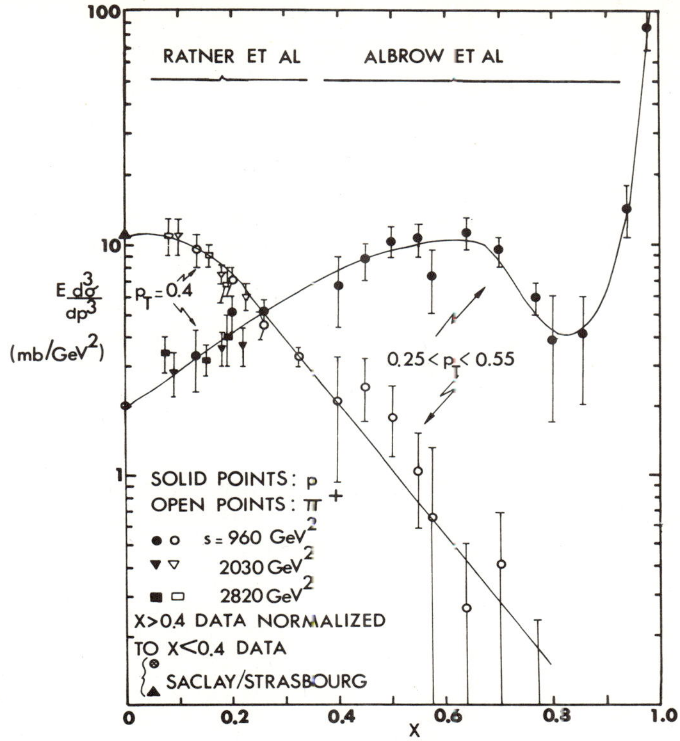

Fig. 8-6 Illustration of the leading particle effect in *pp* inclusive reactions (8SE). Elastic scattering is included.

collisions $f(p_\perp, x \sim +1)$ can be quite different from $f(p_\perp, x \sim -1)$. [In *pp* collisions, symmetry demands that $f(p_\perp, -x) = f(p_\perp, +x)$.] Now let us turn our attention to the $|x| < 1$ region and consider a small interval $\Delta p_\perp^2$, where for $x \neq 0$ we have

$$\frac{d\sigma_{\text{tot}}}{dx} = \left(\frac{\pi p^*_{\max}}{E}\right) f(p_\perp, x)\, \Delta p_\perp^2 \sim \left(\frac{\pi}{|x|}\right) f(p_\perp, x)\, \Delta p_\perp^2 \tag{8-63}$$

Thus plots of $f(p_\perp, x)_{\text{fixed } p_\perp}$ versus x tend to be misleading about the x value of most particles. The additional factor of $|x|$ in Eq. 8-63 shows that most particles have $|x|$ close to zero.

In Secs. 7-2 and 7-3 and in this section we have provided an overall picture of the behavior of many-body, inelastic processes at high energy. We discussed:

1. The dependence of the average multiplicity on the energy (Sec. 7-2).
2. The multiplicity distribution at fixed energy (Sec. 7-2).

3. The sharp cutoff in transverse momentum (Sec. 7-3).
4. The scaling law behavior of $f(p_\perp, x)$ (this section).
5. The dependence of $f(p_\perp, x)$ on $p_\perp$ and x (this section).

The task of the theory of strong interactions is to explain this behavior. At present we cannot carry out this task in a fundamental or rigorous way. We can only use models we know to be crude and incomplete, although we hope that they illustrate or incorporate some basic aspects of many-body, high energy inelastic processes. The remainder of this chapter is devoted to these models.

8-7 The Transverse Momentum Cutoff Phase Space Model. In the transverse momentum cutoff phase space model, we simply introduce into phase space a sharp cutoff in the transverse momentum! We give no explanation for the cutoff; hence this model is more mathematical than physical. But it provides useful insights into the effect on physical observables of such a limited phase space. To simplify our considerations, we take all the particles in the final state to be massless and of the *same* type. As a point of departure we note that the *simple phase space model* using the $|\bar{T}_N|^2$ given by the Fermi statistical model has the form

$$\sigma_N(s) = \frac{B(s)}{N!}\left[\frac{2}{\pi E_{\text{stat}}^2}\right]^N R_N^{(m=0)}(E_{\text{cms}}, 0) \tag{8-64}$$

$$R_N^{(m=0)}(E_{\text{cms}}, 0) = \int \left[\prod_{n=1}^{N} \frac{d^3p_n}{2E_n}\, \delta^4(P_f - P_i)\right] \tag{8-65}$$

derived in Sec. 8-4. The *transverse momentum cutoff phase space model has the form*

$$\sigma_{N,\,\text{cutoff}}(s) = \frac{B'(s)A^N}{N!} R_{N,\,\text{cutoff}}^{(m=0)}(E_{\text{cms}}, 0) \tag{8-66}$$

$$R_{N,\,\text{cutoff}}^{(m=0)}(E_{\text{cms}}, 0) = \int \left[\prod_{n=1}^{N} \frac{e^{-ap_{\perp n}}\, d^3p_n}{2E_n}\, \delta^4(P_f - P_i)\right] \tag{8-67}$$

The constant A in Eq. 8-66 depends in part on the transverse momentum distribution. Also it is required dimensionally and is thus analogous to the factor $2/\pi E_{\text{stat}}^2$ in Eq. 8-64. Hence it incorporates the N dependence of $|\bar{T}_N|^2$. Any s dependence of $|\bar{T}_N|^2$ is incorporated in $B'(s)$, hence $|\bar{T}_N|^2$ does not appear explicitly.

The integral in Eq. 8-67 cannot be done exactly, even for massless particles. There are ways to obtain good approximations, but they are complicated. We proceed in a cruder way, obtaining a crude approximation. But the answers we obtain will be useful, and the very crudeness of the approximation will allow us to understand the basic mechanism. The exponential transverse momentum cutoff means that we can ignore the contribution of $p_{\perp n}{}^2$ to the total energy of particle n, and we can ignore the conservation of the transverse momentum. Then we have

$$R_{N,\,\mathrm{cutoff}}^{(m=0)}(E_{\mathrm{cms}},0) \approx \left(\frac{\pi}{a^2}\right)^N \int \left[\prod_{n=1}^{N} \frac{dp_{\parallel n}}{E_n}\,\delta(E_f - E_{\mathrm{cms}})\,\delta(p_{\parallel f})\right] \qquad (8\text{-}68)$$

The effect of E_n in the denominator is to give most weight to the small $p_\parallel$ region. This mathematical observation is confirmed by the experimental observations on the distributions in x, which were discussed near the end of the previous section. Now if most particles have small energies, their $p_\parallel$ distributions are not affected by the overall energy conservation. Therefore, we take the final bold step of ignoring energy and longitudinal momentum conservation. We write

$$R_{N,\,\mathrm{cutoff}}^{(m=0)}(E_{\mathrm{cms}},0) \approx \left(\frac{2\pi}{a^2}\right)^N \left(\frac{h}{E_{\mathrm{cms}}^2}\right) \int \left[\prod_{n=1}^{N-2} \frac{dE_n}{E_n}\right] \qquad (8\text{-}69)$$

Here we have compensated for our neglect of energy and longitudinal momentum conservation by integrating over the $N-2$ particle phase space. The term (h/E_{cms}^2), with h a constant, is required dimensionally. Alternatively, we can suppose that the E and $p_\parallel$ conservation is accomplished by the two leading particles. Then h is 4. Of course such crude approximations only hold for N greater than 4 or 5. We now have a trivial integral and

$$R_{N,\,\mathrm{cutoff}}^{(m=0)}(E_{\mathrm{cms}},0) \approx \left(\frac{2\pi}{a^2}\right)^N \left(\frac{h}{E_{\mathrm{cms}}^2}\right)\left(\ln\frac{E_{\mathrm{cms}}}{E_{\mathrm{min}}}\right)^{N-2} \qquad (8\text{-}70)$$

where E_{min} is a minimum energy corresponding to $\langle p_\perp{}^2\rangle^{1/2}$. Then

$$\sigma_{N,\,\mathrm{cutoff}}(s) = B(s)\,\frac{[2C\ln(E_{\mathrm{cms}}/E_{\mathrm{min}})]^N}{N!} \qquad (8\text{-}71a)$$

Here we have absorbed some constants into A yielding the constant $2C$. We have extracted a $[\ln(E_{\mathrm{cms}}/E_{\mathrm{min}})]^2$ term from $B'(s)$ to make a simpler looking expression and yield $B(s)$. Setting $s_0 = E_{\mathrm{min}}^2$, we obtain the alternate form

$$\sigma_{N,\,\mathrm{cutoff}}(s) = \frac{B(s)[C\ln(s/s_0)]^N}{N!} \qquad (8\text{-}71b)$$

We easily find

$$\sigma_{\text{tot}}(s) = B(s)\left(\frac{E_{\text{cms}}}{E_{\text{min}}}\right)^{2C} = B(s)\left(\frac{s}{s_0}\right)^{C} \tag{8-72}$$

The average multiplicity is given by the logarithmic energy dependence

$$\langle N \rangle = 2C \ln \frac{E_{\text{cms}}}{E_{\text{min}}} = C \ln \frac{s}{s_0} \tag{8-73}$$

and the multiplicity distribution is given by the Poisson formula

$$P(N) = \frac{\langle N \rangle^N}{N!} e^{-\langle N \rangle} \tag{8-74}$$

Exercise 8-12. Compare Eq. 8-71 with the equivalent formula, Eq. 8-38, for the Fermi statistical model form of the simple phase space model. What are the similarities and differences? Derive Eqs. 8-68 through 8-74. Observe that the explicit energy dependence in Eq. 8-72 is meaningless because $B(s)$ and C are not fixed by the model.

Exercise 8-13. Observe that Eq. 8-73 is the same as the experimental law given in Eq. 7-1b and illustrated in Fig. 7-2. Thus this model gives a better fit than the Fermi statistical model to the energy dependence of the average multiplicity. Use the experimental values of the a and b constants in Eq. 7-1b to replace the C factor and to evaluate E_{min}. Comment on these values.

Exercise 8-14. The Poisson distribution of Eq. 8-74 was used quite frequently until recently to fit experimental multiplicity distributions. Compare it with the data in Fig. 7-4, demonstrating that it is a crude fit for multiplicities in the neighborhood of the average multiplicity but a poor fit for large multiplicities.

Exercise 8-15. Derive Eqs. 8-75 and 8-76. Remember that all the particles in the final state are identical; hence N particles in σ_N contribute to the one-particle distributions.

It is easy to derive the one-particle inclusive distribution by backtracking from Eq. 8-69. We obtain [ignoring $\delta^4(P)$ and the $N \to N - 2$ trick in Eq. 8-69]

$$\sigma_{N,\,\text{cutoff}} = B(s) \frac{A''^N}{N!} \left[\prod_{n=1}^{N} \int \frac{e^{-ap_{\perp n}}\, dp_{\perp n}{}^2\, dp_{\parallel n}}{E_n} \right] \tag{8-75}$$

$$f(p_\perp, x) = \frac{E\, d\sigma_{\text{tot}}}{\pi p^*_{\text{max}}\, dp_\perp{}^2\, dx} = \frac{A''}{\pi} \sigma_{\text{tot}}\, e^{-ap_\perp} \tag{8-76}$$

Thus we have derived Eq. 8-62, a specific form of the scaling law. As required by the scaling law $f(p_\perp, x)$ is independent of s, since σ_{tot} is independent of s.

By analogy we construct a more general form of Eq. 8-75, namely,

$$\sigma_{N,\,\text{cutoff}} = B(s)\,\frac{A''^N}{N!}\left[\prod_{n=1}^{N}\int \frac{f(p_{\perp n}, x_n)\, dp_{\perp n}^2\, dp_{\parallel n}}{E_n}\right] \tag{8-77}$$

and we take this as the final form of the transverse momentum cutoff phase space model. We have seen that this model predicts quite reasonable multiplicity and one-particle distribution properties for inclusive reactions. Therefore, any model leading to a transverse momentum cutoff will have these properties. The remainder of this chapter is devoted to models that contain physical mechanisms leading to a transverse momentum cutoff.

However, there are additional properties not possessed by the transverse momentum cutoff phase space model. It cannot explain correlations between incident and final state particles beyond those caused by the cutoff in $p_\perp$. Hence it cannot explain the leading particle effect. Also it cannot explain higher order correlations among the secondary particles. More physical models must then incorporate these properties.

Exercise 8-16. In connection with Eqs. 8-17 we stated that at very high multiplicity, for fixed s, the simple phase space model gives the correct one-particle distributions. Explain this using the transverse momentum cutoff phase space model; and show that high multiplicity, at fixed s, corresponds to $N \sim E_{\text{cms}}/(\langle p_\perp^2\rangle + m^2)^{1/2}$.

To conclude this section we introduce yet another kinematic variable, the *longitudinal rapidity* (8BY), almost always called simply (and slightly incorrectly) the rapidity:

$$y = \tfrac{1}{2}\ln\left(\frac{E + p_\parallel}{E - p_\parallel}\right) = \ln\left(\frac{E + p_\parallel}{\sqrt{p_\perp^2 + m^2}}\right) \tag{8-78}$$

At first this looks like a rather ungainly and artificial construction. But the rapidity y has two useful features. First, even for particles with nonzero mass,

$$\frac{dp_\parallel}{E} = dy \tag{8-79}$$

Second, under a boost (Lorentz velocity transformation), $\xi = \text{arctanh}\,\beta$, along the axis defining the longitudinal momentum:

$$y \to y' = y + (\tfrac{1}{2})\ln\left(\frac{1+\beta}{1-\beta}\right) = y + \xi \tag{8-80}$$

Exercise 8-17. Derive Eqs. 8-79 and 8-80: ξ is introduced in Eq. 6-88 and discussed in more detail in connection with Eq. 9-20. Show that if $p_\perp$ is zero, the rapidity is a boost parameter (see Eq. 9-20). Use Eq. 8-80 to show that differences in the rapidity of two particles are not changed by boosts along the longitudinal momentum axis. An example of such a boost is the transformation from the laboratory frame to the center of mass frame.

The use of the rapidity variable leads to more elegant forms in many of the equations we have been using. Thus Eq. 8-59 becomes

$$f(p_\perp, y, s) = \frac{d\sigma_{\text{tot}}(s)}{\pi dp_\perp^2 \, dy} \tag{8-81}$$

and Eq. 8-75 becomes

$$\sigma_{N,\,\text{cutoff}} = B(s)\,\frac{A'^N}{N!}\left[\prod_{n=1}^{N}\int e^{-ap_{\perp n}}\, dp_{\perp n}^2\, dy_n\right] \tag{8-82}$$

even when the particles have nonzero mass.

Exercise 8-18. Equation 8-82 has stimulated the idea that many-body reactions can be thought about in analogy to a *one-dimensional gas*, the gas consisting of points in phase space distributed in terms of the rapidity variable y. Thus Eq. 8-82 would predict a uniform distribution in y. Consult the literature (Refs. 8BJ and 8BA, e.g.), for more details.

8-8 Statistical Models with Forward Dominance. In this section we describe models of many-body inelastic reactions in which the statistical concepts, which we have been using in isolation, are combined with the forward dominance concept of Sec. 7-9. This combination is effected by the construction of a physical mechanism for many-body reactions.

1. *The Two-Fireball, Diffractive-Excitation and Related Models.* We now discuss an old model (8CO, 8MI), first developed to explain cosmic ray measurements of inelastic interactions in emulsions and cloud chambers. As with other old models in physics, the contents of the model have changed through the years. We present the basic contents of the model, as used, for example, by Refs. 8AD, 8HW, and 8JAC. We do not distinguish among the various forms of the model—there can be quite large differences (8MU)—and for brevity we refer to all forms collectively as fireball models. The simple two-fireball theory of many-body reactions assumes that the incident hadrons 1 and 2 interact through a forward dominant process. The mechanism is illustrated in Fig. 8-7. For convenience,

the interaction is shown in the center of mass frame. In the forward dominant scattering process, the particles 1 and 2 are excited to higher unstable states 1* and 2*, which are called *fireballs.* We suppose they have masses M_1^* and M_2^*. These are not unique masses; on the contrary, a continuous spectrum of these

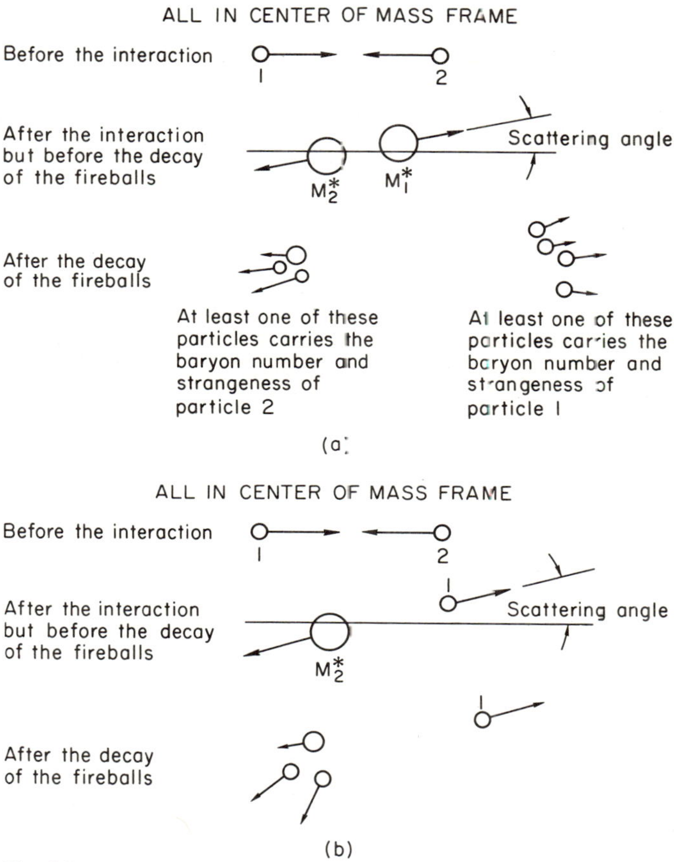

Fig. 8-7 The two-fireball model for many-body reactions.

masses is produced. Included among the masses M_1^* and M_2^* are the initial hadron masses M_1 and M_2, respectively. Thus we include in the model the *single*-fireball excitation process illustrated in Fig. 8-7*b*. These single-fireball excitation processes produce the *leading particle effect.* Thus the probability of the mass M^* being produced is $f(M^*, s)$ where

$$\int_{M_i}^{\infty} dM_i^* \, f(M_i^*, s) = 1, \qquad i = 1, 2 \tag{8-83}$$

As usual, we denote by t the four-momentum squared that is transferred from 1 to 2 during the forward dominant excitation process. Then the excitation process is described by the differential cross section

$$\frac{d^3\sigma_{\text{inel}}}{dt\, dM_1^*\, dM_2^*} = g(t, s)f(M_1^*, s)f(M_2^*, s) \tag{8-84}$$

Here we have made the simple assumptions that $f(M^*, s)$ is independent of t and $g(t, s)$ is independent of M_1^* and M_2^*. The total cross section for the production of all many-body final states is

$$\sigma_{\text{inel}}(s) = \int_{M_1}^{\infty} \int_{M_2}^{\infty} \int_{t_{\text{min}}}^{\infty} dM_1^*\, dM_2^*\, dt[g(t, s)f(M_1^*, s)f(M_2^*, s)] \tag{8-85}$$

where t_{min} depends on M_1^* and M_2^*—an important effect for large M_i^*.

The forward dominant excitation process might be, or at least resemble, a diffractive inelastic process of the type discussed in Sec. 7-8. Hence some forms of this model are called *diffractive-excitation* models. If the diffractive concept is applied rigorously, we must have $f(M_i^*, s)$ independent of s for $s \gg M_i^*$. Such a model can make use of the fragmentation hypothesis developed in the second part of this section. But we continue to discuss the more general model. It is convenient for pedagogical purposes to set

$$g(t, s) = C_{\text{inel}}\, e^{-b|t|} \tag{8-86}$$

$$\sigma_{\text{inel}} = \frac{C_{\text{inel}}}{b} \tag{8-87}$$

and to set t_{min} at zero in Eq. 8-85.

Before discussing how the fireballs decay, we comment on an older two-fireball theory (8MI), which is applied primarily to proton–proton collisions. It assumes that the two fireballs are produced *in addition to* the two incident nucleons. There are assumed to be in the final state two nucleons and two fireballs, the fireballs being thought of as lumps of purely mesonic matter. Thus in a different way the older theory introduced the leading particle concept with respect to the baryons.

Returning to the simpler model, we take up the question of how the fireballs decay. The basic assumption is that the decay of the fireball is *independent* of how the fireball is produced. Thus the fireball decay is quite similar to the decay of a resonant particle. In fact, the only major difference between the fireball concept and the resonant particle concept is that the former has a continuous mass spectrum and the latter has a discrete mass spectrum. A second basic assumption is that the decay of each fireball proceeds *independently* of the decay of the other fireball. All final state interactions are ignored. A third assumption that is almost always made is that each fireball decays *isotropically* in its *own rest frame.* This assumption is not essential; for example, the spin orientation of an incident nucleon might be carried into the associated fireball. But for simplicity most authors assume isotropic decay distributions in the fireball rest frame, and we do the same.

For additional simplicity we assume that M_i^* decays into the incident hadron i, thus ensuring the conservation of baryon number and strangeness, and into n_i pions of mass m. The decays should be treated by the statistical methods to produce multiplicity distribution functions for n_i at each value of M_i^*. But we leave that for an exercise. Here we make three very simple assumptions for the decay represented by

$$M_i^* \to \text{particle } i + n_i \text{ pions}$$

We assume that (1) in the M_i^* rest frame each pion has the same total energy E', (2) the particle i, if a nucleon, has negligible kinetic energy; and (3) E' is a *constant independent of* M^* *and* s. Of course these are very crude assumptions.

Exercise 8-19. The energy E' can be estimated from the average transverse momentum $\langle p_\perp \rangle$, which we may take to be 350 MeV/c. Estimate E', using the fact that $p_\perp$ is the same in the M^* frame and the laboratory frame, and assuming that the decay is isotropic in the M^* frame. For an example of how some authors do this, see Ref. 8HW.

From Ex. 8-19 a reasonable and easily remembered estimate for E' is 400 MeV. The number of pions produced in the decay of the $M_{1,2}^*$ fireball is

$$n(M_i^*) = \langle N(M_i^*) \rangle = \frac{M_i^* - M_i}{E'}, \qquad i = 1, 2 \tag{8-88}$$

and the average multiplicity of the particles produced by a fireball varies linearly with the mass of the fireball. This is quite different from the result that might be yielded if statistical considerations were applied to the fireball. For example, if we applied the Fermi statistical model separately to each fireball, we would obtain

$$\langle N(M_i^*) \rangle \sim c_0 + c_1 \left(\frac{M_i^*}{M_{\text{stat}}} \right)^{2/3}, \qquad i = 1, 2 \tag{8-89}$$

This follows from Eq. 8-42. Once $f(M_i^*, s)$ and the multiplicity distribution from a fireball are fixed, we can calculate the overall multiplicity distributions.

Exercise 8-20. For an example of this theory we turn again to the paper of Hwa and Lam (8HW) on proton–proton interactions. In this system $M_1 = M_2 = M$. They assume

$$f(M_i^*, s) = \frac{\text{constant}}{M_i^{*2}} \tag{8-90}$$

an expression independent of s. They then calculate the average charged multiplicity $\langle N_{\text{ch}} \rangle$. Repeat their work and obtain their formula, which shows that at very high energies, $\langle N_{\text{ch}} \rangle$ increases as ln (s).

Exercise 8-21. As we have mentioned, the assumption of a constant E' is very crude. Certainly the energies of the decay products of the fireball will have a statistical distribution. One way to introduce this is to use the thermodynamic model of Sec. 8-5. Note that E_{cms}, used in Sec. 8-5, must be replaced by M_i^* in discussing the thermodynamics of fireball i. Hagedorn and his colleagues (8HA) have developed a very elaborate and comprehensive model combining the fireball concept and the thermodynamic concept. Many new ideas are introduced, and further details are given in the reference.

In the foregoing discussion and exercises we have emphasized the multiplicity calculations in the fireball model. We now turn to the one-particle momentum distributions. The prediction of the transverse momentum distribution is straightforward. One object of the two-fireball model is to drastically limit $p_\perp$ compared to the available energy. This drastic limitation is achieved in two ways. First, the momentum allowed to each particle in the M^* frame is limited to a few hundred MeV/c. The simple model we have developed accomplishes this by fixing E' in Eq. 8-88 at 400 MeV. Second, b in Eq. 8-86 is taken to be of the order of 10 $(\text{GeV/c})^{-2}$, as in elastic scattering. This means that the transverse velocity of the fireballs is very small. Hence the additional transverse momentum, acquired by the decay products by virtue of the transverse velocity of the fireballs, is also small.

Exercise 8-22. In a many-body, pion–proton interaction, suppose that the pions in the rest frame of either M_i^* have an isotropic momentum distribution given by

$$\rho(p^2)\, dp^2 = \beta\, e^{-\alpha p^2}\, dp^2 \tag{8-91}$$

where p is the magnitude of the three-momentum, and α and β are constants. Take $\langle p \rangle = 370$ MeV/c, and take b in Eq. 8-86 to be $10(\text{GeV/c})^{-2}$. Finally, use Eq. 8-90 and calculate the pion transverse momentum distribution and average transverse momentum in the laboratory frame.

Clearly one can also calculate longitudinal momentum distributions. We cannot carry out such calculations for lack of space, but in Sec. 8-10 we give a general discussion of such distributions using the rapidity variable (not shown here). The scaling law, Eq. 8-60, can be built into fireball models.

2. *The Fragmentation Model.* The fragmentation model is in some ways related to the two-fireball model and is sometimes built into the fireball model. Hence both models suffer from the same deficiencies. Ad hoc mechanisms must be introduced to allow the calculation of multiplicity distributions and momentum spectra. But the fragmentation model has the special virtue of predicting how the momentum spectra will vary with energy, at very high energy. As the next section indicates, the prediction is correct.

The fragmentation model developed by Yang and his colleagues (8BEN) is most easily understood by considering the rest frame of one of the particles in the interaction $a + b$. Let us suppose that a is the incident particle and b the target particle in the laboratory system. Then the rest frame of b is the laboratory frame. As illustrated in Fig. 8-8a, at very high energies, the incident particle a appears as a Lorentz contracted disk. The passage of a through b *fragments* b into particles c, d, e . . . . Of course the term "fragments" is just another way to describe the excitation and decay process of the last section. Let particle c have momentum $\mathbf{p}_c$, and so forth. Think of the energy of a, which is approximately $\sqrt{s}$, as indefinitely increasing. Then the *fragmentation model*, also called the *hypothesis of limiting fragmentation*, proposes that the probability of producing that particle c with momentum $\mathbf{p}_c$ is independent of s for sufficiently high s.

The proposal rests on the following crude idea. By the uncertainty principle, the production of a secondary particle c of momentum $\mathbf{p}_c$ depends on the interaction over a volume whose linear dimensions are at least as large as $L_c = \hbar/\mathbf{p}_c$. But the Lorentz contraction of particle a yields a disk of thickness $T =$ (constant)/$\sqrt{s}$. When

$$T \ll L_c \qquad \text{or} \qquad \sqrt{s} \gg (\text{constant}) \times \mathbf{p}_c \tag{8-92}$$

the production of particle c with momentum $\mathbf{p}_c$ cannot depend on $\sqrt{s}$. Since for any $\mathbf{p}_c$ fixed in the b rest frame

$$|\mathbf{p}_c|/\sqrt{s} \xrightarrow[s\to\infty]{} 0$$

the condition of Eq. 8-92 will be attained, at sufficiently large s, for any fixed $\mathbf{p}_c$. Thus the fragmentation model predicts that the distribution in $\mathbf{p}_c$ should approach a limit as s goes to infinity. Explicitly, we write

$$f_b(p_{\perp c}, p_{\parallel c}, s) \xrightarrow[s\to\infty]{} f_b(p_{\perp c}, p_{\parallel c}) \tag{8-93a}$$

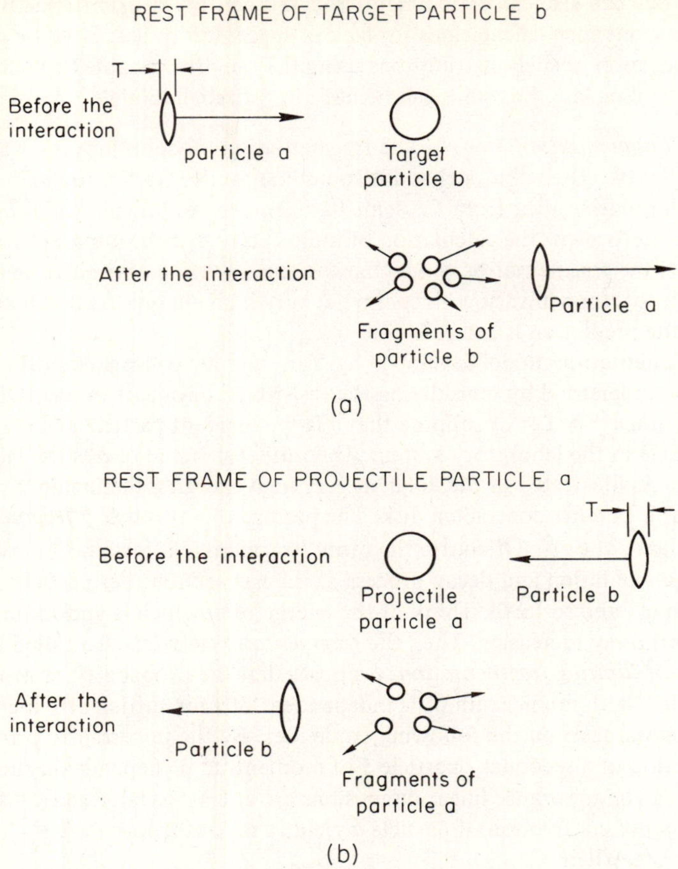

Fig. 8-8 The fragmentation model for many-body reactions.

in the *rest frame of b*. And for those particles produced in the fragmentation of particle *a* (Fig. 8-8*b*)

$$f_a(p_{\perp c}, p_{\parallel c}, s) \xrightarrow[s\to\infty]{} f_a(p_{\perp c}, p_{\parallel c}) \tag{8-93b}$$

in the *rest frame of a.*

Now, there is no way to tell which final state particles came from the breakup of the target *b* and which from the breakup of the projectile *a.* Even theoretically one cannot say if a particular particle came from the fragmentation of *b* rather than from the fragmentation of *a.* But using the principle of forward dominance, it is assumed that the particles whose three-momenta are closest to the three-momenta of particle *i* come from the breakup of *i.* The predictions of Eqs. 8-93 form a *scaling law* called *Yang scaling* or *fragmentation model scaling*. It is easy to show that this scaling law agrees with the Feynman scaling law of Eq. 8-60.

Let us consider $p_{\parallel c}$, the longitudinal momentum of secondary particle c, in the laboratory frame which is the rest frame of particle b. As s increases

$$p_{\parallel c} \underset{s\to\infty}{=} [x + |x|] \frac{s}{4m_b} + \frac{p_\perp^2 + m_c^2 - x^2 m_b^2}{2m_b|x|} \tag{8-94}$$

in the *rest frame of target particle b.*

Here m_i is the mass of the particle i, and for $-1 \leqslant x < 0$:

$$p_{\parallel c} \underset{s\to\infty}{=} \frac{p_\perp^2 + m_c^2 - x^2 m_b^2}{2m_b|x|} \qquad \text{(in rest frame of target } b\text{)} \tag{8-95}$$

Therefore for $-1 \leqslant x < 0$, $p_{\parallel c}$ remains finite in the b rest frame as $s \to \infty$; and Eq. 8-92 is obeyed. Thus $-1 \leqslant x < 0$ corresponds to the kinematic region where the hypothesis of limited fragmentation is valid for the breakup of particle b.

But if $0 < x \leqslant +1$, Eq. 8-94 shows that $p_{\parallel c}$ goes to ∞ as s goes to ∞; and Eq. 8-92 will not be obeyed for this region. *However*, if for those particles with $0 < x \leqslant +1$ we make a Lorentz transformation to the *rest frame* of the projectile a, we obtain

$$p_{\parallel c} \underset{s\to\infty}{=} \frac{p_\perp^2 + m_c^2 - x^2 m_a^2}{2m_a|x|} \qquad \text{(in rest frame of projectile } a\text{)} \tag{8-96}$$

Therefore, $0 < x \leqslant +1$ corresponds to the kinematic region where the hypothesis of limiting fragmentation is valid for the breakup of particle a.

Exercise 8-23. Fill in the steps leading to Eqs. 8-94 and 8-95.

Thus at sufficiently high energies there is an energy-*independent* relationship between x and either $p_{\parallel c}$ in the b rest frame or $p_{\parallel c}$ in the a rest frame. Therefore, at sufficiently high energies, the Feynman scaling prediction Eq. 8-60, agrees with Eqs. 8-93, the fragmentation model prediction. With one exception, we take the predictions as equivalent. The exception is the point at which x is zero, where there is *no* prediction from the simple fragmentation model.

8-9 Field Theory Models, Multiperipheral Models, and Elastic Scattering. If we could, we would like to calculate the matrix element T_N in Eq. 8-1 starting from quantum field theory. But we do not know how to do this in a fundamental or rigorous way. Indeed we do not know if it can be done, for in the end we do not know if field theory, as we know it, can be applied to strong interactions.

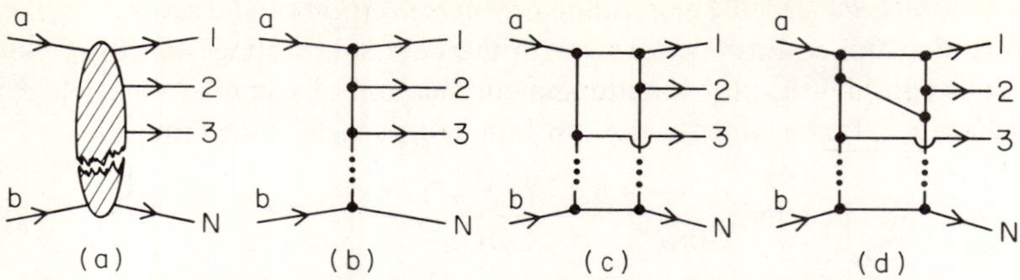

Fig. 8-9 Examples of Feynman diagrams contributing to many-body reactions.

But let us maintain our hopes. Then the program would be to resolve the inelastic reaction in Fig. 8-9*a* into a sum over infinite sets of Feynman diagrams. In order of increasing complexity, some of these diagrams are shown in Fig. 8-9*b* through 8-9*d*.

Now if we know how to do such sums we should also be able to sum over the diagrams in Fig. 8-10, which contribute to elastic scattering. Therefore, in addition to demanding that field theory models explain the behavior of high energy inelastic reactions; we can also demand that such models explain the behavior of high energy diffractive scattering as described in Ch. 2. In particular, such models should lead to a field theory derivation of the eikonal approximation to the optical model, Sec. 2-7.

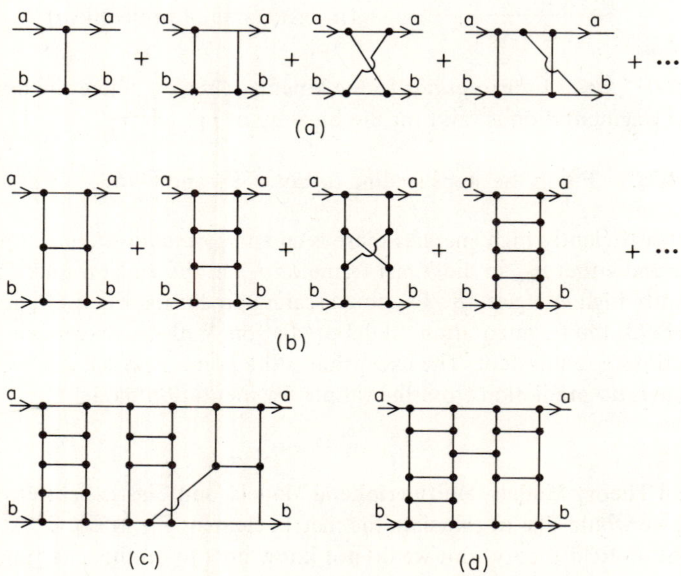

Fig. 8-10 Examples of Feynman diagrams contributing to diffractive elastic scattering.

Diagrams such as those in Figs. 8-9 and 8-10 reveal the connection between diffractive elastic scattering and inelastic processes—a connection treated semiclassically by the optical model (Ch. 2). For example, consider the elastic diagram in Fig. 8-11*a* split as shown in Fig. 8-11*b*. The elastic amplitude in Fig. 8-11*a* is then closely related to the two 2-body to N-body inelastic amplitudes in Fig. 8-11*b*. We say closely related but not equivalent, because the inner particles c, d, e, f in Fig. 8-11*a* are virtual and those in Fig. 8-11*b* are real.

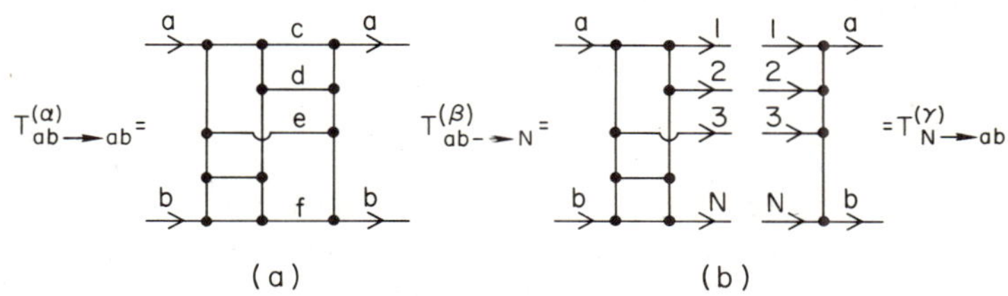

Fig. 8-11 The elastic scattering Feynman diagram in (*a*) is closely related to the inelastic scattering Feynman diagrams in (*b*).

Of course if we understand how the inelastic amplitudes will vary if the inner particles become virtual, we can write

$$T^{(\alpha)}_{ab\to ab}(s,t) = (\text{kinematic factors}) \int \left[d^3p_1 \cdots d^3p_N T^{(\beta)}_{ab\to N}(s;p_1\cdots p_N) \times T^{(\gamma)\dagger}_{ab\to N}(s;p_1\cdots p_N) \left(\frac{1}{p_1^2 - m_1^2} \cdots \frac{1}{p_N^2 - m_N^2} \right) \right] \tag{8-97}$$

If we sum over all types of inelastic diagrams β and γ, and if we sum over all N, we will obtain the *complete* elastic amplitude. Thus we obtain a generalization of the optical theorem, Sec. 6-7, to nonforward elastic amplitudes.

The more general field theory models are not discussed in this book. First, we do not have the space to derive their main results. And in the spirit of this book we are reluctant to discuss theories that we cannot present in a mathematically simple form. Second, we do not known if these models are correct. For in these calculations the most complicated sets of diagrams are ignored or approximated, and questions of basic divergencies in field theory are answered phenomenologically.

Therefore, we restrict our discussion to a special type of field theory model—the multiperipheral model. But before making this restriction we note that general field theory models have been used to derive the eikonal approximation to the optical model. In particular, the characteristic form of that approximation (Eq. 2-69) has been obtained. Extensive work has been done by Cheng and Wu

(8CHE), Cicuta and Sugar (8CI), and Chang and Yan (8CH). They summed over all sets of diagrams of the forms shown in Fig. 8-10*b* and 8-10*c*, which are called *ladder* or *tower* diagrams.

Exercise 8-24. Abarbanel and Itzykson (8AB) have given a relatively simple field theory derivation of the eikonal approximation using just the diagrams in Fig. 8-10*a* in which a neutral spin 1 particle is exchanged. Consult their paper to learn how such calculations are made.

The references just given exclude diagrams such as Fig. 8-10*d* where there is particle exchange between the ladders. But such diagrams are just the ones that when split, as in Fig. 8-11*b*, may contribute most to many-body inelastic reactions (8AA), and such diagrams may contribute most to diffractive inelastic scattering. Reference 8AA discusses models that include these more complicated diagrams.

Having discussed the problems and uncertainties of general field theory models as applied to diffractive elastic scattering, we return to many-body inelastic reactions and a special field theory model of these reactions. This is the *simple multiperipheral* model based on the diagram in Fig. 8-9*b* and shown in more detail in Fig. 8-12. A useful first reference is 8DE. Before proceeding, we note that two of these diagrams together (as in Fig. 8-11*b*) are closely related to the ladder diagrams for diffractive elastic scattering in Fig. 8-10*b*. Therefore, the multiperipheral model calculation can be used with straightforward modification to calculate one of the infinite sets of diagrams for diffractive elastic scattering.

As we consider the path in Fig. 8-12 from vertex 1 to vertex N, we realize that we have a succession of one-particle exchange processes. Each of these one-particle exchange processes is of the type discussed in a preliminary way in

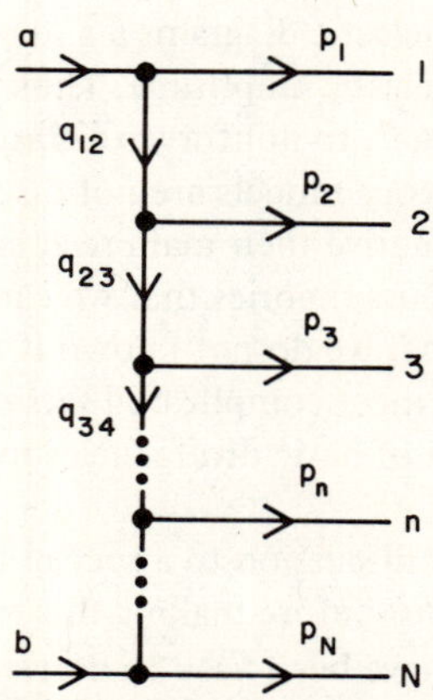

Fig. 8-12 The Feynman diagram for the simple multiperipheral model of many-body reactions.

Sec. 7-5, and discussed extensively in Chs. 12, 13, 17, and 18. As noted in Ch. 18, such processes are often called *peripheral.* Hence the term *multiperipheral* describes the entire diagram. For simplicity we assume that all particles have spin zero.

In Fig. 8-12 the four-momentum carried by outgoing particle n is p_n, and the four-momentum carried from vertex n to vertex $n+1$ is $q_{n,n+1}$. Examining a two-vertex region in Fig. 8-12, we see that we can regard that region as consisting of a two-body reaction, although the two incident particles are virtual. Then we define the typical two-body reaction quantities

$$s_{n,n+1} = (p_n + p_{n+1})^2, \qquad t_{n,n+1} = (q_{n,n+1})^2 \tag{8-98}$$

Then the matrix element T_N in Eq. 8-1 has the form

$$\begin{aligned} T_N(s, p_1, \ldots, p_N) = g^N R(s_{12}, t_{12}) R(s_{23}, t_{23}) \cdots R(s_{N-1,N}, t_{N-1,N}) \\ \times F(t_{12}) F(t_{12}, t_{23}) \cdots F(t_{N-2,N-1}, t_{N-1,N}) F(t_{N-1,N}) \end{aligned} \tag{8-99}$$

Here $R(s_{n,n+1}, t_{n,n+1})$ is the propagator from vertex n to vertex $n+1$, g is the vertex coupling constant, and F is the vertex form factor (see Sec. 13-2). For all vertices except those at the ends of the chain, F is a function of both the t carried into the vertex and the t carried out of the vertex. Since all particles have spin zero, if we assume *elementary* one-particle exchange (Sec. 7-5 and Ch. 12), we have simply $R(s_{ij}, t_{ij}) = (t_{ij} - m^2)^{-1}$. But as discussed in Ch. 17, a more realistic form for the propagator is

$$R_{ij} = f(s_{ij}, t_{ij})\, e^{-bt_{ij}} \tag{8-100}$$

where most of the dependence on $t_{n,n+1}$ is given by the exponential.

Equations 8-99 and 8-100 enable us to describe the overall effect of the matrix element T_N on the inelastic process.

1. T_N will be proportional to g^N.

2. The individual propagators will be a maximum when the corresponding $t_{n,n+1}$ is a minimum. With the z axis defined along the directions of motion of particles a and b, we define

$$\begin{aligned} q_{n,n+1} &= (q^0_{n,n+1}, \mathbf{q}_{n,n+1}); \qquad q_{\parallel,n,n+1} = q_{z,n,n+1}; \\ q_{\perp,n,n+1} &= (q^2_{x,n,n+1} + q^2_{y,n,n+1})^{1/2} \end{aligned} \tag{8-101}$$

The tendency to minimize $t_{n,n+1}$ will result in a tendency to minimize $q_{\perp,n,n+1}$; and here we have an explanation of the sharp cutoff in transverse momentum.

3. Similarly, there will be a tendency to minimize both $q^0_{n,n+1}$ and $q_{\parallel,n,n+1}$ consistent with the requirement that $(\Sigma_n p_n)^2 = s$.

Thus we expect that particle 1, and those close to it on the chain, would form the group of particles tending to carry the three-momentum and internal quantum numbers of incident particle 1. Similarly, particle N and the particles close to it on the chain would tend to carry the three-momentum and internal quantum numbers of incident particle b. Thus forward dominance would occur. The particles produced midway along the chain would have smaller momenta than those produced at either end of the chain, if the momenta are measured in the center of mass frame.

Exercise 8-25. Amplify the observations we have just made.

Thus the multiperipheral model directly generates the major characteristics of many-body inelastic reactions. To make further progress we need a reasonable guess for $F(s_{ij}, t_{ij})$ in Eq. 8-100. Regge theory provides such a guess, but since we do not develop that theory until Ch. 17, we postpone further quantitative discussion of the multiperipheral model to that chapter. We only note that a reasonable guess for F leads to the scaling law for inclusive distributions (Eq. 8-60) and the logarithmic energy dependence of the average multiplicity (Eq. 8-73 or 7-1b).

Now we must confess that the diagram of Fig. 8-12 poses a problem when we demand that most of the outgoing particles be pions. Because G-parity conservation requires that the adjacent propagators have alternate G-parities, at least two kinds of virtual particles–pions and ρ mesons, for example–must be involved. Once we allow this, we certainly should allow the diagram in Fig. 8-13*a* in which all the propagators are pions and all the intermediate secondary particles are ρ mesons. And certainly the diagram in Fig. 8-13*b* must be allowed. For the only difference between the two diagrams is that in Fig. 8-13*a* we assume that the 4-pion vertex is dominated by ρ production. One might go further and consider the diagram in Fig. 8-13*c*. The philosophy of some multiperipheral models is that such a diagram can always be resolved into simpler multiperipheral diagrams, as is indicated in Fig. 8-13*d*. Other models allow three or more particle clusters, as in Fig. 8-13*c*.

In concluding this section we must remember that the simple multiperipheral model considers only one type of field theory diagram; as illustrated in Fig. 8-9, there is an infinite number of other types. Therefore, we cannot demand too much from the simple multiperipheral model. We cannot expect it to reproduce reality; but we hope that it is a reasonable illustration of what would be obtained from a complete field theory calculation of many-body inelastic reactions. For further information, refer to Refs. 8FR, 8ZA, 8MU, and 8DE.

8-10 Energy and Longitudinal Momentum Distributions. In this final section we make some general remarks about energy and longitudinal momentum distributions, considering the energy first. In high energy, hadron–hadron reactions, most of the energy E_{cms} is *not* available for the production of particles; and speaking very crudely, most of the E_{cms} is not involved intimately in the reaction mechanism. Apparently most of the E_{cms} is carried through the reaction simply as kinetic energy. To amplify this vague idea, let us contrast the Fermi statistical model with the models presented in the last two sections. In the Fermi model, *all* the E_{cms} is "dumped" into the volume Ω, and statistical

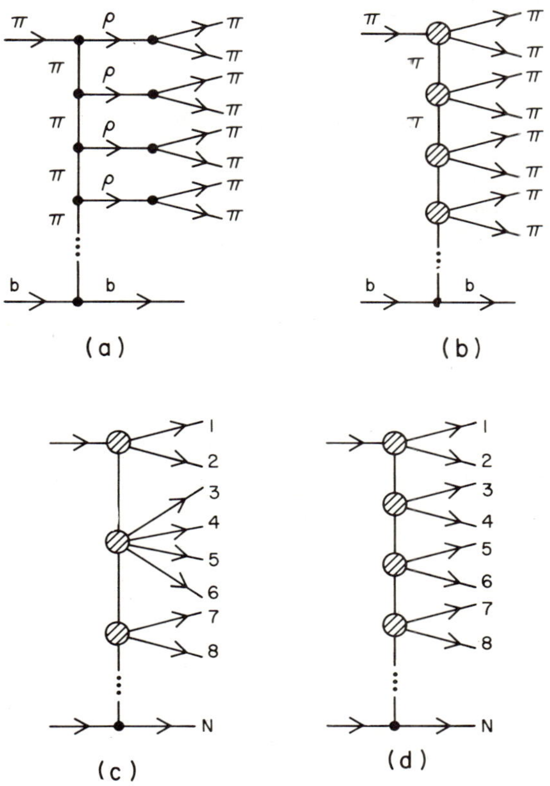

Fig. 8-13 More complicated forms of the multiperipheral model.

equilibrium is reached. Hence all the E_{cms} plays an intimate part in the interaction and is available for particle production. Contrast this situation with the situation in the fireball models; where only some of the E_{cms} is used to excite the incident hadrons into fireballs and most of the E_{cms} simply changes from incident particle kinetic energy to fireball kinetic energy. The fragmentation model inherently includes a similar idea in the form of the limited fragmentation hypothesis. The multiperipheral model is based on the idea that a many-body reaction consists of a series of two-body reactions. And as discussed in Secs. 7-4 and 7-5, the t-channel exchange process in these two-body collisions involves very little transfer of energy or three-momentum between incident particles. Therefore, using a concept introduced in Sec. 1-4, we say that most hadron-hadron reactions involve one or more *glancing* collisions between the hadrons. *Violent* collisions are rare. Of course there is no rigorous way to determine the degree of violence of a particular reaction from its kinematics. But we tend to believe that collisions are more violent when one or more of the secondary particles have large transverse momentum, $p_\perp > 1$ or 2 GeV/c. In two-body reactions this corresponds to large $|t|$. We return to this topic in Ch. 22.

The foregoing discussion provides an explanation of the puzzle mentioned in the discussion following Ex. 7-4. Why are most secondary particles pions, even at very high energies? The answer is that the total E_{cms} is only rarely available to produce secondary particles. Usually only a small fraction of E_{cms} is available, a pion production is much favored because of the low mass of the pion.

Exercise 8-26. Use the two-fireball model to discuss this point.

To conclude this section and this very long chapter we discuss one-particle longitudinal momentum distributions using the rapidity variable. First let us consider the transverse momentum cutoff phase space model (Sec. 8-7). From Eq. 8-82 we see that this model will predict the rapidity distribution shown in Fig. 8-14*a*. Here we have also used our physical intuition to argue that the

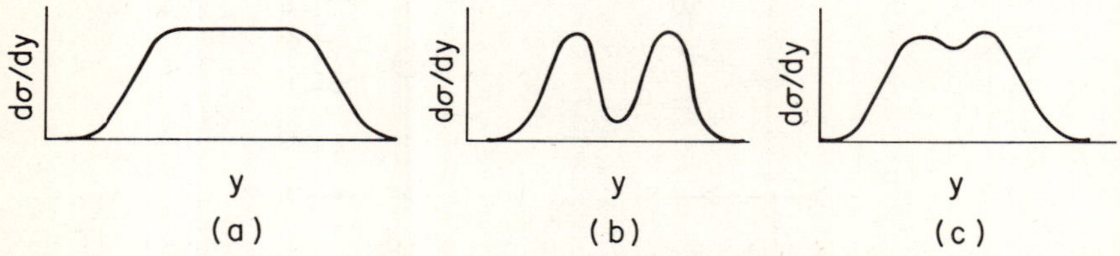

Fig. 8-14 Various predictions for the behavior of the invariant inclusive cross section versus the rapidity y. We can either take $p_\perp$ as fixed or suppose that the inclusive cross section has been integrated over $p_\perp$.

distribution cannot cut off sharply at the extremes of the rapidity range. Figure 8-14*a* is a good representation of what actually happens, for it is similar to the experimental rapidity distribution shown in Fig. 8-15.

Exercise 8-27. Take all the secondary particles to be pions and use Eq. 8-78 to explain the extremes of the rapidity distribution in Fig. 8-15.

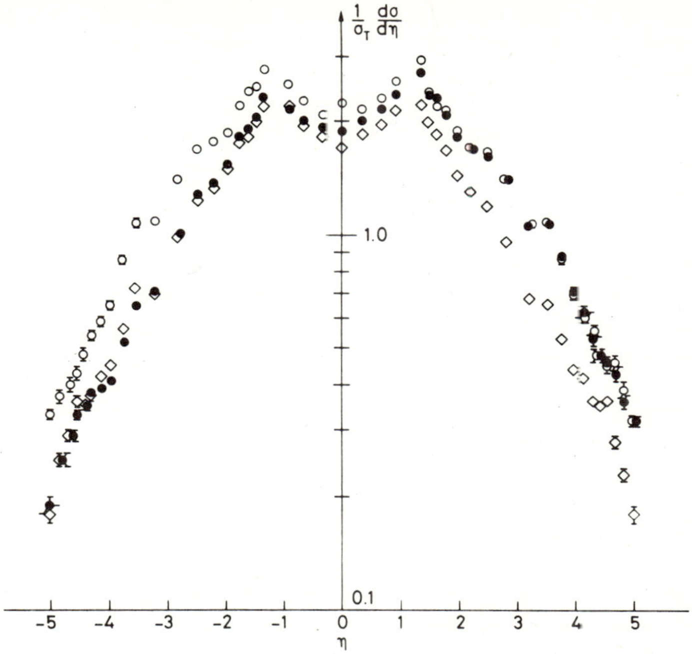

Fig. 8-15 The $(1/\sigma_{tot})$ $(d\sigma/d\eta)$ versus η, where η is the rapidity, in pp collisions (8BEL). The data come from a pp colliding beam experiment; the energy of each beam and the rapidity are all in the actual laboratory frame. The magnitude of the laboratory three-momenta of the two beams p_1 and p_2 are: diamonds, $p_1 = p_2 = 15.4$ GeV/c; open circles, $p_1 = p_2 = 26.6$ GeV/c; closed circles, $p_1 = 15.4$ GeV/c, $p_2 = 26.6$ GeV/c. Our main interest is the distribution when both beams have the same energy. See the reference for an explanation of the distribution when the energies are not equal.

The multiperipheral model calculation in Ch. 17 shows that Fig. 8-14*a* also represents the usual prediction of the multiperipheral model (see, however, Ref. 8DE).

The rapidity distributions predicted by fireball and related models depends on the function of $f(M^*, s)$ and the statistical manner in which the fireball decays. At one extreme, suppose that

$$f(M^*, s) = \delta(M^* - M_0), \qquad M_0 > M_{\text{nucleon}}; \qquad E' = \text{constant} = 400 \text{ MeV} \tag{8-102}$$

using the notation of Sec. 8-8. Then at sufficiently high energy one obtains the double-peak rapidity distribution of Fig. 8-14*b*. If the energy is not high enough, the two peaks will overlap as in Fig. 8-14*c*.

Exercise 8-28. Justify Fig. 8-14*b* and 8-14*c*. Discuss the effect on the width of each peak if the fireball decays into resonances such as ρ mesons, which then decay into pions.

If we relax the conditions of Eq. 8-102, the peaks in Fig. 8-14*b* will broaden. For example, if $f(M^*, s)$ is not a δ function, the different M^*'s that are produced will have different rapidities—and this will broaden the peaks.

Exercise 8-29. Illustrate the last point numerically by setting $E_{\text{cms}} = 50$ GeV and calculating the rapidity difference between fireballs with $M^* = 1, 2, 3, \ldots,$ GeV. Then retaining for all the fireballs, $E' =$ constant $= 400$ MeV, superimpose the peaks from the individual fireballs to obtain the total peak.

The foregoing argument indicates that it is possible in fireball models to eliminate the valley between the peaks in Fig. 8-14*b* and to obtain again Fig. 8-14*a*. Thus on the basis of one-particle rapidity distributions it is not possible to conclusively distinguish between the various models of many-body reactions. And one cannot use the one-particle transverse momentum distribution, the scaling law, or the energy dependence of N to distinguish between models, because all the models have been designed to explain these observations.

Therefore we turn to higher order effects (8JA, 8MU). For example, fireball models predict correlations between all the particles produced in a single fireball—which amounts to roughly half of the particles in the reaction. On the other hand, multiperipheral models predict correlations only between particles produced close to each other along the multiperipheral chain. But one cannot expect by such studies to completely vindicate any of these models, which are too simple to completely reproduce the reality of many-body, strong interaction dynamics. And probably the most we can expect from the different models is that they emphasize different aspects of this reality. For example, diffractive models, such as the fragmentation model or a diffractive-excitation fireball model, may only be applicable to the class of events with low multiplicities—2, 4, or 6. The higher multiplicity events would then be explained by a nondiffractive mechanism such as the multiperipheral model or a nondiffractive fireball model. In terms of rapidity plots such as Fig. 8-15, we might expect the diffractive mechanism to produce events that mostly populate the end regions of the plot. And the nondiffractive mechanism would produce events that mostly populate the central region.

REFERENCES

8AA S. Auerbach et al., *Phys. Rev.* **D6**, 2216 (1972).

8AB H. O. I. Abarbanel and C. Itzykson, *Phys. Rev. Lett.* **23**, 53 (1969).

8AD R. K. Adair, *Phys. Rev.* **D5,** 1105 (1972).

8BA M. Bander, *Phys. Rev.* **D6,** 164 (1972).

8BE A. Bertin et al., *Phys. Lett.* **42B,** 493 (1972).

8BEA J. V. Beaupre et al., *Phys. Lett.* **37B,** 432 (1971).

8BJ J. Bjorken, in *Proceedings of the Rochester Meeting on Particles and Fields–1971,* A. C. Melissinos and P. F. Slattery Eds. (American Institute of Physics, New York, 1971).

8BEL G. Bellettini et al., *Phys. Lett.* **45B,** 69 (1973).

8BEN J. Benecke et al., *Phys. Rev.* **188,** 2159 (1969).

8BY E. Byckling and K. Kajanti, *Particle Kinematics* (Wiley, New York, 1973).

8CH S. J. Chang and T. M. Yan, *Phys. Rev. Lett.* **25,** 1586 (1970).

8CHE H. Cheng and T. T. Wu, *Phys. Rev. Lett.* **24,** 1456 (1970).

8CI G. M. Cicuta and R. L. Sugar, *Phys. Rev.* **D3,** 970 (1971).

8CO G. Cocconi, *Phys. Rev.* **111,** 1699 (1958).

8CZ O. Czyzewski in *Proceedings of the International School of Elementary Particle Physics, Herceg–Novi,* M. Nikolic, Ed. (Gordon and Breach, New York, to be published).

8DE C. E. De Tar, *Phys. Rev.* **D3,** 128 (1971).

8FE E. Fermi, *Progr. Theoret. Phys.* **5,** 570 (1950).

8FER M. Ferro-Luzzi et al., *Nuovo Cimento* **36,** 1101 (1965).

8FEY R. P. Feynman, *Phys. Rev. Lett.* **23,** 1415 (1969).

8FR W. R. Frazer et al., *Rev. Mod. Phys.* **44,** 284 (1972).

8HA R. Hagedorn in *Proceedings of Colloquium on High Multiplicity Hadronic Interactions,* École Polytechnique, Paris, May 1970. Also issued as a CERN Preprint TH.1174; and references contained therein.

8HW R. C. Hwa and C. S. Lam, *Phys. Rev. Lett.* **27,** 1098 (1971).

8IC Good sources for data on many-body reactions and references to particular experiments are the review articles in the recent *Proceedings* of the International Conferences on High Energy Physics. These conferences are held every two years.

8JA M. Jacob, in *Proceedings of the Sixteenth International Conference on High Energy Physics, 1972* (National Accelerator Laboratory, Batavia, Ill., 1973).

8JAC M. Jacob and R. Slansky, *Phys Rev.* **D5,** 1847 (1972).

8KA G. Kallen, *Elementary Particle Physics* (Addison-Wesley, Reading, Mass., 1964).

8KR M. Kretzschmar, *Ann. Rev. Nucl. Sci.* **11,** 1 (1961).

8LA D. Q. Lamb, in *Proceedings of the Colloquium on High Multiplicity Hadronic Interactions, 1970* (École Polytechnique, Paris, 1970).

8MI M. Miesowicz, *Progr. Elem. Part. Cosmic Ray Phys.* **10,** 101 (1971).

8MIL R. H. Milburn, *Rev. Mod. Phys.* **27,** 1 (1955).

8MU A. H. Mueller, in *Proceedings of the Sixteenth International Conference on High Energy Physics, 1972* (National Accelerator Laboratory, Batavia, Ill., 1973).

8NY P. Nyborg and O. Skjeggestad, in *Kinematics and Multiparticle Systems,* M. Nikolic, Ed. (Gordon and Breach, New York, 1968).

8SE J. C. Sens, in *Proceedings of the Fourth International Conference on High Energy Collisions, 1972,* J. R. Smith, Ed. (Rutherford High Energy Laboratory, Chilton, England, 1973).

8ZA F. Zachariasen, *Phys. Rep.* **2C,** 1 (1971).

CHAPTER

9

Rotations, Boosts, and the Helicity Formalism

9-1 Introduction. In this chapter we introduce and develop the helicity formalism, a formalism that is used extensively in this book. The chapter begins with a summary of the properties of the rotation matrices – the D-matrices. The D-matrices are used widely both in the construction of the helicity formalism and in its applications.

The D-matrices are first used to define one-particle helicity states. This allows the construction of two-particle, plane wave helicity states and angular momentum helicity states. The relations between these two types of helicity states are then established. References 9RO, 9WE, 9MA, 9JA supply more formal and extensive discussions.

We also use this chapter to develop some practical applications of Lorentz velocity transformations (Secs. 9-4, 9-8, and 9-9).

A large amount of formal manipulation makes reading the chapter tedious, but there is no way to avoid the mathematics. It is worth following in detail, if only as a way of becoming familiar with rotations, boosts, and helicity amplitudes.

9-2 The Rotation Matrices. Rotations in three-dimensional space are represented by the unitary operator $U(\theta, \hat{n})$ defined in Sec. 6-10. For the remainder of the book we use the mnemonic notation $R(\theta, \hat{n})$ for this three-dimensional rotation operator. Thus Eq. 6-85 becomes

$$R(\theta, \hat{n}) = e^{-i\theta \hat{n}\cdot \mathbf{J}} \tag{9-1}$$

In classical mechanics the rotations of an object are described by the Euler angles α, β, γ. There are unfortunately various ways to define these angles. Our definition uses a *right-handed* coordinate system. We take the active view (Sec. 6-8); the coordinate axes are *fixed* in space and α, β, γ define the rotation of the physical system with reference to those fixed axes. There is *first* a rotation by the angle γ about the z axis, *next* a rotation by the angle β about the y axis, and *third* a rotation by the angle α about the z axis. This convention has been carried over into quantum mechanics, and the general rotations of a physical system are described by the notation

$$R(\alpha\beta\gamma) = R(\alpha, \hat{n}_z)R(\beta, \hat{n}_y)R(\gamma, \hat{n}_z) \tag{9-2}$$

where $\hat{n}_y$ and $\hat{n}_z$ are unit vectors along the y and z axes fixed in space. The two rotations about the z axis are less necessary here than in classical mechanics, because in quantum mechanics we never need to explicitly describe the angular velocity of a body spinning about an axis fixed in that body. However, as the reader may expect, there are phase considerations here which replace the angular velocity considerations in classical mechanics. Eq. 9-2 can be rewritten as

$$R(\alpha\beta\gamma) = e^{-i\alpha J_z} e^{-i\beta J_y} e^{-i\gamma J_z} \tag{9-3}$$

In elementary quantum mechanics special methods are developed for treating the rotations of spin $\frac{1}{2}$ and spin 1 systems. But it is useful to have a general method for describing and calculating the effects of rotations on a system with arbitrary spin. To develop such a method we observe that the rotation operator R commutes with the angular momentum operator $\mathbf{J}^2$. Or in more physical terms, we observe that a rotation of a physical system does not change the magnitude of the total angular momentum of that system. Therefore, a rotation cannot change the eigenvalue j of a $|jm\rangle$ state.

For the exact description of the effect of a rotation $R(\alpha, \beta, \gamma)$ on a state $|jm\rangle$ we take the active view of the transformation as defined in Sec. 6-8. That is, we rotate the physical system by the Eulerian angles α, β, γ; we do *not* rotate the coordinate axes. Then under the rotation $R(\alpha\beta\gamma)$ we have

$$|jm\rangle \rightarrow R(\alpha\beta\gamma)|jm\rangle = \sum_{m'=-j}^{+j} D^j_{m'm}(\alpha\beta\gamma)|jm'\rangle \tag{9-4}$$

Thus the state $|jm\rangle$ is now described by a linear combination of $|jm'\rangle$ states, and the coefficients in this linear combination are functions of α, β, γ; they form the elements of the *rotation matrix* or *D-matrix*. Thus the D-matrix describes how under a rotation the state $|jm\rangle$ is distributed over the jth multiplet. The amplitude $D^j_{m'm}(\alpha\beta\gamma)$ describes how much of the amplitude of the physical system is in the state $|jm'\rangle$.

We call $D^j_{m'm}(\alpha\beta\gamma)$ a D-matrix element or a D function. It is a complex trigonometric function of α, β, and γ. For example,

$$D^1_{11}(\alpha\beta\gamma) = e^{-i\alpha} e^{-i\gamma} \cos^2 \tfrac{1}{2}\beta$$

The D function has very simple dependence on α and γ, but very complicated dependence on β. We first look at the α and γ dependence. Take the scalar product of both sides of Eq. 9-4 with $\langle jm'|$ and use Eq. 9-3 to express R in terms of J. Then

$$D^j_{m'm}(\alpha\beta\gamma) = e^{-i\alpha m'} \langle jm'| e^{-i\beta J_y} |jm\rangle e^{-i\gamma m} \tag{9-5}$$

We next define the d-matrix by

$$\langle jm'| e^{-i\beta J_y} |jm\rangle = d^j_{m'm}(\beta) \tag{9-6}$$

so that

$$D^j_{m'm}(\alpha\beta\gamma) = e^{-i\alpha m'} d^j_{m'm}(\beta)\, e^{-i\gamma m} = e^{-i\alpha m' - i\gamma m}\, d^j_{m'm}(\beta) \tag{9-7}$$

Although $d^j_{m'm}(\beta)$ is a function of only β, it is a complicated function. There are various explicit ways to write out $d^j_{m'm}$. Perhaps the most transparent formula is

$$d^j_{m'm}(\beta) = \sum_n \left[\frac{(-1)^n [(j+m)!(j-m)!(j+m')!(j-m')!]^{1/2}}{(j-m'-n)!(j+m-n)!(n+m'-m)!n!} \times (\cos \tfrac{1}{2}\beta)^{2j+m-m'-2n} (-\sin \tfrac{1}{2}\beta)^{m'-m+2n} \right] \tag{9-8}$$

In Eq. 9-8 the limits of summation over n are such that none of the factorials have negative arguments.

Exercise 9-1. Use Eq. 9-8 to calculate some d-matrix elements for $j = 1$ and compare your answers with the table of the d-matrix elements given in App. C. Show $d^j_{00}(\theta) = P_j(\cos\theta)$ where P_j is a Legendre polynomial and $j = 0, 1, 2$.

Even though Eq. 9-8 is transparent, it is tedious to use it to calculate $d^j_{m'm}$ if $|m'|$ or $|m|$ is greater than 1. It is best and easiest to use tables of the d-matrix elements and to use specific relationships for special values of these elements, as opposed to calculating these elements directly from Eq. 9-8. Such tables and relationships are collected in App. C. However, Eq. 9-8 is useful for the derivation of some very commonly used symmetry relationships, some of which are worked out in the next few paragraphs. We note that $d^j_{m'm}(\beta)$ is real.

If $-\beta$ is inserted in Eq. 9-8 only the sin $\beta/2$ term is affected. This yields

$$d^j_{m'm}(-\beta) = (-1)^{m'-m+2n}\, d^j_{m'm}(\beta) = (-1)^{m'-m}\, d^j_{m'm}(\beta) \tag{9-9}$$

Another relation is obtained by noting that because $R(\alpha\beta\gamma)$ is a unitary operator, $D(\alpha\beta\gamma)$ is a unitary matrix. Therefore, we have

$$D^\dagger(0\beta 0) = D^{-1}(0\beta 0) = [e^{-iJ_y\beta}]^{-1} = e^{iJ_y\beta} = D(0, -\beta, 0)$$

Isolating the d-matrix, we obtain

$$d^j_{m'm}(-\beta) = d^j_{mm'}(\beta)$$

which combined with Eq. 9-9 yields

$$d^j_{mm'}(\beta) = (-1)^{m'-m}\, d^j_{m'm}(\beta) \tag{9-10}$$

Although Eqs. 9-9 and 9-10 have been written in terms of the d-matrix, they all can be directly applied to the D-matrix using Eq. 9-7. As an example, we observe that

$$\begin{aligned} D^j_{m'm}(\alpha\beta\gamma) &= e^{-im'\alpha}\, d^j_{m'm}(\beta)\, e^{-im\gamma} = e^{-im\gamma}\, d^j_{mm'}(-\beta)\, e^{-im'\alpha} \\ &= D^j_{mm'}(\gamma, -\beta, \alpha) \end{aligned} \tag{9-11}$$

Our final observation (see, e.g., Ref. 9RO) is

$$\begin{aligned} &\int_0^{2\pi} d\alpha \int_0^{2\pi} d\gamma \int_0^{2\pi} \sin\beta\, d\beta\, [D^{j*}_{mn}(\alpha\beta\gamma) D^{j'}_{m'n'}(\alpha\beta\gamma)] \\ &\qquad = \delta_{mm'}\, \delta_{nn'}\, \delta_{jj'} \left[\frac{8\pi^2}{2j+1}\right] \end{aligned} \tag{9-12}$$

a very nice result. Thus two different D-matrix elements are orthogonal when integrated over all angles.

9-3 Cartesian and Spherical Tensors. In this section we clarify the relationship between two kinds of tensor quantities—spherical and Cartesian. We first define each type and then investigate the relation between the two types. In the previous section we were discussing the angular momentum eigenfunctions $|jm\rangle$. Under a rotation of the physical system by the operator $R(\alpha\beta\alpha)$, $|jm\rangle$ transforms according to the rule

$$|jm\rangle \to R(\alpha\beta\gamma)|jm\rangle = \sum_{m'} D^j_{m'm}(\alpha\beta\gamma)\ jm'\rangle \tag{9-13}$$

Any set of $2j + 1$ functions identified by the label ϕ_{jm}, which transforms under rotation according to this rule, is called a spherical tensor function of rank j. Thus we have

$$\phi_{jm} \to R(\alpha\beta\gamma)\phi_{jm} = \sum_{m'} D^j_{m'm}(\alpha\beta\gamma)\phi_{jm'} \tag{9-14}$$

When j is integral, the spherical harmonics provide an example:

$$Y_l^m(\theta, \phi) \to R(\alpha\beta\gamma)Y_l^m(\theta, \phi) = \sum_{m'} D^j_{m'm}(\alpha\beta\gamma)Y_l^{m'}(\theta', \phi') \tag{9-15}$$

The spherical tensors are quantities with a specific behavior under rotation, Eq. 9-14. The Cartesian tensors in three dimensions also have a specific behavior under rotation, but the behavior is different from that of the spherical tensors. The description of the Cartesian tensors starts from the consideration of a displacement vector $\mathbf{r}$.

In analogy to the discussion in Sec. 6-8, under a rotation of the physical system containing $\mathbf{r}$ we have

$$\mathbf{r} \to \mathbf{r}' = A(\alpha\beta\gamma)\mathbf{r} \tag{9-16a}$$

$$r_i' = \sum_j A_{ij}(\alpha\beta\gamma)r_j \tag{9-16b}$$

where

$$A(\alpha\beta\gamma) = A_z(\alpha)A_y(\beta)A_z(\gamma) \tag{9-17a}$$

$$A_y(\beta) = \begin{pmatrix} \cos\beta & 0 & \sin\beta \\ 0 & 1 & 0 \\ -\sin\beta & 0 & \cos\beta \end{pmatrix}, \quad A_z(\alpha) = \begin{pmatrix} \cos\alpha & -\sin\alpha & 0 \\ \sin\alpha & \cos\alpha & 0 \\ 0 & 0 & 1 \end{pmatrix} \tag{9-17b}$$

Here again, the rotations of the point $\mathbf{r}$ in the physical system are carried out with respect to a fixed set of coordinate axes.

Any three-component quantity (v_1, v_2, v_3) that behaves under a rotation according to these rules is called a Cartesian tensor of the first rank. A quantity with nine components specified by the double subscript w_{ij} $(i = 1, 2, 3; j = 1, 2, 3)$ is a Cartesian tensor of the second rank if it transforms under rotation according to the rule

$$w_{ij}' = \sum_{ij} A_{ik}A_{jl}w_{kl} \tag{9-18}$$

The extension to higher order Cartesian tensors is straightforward.

We now turn to the relation between spherical and Cartesian tensors, which exists only for spherical tensors of integral rank. A spherical tensor of rank 0 and Cartesian tensor of rank 0 are both scalars, hence are equivalent. The first nontrivial case is for first-rank tensors. We find the relationship by writing the spherical harmonics for $l = 1$ (which represent the spherical tensors of rank 1) in the form

$$Y_1^{\pm 1} = \mp \left(\frac{3}{8\pi}\right)^{1/2} \frac{x \pm iy}{r}, \qquad Y_1^{\,0} = \left(\frac{3}{4\pi}\right)^{1/2} \frac{z}{r}$$

Therefore, a first-rank spherical tensor ϕ_{lm} and a Cartesian vector (v_1, v_2, v_3) are related by

$$\phi_{11} = -\left[\frac{v_1 + iv_2}{\sqrt{2}}\right], \qquad \phi_{10} = v_3, \qquad \phi_{1-1} = \left[\frac{v_1 - iv_2}{\sqrt{2}}\right] \tag{9-19}$$

For higher rank Cartesian tensors the situation is more complicated. Thus the second-rank Cartesian tensor w_{ij} has nine components, whereas the second-rank spherical tensor has only five. In fact, w_{ij} is equivalent to a linear combination of the spherical tensors $j = 2, j = 1$, and $j = 0$. It is tedious to work out the relationship; we leave that for Ex. 9-2, only quoting the results here. Three different objects are formed from w_{ij}.

1. $\omega = \mathrm{Tr}\,(w_{ij})$ is a scalar and of course equivalent to a zero-tank spherical tensor.
2. $\omega_k = \frac{1}{2}(w_{ij} - w_{ji})$, ($ijk$ cyclic) is an axial vector.
3. $\omega_{ij} = \frac{1}{2}[w_{ij} + w_{ji} - \frac{2}{3}\omega]$ is a linear combination of the elements of a second-rank spherical tensor; ω_{ij} is a zero trace, symmetric, second-rank Cartesian tensor.

Thus Cartesian tensors of rank 2 and above have a complicated relationship to spherical tensors. For many applications, the spherical tensors with their simple rotational properties are most convenient. Why do we bother with the Cartesian tensors in this book? The reason is that in the construction of matrix elements, which must be Lorentz scalars (see the discussion in Ch. 12), we naturally use four-dimensional Cartesian tensors. In a specific Lorentz frame these quantities reduce to three-dimensional Cartesian tensors; and so one must be prepared to handle these latter quantities, relating them to the spherical tensors. Similarly, we use Cartesian four-vectors to construct the spin functions of relativistic spin 1 particles.

Exercise 9-2. Derive the relationship just given between a second-rank Cartesian tensor and the spherical tensors.

9-4 Lorentz Velocity Transformations. The construction of helicity states—the main purpose of this chapter—depends on the use of both spatial rotations and Lorentz velocity transformations. Since we tend to be less familiar with Lorentz velocity transformations than with spatial rotations, we devote some space to the development of useful formulas and to our notation. We use the rather ugly word *boost* and the phrase *Lorentz velocity transformation* interchangeably. The very useful variable

$$\xi = \operatorname{arctanh} v = \tfrac{1}{2} \ln \left(\frac{E + |\mathbf{p}|}{E - |\mathbf{p}|} \right) \tag{9-20}$$

appears to have no universally used name. We call ξ the *boost parameter* or *velocity parameter* (9TA). ξ is also called the *rapidity*, but we use that term only in connection with the longitudinal rapidity in inclusive reactions (Eq. 8-78). For the Lorentz velocity transformation operator $U(\xi)$ we almost always use $L(\xi)$ as a mnemonic aid.

A boost need not be along the x, y, or z axes, it can be along an arbitrary direction described by the unit vector $\hat{n}$. Unfortunately, there is no unambiguous conventional notation for such a boost, and we often use our own notation $L(\xi, \hat{n})$. Thus Eq. 6-93 generalizes to

$$L(\xi, \hat{n}) = e^{-i\xi\hat{n}\cdot\mathbf{K}} \tag{9-21}$$

Even more unfortunately, there is one often-used notation that can be confusing. We must explain and sometimes use this notation, to enable the reader to follow the literature. Suppose that a particle of mass m at *rest* receives a boost which gives the particle a three-momentum $\mathbf{p}$. That boost is denoted by $L(\mathbf{p})$. But a Lorentz transformation has nothing to do with the mass of the system being transformed. Therefore, what is actually meant is

$$L(\mathbf{p}) = L\left(\xi = \operatorname{arcsinh} \frac{|\mathbf{p}|}{m},\ \hat{n} = \frac{\mathbf{p}}{|\mathbf{p}|} \right) \tag{9-22}$$

But it is even more confusing when a boost $L(\mathbf{p})$ is applied to a particle that already had some momentum $\mathbf{p}'$. What we mean by such a boost and what is generally meant in the literature by such a boost is the following: $\mathbf{p}$ is the momentum that a particle of mass m at rest would acquire by the boost $L(\mathbf{p})$; $\mathbf{p}$ is *not* the momentum attained by the particle that originally had the momentum $\mathbf{p}'$ – that particle attains a momentum $\mathbf{p}''$, which is calculated by applying the boost $L(\mathbf{p})$ to $\mathbf{p}'$.

Next we would like to have a general formula for the boost operator $L(\xi, \hat{n})$ in terms of the boost operator $L(\xi_z)$ along the $+z$ axis. Let the spherical angles of $\hat{n}$ be (θ, ϕ). In the passive view, we rotate the coordinate axes so that the $+z$ axis coincides with $\hat{n}$, we carry out the boost $L(-\xi_z)$ on the coordinate axes, and we rotate the coordinate axes back to their original orientation. Although it is easier to understand the process in the passive view, we use the active view operator, which boosts the physical system, namely

$$L(\xi, \hat{n}) = R(\phi, \theta, 0) L(\xi_z) R^{-1}(\phi, \theta, 0) \tag{9-23}$$

We try to avoid having to work out explicit forms for the boost operators, but sometimes we must use explicit expressions. For example, we often need the matrix form in four-space of $L(\eta, \hat{n})$. For a boost along the $+z$ axis (see Eq. 6-90)

$$L(\eta_z) = \begin{bmatrix} \cosh\eta & 0 & 0 & \sinh\eta \\ 0 & 1 & 0 & 0 \\ 0 & 0 & 1 & 0 \\ \sinh\eta & 0 & 0 & \cosh\eta \end{bmatrix} \tag{9-24}$$

and for a boost of arbitrary direction in the xz plane

$$L(\eta, \hat{n}) = \begin{bmatrix} \cosh\eta & \sin\theta\sinh\eta & 0 & \cos\theta\sinh\eta \\ \sin\theta\sinh\eta & [1+\sin^2\theta(\cosh\eta - 1)] & 0 & [\cos\theta\sin\theta(\cosh\eta - 1)] \\ 0 & 0 & 1 & 0 \\ \cos\theta\sinh\eta & [\cos\theta\sin\theta(\cosh\eta - 1)] & 0 & [1+\cos^2\theta(\cosh\eta - 1)] \end{bmatrix} \tag{9-25}$$

where θ is the angle between $\hat{n}$ and the $+z$ axis.

Exercise 9-3. Derive Eq. 9-25 from Eqs. 9-24 and 9-23.

To conclude this section we consider the boost operator for four-spinors. For our purposes, a sufficiently general case is again a boost of arbitrary direction in the xz plane. It is interesting to begin with Eqs. 6-85 and 6-93, using the generators J_y and K_z given in App. E, giving

$$R(0, \theta, 0) = \begin{bmatrix} \cos\frac{1}{2}\theta & -\sin\frac{1}{2}\theta & 0 & 0 \\ \sin\frac{1}{2}\theta & \cos\frac{1}{2}\theta & 0 & 0 \\ 0 & 0 & \cos\frac{1}{2}\theta & -\sin\frac{1}{2}\theta \\ 0 & 0 & \sin\frac{1}{2}\theta & \cos\frac{1}{2}\theta \end{bmatrix} \tag{9-26}$$

and

$$L(\eta_z) = \begin{bmatrix} \cosh\frac{1}{2}\eta & 0 & \sinh\frac{1}{2}\eta & 0 \\ 0 & \cosh\frac{1}{2}\eta & 0 & -\sinh\frac{1}{2}\eta \\ \sinh\frac{1}{2}\eta & 0 & \cosh\frac{1}{2}\eta & 0 \\ 0 & -\sinh\frac{1}{2}\eta & 0 & \cosh\frac{1}{2}\eta \end{bmatrix} \tag{9-27}$$

Exercise 9-4. Derive Eq. 9-26 from Eq. 6-85, and Eq. 9-27 from Eq. 6-93.
Finally, using Eq. (9-23), we write

$$L(\eta, \hat{n}) = \begin{bmatrix} \cosh \frac{1}{2}\eta & 0 & \cos\theta \sinh \frac{1}{2}\eta & \sin\theta \sinh \frac{1}{2}\eta \\ 0 & \cosh \frac{1}{2}\eta & \sin\theta \sinh \frac{1}{2}\eta & -\cos\theta \sinh \frac{1}{2}\eta \\ \cos\theta \sinh \frac{1}{2}\eta & \sin\theta \sinh \frac{1}{2}\eta & \cosh \frac{1}{2}\eta & 0 \\ \sin\theta \sinh \frac{1}{2}\eta & -\cos\theta \sinh \frac{1}{2}\eta & 0 & \cosh \frac{1}{2}\eta \end{bmatrix} \tag{9-28}$$

Exercise 9-5. Derive Eq. 9-28.

9-5 One-Particle Helicity States. In the helicity formalism, the spin of the particle is quantized along the direction of the three-momentum of the particle (a more formal definition is presented in the next paragraph). The spin component along this axis is called the helicity (μ) and the state is called a helicity state. A particle with nonzero mass and spin s has $2s + 1$ helicity states with μ going from $+s$ to $-s$. A zero mass particle has just two helicity states: $\mu = \pm s$ (9JA, 9WE, 9MA). The helicity formalism is of great use in elementary particle theory because the spin states are defined directly for relativistic particles. There is no need to take each particle to its rest system to calculate its spin state, as is required in the spin-orbit formalism. In two-particle systems the orbital angular momentum concept is avoided, simplifying equations and calculations.

The formal construction of an helicity state of a nonzero mass particle of spin s begins with the particle in its rest frame. First the $+z$ axis is taken as the axis of spin quantization and the component of the spin along that axis is called μ. Next the particle is given a boost $L(\xi_z)$ along the z axis in the $+z$ direction. It acquires a momentum

$$p_z = m \sinh \xi_z \tag{9-29}$$

Thus in the conventional notation the boost is $L(p_z)$, as defined in Eq. 9-22. During the boost $L(p_z)$ the $+z$ axis remains the axis of spin quantization. The particle is then given an arbitrary direction of motion by carrying out an arbitrary spatial rotation. The three-momentum is now $\mathbf{p}$. The direction of $\mathbf{p}$ is arbitrary, but $|\mathbf{p}| = p_z$. During and after this rotation, the axis of spin quantization is defined as the direction of the three-momentum $\mathbf{p}$. That is, the axis of spin quantization has the direction of the unit vector $\mathbf{p}/|\mathbf{p}|$. This is the crucial difference between the spin-orbit formalism and the helicity formalism. A *helicity state* is *denoted* by $|\mathbf{p}, \mu\rangle$ or $|\mathbf{p}\mu\rangle$. We almost always use a Greek letter to denote the helicity. This helps to distinguish helicity states from rest frame spin states (Sec. 3–5).

When the particle is at rest, the helicity state, now denoted by $|0, \mu\rangle$, is congruent to the rest frame spin state $|s, m\rangle$ with $m = \mu$. The z axis is taken as the axis of quantization for both $|0, \mu\rangle$ and $|s, m\rangle$.

The final momentum $\mathbf{p}$ is along the direction specified by the angles θ and ϕ. For the rotation operator $R(\alpha\beta\gamma)$ we follow Jacob and Wick (9JA) and use $R(\alpha\beta\gamma) = R(\phi, \theta, -\phi)$. This choice is arbitrary; we remarked earlier that in quantum mechanics, unlike classical mechanics, the use of both γ and α was related to phase question. The Jacob and Wick $R(\phi, \theta, -\phi)$, which we use, has the property that at $\theta = 0$ the phase introduced by the rotation is identically zero for all ϕ. This simplifies some equations but complicates others. We summarize the formal definition by

$$|\mathbf{p}\mu\rangle = R(\phi, \theta, -\phi)|p_z\mu\rangle = R(\phi, \theta, -\phi)L(p_z)|0, \mu\rangle \tag{9-30}$$

The state $|\mathbf{p}\mu\rangle$ is called a one-particle helicity state of momentum $\mathbf{p}$ and helicity μ.

The state $|\mathbf{p}\mu\rangle$ is an eigenstate of the three-dimensional momentum operator $\mathbf{P}$. Thus we have

$$\mathbf{P}|\mathbf{p}\mu\rangle = \mathbf{p}|\mathbf{p}\mu\rangle \tag{9-31}$$

These states are plane wave states of infinite extent, as discussed in Ch. 6. We use the normalization described in Ch. 6, that is,

$$\langle \mathbf{p}'\mu'|\mathbf{p}\mu\rangle = (2\pi)^3 2E\delta^3(\mathbf{p}' - \mathbf{p})\delta_{\mu'\mu} \tag{9-32}$$

where $|\mathbf{p}\mu\rangle$ is also an eigenstate of helicity. To clarify this statement we must define a helicity operator h. The definition is

$$h = \frac{\mathbf{J} \cdot \mathbf{P}}{|\mathbf{P}|} = \mathbf{J} \cdot \hat{P} \qquad \text{(particle in motion)} \tag{9-33a}$$

where $\mathbf{P}$ is the three-momentum operator, as in Eq. 9-31. In the particle's rest frame we use the special definition,

$$h = J_z \qquad \text{(particle at rest)} \tag{9-33b}$$

Thus in the particle's rest frame the *helicity states coincide* with the *rest frame spin states* defined in Sec. 3-5. We rarely use the helicity operator h explicitly. But it is instructive to work with h for a while to develop a physical feeling for the helicity concept.

The operator h is invariant under spatial rotations. Therefore, h and the rotation operator R commute. Thus we have

$$[h, R(\theta, \hat{n})] = 0 \tag{9-34}$$

To understand how h behaves under an arbitrary boost, we need only consider an infinitesimal boost

$$L(\delta\xi, \hat{n}) = I - i(\delta\xi)\hat{n} \cdot \mathbf{K}$$

and use the commutation relations among **J**, **P**, and **K** (Eq. 6-94). Inspection of these commutation relations shows that in general $L(\delta\xi, \hat{n})$ and h do *not* commute. Thus h is not invariant under an arbitrary boost. But if $\hat{n}$ and **p** are in the same direction, we write

$$[\mathbf{J} \cdot \mathbf{p}, L(\delta\xi, \hat{n})] = 0$$

Here **p** is *not* an operator.

Therefore

1. The helicity state of a particle does not change under an arbitrary spatial rotation.
2. The helicity state of a particle does not change under a boost along the direction of the three-momentum of the particle.
3. The helicity state of a particle with mass may change under an arbitrary boost. In fact, as shown in Sec. 9-9, it will change if the boost is not along the direction of the three-momentum of the particle.

Exercise 9-6. Use the commutation rules to verify the last equation.

Exercise 9-7. Use the discussion of the last paragraph to show that $|\mathbf{p}\mu\rangle$ is an eigenstate of h.

It is instructive to rewrite Eq. 9-30, with $\phi = 0$ for simplicity, in the form

$$|\mathbf{p}\mu\rangle = R(0, \theta, 0)L(p_z)R^{-1}(0, \theta, 0)R(0, \theta, 0)|0, \mu\rangle$$

Then using Eq. 9-23

$$|\mathbf{p}\mu\rangle = L(\mathbf{p})R(0, \theta, 0)|0, \mu\rangle \tag{9-35}$$

Equation 9-35 says that the helicity state $|\mathbf{p}\mu\rangle$ can be constructed by rotating the state $|0, \mu\rangle$ to "point" along the direction of **p** and giving it a boost to momentum **p**. Comparing with Eq. 9-30, we see that Eq. 9-35 is an alternate, but equivalent, way of preparing a helicity. For most purposes it is simpler to use Eq. 9-30. But Eq. 9-35 perhaps gives more physical insight into the nature of a helicity state.

To conclude this section we note that we cannot use Eq. 9-30 or 9-35 for massless particles such as photons. We must always start from some state $|p_z\mu\rangle$ in which $p_z \neq 0$. This is not a problem as long as we recall that μ is restricted to $\pm s$ and pay attention to phase (see Ref. 9WE).

9-6 Two-Particle, Plane Wave Helicity States. In this section we consider states made up of two free particles called 1 and 2. Their masses are m_1 and m_2; their spins are s_1 and s_2. The particles may be different or identical. The two-particle states can be constructed as direct products with the obvious notation

$$|\mathbf{p}_1\mathbf{p}_2\mu_1\mu_2\rangle = |\mathbf{p}_1\mu_1\rangle|\mathbf{p}_2\mu_2\rangle \tag{9-36}$$

We are particularly interested in two-particle states in the center of mass of the two particles, where $\mathbf{p}_1 + \mathbf{p}_2 = 0$. Our first task is to construct such states, being mindful of phase considerations. We define $\mathbf{p} = \mathbf{p}_1 = -\mathbf{p}_2$; and we say that particle 1 has the direction given by the angles θ and ϕ. We denote a center of mass, two-particle state by

$$|p\theta\phi\mu_1\mu_2\rangle = |\mathbf{p}_1, -\mathbf{p}_1\mu_1\mu_2\rangle \tag{9-37}$$

where $p = |\mathbf{p}|$. The helicity of each single-particle state is defined as positive along the momentum $\mathbf{p}_1$ or $\mathbf{p}_2$ of that state. For example, in Fig. 9-1*a* both μ_1 and μ_2 are positive, but in Fig. 9-1*b* μ_1 is positive and μ_2 is negative. We define the total helicity

$$\mu = \mu_1 - \mu_2$$

As an illustration, consider the state made up of two spin $\frac{1}{2}$ particles.

$$\begin{aligned} \mu_1 &= +\tfrac{1}{2} \quad +\tfrac{1}{2} \quad -\tfrac{1}{2} \quad -\tfrac{1}{2} \\ \mu_2 &= +\tfrac{1}{2} \quad -\tfrac{1}{2} \quad +\tfrac{1}{2} \quad -\tfrac{1}{2} \\ \mu &= 0 \qquad 1 \quad -1 \qquad 0 \end{aligned}$$

Exercise 9-8. Make a similar table for a state consisting of a spin $\frac{1}{2}$ and a spin 1 particle. Also construct such a table if the spin 1 particle is a photon.

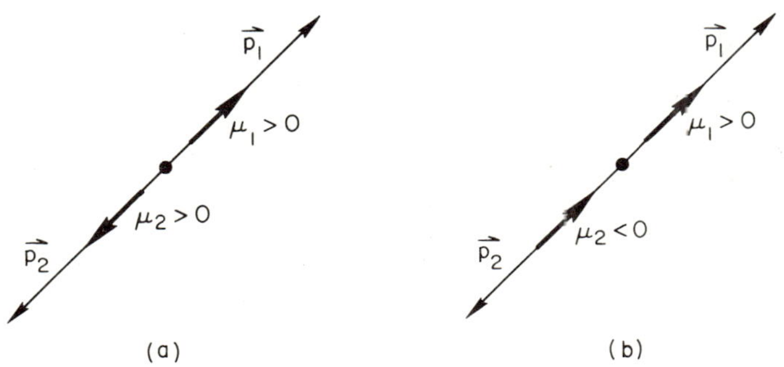

Fig. 9-1 Two-particle, plane wave helicity states.

A phase question arises in the definition of these two-particle, center of mass states when we consider particle 2. We can see the problem most easily by looking at the special case $\mathbf{p}_1 = \mathbf{p}_z$, $\mathbf{p}_2 = -\mathbf{p}_z$; that is, $\theta = 0$, $\phi = 0$. We could define

$$|-p_z\mu_2\rangle = L(-p_z)|0, -\mu_2\rangle \tag{9-38a}$$

Or we could first produce the state $|p_z\mu_2\rangle$ and then rotate it by 180° about the y axis. If we designate the second definition by a prime, we have

$$|-p_z\mu_2\rangle' = e^{-i\pi J_y}L(p_z)|0, \mu_2\rangle \tag{9-38b}$$

If in both these equations we let $p_z \to 0$, Eq. 9-38a yields $|-p_z\mu_2\rangle \to |0, -\mu_2\rangle$ and Eq. 9-38b yields $|-p_z\mu_2\rangle \to (-1)^{s_2-\mu_2}|0, -\mu_2\rangle$. Here we have used

$$d^s_{\mu',\mu}(\pi) = (-1)^{s+\mu'}\delta_{-\mu',\mu}$$

Thus the two definitions differ by the phase $(-1)^{s_2-\mu_2}$. We can adopt either definition or change one definition to make the two definitions consistent. We follow Jacob and Wick, changing Eq. 9-38b to

$$|-p_z\mu_2\rangle = (-1)^{s_2-\mu_2}e^{-i\pi J_y}L(p_z)|0, \mu_2\rangle \tag{9-38c}$$

Then the phase that appears when p_z goes to zero will cancel. Other authors use other conventions.

The orthonormal properties of the two-particle, plane wave helicity state follow directly from Eq. 9-32. In any Lorentz frame we have

$$\begin{aligned}\langle \mathbf{p}'_1\mathbf{p}'_2\mu'_1\mu'_2|\mathbf{p}_1\mathbf{p}_2\mu_1\mu_2\rangle &= \langle \mathbf{p}'_1\mu'_1|\mathbf{p}_1\mu_1\rangle\langle \mathbf{p}'_2\mu'_2|\mathbf{p}_2\mu_2\rangle \\ &= (2\pi)^6 2E_1 2E_2\delta^3(\mathbf{p}'_1 - \mathbf{p}_1)\delta^3(\mathbf{p}'_2 - \mathbf{p}_2)\delta_{\mu'_1\mu_1}\delta_{\mu'_2\mu_2}\end{aligned} \tag{9-39}$$

But we are restricting our considerations to center of mass, two-particle, plane wave helicity states. The six-dimensional Dirac δ function in Eq. 9-39 can then be rewritten in the following way. If P is the total four-momentum of the two particles, in the center of mass we have

$$\begin{aligned}2E_1 2E_2\delta^3(\mathbf{p}'_1 - \mathbf{p}_1)\delta^3(\mathbf{p}'_2 - \mathbf{p}_2) &= \frac{4\sqrt{s}}{p}\,\delta^4(P' - P) \\ &\quad \times \delta(\cos\theta' - \cos\theta)\delta(\phi' - \phi)\end{aligned} \tag{9-40}$$

where $p = |\mathbf{p}_1| = |\mathbf{p}_2|$.

Exercise 9-9. Derive Eq. 9-40.

Thus in the center of mass Eq. 9-39 becomes

$$\langle p'\theta'\mu_1'\mu_2'|p\theta\phi\mu_1\mu_2\rangle = (2\pi)^6 \frac{4\sqrt{s}}{p}\,\delta^4(P'-P)\delta(\cos\theta'-\cos\theta) \times \delta(\phi'-\phi)\delta_{\mu_1'\mu_1}\delta_{\mu_2'\mu_2} \qquad (9\text{-}41)$$

The four-dimensional Dirac δ function in Eq. 9-41 is closely related to the Dirac δ function that appears in Eq. 6-24 in the definition of the T-matrix. It is thus significant. However, since it is convenient not to carry it through the calculations that follow, we write $|p\theta\phi\mu_1\mu_2\rangle$ as the direct product of two states $|\theta\phi\mu_1\mu_2\rangle$ and $|P\rangle$ where $|P\rangle$ depends only on the four-momentum P:

$$|p\theta\phi\mu_1\mu_2\rangle = (2\pi)^3 \left[\frac{4\sqrt{s}}{p}\right]^{1/2} |\theta\phi\mu_1\mu_2\rangle|P\rangle \qquad (9\text{-}42)$$

The orthonormal properties are

$$\langle P'|P\rangle = \delta^4(P'-P) \qquad (9\text{-}43)$$

and

$$\langle\theta'\phi'\mu_1'\mu_2'|\theta\phi\mu_1\mu_2\rangle = \delta(\cos\theta'-\cos\theta)\delta(\phi'-\phi)\delta_{\mu_1'\mu_1}\delta_{\mu_2'\mu_2} \qquad (9\text{-}44)$$

This notation is a little confusing because $|\theta\phi\mu_1\mu_2\rangle$ depends implicitly on $p = |\mathbf{p}_1| = |\mathbf{p}_2|$ in the center of mass. The absence of a momentum parameter in the notation indicates only that the motion of the center of mass has been removed. The normalizations constants in Eq. 9-42 are not unique or profound, they are only convenient and conventional.

9-7 Two-Particle, Angular Momentum Helicity States. The two-particle plane wave helicity states $|\theta\phi\mu_1\mu_2\rangle$ are not eigenstates of $\mathbf{J}^2$ or even of J_z. We are going to construct in the center of mass a two-particle helicity state which is an eigenfunction of $\mathbf{J}^2$ and J_z. We first observe that p, μ_1, μ_2, and μ are invariant to a rotation in the center of mass. Therefore, they can be fixed along with the quantum numbers J and M to specify a state $|JM\mu_1\mu_2\rangle$. The state is defined by

$$\mathbf{J}^2|JM\mu_1\mu_2\rangle = (J)(J+1)|JM\mu_1\mu_2\rangle$$

$$J_z|JM\mu_1\mu_2\rangle = M|JM\mu_1\mu_2\rangle$$

$$R(\alpha\beta\gamma)|JM\mu_1\mu_2\rangle = \sum_{M'} D^J_{M'M}(\alpha\beta\gamma)|JM'\mu_1\mu_2\rangle \qquad (9\text{-}45)$$

and the normalization

$$\langle J'M'\mu_1'\mu_2'|JM\mu_1\mu_2\rangle = \delta_{JJ'}\delta_{MM'}\delta_{\mu_1\mu_1'}\delta_{\mu_2\mu_2'} \qquad (9\text{-}46)$$

Our problem is to relate this state to the plane wave state $|\theta\phi\mu_1\mu_2\rangle$. But the manipulations that must be carried out to solve this problem are tedious. Therefore, to save space, we refer the reader to Refs. 9WE and 9MA, and we give only the solution to the problem in the form of the next two equations.

$$|\theta\phi\mu_1\mu_2\rangle = \sum_{JM} \left[\frac{2J+1}{4\pi}\right]^{1/2} D^J_{M\mu}(\phi, \theta, -\phi)|JM\mu_1\mu_2\rangle \tag{9-47}$$

$$|JM\mu_1\mu_2\rangle = \left[\frac{2J+1}{4\pi}\right]^{1/2} \int_0^{2\pi} d\phi \int_0^{\pi} \sin\theta\, d\theta\, [D^{J*}_{M\mu}(\phi, \theta, -\phi)|\theta\phi\mu_1\mu_2\rangle] \tag{9-48}$$

$$\mu = \mu_1 - \mu_2; \qquad D^{J*}_{M\mu} = \text{complex conjugate of } D^J_{M\mu}$$

Thus the two-particle plane wave helicity state $|\theta\phi\mu_1\mu_2\rangle$ is an infinite sum of two-particle angular momentum helicity states. The summation is over all integral or half-integral values of J. And for a particular value of J, there is a sum over the M values allowed by that J. But the helicity of each particle is the same throughout the summation and is equal to the particle's helicity in the plane wave state. Thus Eq. 9-47 is a more general form of the partial wave expansion of a plane wave, introduced in Ch. 3. Now turning to Eq. 9-48 we see that the angular momentum state $|JM\mu_1\mu_2\rangle$ is an integral over the spectrum of plane wave states. Thus the integral is weighted by $D^{J*}_{M\mu}(\phi, \theta, -\phi)$, and $D^{J*}_{M\mu}$ is a generalization of the spherical harmonic functions.

As a simple example, consider two spin zero particles in a plane wave state moving along the $+z$ axis. Then $|\phi\theta\mu_1\mu_2\rangle = |0000\rangle$ and Eq. 9-47 reduces to

$$|0000\rangle = \sum_J \left[\frac{2J+1}{4\pi}\right]^{1/2} |J000\rangle, \qquad J \text{ integral} \tag{9-49}$$

Comparing this with

$$e^{ikz} = \sum_l i^l j_l(kr)[2l+1]P_l(\cos\theta)$$

and ignoring the radial behavior, we see that there is a great similarity. The state $|J000\rangle$ represents the Jth partial wave implicitly, just as the Legendre polynomial $P_J(\cos\theta)$ represents it explicitly. Thus each $|J000\rangle$ state represents a partial wave.

Exercise 9-10. Work out the equivalent of Eq. 9-49 for a system consisting of a spin $\frac{1}{2}$ and a spin zero particle.

One final remark about this formalism. As the next chapter reveals, we never represent the $|JM\mu_1\mu_2\rangle$ states by explicit functions of θ and ϕ. Rather, the entire θ and ϕ dependence appears in the D functions through the use of Eqs. 9-47 and 9-48.

9-8 The Wigner Rotation. We now turn to the effect on a one-particle helicity state of an arbitrary *restricted homogeneous Lorentz transformation*. As discussed in Sec. 6-9, such a transformation is composed of a sequence of spatial rotations and boosts. We know that a one-particle helicity state is invariant under a spatial rotation. Therefore, the matter reduces to the effect of an arbitrary boost on a one-particle helicity state. To study this effect we must first carry out a calculation

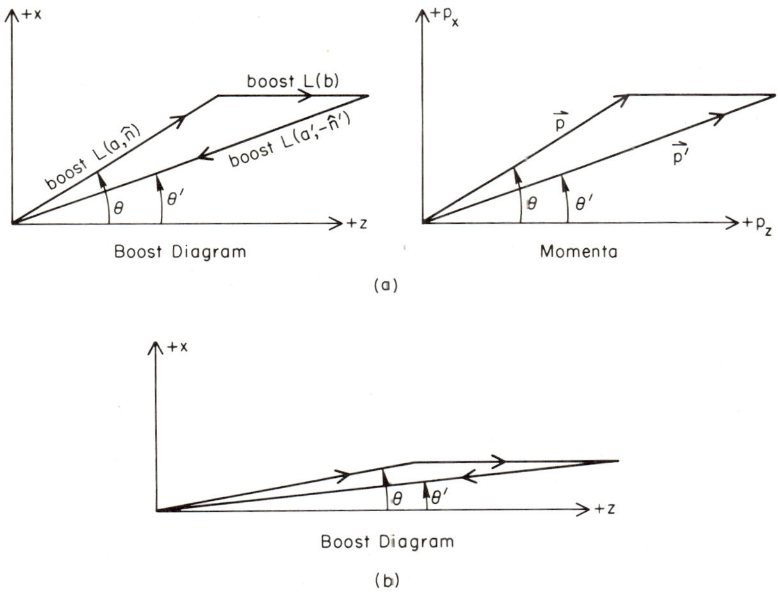

Fig. 9-2 The Wigner rotation.

in classical relativistic kinematics, which leads to the subject of this section—the concept of the Wigner rotation. In the next section we apply this concept to the helicity formalism.

For the following discussion we can restrict our considerations to boosts in the x–z plane without loss of generality. We begin with a classical object, of mass m, at rest. We then carry out a sequence of boosts illustrated in momentum space in Fig. 9-2a. First the boost $L(a, \hat{n})$ gives the object a momentum $\mathbf{p}$, where θ is the angle between $\hat{n}$ and the $+z$ axis. Next the object is given a boost $L(b)$ *parallel* to the z axis. The momentum is now $\mathbf{p}'$. Finally, the object is returned to rest by a boost $L(a', -\hat{n}) = L(-\mathbf{p}')$, where θ' is the angle between $\hat{n}'$ and the $+z$ axis.

Exercise 9-11. Show that any sequence of three boosts that take an object from rest back to rest can be oriented as in Fig. 9-2a.

If all the boosts are nonrelativistic, when the object is returned to rest it will have the same spatial orientation it originally possessed. But if the boosts are relativistic, the object when returned to rest will in general be rotated with respect to its original orientation. This is often called a Wigner rotation (9WIG). To demonstrate this effect without too much messy algebra involving hyperbolic and trigonometric functions, we consider the special case when the angles θ and θ' are infinitesimally small (Fig. 9-2*b*). That is, we calculate to first order in θ and θ'.

We wish to study the transformation of the four-momentum P

$$P \to P' = L(a', -\hat{n}')L(b)L(a, \hat{n})P \tag{9-50}$$

Equation 9-50 can be rewritten as

$$L(a', \hat{n}')P' = L(b)L(a, \hat{n})P \tag{9-51}$$

From Eqs. 9-24 and 9-25 we obtain to first order in θ and θ'

$$L(a'\hat{n}') = \begin{pmatrix} \cosh a' & \theta' \sinh a' & 0 & \sinh a' \\ \theta' \sinh a' & 1 & 0 & \theta'(\cosh a' - 1) \\ 0 & 0 & 1 & 0 \\ \sinh a' & \theta'(\cosh a' - 1) & 0 & \cosh a' \end{pmatrix} \tag{9-52}$$

$$L(b)L(a, \hat{n})$$

$$= \begin{pmatrix} \cosh(a+b) & \theta[\sinh(a+b) - \sinh b] & 0 & \sinh(a+b) \\ \theta \sinh a & 1 & 0 & \theta(\cosh a - 1) \\ 0 & 0 & 1 & 0 \\ \sinh(a+b) & \theta[\cosh(a+b) - \cosh b] & 0 & \cosh(a+b) \end{pmatrix} \tag{9-53}$$

To obtain the condition for returning the object to rest, we set $P' = P = (m, 0, 0, 0)$. Then we have

$$a' = a + b \tag{9-54}$$

$$\theta' \sinh a' = \theta \sinh a \tag{9-55}$$

Equation 9-54 is simply the rule that the boost parameters add for parallel boosts.

Exercise 9-12. Derive Eqs. 9-51 through 9-55.

But even though Eqs. 9-54 and 9-55 make the elements of the first column of the matrices in Eqs. 9-52 and 9-53 equal, they do *not* make all the elements equal. For example, comparing the elements L_{01} gives

$$\theta' \sinh a' = \theta \sinh a \neq \theta[\sinh(a+b) - \sinh b]$$

Exercise 9-13. Show that the two matrices are completely equal if all the boosts are nonrelativistic.

The difference between the matrices amounts to a spatial rotation—the Wigner rotation. To see this explicitly, consider an infinitesimal rotation $R(0, \omega, 0)$ and replace Eq. 9-50 by

$$R(0, \omega, 0)P' = L(a', -\hat{n}')L(b)L(a, \hat{n})P$$

Then if we demand $P' = P$ for *all* P, we obtain

$$R(0, \omega, 0) = L(a', -\hat{n}')L(b)L(a, \hat{n}) \tag{9-56}$$

$$L(a', \hat{n}')R(0, \omega, 0) = L(b)L(a, \hat{n}) \tag{9-57}$$

To first order in ω and θ', the left-hand side of Eq. 9-57 is

$$L(a', \hat{n}')R(0, \omega, 0) =$$

$$\begin{bmatrix} \cosh a' & (\theta' - \omega)\sinh a' & 0 & \sinh a' \\ \theta' \sinh a' & 1 & 0 & \omega + \theta'(\cosh a' - 1) \\ 0 & 0 & 1 & 0 \\ \sinh a' & -\theta' + (\theta' - \omega)\cosh a' & 0 & \cosh a' \end{bmatrix} \tag{9-58}$$

Equating the matrices in Eqs. 9-58 and 9-53, we obtain the rotation angle

$$\omega = \theta \frac{[\sinh a + \sinh b - \sinh (a + b)]}{\sinh (a + b)} \tag{9-59}$$

Exercise 9-14. Show that Eqs. 9-54, 9-55, and 9-59 make the matrices in Eqs. 9-53 and 9-58 completely equal.

Thus the sequence of boosts on the right-hand side of Eq. 9-56 is equivalent to a simple rotation by an angle ω about the y axis. For further discussion, refer to Refs. 9WIG and 9TA.

9-9. Behavior of One-Particle Helicity States Under an Arbitrary Boost. We now apply the concept of the Wigner rotation to one-particle helicity states. The manipulations to be performed are actually simpler than those of the previous section because we deal with the boost $R(0, \theta, 0)L(a)$ rather than with the boost $R(0, \theta, 0)L(a)R^{-1}(0, \theta, 0)$. Therefore, we need no longer restrict ourselves to very small angles.

Returning to Fig. 9-2a, we prepare the helicity state

$$|\mathbf{p}\mu\rangle = R(0, \theta, 0)L(p_z)|0, \mu\rangle, \qquad |\mathbf{p}| = m \sinh a \tag{9-60}$$

where the angle θ between **p** and the $+z$ axis is arbitrary. Next we give the state $|\mathbf{p}\mu\rangle$ a boost $L(b)$ parallel to the z axis. Then we have

$$|x\rangle = L(b)|\mathbf{p}\mu\rangle = L(b)R(0, \theta, 0)L(p_z)|0, \mu\rangle \tag{9-61}$$

and the momentum is

$$\mathbf{p}' = L(b)\mathbf{p}, \qquad |\mathbf{p}'| = m \sinh a' \tag{9-62}$$

If not for the Wigner rotation, we could prepare the state $|x\rangle$ simply by $R(0, \theta', 0)L(p_z')|0, \mu\rangle$. But the Wigner rotation requires

$$|x\rangle = R(0, \theta', 0)L(p_z')R(0, \omega, 0)|0, \mu\rangle \tag{9-63}$$

Then to find ω we require (cf. Eq. 9-57)

$$R(0, \theta', 0)L(p_z')R(0, \omega, 0) = L(b)R(0, \theta, 0)L(p_z) \tag{9-64}$$

Using Eq. 9-24 we calculate the matrices on both sides of Eq. 9-64 and equate the respective elements. First we find the two conditions that satisfy Eq. 9-62.

$$\cosh a' = \cosh b \cosh a + \cos\theta \sinh b \sinh a \tag{9-65}$$

$$\sin\theta' \sinh a' = \sin\theta \sinh a \tag{9-66}$$

Equations 9-65 and 9-66 are the extensions of Eqs. 9-54 and 9-55 to non-infinitesimal angles. Finally, the Wigner rotation angle is given by

$$\sin\omega = \frac{\sinh b}{\sinh a'} \sin\theta = \frac{m \sinh b}{|\mathbf{p}'|} \sin\theta \tag{9-67}$$

Exercise 9-15. Derive Eqs. 9-65 to 9-67.

Exercise 9-16. Discuss the dependence of ω on θ and on b. Draw pictures and work out numerical examples illustrating this dependence. Note that ω for a one-particle helicity state (Eq. 9-67) differs from ω for a four-vector (Eq. 9-59). Why?

To understand the significance of those considerations, we write Eqs. 9-61 and 9-63 in the forms

$$|0, \mu\rangle \to |x\rangle = [L(b)R(0, \theta, 0)L(p_z)]|0, \mu\rangle = L(b)|\mathbf{p}\mu\rangle$$

$$\begin{aligned}|0, \mu\rangle \to |x\rangle &= R(0, \theta', 0)L(p_z')R(0, \omega, 0)|0, \mu\rangle \\ &= R(0, \theta', 0)L(p_z') \sum_{\mu'} d^s_{\mu'\mu}(\omega)|0, \mu'\rangle \\ &= \sum_{\mu'} d^s_{\mu'\mu}(\omega)|p'\mu'\rangle\end{aligned}$$

Thus

$$L(b)|\mathbf{p}\mu\rangle = \sum_{\mu'} d^s_{\mu'\mu}(\omega)|p'\mu'\rangle \tag{9-68}$$

Equation 9-68 defines how a one-particle helicity state is changed by an arbitrary boost $L(b)$. We see that the state is transformed from a single helicity state to a "mixed" helicity state. The coefficients in the "mixed" helicity state are given by the function $d^s_{\mu'\mu}(\omega)$ where ω is the Wigner rotation produced by the arbitrary boost. There is also a change in the three-momentum given by Eq. 9-62. But the interesting change is the conversion of a single helicity state into a "mixed" helicity state. Thus we have finally demonstrated explicitly the truth of the statement made immediately before Ex. 9-6. Under an arbitrary Lorentz transformation, the helicity of a particle may change. For an interesting non-Euclidian geometric interpretation see Refs. 9MA, 9WE, 9WI.

Exercise 9-17. Use Eq. 9-67 to show that ω is always zero for a massless particle. Hence show that the helicity of a massless particle is preserved under an arbitrary boost.

To conclude this section we present a common application of Eq. 9-68 to elastic scattering. In the laboratory frame 1 is the incident particle and 2 is the target particle. In the center of mass frame the magnitude of the three-momentum is p^* and the scattering angle θ^*. Suppose that *after* the scatter, particle 1 in the *center of mass* system is in the helicity state $|\mathbf{p}^*\mu\rangle$. The problem is to calculate the helicity state of particle 1 after the scatter in the *laboratory system*. The transformation from the center of mass system to the laboratory system is accomplished by a boost $L(b)$. Therefore, by Eq. 9-68, we have

$$L(b)|\mathbf{p}^*\mu\rangle = \sum_{\mu'} d^s_{\mu'\mu}(\omega_1)|\mathbf{p}'\mu'\rangle$$

The value of ω_1 is given by

$$\sin\omega_1 = \frac{m_1 p^*}{m_2 p_1}\sin\theta^* \tag{9-69}$$

where $\mathbf{p}_1$ is the laboratory momentum of particle 1 after the scatter and m_i is the mass of particle i. If θ^* is not close to 180° at high energies, Eq. 9-69 simplifies to

$$\sin\omega_1 = \frac{2m_1}{\sqrt{s}}\tan\tfrac{1}{2}\theta^* \qquad s\to\infty \tag{9-70}$$

a beautiful result. Finally when $s \gg t$ we obtain

$$\sin \omega_1 = \frac{2m_1\sqrt{|t|}}{s} \qquad s \to \infty \tag{9-71}$$

Equations 9-69 through 9-71 indicate the very important fact that if $\theta^* \neq 0$, the helicity of the scattered particle changes in the transformation from the center of mass frame to the laboratory frame. Equation 9-71 shows that for fixed t, as the energy increases the Wigner rotation, ω_1 decreases. And if $[m_1\sqrt{|t|}/s]$ is sufficiently small, the helicity is approximately invariant.

Exercise 9-18. Derive Eqs. 9-69 through 9-71.

Exercise 9-19. Suppose that $m_1\sqrt{t}/s = 0.01$. Calculate the coefficients in Eq. 9-68 for a spin 1 particle and for a spin 3/2 particle.

Exercise 9-20. Show that the Wigner rotation ω_2 for the transformation of the helicity states of the target particle (particle 2) from the center of mass frame to the laboratory frame is

$$\sin \omega_2 = \frac{2m_2}{(4m_2{}^2 + |t|)^{1/2}}, \qquad s \to \infty \tag{9-72}$$

Hence even at very high energies the helicity of particle 2 under this transformation is not invariant.

REFERENCES

9JA M. Jacob and G. C. Wick, *Ann. Phys. (N.Y.)* 7, 404 (1959). This paper is condensed, hence difficult to read as an introductory paper.

9MA A. D. Martin and T. D. Spearman, *Elementary Particle Physics* (North-Holland, Amsterdam, 1970).

9RO M. E. Rose, *Elementary Theory of Angular Momentum* (Wiley, New York, 1957).

9TA E. F. Taylor and J. W. Wheeler, *Spacetime Physics* (Freeman, San Francisco, 1966).

9WE J. Werle, *Relativistic Theory of Reactions* (North-Holland, Amsterdam, 1966).

9WI G. C. Wick, *Ann. Phys. (N.Y.)* 18, 65 (1962) (Appendix).

9WIG E. P. Wigner, *Rev. Mod. Phys.* **29**, 255 (1957).

CHAPTER

10

Properties of Helicity Amplitudes

10-1 Introduction. In this chapter we derive the general properties of two-body helicity amplitudes that depend on the invariance of the T-matrix under spatial rotation, space inversion, and time reversal. No extensive applications of these properties are made in this chapter. Rather, those applications are discussed in Chs. 11 through 13; other applications appear later in the book. In Sec. 10-2 the general equations for the two-body helicity amplitudes and differential cross sections are derived. The limitations imposed on these amplitudes by invariance under space inversion and time reversal are developed in Secs. 10-3 and 10-4 and applied in Sec. 10-5. In Secs. 10-6 and 10-7 we discuss the simple and beautiful properties of helicity amplitudes for a scattering angle of zero and for small scattering angles.

In Sec. 10-8 we use the helicity formalism to generalize the eikonal approximation to elastic scattering, an approximation introduced in Ch. 2 and in Sec. 3-4.

We extend the impact parameter representation to particles with spin and to inelastic as well as elastic two-body processes. This extended impact parameter representation is used in Ch. 13 and in later chapters to correct inelastic scattering calculations for initial and final state absorption.

10-2 Helicity Amplitudes and the Partial Wave Expansion. The general two-body reaction is $a + b \to c + d$. Particle i has spin s_i and mass m_i, where m_i may be zero. This reaction includes the most general two-body inelastic process because particles a, b, c, and d may all differ from each other. Of course elastic scattering, $a + b \to a + b$ is included as a special case. Most important we also include in this treatment quasi-two-body reactions in which particles c or d, or both c and d, are strong interaction resonances. An example of such a reaction is $K^+ + p \to K^{*+} + p$.

Our object is to apply the helicity formalism to the matrix element T_{fi} introduced in Eq. 6-24; T_{fi} is related to the two-body differential cross section through equations like 6-39 and 6-43. We can make maximum use of the helicity formalism if we consider only the *center of mass* system. Unless we specifically state otherwise, all the considerations in this chapter are restricted to the cms.

The initial state $|i\rangle$ of particles a and b is described by a center of mass, two-particle, plane wave helicity state in which particle a moves in the $+z$ direction and particle b moves in the $-z$ direction. The magnitude of the three-momentum of a or b is p_i. Hence, using the notation of Eq. 9-42, we have

$$|i\rangle = |p_i 0 0 \mu_a \mu_b\rangle \tag{10-1a}$$

The final state $|f\rangle$ of particles c and d is described by a similar state except that particle c is moving in the direction given by the angles (θ, ϕ), and particle d is moving in the opposite direction. The magnitude of the center of mass three-momentum of c or d is p_f. Hence

$$\langle f| = \langle p_f \theta \phi \mu_c \mu_d| \tag{10-1b}$$

For later use we define

$$\mu_i = \mu_a - \mu_b, \qquad \text{and} \qquad \mu_f = \mu_c - \mu_d \tag{10-2}$$

To evaluate $\langle f|T|i\rangle$ in Eq. 6-24 we must extract $(2\pi)^4\delta^4(P' - P)$. This is easily done using Eqs. 9-43 and 9-44. Thus we obtain

$$T_{fi} = \frac{(2\pi)^2 4\sqrt{s}}{[p_i p_f]^{1/2}} \langle \theta \phi \mu_c \mu_d | T | 0 0 \mu_a \mu_b \rangle \tag{10-3}$$

Here T_{fi} is expressed in terms of two-particle, plane wave helicity states, but to make use of angular momentum considerations we must express T_{fi} in terms of the two-particle, angular momentum helicity states discussed in Sec. 9-7. This can be done through Eq. 9-47, and we have

$$T_{fi} = \frac{(2\pi)^2 4\sqrt{s}}{[p_i p_f]^{1/2}} \sum_{J_f M_f} \sum_{J_i M_i} \left\{ \left[\frac{(2J_f + 1)(2J_i + 1)}{16\pi^2} \right]^{1/2} D^{J_f *}_{M_f \mu_f}(\phi, \theta, -\phi) \right. \tag{10-4}$$

$$\left. \times\, D^{J_i}_{M_i \mu_i}(0, 0, 0) \langle J_f M_f \mu_c \mu_d | T | J_i M_i \mu_a \mu_b \rangle \right\}$$

We now obtain our reward for having introduced angular momentum helicity states. In the matrix element of Eq. 10-4, the conservation of total angular momentum demands that $J_f = J_i = J$ and $M_f = M_i = M$. Also, the matrix element cannot depend on M if the S-matrix is rotationally invariant.

Exercise 10-1. Justify the last statement. Sketch a scattering problem in non-relativistic quantum mechanics in which this statement would not be true.

Thus the matrix element in Eq. 10-4 reduces to $\langle JM\mu_c\mu_d | T | JM\mu_a\mu_b\rangle$, which we write in the form $\langle \mu_c\mu_d | \tilde{T}^J(s) | \mu_a\mu_b\rangle$. We emphasize that this matrix element depends on μ_a, μ_b, μ_c, and μ_d as well as on J. It does *not* depend on M. We have placed the tilde over T^J because there is an overall constant associated with the matrix element which we do not determine until the next paragraph. Returning to Eq. 10-4, the J and M equalities we have just discussed lead to

$$T_{fi} = \frac{4\pi\sqrt{s}}{[p_i p_f]^{1/2}} \sum_J \left[(2J + 1)\langle \mu_c\mu_d | \tilde{T}^J(s) | \mu_a\mu_b\rangle\, e^{i(\mu_i - \mu_f)\phi}\, d^J_{\mu_i \mu_f}(\theta) \right] \tag{10-5}$$

The derivation of Eq. 10-5 started from Eq. 6-24 in which the T-matrix is a function of continuous parameters such as the momenta and angles. Yet in arriving at Eq. 10-5 we have obtained a T-matrix that is a function of discrete parameters—the total angular momentum and the helicities—as well as a function of the total energy. This discrete parameter T-matrix was introduced in Ch. 3, used in Ch. 5, and treated in a general way in Eq. 6-8. But we have made no effort to define the continuous parameter T-matrix of Eq. 6-24 and the discrete parameter T-matrix of Eq. 6-8 to normalize them in the same way. Therefore, we will not be surprised if the discrete parameter $\tilde{T}^J$-matrix is not normalized in accord with our usage of T in Ch. 3. To bring them into agreement we use Eq. 6-42 to obtain the center of mass f amplitude when all particles have spin zero. Then

$$f_{\text{el}}(s, \theta, \phi) = \frac{1}{2p_i} \sum_J (2J + 1)\langle 00 | \tilde{T}^J(s) | 00\rangle P_J(\cos\theta) \tag{10-6}$$

where we have used Ex. 9-1. Now in the notation of Ch. 3 we have

$$f_{\text{el}}(s, \theta, \phi) = \frac{1}{p_i} \sum_J (2J + 1) T_J P_J(\cos \theta) \tag{10-7}$$

where l has been replaced by J. Comparing Eqs. 10-6 and 10-7 we see that we can attain the notation of Ch. 3 if we define

$$\langle \mu_c \mu_d | T^J(s) | \mu_a \mu_b \rangle = 1/2 \langle \mu_c \mu_d | \tilde{T}^J(s) | \mu_a \mu_b \rangle \tag{10-8}$$

With the normalization adjustment of Eq. 10-8, we can now write

$$f_{\mu_c \mu_d; \mu_a \mu_b}(s, \theta, \phi) = \left(\frac{1}{p_i^*}\right) e^{i(\mu_i - \mu_f)\phi} \sum_J [(2J + 1) \langle \mu_c \mu_d | T^J(s) | \mu_a \mu_b \rangle$$

$$\times\ d^J_{\mu_i \mu_f}(\theta)]; \qquad \mu_f = \mu_c - \mu_d, \qquad \mu_i = \mu_a - \mu_b \tag{10-9a}$$

$$T_{\mu_c \mu_d; \mu_a \mu_b}(s, \theta, \phi) = 8\pi\sqrt{s} \left(\frac{p_i^*}{p_f^*}\right)^{1/2} [f_{\mu_c \mu_d; \mu_a \mu_b}(s, \theta, \phi)] \tag{10-9b}$$

In Eq. 10-9a, the summation begins at $J = \mu_i$ or $J = \mu_f$, depending on which is larger. Equation 6-39 becomes

$$\left(\frac{d\sigma}{d\Omega}\right)_{\mu_c \mu_d; \mu_a \mu_b}(s, \theta, \phi) = \frac{1}{64\pi^2 s} \left(\frac{p_f^*}{p_i^*}\right) |T_{\mu_c \mu_d; \mu_a \mu_b}(s, \theta, \phi)|^2 \tag{10-10}$$

All quantities in this set of equations are in the cms.

We conclude this section by writing the discrete parameter S-matrix in the helicity formalism. Using Eq. 6-8 gives

$$\langle \mu_c \mu_d | T^J(s) | \mu_a \mu_b \rangle = \frac{\langle \mu_c \mu_d | S^J(s) - I | \mu_a \mu_b \rangle}{2i} \tag{10-11a}$$

$$= \frac{\langle \mu_c \mu_d | S^J(s) | \mu_a \mu_b \rangle - \delta_{fi}}{2i}$$

where

$$\delta_{fi} = \delta_{ca}\, \delta_{db}\, \delta_{\mu_c \mu_a}\, \delta_{\mu_d \mu_b} \tag{10-11b}$$

In the last expression $\delta_{ca}\, \delta_{db}$ means that for this term to be nonzero, c and a must be the same type of particle and d and b must be the same type of particle. Thus δ_{fi} can only be nonzero for elastic scattering.

10-3 Space Inversion Properties of Helicity Amplitudes. In this section we want to determine the behavior under space inversion of helicity states and helicity amplitudes. The calculation is somewhat tedious and we carry out only the first part of the calculation as an example of the method. We use $\mathscr{P}$ to denote the parity operator, and η_i denotes the intrinsic parity of particle i. Let us first recall (Sec. 3-5) how space inversion affects nonrelativistic spin states. For a rest frame spin state $|s, m\rangle$, we have simply

$$\mathscr{P}|s, m\rangle = \eta|s, m\rangle \tag{10-12a}$$

For two nonrelativistic particles coupled through the spin-orbit formalism, we have

$$\mathscr{P}|s_1 s_2 slJM\rangle = \eta_1\eta_2(-1)^l|s_1 s_2 slJM\rangle \tag{10-12b}$$

Thus we see that the angular momentum states of particles at rest or in nonrelativistic motion transform very simply under space inversion. In particular, these states are eigenstates of the parity operator $\mathscr{P}$. However, we shall find that helicity states do *not* transform as simply under space inversion. And in particular, the simplest helicity states are *not* eigenstates of $\mathscr{P}$. This is the price we have to pay for the convenience of using the helicity states.

Now let us consider the helicity state $|\mathbf{p}\mu\rangle$. Under $\mathscr{P}$, $\mathbf{p} \to -\mathbf{p}$, but spin operators and spin components do *not* change sign. Therefore $\mu \to -\mu$.

Exercise 10-2. Justify these last statements.

Thus, to within an undetermined phase ξ, we write

$$\mathscr{P}|\mathbf{p}\mu\rangle = \xi|-\mathbf{p}, -\mu\rangle \tag{10-13}$$

To determine ξ we must use more formal considerations. Since ξ does not depend on the direction of $\mathbf{p}$, we can work with the helicity state $|p_z, \mu\rangle$. We use the equalities

$$L(p_z) = \mathscr{P}L(-p_z)\mathscr{P}, \qquad L(p_z) = e^{-i\pi J_y}L(-p_z)\, e^{i\pi J_y} \tag{10-14}$$

and some relations from App. C. Then we have

$$\begin{aligned}
\mathscr{P}|p_z, \mu\rangle &= \mathscr{P}L(p_z)|0, \mu\rangle \\
&= L(-p_z)\mathscr{P}|0, \mu\rangle \\
&= \eta\, e^{i\pi J_y}L(p_z)\, e^{-i\pi J_y}|0, \mu\rangle \\
&= \eta\, e^{i\pi J_y}L(p_z)(-1)^{s-\mu}|0, -\mu\rangle \\
&= \eta(-1)^{s-\mu}\, e^{i\pi J_y}|p_z, -\mu\rangle \\
&= \eta(-1)^{s-\mu}|-p_z, -\mu\rangle
\end{aligned} \tag{10-15}$$

Exercise 10-3. Fill in the missing steps in the derivation of Eq. 10-15.

Thus the additional phase change is $(-1)^{s-\mu}$. Finally, it is important to note from Eq. 10-13 or 10-15 that a single-particle helicity state is *not* a parity eigenstate. Contrast this statement with Eq. 10-12a.

For the application of parity considerations to helicity amplitudes, we need to know the effect of space inversion on the two-particle, angular momentum helicity state $|JM\mu_1\mu_2\rangle$. This is calculated from Eqs. 10-15 and 9-48. But the calculation is tedious (10WE, 10MA), and we only give the result:

$$\mathscr{P}|JM\mu_1\mu_2\rangle = \eta_1\eta_2(-1)^{s_1+s_2-J}|JM, -\mu_1, -\mu_2\rangle \tag{10-16}$$

The change of μ_i to $-\mu_i$ and the phase $(-1)^{s_i}$ comes from Eq. 10-15; J and M do not change under $\mathscr{P}$ because the total angular momentum operator $\mathbf{J}$ commutes with $\mathscr{P}$. Thus we can understand all of Eq. 10-16 except for the phase $(-1)^{-J}$, which comes from the detailed calculation where a $d^J_{M\mu}(\pi)$ appears.

Although the state $|JM\mu_1\mu_2\rangle$ is not an eigenstate of $\mathscr{P}$, we can easily construct eigenstates of $\mathscr{P}$ by the definition

$$|JM\mu_1\mu_2\rangle_\pm = \frac{1}{\sqrt{2}}\left[|JM\mu_1\mu_2\rangle \pm \eta_1\eta_2(-1)^{s_1+s_2-J}|JM, -\mu_1, -\mu_2\rangle\right] \tag{10-17a}$$

$$\mathscr{P}|JM\mu_1\mu_2\rangle_\pm = \pm|JM\mu_1\mu_2\rangle \tag{10-17b}$$

Exercise 10-4. Verify Eqs. 10-17. Consider the case of two different spin $\frac{1}{2}$ particles combined to form a two particle, angular momentum, helicity states. An example of such a pair is a proton and a Λ^0 hyperon. We demand that the two particles be different to avoid the symmetry conditions required by Fermi-Dirac statistics.

Our final tasks in this section are to use Eq. 10-16 to calculate the effects of the parity transformation on the helicity amplitudes. We restrict our considerations to strong and electromagnetic interactions. Then, as discussed in Sec. 6-12, the S- and T-matrices are invariant under the parity transformation. Explicitly, we have

$$\begin{aligned}\langle\mu_c\mu_d|T^J|\mu_a\mu_b\rangle &= \langle\mu_c\mu_d|\mathscr{P}^{-1}T^J\mathscr{P}|\mu_a\mu_b\rangle \\ &= \eta_a\eta_b\eta_c\eta_d(-1)^{s_a+s_b+s_c+s_d-2J}\langle-\mu_c, -\mu_d|T^J|-\mu_a, -\mu_b\rangle\end{aligned}$$

so that

$$\langle-\mu_c, -\mu_d|T^J|-\mu_a, -\mu_b\rangle = \eta_a\eta_b\eta_c\eta_d(-1)^{s_c+s_d-s_a-s_b}\langle\mu_c\mu_d|T^J|\mu_a\mu_b\rangle \tag{10-18}$$

We have used Eq. 10-16 and we have realized that $(-1)^{s_a+s_b-J} = (-1)^{J-s_a-s_b}$ because $s_a + s_b - J$ is an integer. Thus the T^J matrix transforms simply under space inversion. Next it is easy to show, using Eqs. 10-9, that for $\phi = 0$

$$T_{-\mu_c,-\mu_d;-\mu_a,-\mu_b}(s,\theta,0) = \eta_a\eta_b\eta_c\eta_d(-1)^{s_c+s_d-s_a-s_b}(-1)^{\mu_i-\mu_f} \times T_{\mu_c\mu_d;\mu_a\mu_b}(s,\theta,0) \tag{10-19a}$$

An alternate form of Eq. 10-19a for $\phi \neq 0$ is

$$T_{-\mu_c,-\mu_d;-\mu_a,-\mu_b}(s,\theta,\phi) = \eta_a\eta_b\eta_c\eta_d(-1)^{s_c+s_d-s_a-s_b} \times T_{\mu_c\mu_d;\mu_a\mu_b}(s,\theta,\pi-\phi) \tag{10-19b}$$

Exercise 10-5. Complete the derivation leading to Eqs. 10-19.

Thus the complicated manipulations of this section have rewarded us with a simple yet powerful result. The number of independent helicity amplitudes for a reaction is reduced by a factor of roughly 2! This result holds for particles of any spin. Thus it is a very general result, and we use it often. The same result can of course be obtained using the nonrelativistic spin-orbit formalism, but the relationships between the related amplitudes are not so transparent and the relativistic extension must still be made.

10-4 Time-Reversal Properties of Helicity Amplitudes. In this section we summarize the behavior of the helicity states and the helicity amplitudes under time reversal, just as we summarized their behavior under space inversion in the previous section. The manipulations are equally complicated and we give only the results (10WE, 10MA). In this section we use the notation T for the time-reversal operator. Under T, $\mathbf{p} \to -\mathbf{p}$ and $\mathbf{J} \to -\mathbf{J}$. Therefore, the rest frame spin state $|s, m\rangle \to |s, -m\rangle$. There is also a phase introduced here and explicitly

$$\mathrm{T}|s, m\rangle = (-1)^{s-m}|s, -m\rangle \tag{10-20}$$

Other phase choices are possible.

We now take the same route that we followed in our treatment of space inversion. Using $\mathrm{T}L(p_z)\mathrm{T}^{-1} = L(-p_z)$, we obtain

$$\mathrm{T}|p_z, \mu\rangle = (-1)^{2s}|-p_z, \mu\rangle \tag{10-21}$$

Exercise 10-6. Derive Eq. 10-21. Why does μ go to μ under T?

Next we give the effect of T on two-particle, angular momentum helicity states.

$$\mathrm{T}|JM\mu_1\mu_2\rangle = (-1)^{J-M}|J, -M\mu_1\mu_2\rangle \tag{10-22}$$

Equation 10-22 is transparent and we almost could have written it down immediately. The phase change is analogous to that in Eq. 10-20; and the change of M to $-M$ is analogous to the change of m to $-m$ in Eq. 10-20.

Finally, we can calculate the effect of time reversal on the helicity amplitudes. We consider first T^J, writing out the next step in detail to illustrate how the *antiunitary* operator T must be manipulated.

$$\begin{aligned}\langle \mu_c\mu_d|T|\mu_a\mu_b\rangle &= \langle JM\mu_c\mu_d|(\mathrm{T}^{-1}T^\dagger\mathrm{T}\,|JM\mu_a\,\mu_b\rangle)\\ &= (-1)^{J-M}\langle JM\mu_c\mu_d|(\mathrm{T}^\dagger T^\dagger\,|J, -M\mu_c\mu_b\rangle)\\ &= (-1)^{J-M}\langle J, -M\mu_a\mu_b|(T\mathrm{T}|JM\mu_c\mu_d\rangle)\\ &= \langle J, -M\mu_a\mu_b|T|J, -M\mu_c\mu_d\rangle\end{aligned}$$

Therefore

$$\langle \mu_c\mu_d|T^J|\mu_a\mu_b\rangle = \langle \mu_a\mu_b|T^J|\mu_c\mu_d\rangle \tag{10-23a}$$

$$\langle \mu_c\mu_d|S^J|\mu_a\mu_b\rangle = \langle \mu_a\mu_b|S^J|\mu_c\mu_d\rangle \tag{10-23b}$$

Equation 10-23 is a very simple and pleasing result. It says that T^J, hence S^J, are symmetric under time reversal. To put it another way, we have shown that it is possible to select a phase convention in Eq. 10-20 such that the T^J- and S^J-matrices are symmetric under time reversal. Finally, it is easy to show

$$T_{\mu_c\mu_d;\mu_a\mu_b}(s, \theta, 0) = (-1)^{\mu_i-\mu_f}T_{\mu_a\mu_b;\mu_c\mu_d}(s, \theta, 0), \qquad \text{for} \qquad \phi = 0$$

$$T_{\mu_c\mu_d;\mu_a\mu_b}(s, \theta, \phi) = T_{\mu_a\mu_b;\mu_c\mu_d}(s, \theta, \pi - \phi), \qquad \text{for} \qquad \phi \neq 0 \tag{10-24}$$

In Eq. 10-24 it is important to remember that the amplitudes on the right-hand side refer to the reaction

$$c + d \to a + b$$

and the amplitudes on the left-hand side refer to the *time-reversed* reaction

$$a + b \to c + d$$

For a *single* reaction, Eq. 10-24 reduces the number of independent amplitudes only in the case of elastic scattering.

10-5 The Number of Independent Helicity Amplitudes. For the general, two-body inelastic interaction, $a + b \to c + d$, there are

$$N = (2s_a + 1)(2s_b + 1)(2s_c + 1)(2s_d + 1) \tag{10-25}$$

$T_{\mu_c\mu_d;\mu_a\mu_b}(s, \theta, \phi)$ amplitudes if all the particles have nonzero mass. If the ith particle is a photon replace $(2s_i + 1)$ by (2). Parity conservation reduces N to N_{ind} independent amplitudes where

$$N_{\text{ind}} = \frac{N + g}{2} \tag{10-26}$$

Here $g = 1$ if all $(2s_i + 1)$ factors are odd; otherwise $g = 0$. Thus in the associated production reaction, $\pi^- + p \to K^+ + \Sigma^-$, $N_{\text{ind}} = 2$. And in the photoproduction reaction, $\gamma + p \to \pi^+ + n$, $N_{\text{ind}} = 4$. These considerations are of course restricted to parity conserving interactions.

Exercise 10-7. Derive Eqs. 10-25 and 10-26. Calculate N_{ind} for the following reactions:

$$\gamma + p \to \rho + p,\ p + p \to \Delta(1236) + p,\ \pi + p \to \rho + \Delta(1236)$$

For the elastic interaction $a + b \to a + b$, there is a further reduction in the number of independent helicity amplitudes if $(s_a + s_b) > 1/2$. This reduction comes about from time reversal. Remember, we are only concerned with strong and electromagnetic interactions here.

If particles a and b in the elastic interaction are identical or lie in the same isospin multiplet, there is a further reduction in the number of independent helicity amplitudes. This reduction is necessary if the amplitudes are to meet the symmetry requirements of Fermi–Dirac or Bose–Einstein statistics. We do not discuss these questions in this book, and we refer the reader to the literature (10JA, 10WE, 10RO, 10MA).

10-6 Forward Helicity Amplitudes, the Optical Theorem, and Total Cross Sections. In two-body interactions, $a + b \to c + d$, there is a special simplicity to the scattering amplitudes when the center of mass angle between a and c, or between a and d, is zero. Such amplitudes have no angular dependence and have a number of uses in elementary particle physics. They are particularly useful in elastic scattering, because such amplitudes are used in the optical theorem (Sec. 6-7) and in forward dispersion relations (Ch. 15).

We start with the general interaction $a + b \to c + d$ and set the center of mass angle θ, between a and c, to zero. From Eq. 10-9 we write

$$f_{\mu_c\mu_d;\mu_a\mu_b}(s, 0) = \frac{1}{p_i^*}\sum_J \{(2J+1)\langle \mu_c\mu_d | T^J | \mu_a\mu_b\rangle \delta_{\mu_i\mu_f}\} \qquad (10\text{-}27)$$

an equation in which there is no ϕ dependence. Therefore, a necessary condition for

$$f_{\mu_c\mu_d;\mu_a\mu_d}(s, 0) \neq 0 \qquad (10\text{-}28a)$$

is

$$\mu_i = \mu_f \qquad \text{or} \qquad \mu_a - \mu_b = \mu_c - \mu_d \qquad (10\text{-}28b)$$

Equation 10-28b simply means that the total helicity can not change at $\theta = 0$. Of course, even if $\mu_i = \mu_f$, dynamics could still lead to $f_{\mu_c\mu_d,\mu_a\mu_b}(s, 0)$ being zero. Equations 10-28 merely give the maximum number of nonzero amplitudes at $\theta = 0$. Moreover, not all these nonzero amplitudes will be independent because of parity conservation as discussed in Sec. 10-5.

Exercise 10-8. Determine which amplitudes may be nonzero at $\theta = 0$ for the reactions listed in Ex. 10-7.

Exercise 10-9. Derive Eqs. 10-28 for nonrelativistic particles by considering the angular momentum of all the particles to be quantized along the same z axis, that z axis being along the direction of motion of all the particles. Use the fact that the component of the orbital momentum along this z axis is zero.

We now restrict our considerations to elastic scattering, and we use an example in which a and b are two different spin $\frac{1}{2}$ particles. An example of such a system is Λ^0-p elastic scattering. For an arbitrary angle there are six independent amplitudes, listed below. We also show which *forward* amplitudes must be zero by Eq. 10-28.

Amplitude	μ_f	μ_i	Forward Amplitude Must Be Zero
$f_{++,++}$	0	0	No
$f_{+-,++}$	1	0	Yes
$f_{-+,++}$	-1	0	Yes
$f_{--,++}$	0	0	No
$f_{+-,+-}$	1	1	No
$f_{-+,+-}$	-1	1	Yes

Thus there can be only three nonzero, independent forward amplitudes; namely, $f_{++,++}(s, 0)$, $f_{--;++}(s, 0)$ and $f_{+-;+-}(s, 0)$.

Exercise 10-10. Calculate the maximum number of independent forward helicity amplitudes that may be nonzero in ρ meson–proton elastic scattering.

We now apply these considerations to the optical theorem (Sec. 6-7). As discussed there, the optical theorem applies to amplitudes in which the final state is exactly the same as the initial state. In particular, each particle must preserve its own helicity state. We use the phrase *helicity flip* to indicate that the helicity of a particle has changed in an interaction. Then the optical theorem applies to amplitudes in which neither particle undergoes helicity flip. This is an example of the *no spin flip* condition given in Sec. 6-7. Now we can write Eqs. 6-46c and 6-46d in the forms

$$\operatorname{Im} T_{\mu_a\mu_b;\mu_a\mu_b}(s, 0) = 2p^*\sqrt{s}\,\sigma_{\text{tot},\mu_a\mu_b}(s)$$

$$\operatorname{Im} f_{\mu_a\mu_b;\mu_a\mu_b}(s, 0) = \frac{p^*}{4\pi}\,\sigma_{\text{tot},\mu_a\mu_b}(s) \tag{10-29}$$

We have put helicity subscripts on σ_{tot} because there is *no* argument from first principles for $\sigma_{\text{tot},\mu_a\mu_b}$ to be independent of μ_a and μ_b. That is, $\sigma_{\text{tot},\mu_a\mu_b}$ may indeed depend on μ_a and μ_b within the symmetry requirements of space inversion and time reversal invariance. But it is generally assumed that $\sigma_{\text{tot},\mu_a\mu_b}$ is *independent* of μ_a and μ_b at *high* energies, since at high energies the total cross section consists of the summation over a very large number of final states. Many of these final states are of high multiplicity. It is unlikely that the initial total helicity $(\mu_a - \mu_b)$ controls the total production of these final states in a marked way. On the other hand, there are no experimental studies on the equality of the independent $\sigma_{\text{tot},\mu_a\mu_b}$ when two or more $\sigma_{\text{tot},\mu_a\mu_b}$ are possible. We note that two independent $\sigma_{\text{tot},\mu_a\mu_b}$ occur in the proton–proton system and in the photon–proton system. When experimental total cross sections are presented for these systems they are actually an average over the various independent total cross sections. Thus for the photon–proton system we have

$$\sigma_{\text{tot}}(s) = (\tfrac{1}{2})[\sigma_{\text{tot},+1/2,+1}(s) + \sigma_{\text{tot},-1/2,+1}(s)]$$

Of course for systems consisting of a spin zero and a spin $\frac{1}{2}$ particle there is only one independent $f_{\mu_a,0;\mu_a,0}(s, 0)$ and hence only one independent $\sigma_{\text{tot},\mu_a,0}(s)$.

To conclude this section, we discuss the relation of the optical theorem to the forward elastic differential cross section $(d\sigma/d\Omega)_{\theta=0}$. For systems consisting of a spin zero particle and a spin $\frac{1}{2}$ particle the relationship is simple.

$$\left.\frac{d\sigma}{d\Omega}\right|_{\theta=0} = \left[\frac{p^*}{4\pi}\,\sigma_{\text{tot}}(s)\right]^2 + [\operatorname{Re} f_{+1/2,0;+1/2,0}(s, 0)]^2 \tag{10-30}$$

This is a derivation of Eqs. 1-10 to 1-13, which we used in Sec. 1-7 to discuss the real and imaginary parts of the elastic scattering amplitude. Equation 10-30 provides a method for determining $\operatorname{Re} f(s, 0)$ from measurements of $\sigma_{\text{tot}}(s)$ and $(d\sigma/d\Omega)_{\theta=0}$. This method is used, but it is not precise because it requires the comparison of quantities obtained from two different experiments. The more usual method of determining $\operatorname{Re} f(s, 0)$, when both particles are charged, is to measure the interference of the hadronic elastic scattering with the electromagnetic elastic scattering. Since the second method brings in considerations that are peripheral to the subject under discussion, we leave an analysis of this method to the exercise at the end of the section. As is illustrated in that exercise, the interference method is applicable only to a small range of angles close to

$\theta = 0$. In fact, it is conventional to say that the interference method separates Re $f(s, \theta)$ and Im $f(s, \theta)$ at $\theta = 0$.

The extension of Eq. 10-30 to higher spin systems is straightforward, but the result is complicated. We give the result and leave the derivation to Ex. 10-11. All equations are in the center of mass.

$$\text{if} \quad \mu_a' - \mu_b' \neq \mu_a - \mu_b, \left(\frac{d\sigma}{d\Omega}\right)_{\mu_a' \ldots, \theta=0} = 0 \tag{10-31a}$$

$$\text{if} \quad \mu_a' - \mu_b' = \mu_a - \mu_b \qquad \text{but} \quad \mu_a' \neq \mu_a$$

$$\left(\frac{d\sigma}{d\Omega}\right)_{\mu_a' \ldots, \theta=0} = [\operatorname{Im} f_{\mu_a' \ldots}(s, 0)]^2 + [\operatorname{Re} f_{\mu_a' \ldots}(s, 0)]^2 \tag{10-31b}$$

$$\text{if} \quad \mu_a' = \mu_a \qquad \text{and} \qquad \mu_b' = \mu_b$$

$$\left(\frac{d\sigma}{d\Omega}\right)_{\mu_a \ldots, \theta=0} = \left[\frac{p^*}{4\pi}\, \sigma_{\text{tot},\mu_a\mu_b}(s)\right]^2 + [\operatorname{Re} f_{\mu_a' \ldots}(s, 0)]^2 \tag{10-31c}$$

Exercise 10-11. Verify Eqs. 10-31.

Although these equations present a complicated situation, we usually simplify them considerably at high energies with two assumptions.

1. As we have already discussed, we assume that all the independent $\sigma_{\text{tot},\mu_a\mu_b}$ are equal to one another.

2. We assume that all the amplitudes in Eq. 10-31b are very small compared with the amplitudes in Eq. 10-31c. This is based on the recognition that all the amplitudes in Eq. 10-31b are *double* helicity flip amplitudes; that is, the helicity of both particles is changed. Experimentally we find that these double helicity flip forward amplitudes are much smaller than the helicity nonflip forward amplitudes. However, the evidence is incomplete because it depends on polarization measurements that exist only for a few systems in limited energy ranges. Perhaps our strongest reason for this assumption is the diffraction model explanation of elastic scattering (Ch. 2). Diffractive elastic scattering is caused by absorption of part of the incident wave; the unabsorbed part of the incident wave then partially appears as the elastically scattered wave. But this unabsorbed part has not undergone any interaction. Therefore its helicity cannot change. Since the forward elastic scattering is almost totally diffractive at high energies, the non-helicity flip forward amplitudes must be much larger than the helicity flip forward amplitudes. Of course most of the latter forward amplitudes are exactly zero by Eq. 10-28. It is only the double helicity flip amplitudes in Eq. 10-31b that must be shown to be relatively small by this diffraction scattering argument.

Exercise 10-12. To study the method for determining the real part of the forward elastic scattering amplitude, we use the simplest example—the elastic scattering of charged pions on protons. Fill in the steps outlined below.

1. Since the interference takes place at very small angles, the proton form factor (see Ch. 19) in electromagnetic scattering is unity, and the electromagnetic scattering amplitude is

$$f_{\text{em}}(\theta) = \mp \frac{2p^*\alpha}{|t|} e^{-i2\delta} = \mp g(\theta)\, e^{-i2\delta} \tag{10-32a}$$

where α is the fine structure constant and p^* is the momentum in the center of mass.

2. The phase δ is given approximately at high energies by (10FO)

$$\delta = \mp\alpha[\ln(R\sqrt{|t|}) + 0.29] \tag{10-32b}$$

The upper (lower) sign is for $\pi^+p(\pi^-p)$. R is roughly the size of the nucleon plus pion charge distribution.

3. With respect to the strong interactions, only the amplitude $f_{0+,0+}(s, \theta, \phi)$ need be considered. By the arguments in this section, $f_{0-,0+}(s, \theta, \phi)$ can be ignored for small θ. Denote $f_{0+,0+}(s, \theta, \phi)$ by $f_{++}(\theta)$; ignore ϕ because the target protons are unpolarized.

4. The complete scattering amplitude is

$$f(\theta) = f_{++}(\theta) + g(\theta)\, e^{-i2\delta} \tag{10-32c}$$

$$|f(\theta)|^2 = [1 + r^2][\text{Im}\, f_{++}(\theta)]^2 + [g(\theta)]^2 \mp 2g(\theta)[\text{Im}\, f_{++}(\theta)][r \cos 2\delta \pm \sin 2\delta] \tag{10-32d}$$

where

$$r = \frac{\text{Re}\, f_{++}(\theta)}{\text{Im}\, f_{++}(\theta)} \tag{10-32e}$$

5. The maximum interference occurs at $|f_{++}(\theta)| \approx g(\theta)$. To estimate where this interference occurs, use

$$|f_{++}(\theta)|^2 \approx \frac{p^{*2}}{\pi} \frac{d\sigma}{dt} = \frac{p^{*2}}{\pi} A\, e^{-b|t|} \tag{10-32f}$$

where $A \approx 32$ mb/(GeV/c)^2 for pion-proton elastic scattering in the 10 to 30 GeV/c laboratory momentum range. The maximum interference then turns out to occur at about $|t| = 0.003$ (GeV/c)^2.

6. This interference in pion-proton elastic scattering (Fig. 10-1) extends from $|t| = 0.003$ to $|t| = 0.02$ (GeV/c)^2. Verify that the sign of the interference is as calculated in this exercise. Why is the t range of this method so limited?

7. For $R \approx 10^{-13}$ cm and this t range show that $\delta \approx 0$. Therefore, $f_{\text{em}}(\theta)$ is

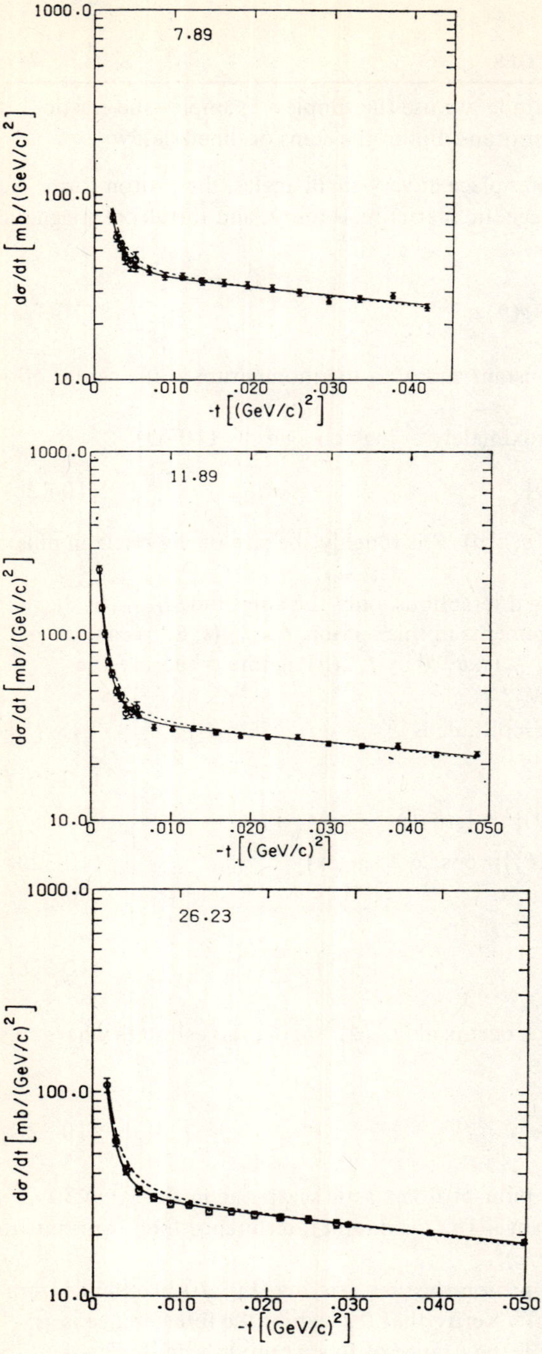

(a)

Fig. 10-1 The elastic differential cross section in the Coulomb interference region for *(a)* $\pi^- p$, *(b)* $\pi^+ p$ (10FO). The solid lines are the best fits to the data. The broken lines are fits if $r = 0$ in Eq. 10-32d. Since the Coulomb scattering amplitude is mostly real (Ex. 10-12), the differences between the lines indicate the interference.

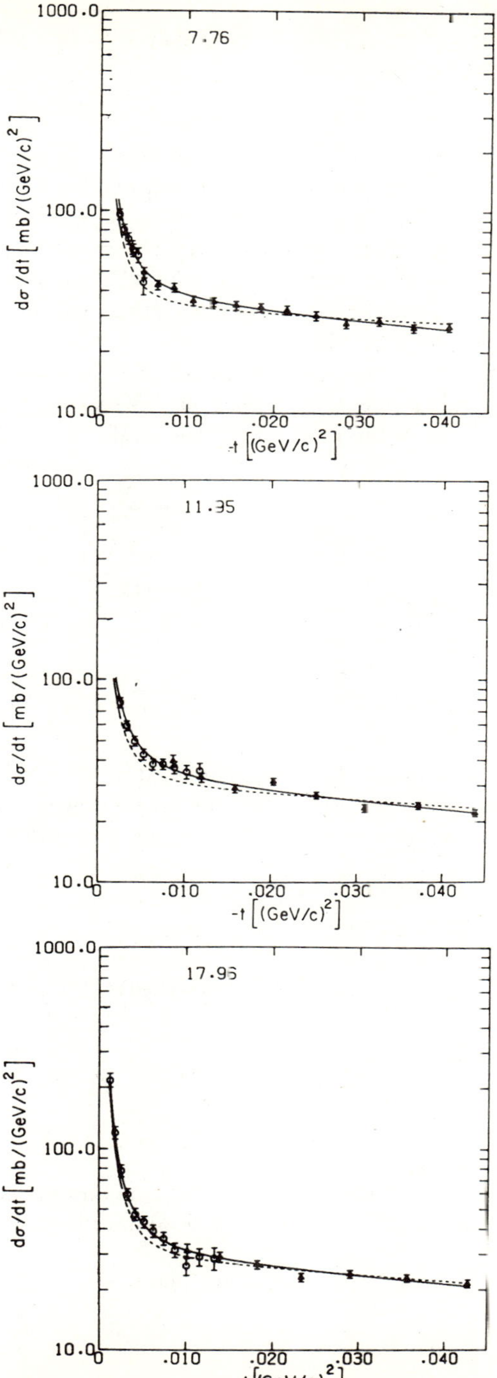

1000.0
7.76
100.0
10.0
dσ/dt [mb/(GeV/c)²]
0
.010
.020
.030
.040
-t [(GeV/c)²]
11.35
17.96

(b)

almost completely real and $f_{\text{em}}(\theta)$ interferes only with Re $f_{++}(\theta)$. For further details, see Ref. 10FO.

10-7 Small Angle Behavior of Helicity Amplitudes. In the last section we found that the center of mass helicity amplitudes exhibit a simple and useful structure when $\theta = 0$. This observation leads us to ask whether the center of mass helicity amplitudes at small but nonzero angles also exhibit some simple and useful structure. These small angle helicity amplitudes do indeed have some rather simple properties. They are developed in this section, and they have a multitude of applications later in this book.

We start with Eq. 10-9, and we wish to study Eq. 10-9 when θ is small. Since all the θ dependence is in the $d^j_{\mu'\mu}(\theta)$ functions, we seek an expression for these functions that emphasizes the small θ behavior. Such an expression is (10DU, 10RO)

$$d^j_{m'm}(z) = (-1)^{m'-m}\left[\frac{\Gamma(j+m'+1)\Gamma(j-m+1)}{\Gamma(j+m+1)\Gamma(j-m'+1)}\right]^{1/2}\frac{1}{(m'-m)!} \tag{10-33}$$

$$\times\left[\frac{1-z}{2}\right]^{(m'-m)/2}\left[\frac{1+z}{2}\right]^{(m'+m)/2}$$

$$\times F\left(-j+m';j+m'+1;m'-m+1,\frac{1-z}{2}\right)$$

where $z = \cos\theta$. Equation 10-33 is very similar to Eq. 9-8; the advantage of Eq. 10-33 for our present purpose is that the lowest power of $\sin(\theta/2) = \sqrt{[(1-z/2]^{1/2}}$ is exhibited explicitly. These expressions hold when

$$j \geqslant |m'|, \qquad j \geqslant |m|, \qquad m' \geqslant m \tag{10-34}$$

Inserting Eq. 10-33 into Eq. 10-9, we obtain

$$f_{\mu_c\mu_d;\mu_a\mu_b}(s,\theta,\phi) = \frac{1}{p_i^*}(\sin\tfrac{1}{2}\theta)^{|\mu_i-\mu_f|}(\cos\tfrac{1}{2}\theta)^{|\mu_i+\mu_f|}\,e^{i(\mu_i-\mu_f)\phi}$$

$$\times\sum_{n=0}^{\infty}a_n(\sin^2\tfrac{1}{2}\theta)^n, \qquad \mu_i = \mu_a-\mu_b,$$

$$\mu_f = \mu_c - \mu_d \tag{10-35}$$

The a_n coefficients depend on the $\langle\mu_c\mu_d|T^J(s)|\mu_a\mu_b\rangle$ matrix elements and on J, μ_i, and μ_f. For θ sufficiently close to 0, $f(s,\theta,\phi)$ has the approximate form

$$f_{\mu_c\mu_d;\mu_a\mu_b}(s,\theta,\phi) \approx \frac{a_0}{p_i^*}(\sin\tfrac{1}{2}\theta)^{|\mu_i-\mu_f|}\,e^{i(\mu_i-\mu_f)\phi}, \qquad \theta\to 0 \tag{10-36}$$

Thus for θ sufficiently close to zero, the angular dependence of $f(s, \theta, \phi)$ depends only on $\mu_i - \mu_f$—*the change in the total helicity*. Thus, as promised, we have found a simple and useful structure in the small angle helicity amplitudes. In the derivation of Eq. 10-36 we have assumed that there is some θ' below which $a_0 \gg a_1(\theta/2) + a_2(\theta/2)^2 + \cdots$. This would not be possible if a_0 were zero. We also note that Eq. 10-36 is not a useful approximation if there is a dominant resonant amplitude in a J state which does not contribute to a_0. Equation 10-36 is particularly instructive when written in the form

$$f_{\mu_c\mu_d;\mu_a\mu_b}(s, t, \phi) \approx A(s)|t - t_{\min}|^{|\mu_i - \mu_f|/2}\, e^{i(\mu_i - \mu_f)\phi}; \qquad \theta \to 0 \qquad (10\text{-}37)$$

where $A(s)$ is a function only of s. Equations 10-35 through 10-37 are for $\mu_i \geqslant \mu_f$; for $\mu_i < \mu_f$ use Eqs. 10-19 first.

Exercise 10-13. Apply Eq. 10-36 to the reactions $\pi^- + p \to \rho^0 + n$, $\gamma + p \to \pi^+ + \Delta(1236)$; and sketch the small θ behavior of the various helicity amplitudes.

10-8 The Impact Parameter Representation for the Two-Body Reactions of Particles with Spin. In Sec. 3-4 we derived, from the partial wave analysis, the eikonal approximations for the elastic scattering of spinless particles. We easily obtained the same result yielded by a much more elaborate analysis in Ch. 2. In this section we employ the same route we followed in Sec. 3-4, but we start from the general expression, Eq. 10-9, for the center of mass, two-body helicity amplitudes. And we obtain powerful generalizations of both the eikonal approximation and the impact parameter representation, including particles of arbitrary spin. Furthermore, we extend the impact parameter representation to two-body inelastic reactions.

We begin with Eq. 10-9 written in the form

$$f_{\mu_c\mu_d;\mu_a\mu_b}(s,z, \phi) = \frac{1}{p_i^*}\, e^{i(\mu_i - \mu_f)\phi} \sum_j \left[(2j + 1)\langle \mu_c\mu_d | T^j(s) | \mu_a\mu_b\rangle \times d^j_{\mu_i\mu_f}(z)\right] \qquad (10\text{-}38)$$

where $z = \cos\theta$. In this derivation the explicit use of z, rather than θ, is conventional and it is slightly more convenient; but it is not necessary. Now through Eq. 10-33 we can extend the $d^j_{m'm}$ from functions of the discrete parameter j to functions of a continuous parameter j. And assuming that $\langle \mu_c\mu_d | T^j(s) | \mu_a\mu_b\rangle$ does not vary too violently with j, we replace it by a continuous function of j,

$\langle \mu_c \mu_d | T(j, s) | \mu_a \mu_b \rangle$. Thus we replace the summation of Eq. 10-38 by the integral of Eq. 10-39

$$f_{\mu_c \mu_d ; \mu_a \mu_b}(s, z, \phi) = \frac{1}{p_i^*} e^{i(\mu_i - \mu_f)\phi} \int_{j_{\min}}^{\infty} dj [(2j + 1) \langle \mu_c \mu_d | T(j, s) | \mu_a \mu_b \rangle \times d^j_{\mu_i \mu_f}(z)] \tag{10-39}$$

The discussion in Sec. 3-4 of the significance of this change from a summation to an integration applies in full to Eq. 10-39. The lower limit of integration $j_{\min}$ is the smallest j that allows the total helicities μ_i and μ_f. Thus $j_{\min}$ is equal to μ_i or μ_f, depending on which is larger. As we might expect, this lower limit leads to some problems because our derivation has most physical significance when j is large. We return in Ch. 13 to a problem associated with $j_{\min}$—the so-called exception terms in the absorption correction to inelastic scattering.

Still following the route of Sec. 3-4, we limit our discussion to the region in which θ^2, hence $(1 - z)$, is small compared to 1. Then the rather obscure function $d^j_{m'm}(z)$ can be approximated by the well-known Bessel function $J_{m'-m}(u)$. To order $(1 - z)$ we have

$$d^j_{m'm}(z) \approx (-1)^{m'-m} \left[\frac{\Gamma(j + m' + 1)\Gamma(j - m + 1)}{\Gamma(j + m + 1)\Gamma(j - m' + 1)} \right]^{1/2} \times [\lambda(j, m', m)]^{-(m'-m)/2} J_{m'-m}[(2(1 - z)\lambda(j, m', m))^{1/2}] \tag{10-40}$$

where

$$\lambda(j, m', m) = j(j + 1) - \tfrac{1}{2}m'(m' + 1) - \tfrac{1}{2}m(m - 1) \tag{10-41}$$

Here $j \geqslant |m'|, j \geqslant |m|, m' \geqslant m$.

Exercise 10-14. Derive Eqs. 10-40 using the expression for the Bessel function given in Eq. B-2 (App. B). Compare Ref. 10DU.

The approximation to the $d^j_{m'm}(z)$ functions provided by Eqs. 10-40 is cumbersome because of the Γ functions and λ factor. Therefore, it is unusual to stop at this stage of the approximation. Rather, we make further use of the philosophy of the approximation—the integration is over values of j that are large compared to 1. One also recognizes that for practical problems, the spins of the particles, hence $|m'|$ and $|m|$, are of the order of 1. Therefore we neglect $|m'|$, $|m|$ and $\frac{1}{2}$ compared to j, and obtain $\lambda(j, m', m) \approx (j + \frac{1}{2})^2$ and

$$d^j_{m'm}(z) \approx (-1)^{m'-m} J_{m'-m} [(j + \tfrac{1}{2})(2(1 - z))^{1/2}]; \qquad m' \geqslant m \tag{10-42}$$

a simple and easily usable approximation.

Inserting Eq. 10-42 into Eq. 10-39, we obtain

$$f_{\mu_c\mu_d;\mu_a\mu_b}(s, z, \phi) = \frac{2}{p_i^*}(-1)^{\mu_i-\mu_f} e^{i(\mu_i-\mu_f)} \int_{j_{\min}}^{\infty} dj$$

$$\times ((j+\tfrac{1}{2})J_{\mu_i-\mu_f}[(j+\tfrac{1}{2})(2(1-z))^{1/2}]\langle\mu_c\mu_d|T(j, s)|\mu_a\mu_b\rangle) \qquad (10\text{-}43)$$

Equations 10-43 and 10-45 are for $\mu_i \geqslant \mu_f$; for $\mu_i < \mu_f$ use Eqs. 10-19.

Finally we make the transition to the impact parameter b; defined in analogy to Eq. 3-26a by

$$p_i^* b = j + \tfrac{1}{2} \qquad (10\text{-}44)$$

The use of p_i^*b rather than p_f^*b or $(p_i^*p_f^*)^{1/2}b$ is arbitrary. However, at high energies $p_i^* \approx p_f^* \approx p^*$ in the center of mass; therefore, the choice is not important for our considerations. Inserting Eq. 10-44 into Eq. 10-43, we obtain the following form for the impact parameter representation of the center of mass helicity amplitudes. In doing this we abandon the $j_{\min}$ lower limit altogether and take the lower limit on b to be zero; we return to this point in Ch. 13. We also replace $(1-z)$ by $2\sin^2 \frac{1}{2}\theta$. For later use we note $(2p^* \sin\frac{1}{2}\theta) \approx \surd(|t - t_{\min}|)^{1/2}$.

$$f_{\mu_c\mu_d;\mu_a\mu_b}(s, \theta, \phi) = 2p^*(-1)^{\mu_i-\mu_f} e^{i(\mu_i-\mu_f)\phi}$$

$$\times \int_0^{\infty} [b\, db\, J_{\mu_i-\mu_f}(2p^*b \sin\tfrac{1}{2}\theta) T_{\mu_c\mu_d;\mu_a\mu_b}(b, s)] \qquad (10\text{-}45)$$

The beautiful and simple result holds for all two-body reactions, inelastic as well as elastic. The derivation depends only on the assumptions of the smoothness of T^j and the contribution of many j states to the amplitude. All the *known* helicity dependence is contained in $J_{\mu_i-\mu_f}$. All the *unknown* helicity dependence is contained in $T_{\mu_c\ldots}(b, s)$, which is only a function of the energy and the impact parameter b. This allows us to think about the interaction using the simple ideas of the impact parameter representation.

To illustrate this, let us fix our attention on inelastic reactions. If we believe that an inelastic reaction involves mostly low angular momentum partial waves, which in turn means small impact parameters, we might write

$$T_{\mu_c\ldots}(b, s) = \lambda_{\mu_c\ldots}(s) \exp \frac{-b}{\beta_{\mu_c\ldots}(s)} \qquad (10\text{-}46a)$$

or

$$T_{\mu_c\ldots}(b, s) = \lambda_{\mu_c\ldots}(s) \exp \frac{-b^2}{\beta_{\mu_c\ldots}(s)} \qquad (10\text{-}46b)$$

Here λ and β would be selected according to our ideas about which helicity amplitudes are most important. Neither Eq. 10-46a or 10-46b lead to a simple integral when inserted in Eq. 10-45. Therefore, when these forms are used, either approximation or numerical integration is necessary to evaluate Eq. 10-45. As a contrasting example, let us suppose, in analogy to Eq. 3-26b, that the inelastic interaction is *peripheral*, taking place mainly at an impact parameter b_0. We expect b_0 to be roughly the hadronic radius. Writing, for simplicity,

$$T_{\mu_c\ldots}(b, s) = \frac{\lambda_{\mu_c\ldots}(s)\,\delta(b - b_0)}{b_0} \tag{10-47}$$

we obtain

$$f_{\mu_c\ldots}(s, t, \phi) = 2p^*\lambda_{\mu_c\ldots}(s)(-1)^{\mu_i-\mu_f}\, e^{i(\mu_i-\mu_f)\phi} \times J_{u_i-u_f}(b_0|t - t_{\min}|^{1/2}) \tag{10-48}$$

This is an important result, for it shows that for a peripheral inelastic interaction, the shape of the forward peak depends directly on $\mu_i - \mu_f$. And $\mu_i - \mu_f$ is simply the change in the total helicity from the initial state to the final state. We also see that an important feature of the forward peak–the position in t of the zeros of J_n–is directly given by b_0. We return to this subject in Ch. 18.

Now let us turn to elastic scattering. We obtain the generalization of the eikonal approximation to the optical model of elastic scattering (Ch. 2 and Sec. 3-4) by writing

$$T_{\mu_{a'}\mu_{b'};\mu_a\mu_b}(b, s) = \frac{\exp\,[i\chi_{\mu_{a'}\mu_{b'};\mu_a\mu_b}(b, s)] - \delta_{\mu_a\mu_{a'}}\delta_{\mu_b\mu_{b'}}}{2i}$$

For high energies we would make $\chi_{\mu_a\mu_b;\mu_a\mu_b}(b, s)$ mostly positive imaginary, hence mostly absorptive. The b dependence of $\chi_{\mu_a\mu_b;\mu_a\mu_b}$ would be such that $\chi_{\mu_a\mu_b;\mu_a\mu_b}$ was a maximum at $b = 0$ and decreased smoothly as b increased. Thus all the physical insight developed in Ch. 2 for the eikonal approximation to the optical model can be applied to $\chi_{\mu_a\mu_b;\mu_a\mu_b}(b, s)$, thus to $f_{\mu_a\mu_b;\mu_a\mu_b}(s, \theta, \phi)$. But we cannot immediately apply the optical model of Ch. 2 to the helicity flip elastic amplitudes $\mu_{a'} \neq \mu_{a'}$, $\mu_{b'} \neq \mu_b$. We would have to introduce potentials that flip the helicity. This can be done and has been done. But it does not provide much physical insight into the helicity flip amplitudes.

Exercise 10-15. As an example that can be used for either elastic or inelastic reactions, assume

$$\begin{aligned} T_{\mu_c\ldots}(b, s) &= \lambda_{\mu_c\ldots}, \qquad 0 \leqslant b \leqslant R \\ T_{\mu_c\ldots}(b, s) &= 0, \qquad R < b \end{aligned} \tag{10-49}$$

This is a generalization of the disk model of elastic scattering introduced in Eq. 2-26. Use a Bessel function integral from Ref. 10AB to obtain

$$f_{\mu_c \ldots}(s, t, \phi) = \left[\frac{4p^* R \lambda_{\mu_c \ldots}(-1)^n e^{in\phi}}{(|t|)^{1/2}}\right]$$

$$\times \sum_{l=0}^{\infty} \left[\left(\frac{(n)(n+2l+1)}{(n+2l)(n+2l+2)}\right) J_{n+2l+1}(R\sqrt{|t|}) \qquad (10\text{-}50)$$

where $n = \mu_i - \mu_f$. Show that Eq. 10-50 reduces to Eq. 2-28 for $n = 0$. Sketch the forward peak given by Eq. 10-50 for $n = 1$ and R, the hadronic radius.

Exercise 10-16. Show that the $f_{\mu_c \mu_d; \mu_a \mu_b}(s, \theta, \phi)$ amplitude of Eq. 10-45 has the same small θ behavior as the amplitudes discussed in Sec. 10-7.

Finally, to pull the various notations together from Ch. 2, Sec. 3-4, and this section, we write

$$S_{fi}(b, s) = e^{i\chi_{fi}(b, s)} \qquad (10\text{-}51\text{a})$$

$$T_{fi}(b, s) = \frac{S_{fi}(b, s) - \delta_{fi}}{2i} = \frac{e^{i\chi_{fi}(b, s)} - \delta_{fi}}{2i} \qquad (10\text{-}51\text{b})$$

thus achieving a uniform notation for the S- and T-matrices in the impact parameter representation. The subscripts indicate the type of particles and their helicity states in the initial state i and the final state f.

REFERENCES

10AB *Handbook of Mathematical Functions*, M. Abramowitz and L. A. Stegun, Eds, (National Bureau of Standards AMS 55, Government Printing Office, 1964).

10DU L. Durand and Y. T. Chiu, *Phys. Rev.* **139**, B646 (1965).

10FO K. J. Foley et al., *Phys. Rev.* **181**, 1775 (1969).

10JA M. Jacob and G. C. Wick, *Ann. Phys. (N. Y.)* **7**, 404 (1959).

10MA A. D. Martin and T. D. Spearman, *Elementary Particle Physics* (North-Holland, Amsterdam, 1970).

10RO M. E. Rose, *Elementary Theory of Angular Momentum* (Wiley, New York, 1957).

10WE J. Werle, *Relativistic Theory of Reactions* (North-Holland, Amsterdam, 1966).

CHAPTER

11

The Density Matrix, Polarization, Scattering, and Decay Processes

11-1 Introduction. This chapter is a summary of the ways in which the spin states of particles are used to obtain information about the amplitudes in elastic and two-body inelastic reactions. We begin with a review of the properties of the density matrix, assuming that the reader has been introduced to the density matrix in quantum mechanics. We then make two simple applications of the density matrix—to the description of unpolarized particle states and to the description of particle states with vector polarization.

The principles developed in Secs. 11-2 through 11-4 are next applied to pion-nucleon elastic scattering. This is the simplest elastic scattering system and the only high energy elastic reaction for which we have complete amplitude information.

In Sec. 11-6 we discuss the application to decay processes of the density matrix and the various polarization concepts. As we emphasize in the introduction to Sec. 11-6, this is a vast subject because it is used so frequently in the elucidation of the internal quantum numbers of elementary particles. But fortunately the vastness of the subject comes from the development of formalisms and applications in great detail—not from the introduction of new physical principles. We can therefore restrict our discussion to showing how the principles already developed in this chapter are applied to the decay process. For a full discussion of the spin and angular distribution aspects of decay processes, see Refs. 11JA, 11CH, 11PI, and 11NI.

11-2 The Density Matrix. At this point we are able to describe the spin state of a particle by using spin angular momentum states $|jm\rangle$ in the rest system of the particle, or by using one-particle, plane wave helicity states $|\mathbf{p}\mu\rangle$. But in many cases a particle is not in a simple spin state described by $|jm\rangle$ or $|\mathbf{p}\mu\rangle$. Rather, the spin state of the particle must be described by a combination of $|jm\rangle$ states or of $|\mathbf{p}\mu\rangle$ states. Such a description is most conveniently carried out through the spin *density matrix*. In this section we review the definition and properties of the spin density matrix. Everything that we do applies both to the rest frame spin states $|jm\rangle$ and to the helicity states $|\mathbf{p}\mu\rangle$. To indicate this we adopt the temporary notation $|m\rangle$ to represent either the rest frame spin state $|jm\rangle$ or the helicity state $|\mathbf{p}\mu\rangle$. (For convenience, we suppress the spin j.) Of course, these two different types of spin states cannot be mixed together indiscriminately in describing the spin state of a particle. The states $|m\rangle$ are orthonormal with respect to j and m, and

$$\langle m'|m\rangle = \delta_{m'm}\delta_{j'j} \tag{11-1}$$

The states $|m\rangle$ also form a complete set with respect to the description of the spin state of the particle. We refer to the states $|m\rangle$ as the *basis* states.

A generalization of the state $|m\rangle$ is a state $|\phi\rangle$ which is a linear combination of $|m\rangle$ states. That is,

$$|\phi\rangle = \sum_{m=-j}^{+j} a_m|m\rangle \tag{11-2a}$$

Any state which can be written in the form of Eq. 11-2a is called a *pure* state or a *coherent* state with respect to the spin. We use the term "coherent state" because we find that it yields more physical insight. The coefficients a_m are given by

$$a_m = \langle m|\phi\rangle \tag{11-2b}$$

Finally, to calculate the expectation value $\langle Q \rangle$ of some operator Q for the state, we have

$$\langle Q \rangle = \langle \phi | Q | \phi \rangle = \sum_{m'} \sum_{m} a_{m'}^* a_m \langle m' | Q | m \rangle \tag{11-3}$$

Equation 11-2 is the most general description of a coherent spin state. But Eq. 11-2a does not represent the most general description of the spin state of a particle, for the particle state $|\phi\rangle$ may be an *incoherent* mixture of several states $|\phi_a\rangle, |\phi_b\rangle, \ldots$. Then the expectation value of the operator Q is

$$\langle Q \rangle = \langle \phi_a | Q | \phi_a \rangle + \langle \phi_b | Q | \phi_b \rangle + \cdots \tag{11-4}$$

If in terms of the basis states $|m\rangle$

$$|\phi_a\rangle = \sum a_m |m\rangle, \qquad |\phi_b\rangle = \sum b_m |m\rangle \cdots \tag{11-5}$$

then

$$\langle Q \rangle = \sum_{m'} \sum_{m} \{[a_{m'}^* a_m + b_{m'}^* b_m + \cdots] \langle m' | Q | m \rangle\} \tag{11-6}$$

which is an obvious extension of Eq. 11-3. Since the coefficient $a_{m'}^* a_m + b_{m'm}^* + \cdots$ does not depend on the operator Q, it is very convenient to define the *density matrix* ρ with elements

$$\rho_{mm'} = a_{m'}^* a_m + b_{m'}^* b_m + \cdots \tag{11-7}$$

Then, writing $\langle m' | Q | m \rangle = Q_{m'm}$, Eq. 11-6 becomes

$$\langle Q \rangle = \sum_{m'm} \rho_{mm'} Q_{m'm} = \sum_{m} (\rho Q)_{mm}$$

$$\langle Q \rangle = \mathrm{Tr}(\rho Q) \tag{11-8}$$

The hermitian density matrix ρ requires $(2j + 1)^2$ real parameters for complete definition. This is to be compared with the $2(2j + 1)$ real parameters needed to define the coherent state of Eq. 11-2a.

For many uses it is more convenient to define a *normalized density matrix* ρ_{norm} by the requirement

$$\mathrm{Tr}\, \rho_{\text{norm}} = 1 \tag{11-9}$$

The relation between ρ and ρ_{norm} is

$$\rho_{\text{norm}} = \frac{\rho}{\mathrm{Tr}\, \rho} \tag{11-10}$$

From Eq. 11-7 it follows immediately that ρ is hermitian.

The density matrix shows directly the degree of coherence between the different states $|m\rangle$ which make up the total state $|\phi\rangle$. If the element $\rho_{mm'} = 0$, there is zero coherence between the component of the state $|\phi\rangle$ in the state $|m\rangle$ and the component of the state $|\phi\rangle$ in the state $|m'\rangle$.

Since everything that we have so far derived depends only on the orthonormal property of the basis states $|m\rangle$, this has been a very general discussion of the density matrix concept. But we can now use the rotational properties of the $|m\rangle$ states to show how ρ behaves under spatial rotation. For both the rest frame spin states $|jm\rangle$ and the helicity states $|\mathbf{p}m\rangle$ we have Eq. 9-4. Therefore, under a rotation $R(\alpha\beta\gamma)$, we write

$$\rho \to R(\alpha\beta\gamma)\rho R^{-1}(\alpha\beta\gamma) \tag{11-11a}$$

$$[R(\alpha\beta\gamma)\rho R^{-1}(\alpha\beta\gamma)]_{nm} = \sum_{n'm'} D^{j}_{nn'}(\alpha\beta\gamma)D^{j*}_{mm'}(\alpha\beta\gamma)\rho_{n'm'} \tag{11-11b}$$

Exercise 11-1. Derive Eq. 11-11.

Additional properties of the spin density matrix depend on the specific processes used to prepare the state $|\phi\rangle$. In later sections of this chapter and in Ch. 12 we examine some of these specific processes. But for the present we include a few exercises to illustrate the use of the spin density matrix.

Exercise 11-2. Suppose that a spin 1 particle at rest is in the state $|1, +1\rangle$ when the spin is quantized along an axis defined by the spherical angles θ and ϕ. Calculate the density matrix when the $+z$ axis is taken as the axis of quantization.

Exercise 11-3. Suppose that a spin $\frac{1}{2}$ particle at rest is in a state described by an incoherent mixture of equal amounts of the states $|\frac{1}{2}, +\frac{1}{2}\rangle$ and $|\frac{1}{2}, -\frac{1}{2}\rangle$, the z axis being taken as the axis of quantization. Write out the density matrix. Rotate the physical system by 90° about the y axis and again write out the density matrix. Can you find a single coherent state that gives the same density matrices?

Exercise 11-4. Repeat Ex. 11-3 for a spin 1 particle at rest in a state described by an incoherent mixture of states, such that the probability of the particle being in the state $|1, m\rangle$ is p_m.

11-3 Descriptions of Special Polarization States. For a particle of spin j, the normalized spin density matrix has $(2j + 1)^2 - 1$ real independent parameters. Thus for j greater than $\frac{1}{2}$, the full description of the polarization state of a particle can be quite complicated. We discuss in this section some of the more common and simple polarization states.

First we define the term "unpolarized", used to refer to the spin state of a particle. A particle is said to be unpolarized when the density matrix is diagonal with all diagonal elements equal, this is

$$\rho = cI \tag{11-12}$$

where I is the unit matrix and c is a real constant.

Exercise 11-5. Show that the density matrix of Eq. 11-12 can be produced only by an incoherent mixture of spin states.

Any spin density matrix not identical to that defined in Eq. 11-12 describes a particle polarized in some way. Exercises 11-2 and 11-4 provide examples of such states. Of the many possible types of spin states, one possesses a prevalent and simple polarization, namely, *vector polarization.* We say that a spin state possesses vector polarization when in the rest frame of the particle, the spin operator **J** has a nonzero expectation value. The expectation value of **J**, $\langle \mathbf{J} \rangle$, is given by Eq. 11-8 to be

$$\langle \mathbf{J} \rangle = \mathrm{Tr}\,(\rho \mathbf{J}) \tag{11-13a}$$

It is conventional to define the vector polarization **P** to be a vector proportional to $\langle \mathbf{J} \rangle$ with a maximum magnitude of 1. Explicitly, we put

$$\mathbf{P} = \frac{\langle \mathbf{J} \rangle}{j\,\mathrm{Tr}\,\rho} = \frac{\mathrm{Tr}\,(\rho \mathbf{J})}{j\,\mathrm{Tr}\,\rho} \tag{11-13b}$$

The definitions of Eqs. 11-13 are in the *rest frame of the particle.* We restrict the definitions to the rest frame because operator **J** depends on the selection of a definite axis of quantization, and we have no obvious way to make this selection for relativistic particles. (We return to the selection problem at the end of this section.)

It is straightforward, however, to extend the definitions of Eqs. 11-13 to nonrelativistic particles, using an axis of quantization fixed in space. We discuss this extension because of the physical insight it provides. The component of **P** along the direction of motion of the particle is called the *longitudinal polarization.* The component of **P** perpendicular to the direction of motion is called the *transverse polarization*

Exercise 11-6. Consider a nonrelativistic particle whose three-momentum **p** makes the spherical angles θ and ϕ with a coordinate system x, y, z fixed in space. The particle has spin j, and the z axis is the axis of spin quantization.

Calculate the longitudinal and transverse polarizations if the particle is in the spin state $|jm\rangle$.

From the expressions for the matrix elements of J_i given in App. C it is straightforward to write explicit formulas for P_i. We now do just that, leaving the derivations to Ex. 11-7.

$$P_x = (j \operatorname{Tr} \rho)^{-1} \sum_\mu (\tfrac{1}{2})[\sqrt{(j-\mu)(j+\mu+1)}\rho_{\mu,\mu+1} + \sqrt{(j+\mu)(j-\mu+1)} \times \rho_{\mu,\mu-1}] \tag{11-13c}$$

$$P_y = (j \operatorname{Tr} \rho)^{-1} \sum_\mu \left(-\frac{i}{2}\right)[\sqrt{(j-\mu)(j+\mu+1)}\rho_{\mu,\mu+1} - \sqrt{(j+\mu)(j-\mu+1)}\rho_{\mu,\mu-1}] \tag{11-13d}$$

$$P_z = (j \operatorname{Tr} \rho)^{-1} \sum_\mu [\mu\rho_{\mu,\mu}] \tag{11-13e}$$

Exercise 11-7. Derive Eqs. 11-13c through 11-13e.

Helicity states provide a completely general basis for the spin density matrix of relativistic particles. Yet sometimes it is inconvenient to use helicity states, even when the particles are relativistic. For example, it is inconvenient to use helicity basis when the vector polarization of a particle is predominantly transverse to the direction of motion of the particle. We resolve this problem by defining a special rest frame—a rest frame having a simple relationship to the helicity formalism. Consider a particle with spin j and a nonzero mass. Suppose that in some Lorentz frame with coordinate axes x, y, z the particle has three-momentum $\mathbf{p}$ and helicity state $|\mathbf{p}\mu\rangle$. Let θ and ϕ be the spherical angles of $\mathbf{p}$, and consider a second coordinate system x', y', z' fixed to the coordinate system x, y, x but with its axes rotated with respect to x, y, z. Specifically, the x', y', z' system is produced by applying the rotation $R(\phi, \theta, 0)$ to the x, y, z system. Thus the z' axis points in the direction of $\mathbf{p}$. Now the helicity state of the particle is invariant to boosts along $\mathbf{p}$. Therefore, if we apply the boost $L(-p_{z'})$ to the particle in the x', y', z' system, we obtain the rest frame helicity state $|0\mu\rangle$. And in this x', y', z' rest frame we can easily connect the helicity state of the particle with a transversely polarized rest frame spin state, or indeed with any rest frame spin state.

Exercise 11-8. Suppose that in the x', y', z' rest frame a particle of spin j has spin component m along an axis of quantization described by the angles θ_1 and

ϕ_1. Note that these angles are in the x', y', z' coordinate system. Show that the helicity state of the particle is given by

$$|\psi\rangle_{\text{helicity}} = \sum_{\mu} D^j_{\mu m}(-\phi_1, -\theta_1, \phi_1)|\mathbf{p}\mu\rangle \tag{11-14}$$

The special rest frame just defined allows us to extend the definitions of transverse and longitudinal polarization. When we speak here of the transverse or longitudinal polarization of a relativistic particle, we mean the transverse or longitudinal polarization as calculated by Eqs. 11-13 *in the special rest frame.* Actually for the longitudinal polarization there is no need to go to the special rest frame. We can use the helicity operator h, of Eq. 9-33, to define the longitudinal polarization; namely,

$$P_{\text{long}} = \frac{\text{Tr}\,(\rho_{\text{hel}}h)}{j\,\text{Tr}\,\rho_{\text{hel}}} = \frac{\sum_{\mu} \mu\rho_{\text{hel},\mu\mu}}{j\,\text{Tr}\,\rho_{\text{hel}}} \tag{11-15}$$

As we observed earlier in this section, for particles of *arbitrary* spin the concept of vector polarization represents just one particularly simple kind of polarized state. Much more complicated kinds of polarized states are possible. But for spin $\frac{1}{2}$ particles the specification of the vector polarization completely describes the polarization of the particle. We conclude this section by demonstrating this very useful point. For spin $\frac{1}{2}$ particles $2\mathbf{J} = \sigma$ where $\sigma_x, \sigma_y, \sigma_z$ are the Pauli matrices, we have

$$\sigma_x = \begin{bmatrix} 0 & 1 \\ 1 & 0 \end{bmatrix} \qquad \sigma_y = \begin{bmatrix} 0 & -i \\ i & 0 \end{bmatrix} \qquad \sigma_z = \begin{bmatrix} 1 & 0 \\ 0 & -1 \end{bmatrix} \tag{11-16}$$

The spin $\frac{1}{2}$ density matrix $\rho_{1/2}$ has only four independent real parameters because it is hermitian. Therefore it can be expressed as the linear combination of four independent hermitian matrices. We choose these matrices to be the unit matrix I and the Pauli matrices. Hence we have

$$\rho_{1/2} = a_0 I + \sum_{i=x,y,z} a_i\sigma_i \tag{11-17}$$

The nice thing about this particular linear combination is that the vector polarization is immediately given by

$$P_i = \frac{a_i}{a_0} \tag{11-18}$$

Thus

$$\rho_{1/2} = a_0(I + \mathbf{P}\cdot\sigma) \tag{11-19}$$

Equation 11-19 is the most general form for the density matrix for spin $\frac{1}{2}$ particles. This formalism for spin $\frac{1}{2}$ particles can be extended to particles of higher spin, although the formulas become more complicated.

Exercise 11-9. Fill in the steps leading to Eq. 11-19. What are the restrictions on the values of the a_i's?

11-4 Differential Cross Sections and Vector Polarization in Two-Body Reactions. In this section we use the methods developed in the previous two chapters and in this chapter to discuss the polarization of particles in two-body interactions. This vast subject has been extensively developed in both nuclear physics (11BA) and in elementary particle physics. In this section we present its simpler aspects, thus preparing for the discussion of the next section on the polarization data of high energy, meson–nucleon elastic scattering.

We consider the two-body reaction $a + b \to c + d$ described by the center of mass helicity amplitudes $f_{\mu_c\mu_d;\mu_a\mu_b}(s, \theta, \phi)$ of Eq. 10-9. To shorten the notation we usually suppress the explicit dependence on s, θ, and ϕ. The most general description of the initial helicity states of particles a and b would be provided by a density matrix ρ^{in} of dimensions

$$[(2s_a + 1)(2s_b + 1)]^2$$

where s_i is the spin of particle i. Of course, the $(2s_i + 1)$ factor is replaced by 2 if the particle has zero mass but nonzero spin. The density matrix ρ^{out} for the final state has similar properties. If we label the rows (columns) of ρ^{in} by the number pairs $\mu_a\mu_b(\mu_a'\mu_b')$, and similarly label ρ^{out}, we obtain the very general relationship

$$\rho^{\text{out}}_{\mu_c\mu_d,\mu_c'\mu_d'} = \sum_{\mu_a\mu_b,\mu_a'\mu_b'} f_{\mu_c\mu_d;\mu_a\mu_b} f^*_{\mu_c'\mu_d';\mu_a'\mu_b'} \rho^{\text{in}}_{\mu_a\mu_b;\mu_a'\mu_b'} \tag{11-20a}$$

Or regarding $f_{\mu_c\mu_d,\mu_a\mu_b}$ as a matrix F, we have the matrix equation

$$\rho^{\text{out}} = F\rho^{\text{in}}F^\dagger \tag{11-20b}$$

Exercise 11-10. Write out Eq. 11-20b in explicit matrix form for the reaction $K^- + d \to \Lambda^0 + n$.

Digressing briefly, we can show that if both particles in the initial state are unpolarized, parity conservation leads to a very useful restriction on ρ^{out}. This restriction is most easily expressed if we suppose that the reaction $a + b \to c + d$

takes place in the x–z plane; hence $\phi = 0$. Equation 11-20a, for both initial state particles unpolarized, reduces to

$$\rho^{\text{out}}_{\mu_c\mu_d;\mu'_c\mu'_d} = \sum_{\mu_a\mu_b} f_{\mu_c\mu_d;\mu_a\mu_b} f^*_{\mu'_c\mu'_d;\mu_a\mu_b} \tag{11-21}$$

Now under parity conservation, Eq. 10-19a applied to Eq. 11-21 yields

$$\rho^{\text{out}}_{-\mu_c,-\mu_d;-\mu'_c,-\mu'_d} = (-1)^{\mu_c+\mu_d-\mu'_c-\mu'_d}\rho^{\text{out}}_{\mu_c\mu_d;\mu'_c\mu'_d} \tag{11-22}$$

Thus the number of independent elements in the density matrix ρ^{out} is reduced by about a factor of 2. Remember that this restriction has been derived in the helicity basis. Similar restrictions hold in other bases for the density matrix, but the restriction will be different from that of Eq. 11-22.

Exercise 11-11. Derive Eq. 11-22. Remember that $f^*_{\mu'_c\mu'_a;\mu_a\mu_b}$ not $f_{\mu'_c\mu'_d;\mu_a\mu_b}$ occurs in Eq. 11-21.

Returning to Eqs. 11-20, we can write a number of formulas for the differential cross section. Thus the center of mass differential cross section for particle c having helicity μ_c and particle d having helicity μ_d is

$$\left(\frac{d\sigma}{d\Omega}\right)_{u_c u_d} = \frac{\rho^{\text{out}}_{\mu_c\mu_d;\mu_c\mu_d}}{\text{Tr}\,\rho^{\text{in}}} \tag{11-23a}$$

Exercise 11-12. Why does the term $\text{Tr}\,\rho^{\text{in}}$ appear in Eqs. 11-23?

The center of mass differential cross section summed over all spin states of particles c and d is

$$\frac{d\sigma}{d\Omega} = \frac{\text{Tr}\,\rho^{\text{out}}}{\text{Tr}\,\rho^{\text{in}}} \tag{11-23b}$$

These equations are obviously extensions of Eq. 10-10.

We could also write formulas for the vector polarizations of particles c and d. But the very generality of Eqs. 11-23 makes them difficult to handle. Therefore, for educational purposes, it is better to simplify them by supposing that (1) particles a and b are either unpolarized or have spin 0, and (2) either the polarization of particle d is undetected or particle d has spin 0. Actually such conditions apply to a large class of experiments. When they do not apply, it is better to work with the specific matrices as in Ex. 11-10. Under conditions (1) and (2) the polarization state of particle c is described by the density matrix.

$$\rho^{c}_{\mu_c,\mu'_c} = \sum_{\mu_a\mu_b\mu_d} f_{\mu_c\mu_d;\mu_a\mu_b} f^*_{\mu'_c\mu_d;\mu_a\mu_b} \tag{11-24}$$

Then we have

$$\rho^c_{-\mu_c, -\mu_c'} = (-1)^{\mu_c - \mu_c'} \rho^c_{\mu_c \mu_c'} \tag{11-25}$$

We now use Eq. 11-25 to derive a classic result on the vector polarization $\mathbf{P}^{\text{out}}$ of a particle under conditions (1) and (2). First we must recall that we define $\mathbf{P}^{\text{out}}$ in the special rest frame x', y', z' discussed in the previous section. We have already said that the reaction leading to Eq. 11-25 takes place in the xz plane. As usual, we take the angle between the three-momentum $\mathbf{p}_c$ of particle c and the z axis to be θ. Then the special rest frame x', y', z' is simply obtained from xyz by a rotation by the angle θ around the y axis. Thus the y' axis coincides with the y axis. Let us consider first $P^{\text{out}}_{z'}$. Using Eqs. 11-25 and 11-13e we find that

$$P^{\text{out}}_{z'} = (j \operatorname{Tr} \rho^c)^{-1} \sum_{\mu} \mu \rho^c_{\mu\mu} = 0 \tag{11-26a}$$

Thus there is zero longitudinal polarization. Turning next to $P^{\text{out}}_{x'}$, the transverse polarization in the $x'z'$ plane, Eq. 11-13c leads to the result

$$P^{\text{out}}_{x'} = 0 \tag{11-26b}$$

Thus the transverse polarization in the plane of the reaction is also zero. Indeed, the only component of the vector polarization that need not be zero is $P^{\text{out}}_{y'}$.

Exercise 11-13. Derive Eq. 11-26b. Show that $P^{\text{out}}_{y'}$ need not be zero.

This is the classic result; the vector polarization must be perpendicular to the plane of the reaction if the particles in the initial state are unpolarized, if the spin state of the other particle in the final state is summed over (or undetected), and if parity is conserved. This is a relativistically correct statement—we have been careful to work in the special rest frame.

Exercise 11-14. This is such a simple result that we might expect to find a simpler derivation. Find such a derivation by using the axial vector nature of the polarization vector **P**. Show that **P** can only be constructed by taking the cross product of the three-momenta of one initial state particle and one final state particle.

Exercise 11-15. Show that the proof of Ex. 11-14 can be extended to a reaction in which *many* particles are produced in the final state but only *one* particle is detected. Consider

$$a + b \to c + d + e + \cdots$$

with particles a and b unpolarized and only particle c detected. In the center of mass system of particles a and b we have the same number of independent vectors available as we had in the two-particle final state case. The argument is the same as before, and the vector polarization of particle c must be perpendicular to the plane containing the three-momenta $\mathbf{p}_a$, $\mathbf{p}_b$, and $\mathbf{p}_c$.

At this point we turn from our development of the theories of the density matrix and polarization states back to our study of the data of elementary particle physics. From the educational viewpoint, it is important to begin to apply these concepts to the data. But more important, we want to see if we can gain additional insight into high energy interactions by applying these concepts. In the next section we make the application to elastic scattering.

11-5 Polarization Phenomena and Helicity Amplitudes in Pion–Nucleon Elastic Scattering. The pion–nucleon elastic scattering process involves in the initial and final states one spin $\frac{1}{2}$ particle and one spin zero particle. This is the simplest system involving particles with spin. And perhaps we have developed almost too powerful a formalism to apply to this simplest of cases. Yet once we get into the details of the experimental amplitudes, we shall be pleased that this is such a simple case. In pion–nucleon elastic scattering we take particles a and c to be the nucleons. As discussed in Ch. 10, the two independent helicity amplitudes, with their abbreviated notation, are

$$f_{+1/2,0;+1/2,0}(s,\theta,\phi) \equiv f_{++}$$

$$f_{-1/2,0;+1/2,0}(s,\theta,\phi) \equiv f_{-+}$$

In the cms we take the scattering to occur in the xz plane, and we take the incident nucleon, particle a, to be moving in the $+z$ direction. The special rest frame of particle a, x'_a, y'_a, z'_a, is then the same as the frame x, y, z. Therefore we simply use x, y, z and the unit vectors $\hat{n}_x$, $\hat{n}_y$, $\hat{n}_z$ for the special rest frame of the incident nucleon. The final nucleon, particle c, has three-momentum $\mathbf{p}_c$ making an angle θ with the $+z$ axis. The special rest frame should be designated by x'_c, y'_c, z'_c; but to simplify the notation we use x', y', z' and the unit vectors $\hat{n}'_x$, $\hat{n}'_y$, $\hat{n}'_z$. These coordinate systems are illustrated in Fig. 11-1.

For the most general case we suppose the incident nucleon to have the normalized density matrix of Eq. 11-19

$$\rho^{\text{in}} = \tfrac{1}{2}[I + \mathbf{P}^{\text{in}} \cdot \sigma] \tag{11-27a}$$

This density matrix is in the special rest frame x, y, z of the incident particle. The F-matrix, using parity conservation as expressed in Eq. 10-19a, is

$$F = \begin{bmatrix} f_{++} & -f_{-+} \\ f_{-+} & f_{++} \end{bmatrix}$$

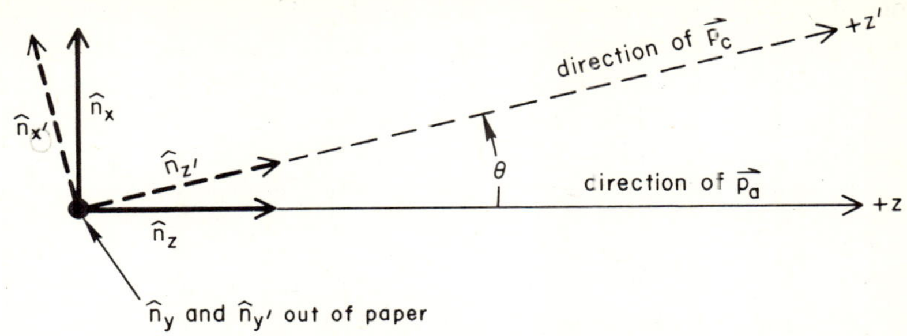

Fig. 11-1 The $\hat{n}_x$, $\hat{n}_y$, $\hat{n}_z$ define the special rest frame for incident nucleon a; $\hat{n}_{x'}$, $\hat{n}_{y'}$, $\hat{n}_{z'}$ define the special rest frame for scattered nucleon c; in πp elastic scattering.

The density matrix of the final nucleon in the x_c', y_c', z_c' special rest frame is given by Eq. 11-20b, namely,

$$\rho^{\text{out}} = F\rho^{\text{in}}F^+$$

$$= \frac{1}{2}\begin{bmatrix} f_{++} & -f_{-+} \\ f_{-+} & f_{++} \end{bmatrix} [I + \mathbf{P}^{\text{in}} \cdot \sigma] \begin{bmatrix} f_{++}^* & f_{-+}^* \\ -f_{-+}^* & f_{++}^* \end{bmatrix} \qquad (11\text{-}27b)$$

Exercise 11-16. Write out explicitly ρ^{out} in Eq. 11-27b. The calculation is straightforward but tedious. Then derive Eqs. 11-28 and 11-29.

From ρ^{out} we calculate the differential cross section in the center of mass and the vector polarization $\mathbf{P}^{\text{out}}$ of the final nucleon in its special rest frame. The results are

$$\frac{d\sigma}{d\Lambda} = I_0(1 + P_0 P_y^{\text{in}}) \qquad (11\text{-}28)$$

$$P_{x'}^{\text{out}} = \frac{I_0(R_0 P_x^{\text{in}} + A_0 P_z^{\text{in}})}{d\sigma/d\Omega} \qquad (11\text{-}29a)$$

$$P_{y'}^{\text{out}} = \frac{I_0(P_0 + P_y^{\text{in}})}{d\sigma/d\Omega} \qquad (11\text{-}29b)$$

$$P_{z'}^{\text{out}} = \frac{I_0(-A_0 P_x^{\text{in}} + R_0 P_z^{\text{in}})}{d\sigma/d\Omega} \qquad (11\text{-}29c)$$

where

$$I_0 = [|f_{++}|^2 + |f_{-+}|^2] \tag{11-30a}$$

$$P_0 = \frac{-2\,\mathrm{Im}(f_{++}f_{-+}^*)}{|f_{++}|^2 + |f_{-+}|^2} \tag{11-30b}$$

$$R_0 = \frac{|f_{++}|^2 - |f_{-+}|^2}{|f_{++}|^2 + |f_{-+}|^2} \tag{11-30c}$$

$$A_0 = \frac{2\,\mathrm{Re}(f_{++}f_{-+}^*)}{|f_{++}|^2 + |f_{-+}|^2} \tag{11-30d}$$

The parameters I_0, P_0, R_0, and A_0 are all functions of the energy and scattering angle θ in the center of mass. (The reaction is confined to the xz plane, thus $\phi = 0$.) However, for brevity of notation we omit the explicit dependence on energy and θ.

To understand Eqs. 11-28 to 11-30, first suppose that $\mathbf{P}^{\text{in}} = 0$. Then as we expect from the results of the last section, the only polarization in the final state is in the $\hat{n}_y$ direction—the direction perpendicular to the reaction plane. Furthermore, since $d\sigma/d\Omega = I_0$, in this case $P_{y'}^{\text{out}} = P_0$. This is the physical significance of P_0. It is the magnitude of the polarization produced by an unpolarized initial state.

Next suppose that the only initial state polarization is in the $\hat{n}_y$ direction such that only P_y^{in} is nonzero. Then $\mathbf{P}^{\text{out}}$ is still perpendicular to the reaction plane. But now the differential cross section, given by Eq. 11-28, depends on P_y^{in} through the coefficient P_0. Thus the coefficient P_0 appears in two different expressions. This is a general result and can be extended to particles of higher spin.

Finally, we turn to the interpretation of A_0 and R_0. First we must note that there are several different definitions of these A and R parameters, some of the differences having to do with the coordinate system. For example A and R are usually measured in the laboratory system and their values given in that system; for simplicity, our parameters are defined in the cms. References 11WO, 11BA, 11BI, and 11JAC give a full discussion. For the physical significance of R_0, suppose that $\mathbf{P}^{\text{in}}$ is in the $\hat{n}_x$ direction. Then $d\sigma/d\Omega = I_0$, and Eqs. 11-29a and 11-29c become

$$P_{z'}^{\text{out}} = R_0 P_x^{\text{in}}$$
$$P_{z'}^{\text{out}} = -A_0 P_x^{\text{in}} \tag{11-31}$$

Thus $R_0 P_x^{\text{in}}$ gives the polarization of the scattered nucleon in the $\hat{n}_{x'}$ direction produced by an initial polarization P_x^{in} in the $\hat{n}_x$ direction. Similarly, $-A_0 P^{\text{in}}$

gives the polarization of the final nucleon in the $n_{z'}$ direction. Thus A_0 and R_0 describe how initial state polarizations in the reaction plane change to final state polarizations in the reaction plane.

Exercise 11-17. If $f_{++} = 0, R_0 = -1$ and $A_0 = 0$. Interpret this result with respect to $\mathbf{P}^{out}$ and show that the same result follows directly from the consideration of the f amplitudes.

The parameters I_0, P_0, R_0, and A_0 are related to f_{++} and f_{-+} by Eqs. 11-30, hence they are not all independent. The simplest measurement—namely, the differential cross section—provides I_0; P_0 is determined by measurement of the differential cross section from a polarized target. Alternatively, P_0 can be determined in a double scattering experiment consisting of the polarization of a nucleon by a primary elastic scatter and the subsequent measurement of that polarization by a secondary elastic scatter. The parameters R_0 and A_0, which require an initial polarization in the plane of the reaction, are much more difficult to measure. The polarization of the final state nucleon must also be measured. Thus the measurement of R_0 and A_0 requires a secondary scattering *after* a primary scattering on a polarized target; or it requires a triple scattering experiment if the original target is unpolarized. If I_0 and P_0 are known, it is not necessary to measure both R_0 and A_0, as Eqs. 11-30 indicate. Once an experiment is set up to measure either one of the latter parameters, however, the other is also measured as an aid in reducing statistical and systematic errors. The measurements of A_0 and R_0 are difficult. And at the time of writing only a few measurements of these parameters have been made in high energy elastic scattering (11DE).

Let us now turn to the data. Typical measurements for I_0, the unpolarized differential cross section, were presented and discussed in Ch. 1. Existing high energy data for P_0 in pion-nucleon elastic scattering appear in Fig. 11-2. At $t = 0$, from Eqs. 11-30b and Sec. 10-6, we must have $P_0 = 0$; but the data do not cover $|t|$ small enough to show this. Inspection of the figure shows that at equal energies $P_0(\pi^+p)$ is rougly the mirror image of $P_0(\pi^-p)$ about the $P_0 = 0$ line. We also observe that P_0 is approximately zero at $|t| \approx 0.6$ (GeV/c)^2. We return to these observations in Sec. 17-9 and Ch. 18.

From the measurements of I_0, P_0, R_0, and A_0 at some angle θ, we would like to determine the two complex quantities f_{++} and f_{-+} at that angle θ, writing

$$f_{++} = |f_{++}(\theta)|\, e^{i\phi_{++}(\theta)}$$
$$f_{-+} = |f_{-+}(\theta)|\, e^{i\phi_{-+}(\theta)} \tag{11-32}$$

Equations 11-30 show that we can determine $|f_{++}(\theta)|$, $|f_{-+}(\theta)|$, and $[\phi_{++}(\theta) - \phi_{-+}(\theta)]$, but we cannot use these equations to determine an absolute phase, say, $\phi_{++}(\theta)$.

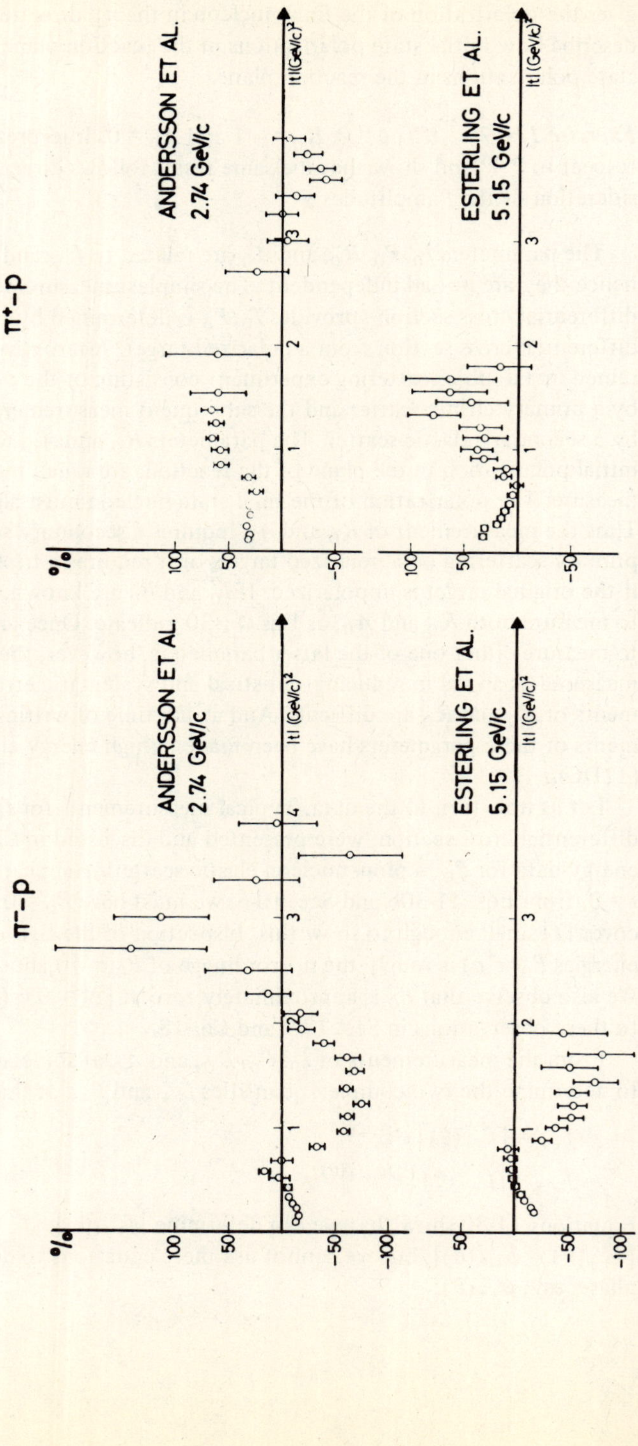
π±-p POLARIZATION
π⁻-p
π⁺-p
%
ANDERSSON ET AL.
2.74 GeV/c
ESTERLING ET AL.
5.15 GeV/c
|t| (GeV/c)²
100
50
-50
-100
1
2
3
4

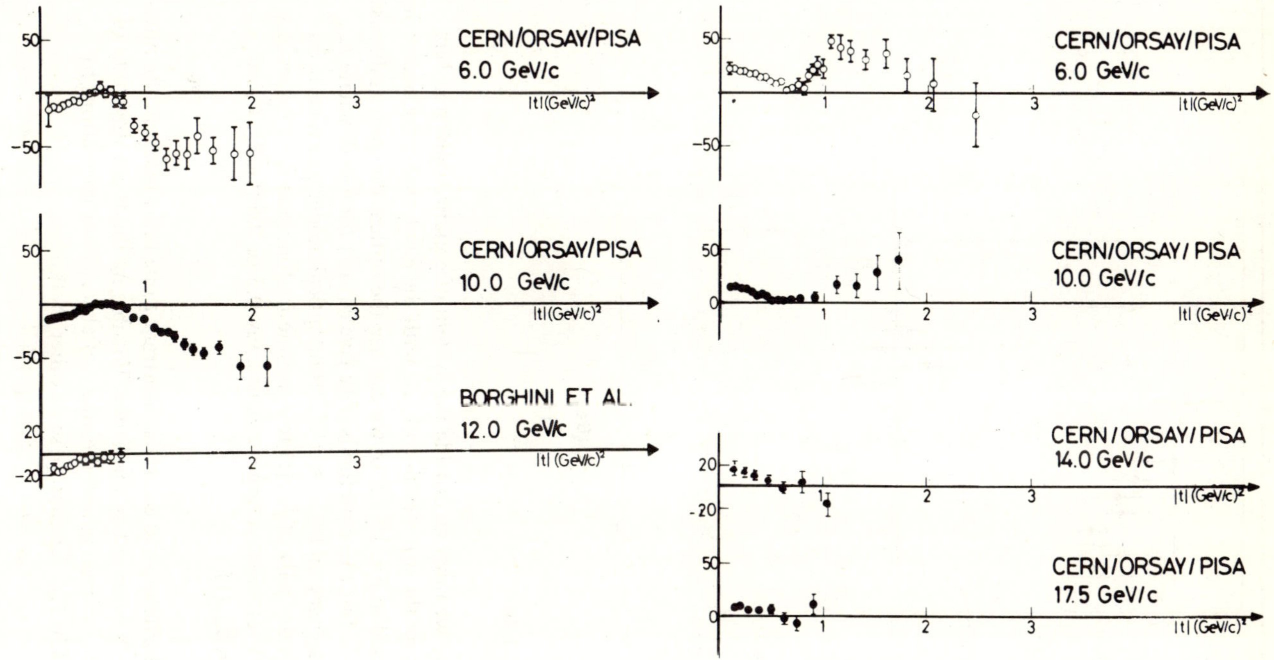

(a)

Fig. 11-2 The polarization P_0 in πp elastic scattering: (*a*) taken from Ref. 11AL, (*b*) taken from Ref. 11GI.

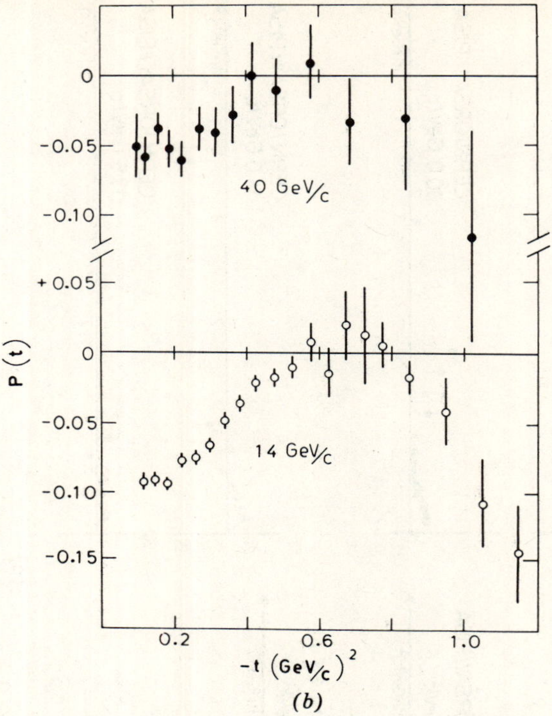

Fig. 11-2b

To determine ϕ_{++} we must go outside the system of equations given in Eqs. 11-30. We can measure the interference of f_{++} with the electromagnetic scattering amplitude as discussed in Ex. 10-12, or we can use the optical theorem as discussed in Secs. 1-7 and 10-6. But these methods only give $\phi_{++}(\theta)$ for $\theta \approx 0$, where f_{++} is almost purely imaginary and $\phi_{++} \approx \pi/2$. From optical model considerations we expect the imaginary part of $f_{++}(t)$ to continue to dominate for t values less than 0.1 or 0.2 $(\text{GeV/c})^2$. But at larger t values, say 0.3 or 0.5 $(\text{GeV/c})^2$, we simply do not know $\phi_{++}(t)$.

Exercise 11-18. To determine $\phi_{++}(t)$ at arbitrary t one must measure the interference of $f_{++}(t)$ with some other amplitude. Possible methods are the study of pion–deuteron elastic scattering and the study of absorptions corrections to inelastic reactions (Ch. 13). Discuss these or other methods.

In the analysis of the elastic scattering amplitudes in the diffraction region, we wish to separate the dominant *diffractive* contribution from the contributions of *other high energy mechanisms.* These other mechanisms are t-channel particle exchanges (Secs. 7-5 and 7-6); or in more modern language, t-channel Regge trajectories (Ch. 17 and Sec. 17-9). This separation can be partially accomplished by analyzing the amplitudes in terms of *t-channel isotopic spin amplitudes.* Then the diffractive contribution, which is common to all hadron–hadron elastic reactions, appears only in the $I_t = 0$ amplitudes (I_t being the total isotopic spin in the t-channel). But we must remember that although the diffractive processes are the dominant contribution to the $I_t = 0$ amplitudes, there can be other contributions. For example, the exchange of an $I = 0$ particle such as the ω will contribute to the $I_t = 0$ amplitudes.

Exercise 11-19. For a further discussion of t-channel exchange and the related idea of crossing, the reader may wish to look ahead to Chs. 15 and 16. In particular, Sec. 16-4 discusses the relation between s-channel and t-channel isotopic spin amplitudes.

Using the method of Sec. 16-4 and the phase space choices of Ref. 11JO we write the elastic scattering helicity amplitudes as follows:

$$\begin{aligned} f_{++}(\pi^{\pm} + p \to \pi^{\pm} + p) &= F^0_{++} \mp F^1_{++} \\ f_{-+}(\pi^{\pm} + p \to \pi^{\pm} + p) &= F^0_{-+} \mp F^1_{-+} \end{aligned} \tag{11-33}$$

The superscript denotes the isotopic spin in the t-channel. But remember that these are still s-channel helicity amplitudes. In principle, Eqs. 11-28, 11-29, 11-30, and 11-33 are sufficient. For the experimental determination of the amplitudes, however, it is very useful to use information from the inelastic charge exchange reaction $\pi^- + p \to \pi^0 + n$, whose amplitude is given by

$$f_{\substack{++\\-+}}(\pi^- + p \to \pi^0 + n) = \sqrt{2}F^1_{\substack{++\\-+}} \tag{11-34}$$

hence F^1 is determined most directly through this reaction.

These amplitudes at 6 GeV/c are given in Fig. 11-3. Here the amplitudes

$$H^I_{\mu\mu'} = \frac{\sqrt{\pi}}{p^*} F^I_{\mu'\mu} \tag{11-35}$$

are used. These amplitudes have the advantage of being Lorentz invariant; for example,

$$\frac{d\sigma}{dt}(\pi^+ p \to \pi^+ p) = |H^0_{++} - H^1_{+-}|^2 \tag{11-36}$$

In $H^I_{\mu\mu'}$ the final helicity is given as the second subscript, which is the opposite of our convention. [As this book is being written, 6 GeV/c is the highest energy at which there is sufficient information for the complete determination of the amplitudes; the difficulty at higher energies is the lack of R_0 and A_0 measurements (11CO).] As we have already noted, the absolute phase of the amplitudes cannot be determined. Therefore, as is customary, H^0_{++} is taken to be a positive imaginary quantity in Fig. 11-3; and all other phases are fixed with respect to this assumption. The assumption itself is justified by the predominantly diffractive nature of H^0_{++}, as previously discussed.

In Fig. 11-3 we see that H^0_{++} is much larger than any of the other amplitudes; and indeed this is what we expect from the optical model (Ch. 2). In particular, the other helicity nonflip amplitude H^1_{++} is a factor of 10 smaller. This amplitude comes from an $I = 1$ t-channel exchange; it does not come from diffractive processes. We also note that both helicity flip amplitudes H^0_{+-} and H^1_{+-} are small. From Eq. 11-30b, we see that the interference with Im H^0_{++} of the real parts of these amplitudes leads to nonzero polarization. And at $t = -0.6$ (GeV/c)^2 where the real parts go to zero, the polarization is also zero. In Sec. 17-9 and Ch. 18 we discuss the present explanations of why these amplitudes go through zero at the t value cited.

Figure 11-2 shows that the polarization in the small t region decreases as the energy increases. This agrees with the following picture of high energy, small t elastic scattering. As discussed in Chs. 1 and 2, the diffractive mechanism leads to an amplitude, namely, H^0_{++}, whose magnitude is either independent of energy or varies very slowly with energy. But the t-channel exchange processes that contribute to the other amplitudes decrease rather rapidly as the energy increases (Secs. 17-6 and 17-9). Therefore, P_0 decreases as the energy increases.

Before concluding this section, it is interesting to consider the partial wave analysis of these amplitudes, Sec. 1-2. Such an analysis is presented in Fig. 11-4 (11BAR): ($[F^I_{\mu\mu'}]_J = \langle \mu' 0 \,|\, T^J(s) \,|\, \mu 0 \rangle$ in Eq. 10-8a). This partial wave analysis is based on a slightly different version of the amplitudes presented in Fig. 11-3. A theoretical model was employed to predict the real part of H^0_{++}, a different fitting method was used, and some of the phase conventions differ by a negative sign. But the main features of the amplitudes are the same. As we expect, the only large amplitude is $[\text{Im } F^0_{++}]_J$; and it has the shape we expect from the discussion in Sec. 3-3. Using impact parameter language, almost all the other amplitudes have a maximum value at nonzero values of the impact parameter $b = (J + \frac{1}{2})/p^*$. They contribute most in the b range of 0.5 to 1.5 fermi. Since the hadronic radius is about 1 fermi, these amplitudes are said to be predominantly *peripheral*, whereas $[\text{Im } F^0_{++}]_J$ is said to be predominantly *central.* We return to this subject in Ch. 18.

Exercise 11-20. By consulting the literature (e.g., Refs. 11AL and 11GI), show that P_0 in elastic scattering in other hadronic systems has a magnitude and energy behavior similar to P_0 in the πN system. Also consult the literature for information

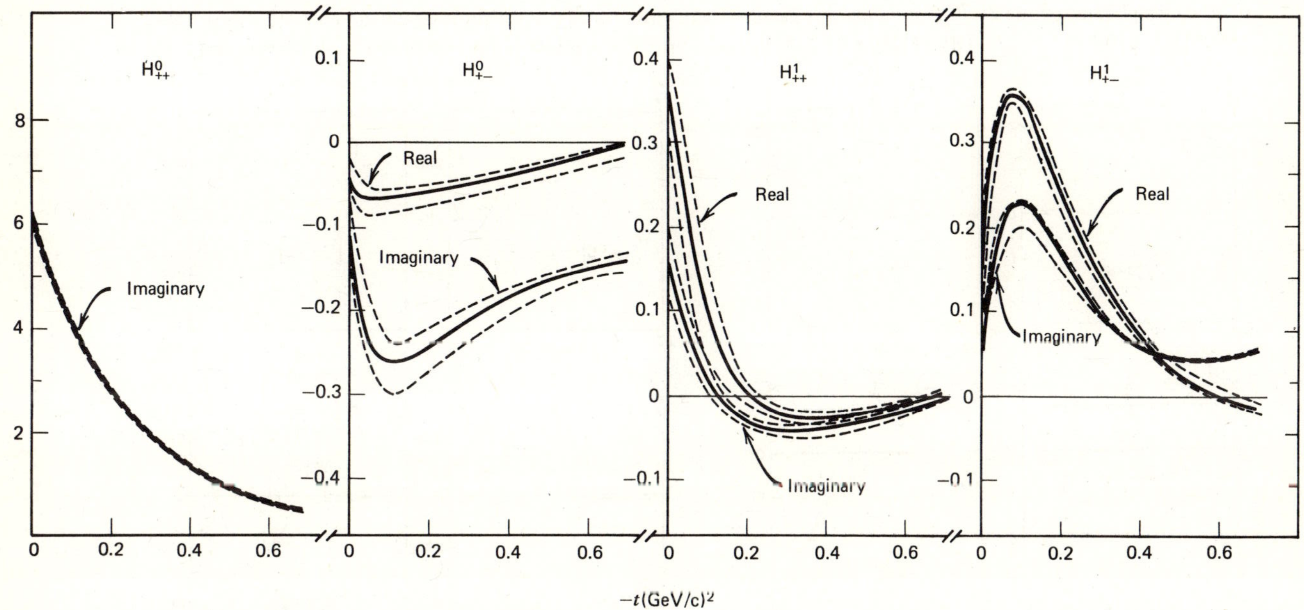

Fig. 11-3 Some *s*-channel helicity amplitudes for pion-nucleon elastic scattering at 6 GeV/c (11JO). Note the different scale used for H^0_{++}. The notation is explained in Eqs. 11-33 through 11-36.

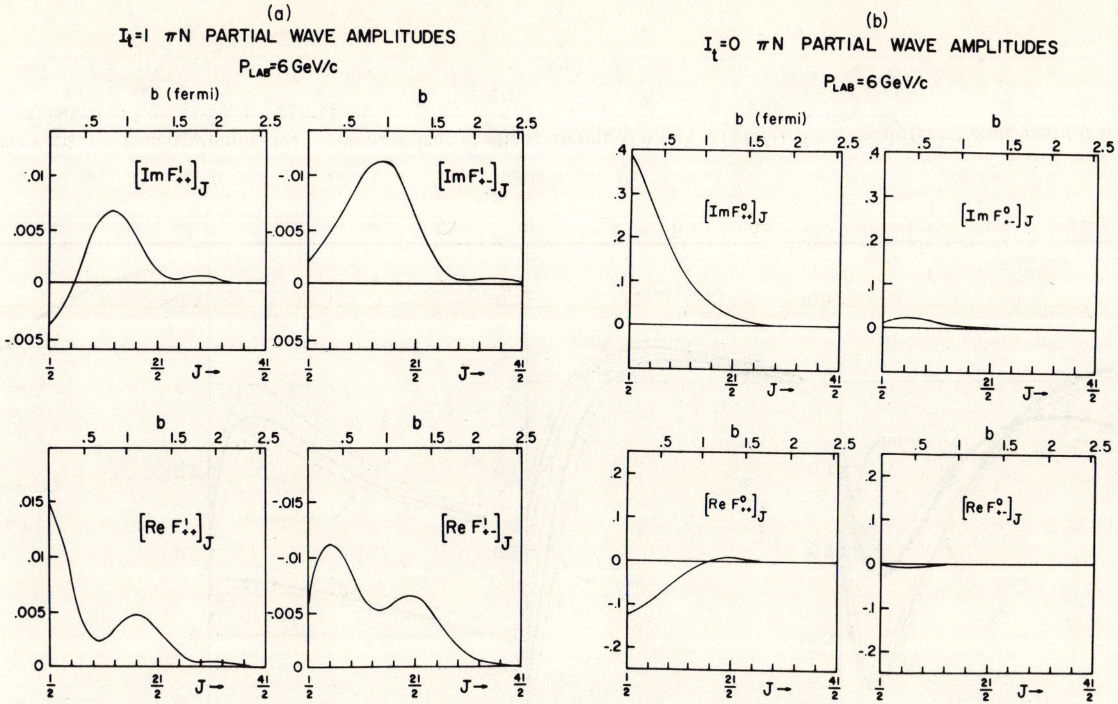

Fig. 11-4 Partial wave amplitudes for pion-nucleon elastic scattering at 6 GeV/c (11BAR): $[F^I_{\mu\mu'}]_J = \langle \mu' 0 | T^J(s) | \mu 0 \rangle$ in Eq. 10-9a; I is the t-channel isotopic spin. The smooth curves connect the values of $[F^I_{\mu\mu'}]_J$ at half-integral values of J.

on the polarization in the backward peak region of elastic scattering. Observe that in this region P_0 does not necessarily decrease as the energy increases. Explain this observation.

Exercise 11-21. Pion–nucleon elastic scattering is often described in terms of the *spin nonflip* and *spin flip* amplitudes $f(s, \theta, \phi)$ and $g(s, \theta, \phi)$. These are defined by again confining the reaction to the xz plane in the center of mass. But now the special rest system x, y, z is used *both* for the incident nucleon and for the scattered nucleon. Therefore, for the scattered nucleon, the rest system x, y, z is obtained from the x', y', z' rest system (which we used for the helicity amplitudes) by rotation about the y axis by the angle θ. Thus the spin of the scattered nucleon, as well as the spin of the incident nucleon, is quantized along the $+z$ axis. The F-matrix is written in the form (other phase conventions are often used)

$$F = fI - ig\sigma \cdot \hat{n}_y = \begin{bmatrix} f & -g \\ g & f \end{bmatrix} \tag{11-37}$$

With the definition of f and g we have produced a relativistic extension of the spin-orbit amplitudes $f_{++}^{(\circ)}$ and $f_{-+}^{(\circ)}$ which we defined much earlier (Sec. 3-5). For *nonrelativistic* particles we have

$$\begin{aligned} f &\equiv f_{++}^{(\circ)} \\ g &\equiv f_{-+}^{(\circ)} \end{aligned} \tag{11-38}$$

The relations between these spin amplitudes and the helicity amplitudes are found by comparing Eq. 11-37 with Eq. 11-27b using Eq. 9-4. For $\phi = 0$, derive the relations

$$f_{++} = f \cos \tfrac{1}{2}\theta + g \sin \tfrac{1}{2}\theta \tag{11-39a}$$

$$f_{-+} = -f \sin \tfrac{1}{2}\theta + g \cos \tfrac{1}{2}\theta \tag{11-39b}$$

11-6 Angular Distributions and Particle Polarizations in Decay Processes. We continue our application of the material developed in Chs. 9 and 10 and in Secs. 11-2 and 11-3 with the study of decay processes. Such processes can be complicated if there is a sequence of decays $a \to 1 + 2$, $1 \to 3 + 4$, or if the final state of the decay process has three or more particles. We therefore restrict our considerations to the simplest decay process, $a \to 1 + 2$ without sequential decays. The emphasis of this book is on interactions at high energy, and the principles used to describe a decay process are sufficiently well illustrated for our purposes by the decay into two bodies. Sequential decays and decays into three or more particles are important in experimental strong interaction physics for finding the internal quantum numbers of resonant particles. But no completely new principles are required to describe those processes, even though the description can be very involved. Therefore we refer the reader to several references on the methods of describing the more complicated decay processes (11NI, 11CH, 11BA, 11JAC).

Having made our apologies for the relative brevity of this section, we consider the process $a \to 1 + 2$ in the rest frame of particle a. The selection of this Lorentz frame greatly simplifies the discussion. The particle a has mass M_a and spin j and is in the spin state $|jm\rangle$ in the rest frame. The final state particles are in the two-particle, plane wave, center of mass helicity state $|\theta\phi\mu_1\mu_2\rangle$ defined in Eq. 9-42. The differential decay rate, from Eq. 6-30, is

$$\left(\frac{d\Lambda}{d\Omega}\right)_{\mu_1\mu_2;m}(\theta,\phi) = N|\langle\theta\phi\mu_1\mu_2|T|jm\rangle|^2 \tag{11-40a}$$

The spherical angles θ and ϕ give the direction of motion of particle 1 relative to the $+z$ axis, which is also the axis of quantization for the state $|jm\rangle$. The N is a proportionality constant that depends on the total decay rate and the normalization conditions. Fortunately we do not need to evaluate this proportionality constant. Rewriting, Eq. 11-40a appears as

$$\left(\frac{d\Lambda}{d\Omega}\right)_{\mu_1\mu_2;m}(\theta,\phi) = |f_{\mu_1\mu_2;m}(\theta,\phi)|^2 \tag{11-40b}$$

$$f_{\mu_1\mu_2;m}(\theta,\phi) = \langle\theta\phi\mu_1\mu_2|f|jm\rangle \tag{11-40c}$$

where we have absorbed into f the numerical factors and normalization constant. From now on we shall concentrate on elucidating the angular dependence of the amplitude $f_{\mu_1\mu_2;m}(\theta,\phi)$.

Next we use Eq. 9-47 and find

$$f_{\mu_1\mu_2;m}(\theta,\phi) = \sum_J \left[\frac{2J+1}{4\pi}\right]^{1/2} D^{J*}_{M\mu}(\phi,\theta,-\phi)\langle JM_{\mu_1\mu_2}|f|jm\rangle \tag{11-41}$$

where $\mu = \mu_1 - \mu_2$. Finally, we use the conservation of angular momentum and the invariance of the S-matrix under spatial rotation to obtain

$$f_{\mu_1\mu_2;m}(\theta,\phi) = \left(\frac{2j+1}{4\pi}\right)^{1/2} D^{j*}_{m\mu}(\phi,\theta,-\phi)\langle\mu_1\mu_2|f|m\rangle \tag{11-42}$$

The matrix element $\langle\mu_1\mu_2|f|m\rangle$ for the decay process is analogous to the matrix element $\langle\mu_c\mu_d|T^J|\mu_a\mu_b\rangle$, which we developed for two-body interactions.

It depends on the helicities of the final states μ_1 and μ_2. Since j is fixed, there is no explicit dependence on j. The matrix element also depends on the total energy in the center of mass $\sqrt{s}$; but since $s = M_a{}^2$, s is not a quantity that can be varied experimentally. It is important to observe that $\langle \mu_1\mu_2 | f | m \rangle$ cannot depend on m because m is equal to j_z. The notation $\langle \mu_1\mu_2 | f | m \rangle$ is thus both unnecessarily complicated and confusing, and we abbreviate it to $f_{\mu_1\mu_2}$ and write Eq. 11-42 in the form

$$f_{\mu_1\mu_2;m}(\theta, \phi) = \left(\frac{2j+1}{4\pi}\right)^{1/2} D^{j*}_{m\mu}(\phi, \theta, -\phi) f_{\mu_1\mu_2} \tag{11-43}$$

$$\mu = \mu_1 - \mu_2$$

Before commenting on Eq. 11-42 and discussing the general use of this result, let us carry a simple calculation to aid our physical understanding of the equation. Suppose that both particles 1 and 2 have spin zero. Then particle a must have integral spin, and we write

$$f_{00;m}(\theta, \phi) = Y_j^m(\theta, \phi) f_{00} \tag{11-44a}$$

$$\left(\frac{d\Lambda}{d\Omega}\right)_{00;m} = |f_{00}|^2 |Y_j^m(\theta, \phi)|^2 \tag{11-44b}$$

Equation 11-44b is just what we expect from more elementary considerations. The angular dependence of the wave function of a particle with integral spin in the rest frame state $|jm\rangle$ is given by the spherical harmonic $Y_j{}^m(\theta, \phi)$. If the particle decays into two spinless particles, the square of that spherical harmonic gives the angular distribution of the decay. Finally, the total decay rate from the state $|jm\rangle$ is

$$\Lambda_{00;m} = |f_{00}|^2 \tag{11-44c}$$

which is independent of m, as it should be. Thus $|f_{00}|^2$ has the direct physical interpretation of being equal to the total decay rate from the initial state $|jm\rangle$.

We now return to Eq. 11-43. The factor $[(2j+1)/4\pi]^{1/2}$ can be absorbed into $f_{\mu_1\mu_2}$, but it is conventional not to do so. As illustrated by the example just given, this convention has the advantage of giving a simple meaning to $f_{\mu_1\mu_2}$.

Following the material in Secs. 11-3 and 11-4, we can write compact formulas using the density matrix $\rho^{(a)}$ for the initial state and the density matrix $\rho^{(1,2)}$ for the final state particles:

$$\rho^{(1,2)}_{\mu_1\mu_2;\mu_1'\mu_2'} = \sum_{mm'} f_{\mu_1\mu_2;m} f^*_{\mu_1'\mu_2';m'} \rho^{(a)}_{mm'} \tag{11-45}$$

The differential decay rate to the final helicity state $\mu_1\mu_2$ is

$$\left(\frac{d\Lambda}{d\Omega}\right)_{\mu_1\mu_2} = \frac{\rho^{(1,2)}_{\mu_1\mu_2;\mu_1\mu_2}}{\mathrm{Tr}\,\rho^{(a)}} \tag{11-46}$$

The total decay rate to the final helicity state $(\mu_1\mu_2)$ is

$$\Lambda_{\mu_1\mu_2} = |f_{\mu_1\mu_2}|^2 \tag{11-47}$$

Equation 11-47 is the generalization of Eq. 11-44c, and its general meaning is discussed in Ex. 11-22. Finally, the total decay rate is

$$\Lambda = \sum_{\mu_1\mu_2} \Lambda_{\mu_1\mu_2} = \sum_{\mu_1\mu_2} |f_{\mu_1\mu_2}|^2 \tag{11-48}$$

Exercise 11-22. Derive Eqs. 11-42 through 11-48. Use Eq. 11-47 to interpret $f_{\mu_1\mu_2}$. When $\mu \neq 0$ in Eq. 11-43, an additional phase $e^{-i\mu\phi}$ comes from our definition of helicity states. This is often removed by a rotation of the rest frame of particle *a*.

We now give the restrictions imposed by parity conservation on the decay process $a \to 1 + 2$. Using the methods of Ch. 10, we have

$$f_{-\mu_1,-\mu_2} = \eta_a\eta_1\eta_2(-1)^{s_1+s_2-j} f_{\mu_1\mu_2} \tag{11-49}$$

where the η's are the intrinsic parities of the particles. Thus in the parity conserving, strong interaction decay of the $\Delta(1236)$ isobar, $\Delta(1236) \to p + \pi$, there is just one independent amplitude, say, $f_{+1/2,0}$. But in the parity nonconserving, weak interaction decay process, $\Lambda^0 \to p + \pi^-$, there are two independent amplitudes $f_{+1/2,0}$ and $f_{-1/2,0}$.

To conclude this section we give an example of the angular distribution produced in the strong interaction decay of a spin 1 particle to two spin zero particles. This example includes the prominent decays

$$\rho \to \pi + \pi \qquad \text{and} \qquad K^*(890) \to K + \pi$$

We assume that particle *a* was produced by a strong interaction process from unpolarized initial state particles α and β, namely

$$\alpha + \beta \to a + b$$

This is almost always true experimentally. We also assume that $\rho^{(a)}$ results from a summation over the spin states of particle *b*. Then from Eq. 11-25 the hermitian, normalized density matrix in the helicity basis is

$$\rho^{(a)} = \begin{bmatrix} \rho_{11} & \rho_{10} & \rho_{1-1} \\ \rho_{10}^* & 1 - 2\rho_{11} & -\rho_{10}^* \\ \rho_{1-1} & -\rho_{10} & \rho_{11} \end{bmatrix}, \qquad \rho_{1-1} \text{ real} \tag{11-50}$$

with just four independent real parameters.

Exercise 11-23. Derive Eq. 11-50.

Exercise 11-24. Using the two assumptions just stated, write out the normalized density matrices for spin $\frac{3}{2}$ and spin 2 particles, and give the minimum number of independent real parameters required in each case.

Using Eq. 11-46 it is now straightforward to write out the differential decay rate, usually called the angular distribution, for the decay of a spin 1 particle to two spin zero particles under the foregoing assumptions. The by now classic result is

$$\frac{d\Lambda}{d\Omega} = \frac{1}{4\pi}\,[1 + (1 - 3\cos^2\theta)(3\rho_{11} - 1)$$
$$-(3\sqrt{2}\sin 2\theta\cos\phi)\,\text{Re}\,\rho_{10} - (3\sin^2\theta\cos 2\phi)\rho_{1-1}] \tag{11-51}$$

Exercise 11-25. Observe that Im ρ_{10} cannot be determined from Eq. 11-51. Investigate the origin of the lack of dependence of the angular distribution of Im ρ_{10}.

REFERENCES

11AL J. Allaby, in *Proceedings of the Fifteenth International Conference on High Energy Physics, 1970* (Naukova Dumka, Kiev, 1972).

11BA H. H. Barschall and W. Haeberli, Eds., *Polarization Phenomena in Nuclear Reactions* (University of Wisconsin Press, Madison, 1970).

11BAR V. Barger and F. Halzen, *Phys. Rev.* **D6**, 1918 (1972).

11BI A. Bialas and B. E. Y. Svensson, *Nuovo Cimento* **42**, 908 (1964).

11CH S. U. Chung, CERN Report 71-8 (CERN, Geneva, 1971).

11CO G. Cozzika et al., *Phys. Lett.* **40B**, 281 (1972).

11DE A. De Lesquen et al., *Phys. Lett.* **40B**, 277 (1972).

11GI G. Giacomelli, in *Proceedings of the Sixteenth International Conference on High Energy Physics, 1972* (National Accelerator Laboratory, Batavia, Ill., 1973).

11JA J. D. Jackson, in *Les Houches Lectures in High Energy Physics, 1965*, C. DeWitt and M. Jacob, Eds. (Gordon and Breach, New York, 1966).

11JAC M. Jacob, in *Proceedings of the International Conference on Polarized Targets and Ion Sources, 1966* (Centre d'Etudes Nucleaires de Saclay, Gif-sur-Yvette, France, 1967).

11JO P. Johnson et al., *Phys. Rev. Lett.* **30**, 242 (1973).

11NI M. Nikolic, Ed, *Analysis of Scattering and Decay* (Gordon and Breach, New York, 1968).

11PI H. Pilkuhn, *The Interaction of Hadrons* (North-Holland, Amsterdam, 1967).

11WO L. Wolfenstein, *Phys. Rev.* **96**, 1654 (1954); *Ann. Rev. Nucl. Sci.* **6**, 43 (1956).

CHAPTER

12

The Elementary One-Particle Exchange Model

12-1 Introduction. In this chapter we return to the one-particle exchange explanation of forward and backward peaks in two-body inelastic interactions—a subject introduced in Ch. 7. We discuss the simplest version of one-particle exchange in which the exchanged particle is a virtual elementary particle having all the internal quantum numbers of its real elementary particle counterpart. We ignore in this chapter the absorption and form factor corrections to the model discussed in Ch. 13. We call this very simple model the *elementary one-particle exchange* model.

Working through the chapter the reader may wonder why we have bothered to present the elementary one-particle exchange model. Although it explains the existence of forward and backward peaks in two-body inelastic interactions, the model usually gives the wrong differential cross-section shape, the wrong energy dependence of the reaction cross section, and the wrong density matrix elements.

In spite of these grave difficulties, however, the model is the foundation for more complicated models that have been developed to give better fits to experiment. Also, it provides insight into some general rules that have been incorporated into our picture of the behavior of two-body inelastic interactions. Finally, this model provides simple illustrations and exercises that involve frequently used manipulations of spin functions. General references for this chapter are 12DO, 12HU, 12JAC, and 12PI.

One-particle exchange models are usually applied to two-body reactions in which one or both of the final state particles are strong interaction resonances. We ignore, however, the resonant nature of these particles and deal with them as though they were stable. This is the conventional approach in one-particle exchange models, and justification for it is given in the literature (12JA).

12-2 Relativistic Spin Functions. In Chs. 9 and 10 we developed the beautiful and general helicity formalism—a formalism specifically designed for relativistic considerations involving spin. Yet as so often happens when we must calculate, we find that a general formalism is too general; we find that we need explicit expressions. This is the case at this point. For spin zero, the spin function φ is simply a Lorentz scalar if the particle has positive intrinsic parity and a Lorentz pseudoscalar if the particle has negative parity. The relativistic spin $\frac{1}{2}$ functions are, of course, more complicated. We review these functions here to connect them with the general helicity formalism and to establish the exact functions used in our calculations.

First we note that the positive energy, rest frame spin states $|0\lambda\rangle$ defined in Sec. 9-5 are

$$|0, +\tfrac{1}{2}\rangle = \sqrt{2m}\begin{bmatrix}1\\0\\0\\0\end{bmatrix}, \qquad |0, -\tfrac{1}{2}\rangle = \sqrt{2m}\begin{bmatrix}0\\1\\0\\0\end{bmatrix} \tag{12-1}$$

It is now easy to construct the helicity states with $\mathbf{p}$ along the z axis

$$|p_z\lambda\rangle = L(p_z)|0\lambda\rangle \tag{12-2}$$

where $L(p_z)$ can be taken directly from Eq. 9-27. We easily obtain

$$u^+(p_z) \equiv |p_z, +\tfrac{1}{2}\rangle = \sqrt{E+m}\begin{bmatrix}1\\0\\p_z/(E+m)\\0\end{bmatrix},$$

$$u^-(p_z) \equiv |p_z, -\tfrac{1}{2}\rangle = \sqrt{E+m}\begin{bmatrix} 0 \\ 1 \\ 0 \\ -p_z/(E+m) \end{bmatrix} \qquad (12\text{-}3)$$

where m is the particle mass, $E = (\mathbf{p}^2 + m^2)^{1/2}$, and in this case $p_z = |\mathbf{p}|$. Note that our spinor normalization is

$$u^{\lambda'\dagger}(\mathbf{p})u^\lambda(\mathbf{p}) = 2E\delta_{\lambda'\lambda}; \qquad \bar{u}^{\lambda'}(\mathbf{p})u^\lambda(\mathbf{p}) = 2m\delta_{\lambda'\lambda} \qquad (12\text{-}4)$$

However, we need more general helicity functions $u^\lambda(\mathbf{p})$, in which $\mathbf{p}$ is along an arbitrary direction in the xz plane given by the angle $\theta = \theta_y$, measured from the $+z$ axis. It is easy to construct this function using

$$u^\lambda(\mathbf{p}) \equiv |\mathbf{p}, \lambda\rangle = R(0, \theta, 0)L(p_z)|0\lambda\rangle \qquad (12\text{-}5)$$

which is simply Eq. 9-30. Then, using Eq. 9-26, we write

$$u^+(\mathbf{p}) \equiv |\mathbf{p}, +\tfrac{1}{2}\rangle = \sqrt{E+m}\begin{bmatrix} \cos\frac{1}{2}\theta \\ \sin\frac{1}{2}\theta \\ \dfrac{p\cos\frac{1}{2}\theta}{E+m} \\ \dfrac{p\sin\frac{1}{2}\theta}{E+m} \end{bmatrix},$$

$$u^-(\mathbf{p}) \equiv |\mathbf{p}, -\tfrac{1}{2}\rangle = \sqrt{E+m}\begin{bmatrix} -\sin\frac{1}{2}\theta \\ \cos\frac{1}{2}\theta \\ \dfrac{p\sin\frac{1}{2}\theta}{E+m} \\ \dfrac{-p\cos\frac{1}{2}\theta}{E+m} \end{bmatrix} \qquad (12\text{-}6)$$

where $p = |\mathbf{p}|$.

Exercise 12-1. Derive Eqs. 12-3 and 12-6.

To conclude this section we develop the relativistic spin functions for spin 1 particles, often called *vector* particles. To fix our ideas we first consider a spin 1 particle in its rest frame. The rest frame spin states are $|0, +1\rangle$, $|0, -1\rangle$, $|0, 0\rangle$, the three components of a first-rank *spherical tensor* (Sec. 9-3). As also discussed in Sec. 9-3, we can describe the spin states in terms of a first-rank *Cartesian tensor* s with $s_x = (1, 0, 0)$, $s_y = (0, 1, 0)$, $s_z = (0, 0, 1)$. It is most convenient, although somewhat confusing, to continue to use the states $|0\lambda\rangle$, however, *and* to express them in a *Cartesian base.* Thus, following the phase convention of Eq. 9-19, we write

$$\epsilon^{(+1)} \equiv |0, +1\rangle = \frac{1}{\sqrt{2}}(-1, -i, 0)$$

$$\epsilon^{(-1)} \equiv |0, -1\rangle = \frac{1}{\sqrt{2}}(+1, -i, 0) \tag{12-7}$$

$$\epsilon^{(0)} \equiv |0, 0\rangle = (0, 0, 1)$$

Our next tasks are to write $\epsilon^{(\lambda)}$ as a four-vector $(\epsilon^{(\lambda)0}, \epsilon^{(\lambda)1}, \epsilon^{(\lambda)2}, \epsilon^{(\lambda)3})$ and to boost these functions out of the rest frame. The first task is easily accomplished by noting (1) that the spin components of a particle must change under a general spatial rotation, and (2) that the time component of a four-vector does not change under spatial rotation. These two conditions can only be satisfied if $\epsilon^{(\lambda)0} = 0$ in the rest frame. Hence Eq. 12-7 becomes

$$\epsilon^{(\pm 1)}(0) \equiv |0, \pm 1\rangle = \frac{1}{\sqrt{2}}(0, \mp 1, -i, 0)$$

$$\epsilon^{(0)}(0) \equiv |0, 0\rangle = (0, 0, 0, 1) \tag{12-8}$$

Our second task is dispatched by using $|p_z\lambda\rangle = L(p_z)|0\lambda\rangle$. We obtain

$$\epsilon^{(\pm 1)}(p_z) \equiv |p_z, \pm 1\rangle = \frac{1}{\sqrt{2}}(0, \mp 1, -i, 0)$$

$$\epsilon^{(0)}(p_z) \equiv |p_z, 0\rangle = (\beta\gamma, 0, 0, \gamma) \tag{12-9}$$

where β is the velocity of the particle and $\gamma = (1 - \beta^2)^{-1/2}$, as usual. Note carefully that the superscripts (± 1) and (0) refer to the *helicity* and *not* to the component of the four-vector. This can be a source of confusion.

Exercise 12-2. Verify the Lorentz condition, Eq. 12-10, using Eq. 12-9.

$$p \cdot \epsilon^{(\mu)}(\mathbf{p}) = 0 \tag{12-10}$$

We could derive the more general spin 1 functions $\epsilon^{(\mu)}(\mathbf{p})$, which correspond to the spin $\frac{1}{2}$ functions of Eq. 12-6, but we use the simplest spin 1 functions that are sufficient to the tasks at hand.

Exercise 12-3. Calculate $\epsilon^{(0)}(\mathbf{p})$ and $\epsilon^{(\pm 1)}(\mathbf{p})$ where $\mathbf{p}$ is at an arbitrary angle in the xz plane. Compare your result with that in Ref. 12AU.

Before leaving this discussion of spin 1 functions, we review some potentially confusing terminology: $\epsilon^{(\pm 1)}(p_z)$ and $\epsilon^{(0)}(p_z)$ of Eq. 12-9 represent helicity states with **p** along the $+z$ axis. In addition, the states $\epsilon^{(\pm 1)}(p_z)$ contain only x and y components, hence are said to represent states of *transverse* polarization. In fact, we can define functions that represent two separate states of transverse polarization, namely,

$$\epsilon^{(x)}(p_z) = (0, 1, 0, 0) \qquad \text{and} \qquad \epsilon^{(y)}(p_z) = (0, 0, 1, 0) \tag{12-11}$$

Finally, the state represented by $\epsilon^{(0)}(p_z) = (\beta\gamma, 0, 0, \gamma)$ is often referred to as a *longitudinally* polarized state, the z component γ representing the longitudinal polarization. However, we note that if $\beta \neq 0$, there is also a nonzero time component in this state.

The photon is a special case of a spin 1 particle with zero mass. The functions of Eq. 12-9 reduce to $\epsilon^{(\pm)}(p_z) = 1/\sqrt{2}(0, \mp 1, -i, 0)$. Thus the photon only has helicity ± 1; or in terms of the previous discussion, it can only be transversely polarized in the two states of Eq. 12-11.

Without discussing further the spin functions representing helicity states for particle with spin greater than 1, we note that such higher spin functions can be constructed by forming products of spin $\frac{1}{2}$ functions and spin 1 functions, the latter functions being the ones we have developed in this section (12AU, 12PI).

12-3 Rules and Philosophy of Elementary One-Particle Exchange Calculations.

In Sec. 7-5 we presented the concepts used in the elementary one-particle exchange model; in this section we list those concepts with a bit more formality. The elementary one-particle exchange model is based on relativistic quantum mechanics and quantum field theory. The calculations are carried out in lowest order in the coupling constants. As shown in Fig. 12-1 each particle i with mass m_i has four-momentum P_i and a spin function $\Phi_i^{(\lambda_i)}$. The calculations are most

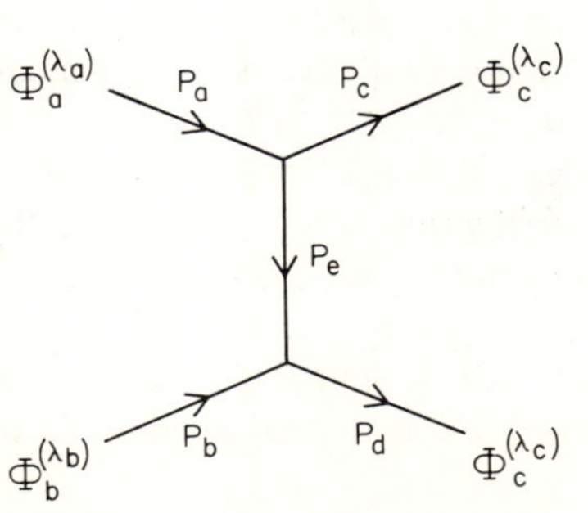

Fig. 12-1 The Feynman diagram for elementary one-particle exchange.

easily carried out in momentum space using the standard methods for Feynman diagrams, as described, for example, in Apps. E and F and in Ref. 12BJ. The rules and philosophy are as follows.

1. Single diagrams in which two or more particles are exchanged are ignored completely. But we can add the amplitudes of two or more one-particle exchange diagrams. Each of the one-particle exchange diagrams obeys rules 2 through 9.

2. For each vertex (iej) we write a *Lorentz scalar* function V_{iej}, which is made up of the spin function Φ_i, Φ_e, Φ_j to the first power, and any combination of the four momenta P_i, P_e, P_j that leads to V_{iej} being a Lorentz scalar. Finally, the strength of the coupling at the vertex is fixed by a coupling constant g_{iej}. Thus *symbolically* we have

$$V_{iej} = g_{iej} h(P_i, P_e, P_j)\Phi_i\Phi_e\Phi_j \tag{12-12}$$

Exercise 12-4. Discuss why the spin function of each particle should not appear more than once in V_{iej}. Give more general arguments for rule 3, using time-reversal invariance.

3. In analogy with quantum electrodynamics, in which the coupling constant is $e = \sqrt{4\pi\alpha}$, we take the coupling constant g_{iej} to be real.

4. The V_{iej} must be invariant under space and time inversion, since we are only concerned with strong and electromagnetic interactions.

5. We must choose the exchanged particle so that baryon number, strangeness, charge, isotopic spin, and G-parity are conserved at the vertex.

6. The exchanged particle is a *virtual*, not a *real* particle, since the square of its four-momentum, $t = P_e^2 = (P_i - P_j)^2 \neq m_e^2$. Although the exchange particle is virtual, we take all its internal quantum numbers, *including its intrinsic spin and parity*, to be the internal quantum numbers of the corresponding real particle.

7. There are usually several hadrons having different masses but the same set of internal quantum numbers. How then is the exchange particle e to be chosen? The philosophy of the model (justified later in the chapter) favors the lowest mass exchange particle that fits rule 5. Sometimes, when we know from other measurements that g_{iej} for the lowest mass exchange particle e is small compared with the $g_{ie'j}$ of the next lowest mass exchange particle e', we use e' in the calculation. Of course as noted in rule 1, we can also use a sum over several diagrams if we think several exchange particles contribute.

8. For a particular exchange particle, rules 2 through 5 do not necessarily limit V_{iej} to a single form. If there are two vertex functions V_{iej} and V'_{iej}, then different, independent coupling constants g_{iej} and g'_{iej} are assigned to the respective vertices.

9. In spite of—or, better, because of—the extreme simplicity of this model, there is yet one more degree of freedom; a vertex can be multiplied by a Lorentz scalar $F(t)$.

Exercise 12-5. Show that because of four-momentum conservation at the vertex (iej) a Lorentz scalar $F(P_i, P_e, P_j)$ can always be reduced to $F(t)$.

There are some restrictions on the form of $F(t)$, such as the analyticity considerations discussed in Ch. 15, but still $F(t)$ is a quite general function. And within the elementary one-particle exchange model there is no way to fix its form. We call $F(t)$ the *vertex form factor*, or simply a *form factor*. This is in analogy to the nucleon form factors, which are measured in electron–nucleon elastic scattering (see Ch. 19). In Ch. 13 we briefly discuss the $F(t)$'s that have been used in purely hadronic one-particle exchange calculations; but here we take $F(t)$ to be constant, and we absorb it into the coupling constant g.

The T amplitude for the process $a + b \to c + d$ with e exchange has the symbolic form (ignoring phases)

$$T = \sum_{e \text{ spin states}} \left[V_{bed} \frac{1}{m_e^2 - t} V_{aec} \right] \tag{12-13}$$

where the propagator term $(m_e^2 - t)^{-1}$ represents the exchange particle except for its spin properties. We have emphasized that the amplitude involves the summation over all the spin states $\Phi_e^{(\lambda_e)}$ of the exchanged particle. We extract $\Phi_e^{(\lambda_e)}$ from V_{iej} by writing

$$V_{aec} = V_{aec}^{(\lambda_c\lambda_a)} \Phi_e^{(\lambda_e)\dagger}, \qquad V_{bed} = V_{bed}^{(\lambda_d\lambda_b)} \Phi_e^{(\lambda_e)} \tag{12-14}$$

Then we put

$$T_{\lambda_c\lambda_d;\lambda_a\lambda_b} = V_{bed}^{(\lambda_d\lambda_b)} R_e V_{aec}^{(\lambda_c\lambda_a)} \tag{12-15}$$

where the full propagator is

$$R_e = \sum_{\lambda_e} \left[\frac{\Phi_e^{(\lambda_e)} \Phi_e^{(\lambda_e)\dagger}}{m_e^2 - t} \right] \tag{12-16}$$

For simplicity, we have ignored phase considerations in Eqs. 12-13 to 12-16.

The propagator R_e is usually a tensor quantity because of the summation over the spin functions. We have emphasized the form given in Eq. 12-16 because it shows how to calculate the propagator for a particle of any spin, once the spin functions are known. Equation 12-16 also provides insight into the meaning of the propagator. The propagators we use are (App. F)

$$\text{spin } 0 \qquad R_e = \frac{-i}{m_e^2 - t} \tag{12-17a}$$

$$\text{spin } \tfrac{1}{2} \qquad R_e = \frac{-i(\not{P}_e + m_e)}{m_e^2 - t} \tag{12-17b}$$

$$\text{spin 1} \qquad R_{\mu\nu} = \frac{-i[P_{e\mu}P_{e\nu}/m_e^2 - g_{\mu\nu}]}{m_e^2 - t} \tag{12-17c}$$

$$\text{photon} \qquad R_{\mu\nu} = \frac{-ig_{\mu\nu}}{t} \tag{12-17d}$$

Other definitions of the propagators may differ by multiplicative constants such as i; see Refs. 12JAC, 12SA, and 12BJ for examples. Such considerations become important when the relative phases of one-particle and two-particle exchange amplitudes are being compared. Incidentally, the same remark holds for the definitions of vertex functions (App. F).

Exercise 12-6. Use Eq. 12-16 to calculate the propagators for spin $\frac{1}{2}$ and spin 1 particles, using the spin functions given in the preceding section. Check that your results are in agreement with Eq. 12-17, taking account of the restricted values of **p** employed in the last section. Use the result of Ex. 12-3.

Once $T_{\lambda_c\lambda_d;\lambda_a\lambda_b}$ is calculated using this set of rules, it is straightforward but often tedious to calculate density matrix elements and cross sections. Most of the remainder of this chapter is devoted to examples of such calculations. We reduce the work as much as possible by tricks and approximations—the most important approximation being the neglect of m_i^2 $(i = a, b, c, d, e)$ compared to s. In many cases these approximations are more than a convenience—they enable us to understand more clearly the basic s and t dependence of the amplitudes.

12-4 Vector Particle Production by Pseudoscalar Exchange—The Differential Cross Section. We consider first the important, yet simple case of the production of a vector (spin 1) particle by the exchange of a pseudoscalar particle. The reaction is

$$\begin{aligned} &0^- + \tfrac{1}{2}^+ \to 1^- + \tfrac{1}{2}^+ \qquad \text{by} \quad 0^- \text{ exchange} \\ &a + b \to c + d \qquad \text{by} \quad e \text{ exchange} \end{aligned} \tag{12-18}$$

A well-studied example of this reaction is $\pi^- + p \to \rho^0 + n$, where the exchanged particle is a π^-; and we shall compare our calculations with the measurements of this reaction.

From the rules given in the last section, the mesonic vertex is

$$V_{aec} = -ig(P_a + P_e) \cdot \epsilon_c^{(\mu)\dagger}\Phi_a\Phi_e^\dagger = -i2gP_a \cdot \epsilon_c^{(\mu)\dagger}\Phi_a\Phi_e^\dagger \tag{12-19}$$

and the baryon vertex is

$$V_{bed} = -iG\bar{u}^{\lambda_d}(\mathbf{p}_d)\gamma_5 u^{\lambda_b}(\mathbf{p}_b)\Phi_e \tag{12-20}$$

The phases come from the conventions given in App. F.

Exercise 12-7. Show that both vertex functions are Lorentz scalars.

The extraction of Φ_e and $\Phi_e^\dagger$ to form the propagator leaves the vertex functions

$$V_{aec}^{(\mu)} = -i2gP_a \cdot \epsilon_c^{(\mu)\dagger}\Phi_a \tag{12-21}$$

$$V_{bed}^{(\lambda_b\lambda_d)} = -iG\bar{u}^{\lambda_d}(\mathbf{p}_d)\gamma_5 u^{\lambda_b}(\mathbf{p}_b) \tag{12-22}$$

Then we have

$$T_{\mu\lambda_d;\lambda_b}(s, t) = \frac{-2gGP_a \cdot \epsilon_c^{(\mu)\dagger}\bar{u}^{\lambda_d}(\mathbf{p}_d)\gamma_5 u^{\lambda_b}(\mathbf{p}_b)}{m_e^2 - t} \tag{12-23}$$

In the future, we ignore scalar or pseudoscalar spin functions such as Φ_a.

We can now calculate the density matrix elements and the differential cross section, but for simplicity we calculate the latter first. Assuming that particle b was unpolarized, we average and sum, respectively, over the spin states of particles b and d using standard trace techniques. We also assume $m_b = m_d = m$, a limitation that is not important at high energies. Then we obtain

$$\left(\frac{d\sigma}{dt}\right)(s, t) = \frac{g^2G^2|t|}{16\pi s(m_e^2 + |t|)^2}\,\frac{\lambda(t, m_a^2, m_c^2)/m_c^2}{\lambda(s, m_a^2, m^2)/s} \tag{12-24}$$

which is written completely as a function of Lorentz scalars, with λ as defined in App. A.

Exercise 12-8. Derive Eq. 12-24. To do this evaluate $|P_a \cdot \epsilon_c^{(\mu)}|^2$ by recognizing that $P_a \cdot \epsilon_c^{(\mu)}$ is a Lorentz scalar, then calculating in the rest frame of particle c.

To aid in the interpretation of Eq. 12-24, we go immediately to high energies $s \gg m_i^2$. Then $\lambda(s, m_a^2, m_b^2) \to s^2$ and

$$\frac{d\sigma}{dt} = \frac{g^2G^2|t|}{16\pi s^2[m_e^2 + |t|]^2}\,\frac{\lambda(t, m_a^2, m_c^2)}{m_c^2} \tag{12-25}$$

Thus the s behavior of this reaction is simply:

1. $d\sigma/dt$ and σ are proportional to s^{-2} at large s.
2. Therefore $|T|_{\text{av}}$ (defined in Eq. 6-44) is independent of s. (12-26)

To understand the t behavior, we observe that at small $|t|$

$$|t| \ll (m_a^2 + m_c^2); \qquad \frac{d\sigma}{dt} = \frac{g^2G^2|t|}{16\pi s^2[m_e^2 + |t|]^2}\,\frac{(m_c^2 - m_a^2)^2}{m_c^2} \tag{12-27}$$

and at large $|t|$

$$|t| \gg (m_a{}^2 + m_c{}^2); \qquad \frac{d\sigma}{dt} = \frac{g^2 G^2 |t|^3}{16\pi s^2 (m_e{}^2 + |t|)^2} \tag{12-28}$$

First we make some general observations. The relationships given in Eq. 12-26 lead to $n = 2$ in Eq. 7-13, quite a reasonable value when compared with the values of n in Table 7-1. But as we soon see, this agreement is an illusion; for if the spin of the exchanged particle is greater than zero, the predicted energy dependence of $|T|_{av}$ is wrong. Next we note the basis for the discussion about exchange particle masses in rule 7 of the previous section. At small $|t|$ the smallest m_e gives the largest $d\sigma/dt$; at large $|t|$ the magnitude of m_e is unimportant as long as there is some m_e with $m_e{}^2 \ll |t|$. Finally, we observe that once t becomes about the same size or larger than $(m_a{}^2 + m_c{}^2)$; $d\sigma/dt$ is predicted to increase with $|t|$. This is a completely wrong prediction, demonstrating that any validity the elementary one-particle exchange model may have is restricted to the small $|t|$ region.

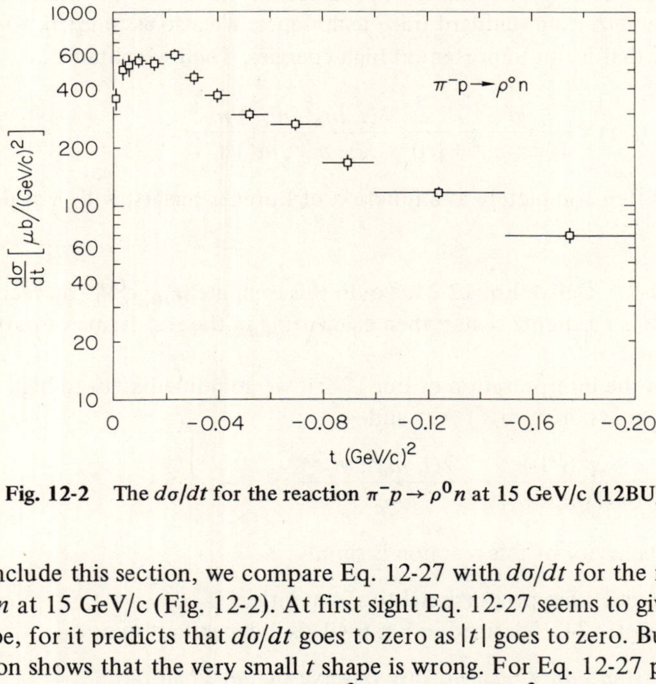

Fig. 12-2 The $d\sigma/dt$ for the reaction $\pi^- p \rightarrow \rho^0 n$ at 15 GeV/c (12BU).

To conclude this section, we compare Eq. 12-27 with $d\sigma/dt$ for the reaction $\pi^- p \rightarrow \rho^0 n$ at 15 GeV/c (Fig. 12-2). At first sight Eq. 12-27 seems to give the right shape, for it predicts that $d\sigma/dt$ goes to zero as $|t|$ goes to zero. But closer observation shows that the very small t shape is wrong. For Eq. 12-27 predicts that the turnover should begin at $|t| \approx m_e{}^2 = 0.02$ (GeV/c)^2, but Fig. 12-2 indicates that the turnover begins at lower $|t|$. In the next chapter we show that the absorption correction can account for this difference.

12-5 Coupling Constants. During the development of the elementary one-particle exchange model in the early 1960s, there was great emphasis on the values of the strong interaction coupling constants required to give the correct magnitude of $d\sigma/dt$. Since the models we use today are quite different from the elementary one-particle exchange model, we do not discuss coupling constants in detail here. The significance and definition of coupling constants from the analytic viewpoint are given in Ch. 15.

The philosophy of the elementary one-particle exchange model is that the coupling constant is independent of the reality or virtuality of the particles at the vertex. Thus the coupling constant $g_{\pi\pi\rho}$ at the mesonic vertex in the reaction $\pi^- + p \to \rho^0 + n$ can be evaluated from the vertex of the ρ decay process, $\rho^0 \to \pi^+ + \pi^-$. We do just this in the next two exercises.

Exercise 12-9. Use Eq. 12-21 to show that the decay rate for $\rho^0 \to \pi^+ + \pi^-$ is given by

$$\Gamma_\rho = \frac{g_{\pi\pi\rho}^2}{6\pi} \frac{[(m_\rho/2)^2 - m_\pi{}^2]^{3/2}}{m_\rho{}^2} \tag{12-29}$$

Exercise 12-10. Identify the Γ_ρ in Eq. 12-29 with the width Γ of the ρ, namely, 135 MeV, as given in the nonrelativistic Breit–Wigner formula; and derive

$$\frac{g_{\pi\pi\rho}^2}{4\pi} = 2.6 \tag{12-30}$$

In the units of this book, where $\hbar = 1$ and $c = 1$; g is dimensionless. In conventional units $g^2/\hbar c$ is dimensionless. Our use of $g^2/4\pi$ means that we are using rationalized units, which is analogous to our use of rationalized electromagnetic units leading to $e^2/4\pi = \alpha$. Also we note that some authors omit the 2 in Eq. 12-21.

The baryon vertex coupling constant $G_{p\pi-n}$ is found more indirectly from a dispersion relations analysis of pion–nucleon scattering as discussed in Ch. 15. We give only the result here

$$\frac{G_{p\pi-n}^2}{4\pi} = 2\left(\frac{G_{p\pi^0 p}^2}{4\pi}\right), \qquad \frac{G_{p\pi^0 p}^2}{4\pi} = 14.9 \tag{12-31}$$

Our major observation regarding the large size of these coupling constants compared to α, and even compared to 1, is that the entire concept of one-particle exchange is a complete paradox. For with such large coupling constants, two-particle and many-particle exchange would certainly be expected to contribute as much, if not more, than one-particle exchange. Yet somehow the one-particle exchange theory has some truth in it, for it correctly predicts when forward peaks will appear in two-body inelastic reactions. And this prediction depends only on the exchange particle–on its internal quantum numbers.

12-6 Vector Particle Production by Pseudoscalar Exchange–The Density Matrix Elements. We can learn more about the elementary one-particle exchange model calculation of the reaction $0^- + \frac{1}{2}^+ \to 1^- + \frac{1}{2}^+$ by 0^- exchange, if we examine the individual helicity amplitudes and density matrix elements. We interrupt the calculation of Sec. 12-4 at Eq. 12-23 and consider the individual $T_{\mu\lambda_d;\lambda_b}$'s. To simplify the calculation, we choose the xz plane as the plane of the reaction, we work in the center of mass, and as illustrated in Fig. 12-3 we place the 1^- particle along the $+z$ axis to use the spin 1 functions of Eq. 12-9.

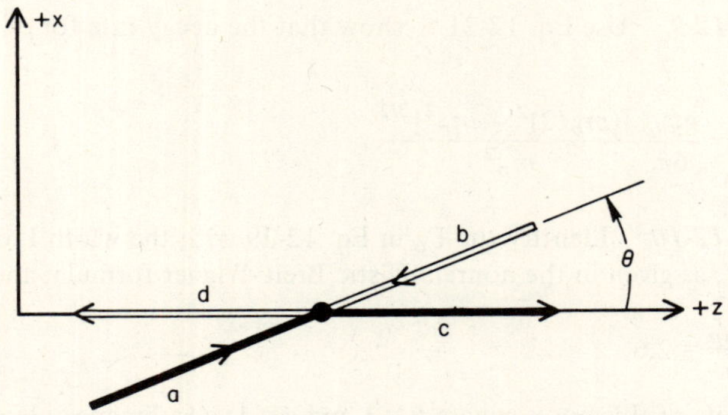

Fig. 12-3 Kinematics of the reaction $a + b \to c + d$ in the x–z plane.

For the mesonic vertex we easily obtain

$$V_{aec}^{(\pm 1)} = \mp i\sqrt{2}gp \sin\theta$$

$$V_{aec}^{(0)} = \frac{-2i\, g[E_a p' - E_c p \cos\theta]}{m_c} \tag{12-32}$$

Here all quantities are in the center of mass, p and p' being the magnitudes of the initial and final three-momenta, respectively.

The calculation of the baryon vertex is a bit more tedious. One annoying feature of the calculation is that we should take account of the baryons being the "second particle" in the two-particle helicity states, defined in Sec. 9-6. (We have already used the 1^- spin function treating that particle as the "first particle.") Ignoring the "second particle" aspects of the baryon spin functions is not disastrous, but it can lead to confusing phase differences. Therefore, we take account of the "second particle" aspect by multiplying the spinors in Eq. 12-6 by $-1^{(s-\lambda)}$ and by replacing θ by $(\pi + \theta)$. Using Fig. 12-3 we obtain

$$u^+(\mathbf{p}_b) = \sqrt{E_b + m}\begin{bmatrix} -\sin\frac{1}{2}\theta \\ \cos\frac{1}{2}\theta \\ \dfrac{-p\sin\frac{1}{2}\theta}{E_b+m} \\ \dfrac{p\cos\frac{1}{2}\theta}{E_b+m} \end{bmatrix} \qquad u^-(\mathbf{p}_b) = \sqrt{E_b + m}\begin{bmatrix} \cos\frac{1}{2}\theta \\ \sin\frac{1}{2}\theta \\ \dfrac{-p\cos\frac{1}{2}\theta}{E_b+m} \\ \dfrac{-p\sin\frac{1}{2}\theta}{E_b+m} \end{bmatrix} \tag{12-33}$$

$$u^+(\mathbf{p}_d) = \sqrt{E_d + m}\begin{bmatrix} 0 \\ 1 \\ 0 \\ \dfrac{p'}{E_d+m} \end{bmatrix} \qquad u^-(\mathbf{p}_d) = \sqrt{E_d + m}\begin{bmatrix} 1 \\ 0 \\ \dfrac{-p'}{E_d+m} \\ 0 \end{bmatrix} \tag{12-34}$$

Then we have

$$\begin{aligned} V_{bed}^{\pm\pm} &= \mp iGZ_- \cos\tfrac{1}{2}\theta \\ V_{bed}^{\pm\mp} &= +iGZ_+ \sin\tfrac{1}{2}\theta \end{aligned} \tag{12-35a}$$

where

$$Z_\pm = \sqrt{2}[E_bE_d \pm pp' - m^2]^{1/2} \tag{12-35b}$$

Exercise 12-11. Derive Eqs. 12-32 and 12-35. Check the relative phases of the space inversion related amplitudes against Eq. 10-19a.

To facilitate understanding of the behavior of these individual amplitudes, we avoid writing out the exact amplitudes, which are a bit cumbersome, but we immediately make the simplification of going to high energy where $s \gg m^2$. Then we have

$$Z_+ \to \sqrt{s},\ Z_- = \sqrt{|t_{\min}|} \to \frac{|m_c{}^2 - m_a{}^2|m}{s} \tag{12-36}$$

$$\begin{aligned} V_{bed}^{\pm\pm} &\to 0 \\ V_{bed}^{\pm\mp} &\to iG\sqrt{s}\sin\tfrac{1}{2}\theta \approx iG\sqrt{|t|} \end{aligned} \tag{12-37}$$

Thus we have the following very general observation: at high energies the coupling of a pseudoscalar exchange particle to spin $\frac{1}{2}$ particles at a vertex leads to the dominance of the spin $\frac{1}{2}$ helicity flip amplitudes over the spin $\frac{1}{2}$ nonhelicity flip amplitudes, except in the very small angle range $\theta \leqslant (|t_{\min}|/s)^{1/2}$.

Exercise 12-12. Fill in the steps in the last set of equations and statements. Show that the last observation is true even if $m_b \neq m_d$.

We now turn to the density matrix elements. First we consider the final spin $\frac{1}{2}$ particle, namely, particle d. If b is *unpolarized* and we sum over the spin states of c, we obtain at high energies (see Ch. 11) the simple result

$$\rho_d{}^H = \begin{bmatrix} \frac{1}{2} & 0 \\ 0 & \frac{1}{2} \end{bmatrix} \tag{12-38}$$

except at very small angles. Here H denotes the helicity frame. Thus particle d is unpolarized; V^{+-} and V^{-+} simply interchange the helicity states of b.

Exercise 12-13. Derive Eq. 12-38. Calculate, with and without the high energy approximation of Eq. 12-36, the density matrix for particle d if particle b is in the pure $+\frac{1}{2}$ helicity state.

Again using the methods of Ch. 11, assuming that b is unpolarized, and summing over the spin states of d, we can calculate the density matrix in the helicity frame for particle c, namely, $\rho_c{}^H$. If we examine the result, Ex. 12-14, we find that *all* the matrix elements are nonzero. Thus in the helicity frame particle c is produced in a mildly complicated state of polarization. In the next section we describe a simpler way to examine the polarization state of particle c.

Exercise 12-14. Calculate $\rho_c{}^H$. Show that it reduces to Eq. 12-39 when $\theta = 0$.

$$\rho_c{}^H(\theta = 0) = \begin{bmatrix} 0 & 0 & 0 \\ 0 & 1 & 0 \\ 0 & 0 & 0 \end{bmatrix} \tag{12-39}$$

12-7 The Gottfried–Jackson Frame. The particle c density matrix elements depend completely on the term $P_a \cdot \epsilon_c^{(\mu)}$ in $V_{aec}^{(\mu)}$. Consider the rest frame spin states of particle c with the z axis defined as the direction of the three-vector $\mathbf{p}_a$ *in the c rest frame*. Using these rest frame spin states and designating by $p_{a,c}$ the magnitude of $\mathbf{p}_a$ *in the c rest frame*, we easily find

$$V_{aec}^{(\pm 1)} = 0, \qquad V_{aec}^{(0)} = 2igp_{a,c} \tag{12-40}$$

a most beautiful result! This special frame for particle c rest frame spin states is used often, and it has a special name—the Gottfried–Jackson frame (12GO). The c density matrix in this frame is

$$\rho_c{}^{GJ} = \begin{bmatrix} 0 & 0 & 0 \\ 0 & 1 & 0 \\ 0 & 0 & 0 \end{bmatrix} \tag{12-41}$$

Equation 12-41 is the same as Eq. 12-39 because the z axis of the helicity frame coincides with the z axis of the Gottfried–Jackson frame when θ is zero.

Such a simple result should have a simple explanation, and it does. In the c rest frame, both $\mathbf{p}_a$ and $\mathbf{p}_e$ lie along the z axis of the Gottfried–Jackson frame. Hence the z component of the orbital angular momentum of the a–e system is zero. Since both a and e have zero spin, the z component of the total angular momentum must be zero. Then by angular momentum conservation, the z component of particle c's spin must be zero, and the density matrix in the Gottfried–Jackson frame is given by Eq. 12-41.

Exercise 12-15. Extend the reasoning used in the last paragraph to a vertex (aec) where the exchanged particle e has spin zero, but particles a and c have any spin. Show that if a *is* unpolarized, then $(\rho_c{}^{GJ})_{00} = 1$ and $(\rho_c{}^{GJ})_{ij} = 0$ otherwise. Compare your reasoning with that first given in Ref. 12TR. What is the definition and significance of the Treiman–Yang angle?

12-8 Charged Lepton–Nucleon Elastic Scattering. To introduce vector particle exchange, we consider first the elastic scattering of charged leptons, electrons, or muons, on nucleons. This partially electromagnetic, partially hadronic process, which proceeds by one-photon exchange, has been the inspiration for much purely hadronic one-vector particle exchange theory.

Unlike purely hadronic one-particle exchange calculations, the charged lepton–proton elastic scattering calculation is quite exact. First, one-photon exchange is the dominant process; we can neglect two-or-more photon exchange and weak interactions. Second, we can use quantum electrodynamics to fix precisely the form of the charged lepton–photon vertex. Third, we can also use quantum electrodynamics to limit the form of the nucleon–photon vertex.

The reaction is

$$\begin{matrix} (e \text{ or } \mu) & + & (p \text{ or } n) & \rightarrow & (e \text{ or } \mu) & + & (p \text{ or } n) \\ a & + & b & \rightarrow & c & + & d \end{matrix} \tag{12-42}$$

and we neglect the charged lepton mass in the high energy approximation. The charged lepton–photon vertex is

$$[V_{a\gamma c}^{\lambda_c\lambda_a}]^\mu = -ie\bar{u}^{\lambda_c}(\mathbf{p}_c)\gamma^\mu u^{\lambda_a}(\mathbf{p}_a) \tag{12-43}$$

and the superscript μ again indicates the component of a four-vector. The propagator is given by Eq. 12-17d.

The nucleon vertex is more complicated because the nucleon is not a point Dirac particle. The nucleon vertex, however, must be a four-vector; thus its contraction with the photon spin (now buried in the propagator) is a Lorentz scalar. The most general form is

$$[V_{b\gamma d}^{(\lambda_d\lambda_b)}]^\nu = -ie\bar{u}^{\lambda_d}(\mathbf{p}_d)[P_b{}^\nu A(t) + P_d{}^\nu B(t) + \gamma^\nu C(t)]u^{\lambda_b}(\mathbf{p}_b) \tag{12-44}$$

Exercise 12-16. Show that Eq. 12-44 is indeed the most general form. The use of the coupling constant e, the unit electric charge, is obvious for the proton. For the neutron, its explicit use is a matter of convention that allows the neutron and the proton to be treated by the same formalism in charged lepton-nucleon elastic scattering.

Quantum electrodynamics sets a further restriction on $V_{b\gamma d}^{(\lambda_d \lambda_b)}$—the restriction of current conservation. Defining the *transition current*, usually simply called the *current*, by

$$[j_{\text{nucleon}}^{(\lambda_d \lambda_b)}] = \frac{[V_{b\gamma d}^{(\lambda_d \lambda_b)}]}{-ie} \tag{12-45}$$

current conservation requires that

$$(P_b - P_d) \cdot j_{\text{nucleon}} = 0 \tag{12-46}$$

Exercise 12-17. Review the subject of current conservation in quantum electrodynamics. Relate Eq. 12-46 to the coordinate space condition $\partial j^\nu(x)/\partial x^\nu = 0$. Discuss the relation of current conservation to gauge invariance.

Exercise 12-18. Discuss the difference between the *transition* current of Eq. 12-45 and the *probability* current of a single particle $j = \bar{u}(\mathbf{p})\gamma u(\mathbf{p})$.

For Eq. 12-46 to hold, Eq. 12-44 must reduce to

$$[j_{\text{nucleon}}^{(\lambda_a \lambda_b)}]^\nu = \bar{u}^{\lambda_d}(\mathbf{p}_d)[(P_b + P_d)^\nu A(t) + \gamma^\nu C(t)]u^{\lambda_b}(\mathbf{p}_b) \tag{12-47}$$

Exercise 12-19. Prove the last statement.

In elementary one-particle exchange calculations, we would make the form factors $A(t)$ and $B(t)$ constants, as discussed in Sec. 12-3. In charged lepton-proton elastic scattering, these form factors are definitely *not* constants; indeed, the determination of their functional form is the chief purpose of charged-lepton proton–elastic scattering measurements. But this is discussed fully in Ch. 19. Now, having obtained Eq. 12-47, the objective of this section, we return to the subject of purely hadronic one-particle exchange calculations.

12-9 Pseudoscalar Particle Production by Vector Exchange. Consider the vector particle exchange process of Eq. 12-48

$$\begin{aligned} 0^- + \tfrac{1}{2}^+ &\to 0^- + \tfrac{1}{2}^+ \qquad \text{by} \quad 1^- \text{ exchange} \\ a + b &\to c + d \qquad \text{by} \quad e \text{ exchange} \end{aligned} \tag{12-48}$$

For convenience we take the spin $\frac{1}{2}$ particles to have $m_b = m_a = m$. An important application of this process is $\pi^- p$ *charge exchange*, $\pi^- + p \to \pi^0 + n$, with ρ^- as the exchange particle.

The spin $\frac{1}{2}$ vertex is constructed in analogy with the spin $\frac{1}{2}$ vertex in charged lepton-nucleon elastic scattering. Following Eq. 12-47 and imposing current conservation, we write

$$[V_{bed}^{(\lambda_d\lambda_b)}]^\nu = -i\bar{u}^{\lambda_d}(\mathbf{p}_d)\left[(G_V + G_T)\gamma^\nu - \frac{G_T}{2m}(P_b + P_d)^\nu\right]u^{\lambda_b}(\mathbf{p}_b) \tag{12-49}$$

The form factors of Eq. 12-47 are here reduced to two *independent* constants as discussed in rule 8 of Sec. 12-3. The *vector coupling constant* G_V and the *tensor coupling constant* G_T, are defined according to the discussion in Ch. 19. Referring to Eqs. 19-16 and 19-17, G_V is the analog of eF_1^N and G_T is the analog of $e\kappa^N F_2^N$. And using Eq. 19-17, we see that G_V and G_T are associated with vector and tensor coupling structures, respectively.

Exercise 12-20. Discuss the meaning of current conservation here.

We also note that current conservation leads to the welcome simplification of the propagator of Eq. 12-17c to the simple form

$$(R_e)_{\mu\nu} = \frac{-ig_{\mu\nu}}{t - m_e^2} \tag{12-50}$$

The spin $\frac{1}{2}$ vertex is complicated, and we can simplify the calculations by evaluating the vertex in the kinematic region of most interest to us–high energy where $s \gg m_i^2$, $i = a, b, c, d$, and $s \gg |t|$. To be precise, we pick out the terms that *at fixed t* vary with *s* as the *highest order in s*. These terms we call the dominant terms at high energy. The spin $\frac{1}{2}$ particles are the "second particles," the spinors of Eqs. 12-33 and 12-34 are used, and the coordinate system is as shown in Fig. 12-3. Since we can use Eq. 10-19 to relate helicity amplitudes connected by space inversion, we calculate only the combination ($\lambda_d = +\frac{1}{2}$, $\lambda_b = +\frac{1}{2}$) and ($\lambda_d = -\frac{1}{2}$, $\lambda_b = +\frac{1}{2}$)

$-i[G_V + G_T]$ coefficients

$$\begin{array}{ll}
\bar{u}_d^{+}\gamma^0 u_b^{+} \approx \sqrt{s}\cos\frac{1}{2}\theta & \bar{u}_d^{-}\gamma^0 u_b^{+} \approx -2m\sin\frac{1}{2}\theta \\
\bar{u}_d^{+}\gamma^1 u_b^{+} \approx -\sqrt{s}\sin\frac{1}{2}\theta & \bar{u}_d^{-}\gamma^1 u_b^{+} \approx \sqrt{|t_{\min}|}\cos\frac{1}{2}\theta \\
\bar{u}_d^{+}\gamma^2 u_b^{+} \approx -i\sqrt{s}\sin\frac{1}{2}\theta & \bar{u}_d^{-}\gamma^2 u_b^{+} \approx -i\sqrt{|t_{\min}|}\cos\frac{1}{2}\theta \\
\bar{u}_d^{+}\gamma^3 u_b^{+} \approx -\sqrt{s}\cos\frac{1}{2}\theta & \bar{u}_d^{-}\gamma^3 u_b^{-} \approx -\sqrt{|t_{\min}|}\sin\frac{1}{2}\theta
\end{array} \tag{12-51a}$$

$+i\left[\frac{G_T}{2m}(P_b + P_d)^\nu\right]$ Coefficients

$$\bar{u}_d{}^+ u_b{}^+ \approx 2m \cos \tfrac{1}{2}\theta \qquad \bar{u}_d{}^- u_b{}^+ \approx -\sqrt{s} \sin \tfrac{1}{2}\theta \tag{12-51b}$$

Exercise 12-21. Derive the relationships given in the last two equations.

We now can make the following important observation, which is supported by experiment. If we have *pure vector coupling*, that is, if $G_T = 0$, then in the high energy approximation

$$[V_{bed}^{++}]^\nu \gg [V_{bed}^{-+}]^\nu; \qquad \theta^2 > \frac{|t_{\min}|}{s} \tag{12-52}$$

Therefore, the *helicity state*, or more briefly the *helicity*, is conserved at the spin $\frac{1}{2}$ vertex.

Exercise 12-22. Show that this important high energy principle of helicity conservation at the spin $\frac{1}{2}$ vertex (if $G_T = 0$) also applies to the charged lepton vertex in charged lepton–nucleon scattering.

Our next task, calculating the full amplitudes, is easily done using the mesonic vertex of the same form as Eq. 12-19. But now the spin function of the vector particle is included in the propagator. Hence we have

$$[V_{aec}]^\mu = -ig(P_a + P_c)^\mu \tag{12-53}$$

Using this equation, the reduced propagator of Eq. 12-50, and Eqs. 12-51 and 12-53, we easily obtain the high energy result

$$T_{++}(s, t) = \frac{2gG_V s}{m_e{}^2 - t}$$

$$T_{-+}(s, t) = \frac{-2gG_T s}{m_e{}^2 - t}\left(\frac{\sqrt{|t|}}{2m}\right) \tag{12-54}$$

where we have taken $\cos \frac{1}{2}\theta \approx 1$, since $s \gg |t|$. Averaging over the initial spin states and summing over the final spin states of the spin $\frac{1}{2}$ particles, we obtain

$$|T|^2_{\text{av}} = \frac{4g^2 s^2}{[m_e{}^2 + |t|]^2}\left[G_V{}^2 + G_T{}^2 \frac{|t|}{4m^2}\right] \tag{12-55}$$

$$\frac{d\sigma}{dt} = \frac{g^2[G_V{}^2 + G_T{}^2(|t|/4m^2)]}{4\pi[m_e{}^2 + |t|]^2} \tag{12-56}$$

The *high energy* s dependence of this exchange process is very simple.

1. $d\sigma/dt$ and σ are independent of s.
2. T_{++} and T_{-+} are proportional to s. (12-57)

First, we observe that this vector exchange energy dependence is very different from that obtained in pseudoscalar exchange, Eq. 12-26. Second, and more important, the energy dependence predicted for vector exchange is certainly physically wrong. For example, for the reaction $\pi^- + p \rightarrow \pi^0 + n$, σ is proportional to $s^{-1.1}$ at high energies; it is *not* independent of s. The vector exchange energy dependence prediction is in fact *wrong* for all two-body inelastic reactions. As discussed in Ch. 7, all these reactions, even those explained by vector particle exchange, have σ's that decrease as s increases. We have now located a fatal flaw of elementary one-particle exchange, which is not corrected even by the modifications described in the next chapter. The flaw is that except for spin zero exchange, one-particle exchange models give the wrong energy dependence of the cross sections. And the rough agreement in the case of spin zero exchange is probably coincidental.

We realize that this incorrect energy dependence is a fundamental problem of the elementary one-particle exchange model when we consider a two-body reaction $a + b \rightarrow c + d$ in which particles a, b, c, and d have spin zero. Suppose the reaction is to be explained by the exchange of a particle of *integral* spin J. As we very briefly mentioned at the end of Sec. 12-2, the spin function $\psi(J)$, for a particle of integral spin J can be constructed out of products of J spin 1 functions. Using the notation of Sec. 12-2, $\psi(J) = \epsilon_1 \epsilon_2 \ldots \epsilon_J$. We are ignoring symmetry questions here. This $\psi(J)$ appears at each "end" of the propagator. Hence rule 2 of Sec. 12-3 requires that V_{aec} contain $P_{a1}P_{a2} \ldots P_{aJ}$, with a similar requirement for V_{bed}. This can also be seen by analogy from Eq. 12-54. The result is that T has the rough form

$$T \sim \frac{(P_a \cdot P_b)^J}{m_e^2 - t} \sim \frac{s^J}{m_e^2 - t} \tag{12-58}$$

This is the generalization of Eqs. 12-26 and 12-57. But it is clearly absurd, for if $J > 1$, σ will increase with energy, and *no* two-body reaction cross section increases with energy at high energy. Having found this fatal flaw, we very briefly mention backward peaks to end the discussion of the elementary one-particle exchange model.

12-10 The Elementary One-Particle Exchange Model and Backward Peaks. As we discussed in Ch. 7, the backward peaks in elastic and inelastic two-body reactions can be qualitatively explained by the one-particle exchange model. Unfortunately, the quantitative predictions made for backward peaks by the elementary one-particle exchange model are very wrong. First, since the backward peak coupling constants are about the same size as those used in calculations of the forward peaks, we would expect the total cross section under a backward peak to be about the same size as that under a forward peak. But as discussed in Ch. 7, forward peaks in a particular reaction are always much larger than backward peaks when both peaks can be produced by the exchange of known hadrons. Second, the cross sections of backward peaks almost always decrease faster with increasing energy than the cross sections of forward peaks. Hence the energy predictions discussed at the end of the previous section will be even more incorrect. Therefore, the elementary one-particle exchange model cannot be applied at all to backward peaks, even when the modifications considered in the next chapter are used. Drastic changes, such as those made by the application of Regge theory (Ch. 17), must be made.

12-11 One-Particle Exchange in the *s*-Channel. This chapter is devoted to the *t*-channel, one-particle exchange mechanism in *s*-channel reactions. As long as we have assembled the required tools, however, let us look at the *s*-channel exchange mechanism in *s*-channel reactions. Consider the Feynman diagram in Fig. 12-4; the external particles for simplicity are assigned spin zero and equal masses m.

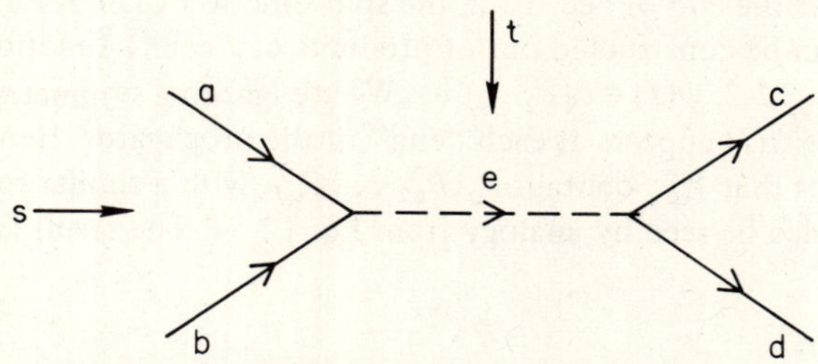

Fig. 12-4 One-particle exchange in the *s*-channel.

The exchanged particle has mass m_e and internal spin J

$$\text{for} \quad J = 0 \qquad T(s, t) = \frac{g^2}{s - m_e^2} \tag{12-59}$$

$$\text{for} \quad J = 1 \qquad T(s, t) = \frac{g^2[s + 2t - 4m^2]}{s - m_e^2} = \frac{4g^2 p^{*2} \cos\theta}{s - m_e^2} \tag{12-60}$$

where p^* are θ are in the center of mass.

Exercise 12-33. Derive these two equations.

We observe that the amplitude contains the $P_J(\cos\theta)$, which is appropriate to the spin of the exchanged particle. This is a general result, but we do not prove it here. For the exchange of a particle with integral spin J, the amplitude has the form

$$T(s, t) = \frac{\text{constant } (p^*)^{2J} P_J(\cos\theta)}{s - m_e^2} \tag{12-61}$$

Exercise 12-24. Equations 12-60 and 12-61 are precise only when all the external particle masses are *equal*. When the masses of the external particles are *unequal*, additional terms appear in the numerator; these terms contain $P_{J-1}(\cos\theta)$, $P_{J-2}(\cos\theta)$, ... Show that this is true for spin 1 by recalculating Eq. 12-60. This effect of unequal masses is important in the one-particle exchange model for Regge trajectories (Sec. 17-3; Refs. 17DU, 17BL, and 17SU).

Exercise 12-25. Use the discussion in Ch. 5 to show that a Breit–Wigner s-channel resonance of mass m_e in partial wave J leads to the amplitude

$$T(s, t) = \frac{\text{constant } (p^*)^{2J} P_J(\cos\theta)}{(s - m_e^2) + i m_e \Gamma} \tag{12-62}$$

Here we have also used the centrifugal barrier penetration factor (Ex. 5-8). Discuss the similarities and differences between Eqs. 12-61 and 12-62.

REFERENCES

12AU P. R. Auvil and J. J. Brehm, *Phys. Rev.* **145**, 1152 (1966).

12BJ J. D. Bjorken and S. D. Drell, *Relativistic Quantum Mechanics* (McGraw-Hill, New York, 1964).

12BL R. Blankenbecler and R. L. Sugar, *Phys. Rev.* **168**, 1597 (1968).

12BU F. Bulos et al., *Phys. Rev. Lett.* **26**, 1457 (1971).

12DO J. T. Donahue, Ph.D. thesis, 1967, University of Illinois.

12DU L. Durand, III, *Phys. Rev.* **154**, 1537 (1967).

12GO K. Gottfried and J. D. Jackson, *Phys. Lett.* 8, 144 (1964).

12HU R. W. Huff, *Phys. Rev.* **133**, B1078 (1964).

12JA J. D. Jackson, *Nuovo Cimento* **34**, 1644 (1964).

12JAC J. D. Jackson and H. Pilkuhn, *Nuovo Cimento* **33**, 906 (1964).

12PI H. Pilkuhn, *The Interaction of Hadrons* (North-Holland, Amsterdam, 1967).

12SA J. J. Sakurai, *Advanced Quantum Mechanics* (Addison-Wesley, Reading, Mass., 1967).

12SU R. L. Sugar and J. D. Sullivan, *Phys. Rev.* **166**, 1515 (1968).

12TR S. Treiman and C. N. Yang, *Phys. Rev. Lett.* 8, 140 (1962).

CHAPTER

13

Modifications of the One-Particle Exchange Model

13-1 Introduction. In this chapter we discuss two modifications of the elementary one-particle exchange model. One modification—the vertex form factor—is only briefly discussed. The vertex form factor is an old concept, but it reappears as the Regge trajectory residue function in Ch. 17.

Most of the chapter is devoted to the other modification, the absorption correction. The concept was first introduced in the early 1960s, and it has remained important to the present. Now we use it in connection with Regge or other models of two-body reactions, as described in Ch. 18. But it is easiest to introduce the absorption correction concept in connection with the elementary one-particle exchange model, and this is what we do in this chapter.

The fundamental motivation for the absorption correction is given in Sec. 13-3, where a partial wave analysis of a simple one-particle exchange amplitude is presented. In Sec. 13-4 the absorption correction for spinless particles is developed in terms of the very useful impact parameter representation. We then deal briefly with the absorption correction, using the full partial wave analysis, for particles with spin.

The chapter concludes with three sections illustrating ways in which the absorption correction modifies the one-particle exchange amplitude. Sections 13-6 and 13-7 contain a general discussion. Section 13-8 applies the theory to one-pion exchange at very small t values. This is the only area in which the theory as presented in this chapter appears to have direct validity.

13-2 Hadronic Vertex Form Factors. The idea of a vertex form factor was introduced in Sec. 12-3. We can associate a function $F_{aec}(t)$ with each vertex (aec) in the one-particle exchange diagram (Fig. 12-1), where t is the four-momentum squared of e, the exchanged particle. The reader who is unacquainted with the vertex form factor concept should jump ahead to Ch. 19 to read about the electromagnetic form factor at the photon–hadron vertex.

The use of hadronic vertex form factors in one-particle exchange has a curious history. In the early 1960s the form factor was introduced and used (13FE). However, during the middle 1960s—the heyday of the application of the absorption correction to elementary one-particle exchange—this factor was often ignored because the narrowing of the forward peak, which was the principal reason for introducing the vertex form factor, could be at least partly achieved by the absorption correction. But during the same period, experimenters who were fitting background spectra under resonances often used for the background calculation a one-particle exchange model modified with an empirical vertex form factor. Finally, with the application of Regge theory to two-body reactions, the vertex form factor was fully reinstated in the form of the residue function of the Regge trajectory.

As discussed in Ch. 19 whenever one of the particles at a vertex has nonzero size, a form factor $F(t)$ must be attached to the vertex. Roughly speaking, $F(t)$ is the manifestation in momentum space of the spatial extent and structure of the particles. The factor $F(t)$ should have the following behavior:

1. $F(0) = 1$.
2. $F(t)$ decreases as t increases.
3. $F(t)$ should decrease to $\sim\frac{1}{2}$ or $1/e$ when $\sqrt{|t|} \sim \hbar/R$, R being the hadronic radius.
4. $F(t)$ can depend on the nature of the particles.

Exercise 13-1. Justify the foregoing statements.

It is not possible at present to develop a fundamental form for $F(t)$. That would require knowledge, which we do not possess, both of the structure of hadrons and of how to translate that structure into $F(t)$. Therefore, various empirical forms for $F(t)$ have been developed. In analogy to the electromagnetic form factor of the proton, we can write

$$F(t) = [1 + a|t|]^{-2} \tag{13-1}$$

or use a propagator form

$$F(t) = [1 + a|t|]^{-1} \tag{13-2}$$

We can also follow the philosophy of Dürr and Pilkuhn (13DUR, 13PI) and base $F(t)$ on angular momentum penetration factors. But for many purposes it is simplest to use the exponential

$$F(t) = e^{-a|t|} \tag{13-3}$$

which is often used for the Regge residue function (Ch. 17). In any case, since we restrict our form factor considerations for present purposes to $|t| \lesssim 0.5\ (\text{GeV/c})^2$, the exact form of $F(t)$ is not important.

Exercise 13-2. Justify the last statement using the restrictions on $F(t)$ given in the previous paragraph.

To conclude this section, we make two remarks. First, within the spirit of the one-particle exchange model, $F(t)$ must be independent of s. Therefore, vertex form factors cannot cure the incorrect energy dependence of the elementary one-particle exchange model. Second, $F_{aec}(t)$ can be dependent on the spin states of particles a and c. Thus we could write $F_{aec}^{(\mu_a, \mu_c)}(t)$, where μ_i is the helicity of particle i. But this allows too much flexibility to the one-particle exchange model, and the model loses all its predictive power. Therefore, we shall always demand that $F_{aec}(t)$ be independent of the spin states of the particles. We now leave the subject of vertex form factors until Ch. 17.

13-3 Partial Wave Analysis of Elementary One-Particle Exchange. To try to understand the origin of some of the flaws of the elementary one-particle exchange model, we return to the partial wave analysis. For the reaction $a + b \to c + d$, which proceeds by e exchange, we write the amplitudes in the high energy approximation

$$T(s, t) = \frac{gG\eta(s)}{m_e^2 - t}, \qquad f(s, t) = \frac{T(s, t)}{8\pi\sqrt{s}} = \frac{gG\eta(s)}{8\pi\sqrt{s}}\left(\frac{1}{m_e^2 - t}\right) \tag{13-4}$$

Here, for simplicity, we have ignored any t dependence due to the spin of the particles; but $\eta(s)$, which is dimensionless, contains the s dependence produced by the spin of particle e. Thus for spin zero exchange $\eta(s)$ is a constant, and for spin

1 exchange $\eta(s)$ is proportional to s. In addition, $t = t_{\text{min}} - 2p_i p_f(1 - \cos\theta)$ still holds in the center of mass.

In the partial wave analysis we have, in the center of mass,

$$f(s, t) = \frac{2}{\sqrt{s}} \sum_{l=0}^{\infty} (2l + 1) T_l(s) P_l(\cos\theta) \tag{13-5}$$

Here we use the high energy approximation, and again the particle spin is ignored. We now define a quantity that appears often in the literature of one-particle exchange

$$z = 1 + \frac{m_e{}^2 - t_{\text{min}}}{2p_i p_f} \underset{s\to\infty}{=} 1 + \frac{2m_e{}^2}{s} \tag{13-6}$$

Then we can put

$$\left(\frac{gG\eta(s)}{8\pi s}\right) \frac{1}{z - \cos\theta} = \sum_{l=0}^{\infty} (2l + 1) T_l(s) P_l(\cos\theta) \tag{13-7}$$

We can easily solve for $T_l(s)$ using the Legendre function of the second kind

$$Q_n(z) = \tfrac{1}{2} \int_{-1}^{+1} \frac{P_n(x)}{z - x}\, dx, \qquad n = \text{integer}, \qquad z > 1 \tag{13-8}$$

We obtain

$$T_l(s) = Q_l(z) \left(\frac{gG\eta(s)}{8\pi s}\right) \tag{13-9}$$

As an example, let us suppose $s \gg m_i^2$, $m_e^2 = 0.5\ \text{GeV}^2$ and $s = 100\ \text{GeV}^2$, such that $z = 1.01$. Using the properties of $Q_n(z)$ given in App. C, we construct the following table showing typical partial wave contributions to the simple propagator $(m_e{}^2 - t)^{-1}$.

l	0	1	2	3	5	10
$Q_l(z)$	2.65	1.68	1.22	0.93	0.58	0.22

Thus at high energies many partial waves contribute to the propagator; the lower partial waves make the largest contributions.

Exercise 13-3. Continue the example by taking the reaction cross section σ to be 0.1 mb. Calculate the values of $T_l(s)$ for $l = 0, 1, 2, 3, 5, 10$.

As we discussed in Sec. 3-3 in connection with the elastic diffraction peak, the shape of a forward peak depends on the order of the partial waves that contribute to the peak. Low order partial waves produce a flattish peak, high order partial waves produce a narrow, sharp peak. From the partial wave viewpoint, therefore, we can easily see why a too-broad forward peak is predicted by the elementary one-particle exchange model when the exchanged particle is other than a pion. The large mass of the exchanged particle leads to too large a contribution from the low partial waves. We could cure this difficulty if we could think of a reason for reducing the contribution of the low partial waves. Just such a reason is provided by the absorption corrections to the one-particle exchange model, and the desire to sharpen the forward peak predicted by the elementary one-particle exchange model was indeed one reason for introducing the absorption corrections here.

However, partial wave analysis reveals another, more fundamental flaw in the elementary one-particle exchange model. As discussed in Ch. 3, unitarity demands that in any inelastic reaction

$$|T_l| \leqslant \tfrac{1}{2} \tag{13-10}$$

Now consider the $l = 0$ partial wave and use the relationships

$$Q_0(z) = \tfrac{1}{2} \ln \left(\frac{z+1}{z-1} \right) \underset{s \to \infty}{\longrightarrow} \tfrac{1}{2} \ln \frac{s}{m_e^{\,2}} \tag{13-11}$$

Then at large s we have

$$T_0(s) = \frac{gG\eta(s) \ln (s/m_e^{\,2})}{16\pi s} \leqslant \tfrac{1}{2} \tag{13-12}$$

and $\eta(s)$ *cannot* increase faster than $s/\ln(s)$ as s increases. In the elementary one-particle exchange model, the exchange of particles with spin 1 or greater will violate this unitarity limit. Thus the elementary one-particle exchange model will in general violate unitarity at sufficiently high energy.

Violation of unitarity by the general elementary one-particle exchange model is a simple example of a problem that still plagues strong interaction theory. As discussed in Sec. 3-2, the total flux F_l in any partial wave l is fixed. As the energy increases, the number of different inelastic channels $N(s)$ increases rapidly. Therefore, unitarity demands that most channels have an amount of flux $\lesssim F_l/N(s)$ in the l partial wave. Suppose for a particular channel we have a correct theory for calculating the amplitudes, hence the partial wave contributions to those amplitudes. For that theory to be correct, the partial wave contributions must meet the unitarity restrictions, which means somehow that unitarity must be built into the theory. In other words, a theory for one inelastic channel must include knowledge about all other inelastic channels that are open at that energy. This is a very complicated requirement for a fundamental theory; and we have no theory of strong interactions that possess the desired property in a fundamental way.

13-4 Introduction to the Absorption Correction of One-Particle Exchange. There are many derivations of the absorption correction to the elementary one-particle exchange model. Some are more general than others, but none are fundamental. Therefore, we choose a somewhat limited but simple derivation that gives much insight into the origin of the correction. We consider an inelastic reaction $a + b \to c + d$ in which, for simplicity, the four particles all have spin 0 and equal masses. Suppose that the one-particle exchange model gives an amplitude $T(s, t)$, which has the impact parameter representation

$$f(s, t) = 2p \int_0^\infty [b\,db\, J_0(b\sqrt{|t|}) T_{ab\to cd}(b, s)] \tag{13-13}$$

as discussed in Sec. 10-8. We can also obtain Eq. 13-13 by the first Born approximation using an inelastic potential $U_{ab\to cd}(s, \mathbf{r})$ with

$$\begin{aligned} \text{ingoing wave} &= e^{i\mathbf{p}\cdot\mathbf{r}} \\ \text{outgoing wave} &= e^{i\mathbf{p}'\cdot\mathbf{r}} \end{aligned} \tag{13-14}$$

Thus we have

$$f(s, t) = -\frac{1}{4\pi} \int d^3r\, e^{i\mathbf{q}\cdot\mathbf{r}} U_{ab\to cd}(s, \mathbf{r}), \qquad \mathbf{q} = \mathbf{p} - \mathbf{p}' \tag{13-15}$$

Now we can improve the approximation by taking into account the elastic scattering of particles a and b on each other in the initial state, and the elastic scattering of particles c and d in the final state. To do this we make use of a tool we have already developed and whose physical significance we feel—the eikonal approximation of Ch. 2. Equation 13-14 is replaced by

$$\begin{aligned} \text{ingoing wave} &= e^{i\mathbf{p}\cdot\mathbf{r}}\, e^{+i\chi_i(s,\mathbf{r})} \\ \text{outgoing wave} &= e^{i\mathbf{p}'\cdot\mathbf{r}}\, e^{-i\chi_f(s,\mathbf{r})} \end{aligned} \tag{13-16}$$

where

$$\begin{aligned} \chi_i(s, \mathbf{r}) &= -\frac{1}{2p} \int_{-\infty}^{z} U_{ab\to ab}(s; x, y, z')\, dz' \\ \chi_f(s, \mathbf{r}) &= -\frac{1}{2p} \int_{z}^{\infty} U_{cd\to cd}(s; x, y, z')\, dz' \end{aligned} \tag{13-17}$$

and $U_{ab\to ab}$ and $U_{cd\to cd}$ are the potentials representing the elastic scattering in the initial and final states, respectively. The meaning of $\chi_{i,f}$ can be reviewed in the discussion of Sec. 2-7. Next Eq. 13-15 is replaced by

$$f_{\text{abs}}(s,t) = -\frac{1}{4\pi}\int [d^3r\, e^{i\mathbf{q}\cdot\mathbf{r}} U_{ab\to cd}(s,\mathbf{r})\, e^{i\chi_i(s,\mathbf{r})}\, e^{i\chi_f(s,\mathbf{r})}] \tag{13-18}$$

This approximation method is also called the *distorted wave Born approximation* because the incoming and outgoing waves have been distorted by their elastic interactions. In Eq. 13-18 we have used the eikonal approximation for the elastic interaction, but we have only used the first Born approximation for the inelastic interaction. Certainly better and more consistent approximations can be devised, but the present approximation is sufficient for our purposes. Equation 13-18 is sometimes justified by assuming that the inelastic interaction is short range compared with the elastic interaction.

To bring Eq. 13-18 into a more useful form, we take advantage of the acknowledged crudeness of the entire method. In almost all two-body inelastic reactions, particles c or d or both will be resonances. We usually have no precise knowledge of the elastic scattering amplitudes for resonances. Therefore, in the end we usually assume that all hadrons have roughly the same elastic scattering amplitudes (we are ignoring spin here), and we write

$$U_{cd\to cd}(s,\mathbf{r}) \approx U_{ab\to ab}(s,\mathbf{r}) \approx U_{\text{el}}(s,\mathbf{r}) \tag{13-19}$$

Thus we obtain the *absorption corrected* inelastic amplitude

$$f_{\text{abs}}(s,t) = 2p\int_0^\infty [bdb\, J_0(b\sqrt{|t|})T_{ab\to cd}(s,b)\, e^{i\chi_{\text{el}}(s,b)}] \tag{13-20}$$

Here we have assumed that $U_{\text{el}}(s,\mathbf{r})$ has cylindrical symmetry, such that

$$\chi_{\text{el}}(s,b) \equiv \chi_{\text{el}}(s,\mathbf{b}) = -\frac{1}{2p}\int_{-\infty}^{+\infty} U_{\text{el}}(s,\mathbf{r})\, dz \tag{13-21}$$

Exercise 13-4. Complete the derivation of Eq. 13-20. Why does Eq. 13-20 reduce to Eq. 13-13 as χ_{el} approaches zero?

We can also write Eq. 13-20 in the particularly instructive form

$$f_{\text{abs}}(s,t) = 2p\int_0^\infty [bdb\, J_0(b\sqrt{|t|})T_{ab\to cd}(s,b)S_{\text{el}}(s,b)] \tag{13-22}$$

where $S_{\text{el}}(s,b) \equiv \exp\,[i\chi_{\text{el}}(s,b)]$, as in Eq. 10-51.

The transparent significance of this result is beautiful. Suppose that the elastic scattering is purely absorptive, a good approximation at high energies. Then $S_{\text{el}}(s,b)$ is simply $\alpha(s,b)$, where $0 \leqslant \alpha(s,b) \leqslant 1$. Now we know that in high energy elastic scattering, when the impact parameter b is small, $\alpha(s,b) \approx 0$. Therefore, in

$f_{\text{abs}}(s, t)$ the contribution of $T_{ab\to cd}(s, b)$ for small b is zero or at least small. As b increases, $\alpha(s, b)$ approaches 1 and the contribution of $T_{ab\to cd}(s, b)$ to $f_{\text{abs}}(s, t)$ approaches the contribution to the nonabsorption corrected $f(s, t)$. Since small impact parameters correspond to low partial waves, the absorption correction reduces the low partial wave contribution of the elementary one-particle exchange amplitude $T_{ab\to cd}(s, b)$ to $f_{\text{abs}}(s, t)$. This is written more formally in the next section, but we can clearly see that the result is the one required in the discussion of the last section.

13-5 Partial Wave Treatment of Absorption Corrections. In this section we discuss the absorption correction using the partial wave expansion, which provides the clearest description of the absorption correction. Indeed, given a modern computer, the partial wave expansion supplies the most straightforward way of calculating the effects of the absorption correction on an inelastic amplitude. Also, employing the partial wave expansion avoids the vagueness associated with the use of the impact parameter representation for low partial waves.

As we have shown several times, the partial wave expansion and the impact parameter representation provide alternative methods for describing elastic and inelastic scattering amplitudes. We have also shown how to go from one method to the other method. Therefore, we could start with Eqs. 13-21 or 13-22 and transform it into a partial wave expansion. However, it is simpler to apply directly the physical insight we gained while developing those equations in the partial wave formalism.

We allow particles a, b, c, and d to have nonzero spin, writing with obvious modifications the uncorrected amplitude for the reaction $a + b \to c + d$

$$f_{\mu_c\mu_d;\mu_a\mu_b}(s, t, \phi) = \frac{1}{p}\sum_J [(2J + 1)T^J_{\mu_c\mu_d;\mu_a\mu_b}(s)D^J_{\mu_i\mu_f}(-\phi, \theta, \phi)] \qquad (13\text{-}23)$$

as given in Eq. 10-9. Similarly, the partial wave expansions for the elastic reactions $a + b \to a + b$ and $c + d \to c + d$ are

$$f_{\text{el},\mu_i\mu_j;\mu_{i'}\mu_{j'}} = \frac{1}{p}\sum_J \left[(2J + 1)\frac{S^J_{\text{el},\mu_i\mu_j;\mu_{i'}\mu_{j'}}(s) - \delta_{\mu_i\mu_{i'}}\delta_{\mu_j\mu_{j'}}}{2i} \times D^J_{\mu_i\mu_f}(-\phi, \theta, \phi)\right] \qquad (13\text{-}24)$$

Then the theory developed in the last section says that each inelastic partial wave amplitude should be corrected by the corresponding elastic partial wave amplitudes. Of the several prescriptions for making this correction, we adopt the one proposed by Sopkovich (13SO, 13DO, 13GO, 13RO). The other prescriptions, when applied, reduce to the same result. The Sopkovich prescription for the absorption corrected inelastic amplitude is

$$f_{\text{abs},\mu_c\mu_d;\mu_a\mu_b}(s,t,\phi)=\frac{1}{p}\sum_J\Bigg\{(2J+1)\sum_{\mu_c'\mu_d'\mu_a'\mu_b'}\left[(S^J_{\text{el},\mu_c\mu_d;\mu_c'\mu_d'}(s))^{1/2}\right.$$
$$\left.\times\, T^J_{\mu_c'\mu_d';\mu_a'\mu_b'}(s)(S^J_{\text{el},\mu_a'\mu_b';\mu_a\mu_b}(s))^{1/2}D^J_{\mu_i\mu_f}(-\phi,\theta,\phi)\right]\Bigg\} \tag{13-25}$$

Exercise 13-5. Discuss the justification of the Sopkovich prescription.

Equation 13-25, with its summation over spin states, is unwieldy. However, our ignorance of the elastic amplitudes in the final state and our general uncertainty about the exactness of the theory convince us that there is no loss in making the following simplification. We assume (1) that we can ignore the elastic helicity *flip* amplitudes and that all elastic helicity *nonflip* amplitudes for a particular elastic reaction are equal, (2) that the elastic amplitudes are roughly the same in the initial and final states, and (3) that the elastic amplitude is purely absorptive, namely,

$$S^J_{\text{el},\mu_i\mu_j;\mu_i\mu_j}(s)=S^J_{\text{el}}(s)=\alpha^J(s);\qquad \alpha^J\text{ real},\qquad 0\leqslant\alpha^J\leqslant 1 \tag{13-26}$$

Then we have

$$f_{\text{abs},\mu_c\mu_d;\mu_a\mu_b}(s,t,\phi)=\frac{1}{p}\sum_J\left[(2J+1)S^J_{\text{el}}(s)T^J_{\mu_c\mu_d;\mu_a\mu_b}(s)\right.$$
$$\left.\times\, D^J_{\mu_i\mu_f}(-\phi,\theta,\phi)\right] \tag{13-27}$$

Thus each partial wave amplitude is reduced by the total absorption in that partial wave, the absorption being measured by the elastic amplitude, which we have assumed to be purely absorptive.

We often use for $\alpha^J(s)$ the form first given in Eq. 3-21

$$S^J_{\text{el}}(s)=\alpha^J(s)=1-C(s)\,e^{-J^2/\tilde{J}^2(s)} \tag{13-28}$$

where $\tilde{J}$ and C are functions of s. Usually C must be of order 1. Rather than discussing the explicit use of Eq. 13-27, we return to the impact parameter formalism, which is more convenient for a general consideration of the effects of absorption corrections on amplitudes.

The impact parameter representation of Eq. 13-27 is

$$f_{\text{abs},\mu_c\mu_d;\mu_a\mu_b}(s,t,\phi)=2p(-1)^n\,e^{in\phi}\int_0^\infty\left[b\,db\,J_n(b\sqrt{|t|})\right.$$
$$\left.\times\, T_{\mu_c\mu_d;\mu_a\mu_b}(s,b)S_{\text{el}}(s,b)\right] \tag{13-29a}$$

where

$$f_{\mu_c\mu_d;\mu_a\mu_b}(s, t, \phi) = 2p(-1)^n e^{in\phi} \int_0^\infty [bdb\, J_n(b\sqrt{|t|}) \times T_{\mu_c\mu_d;\mu_a\mu_b}(s, b)] \qquad (13\text{-}29b)$$

and

$$n = \mu_i - \mu_f = (\mu_a - \mu_b) - (\mu_c - \mu_d)$$

Exercise 13-6. Justify Eqs. 13-29 using Eqs. 13-27, 13-22, and 10-45. Discuss the reason for using helicity amplitudes in Eqs. 13-25, 13-27, and 13-29.

13-6 Some Properties of Absorption Corrected Amplitudes. In this section we exhibit some properties of absorption corrected amplitudes, using for convenience the impact parameter representation, Eq. 13-29, and the elementary one-particle exchange model (to provide the uncorrected inelastic amplitude). But the arguments in this section are quite general, and the properties we demonstrate also apply to inelastic amplitudes obtained from other models. In particular, the arguments apply to the amplitudes obtained by Regge theory, as discussed in Ch. 18.

First we examine how the absorption correction changes the shape of the forward peak for the amplitude of Eq. 13-4, where any spin induced t dependence has been suppressed but a general s dependence $\eta(s)$ has been allowed. We restrict all particle spins to 0 and use (App. B)

$$\frac{1}{\alpha^2 + \beta^2} = \int_0^\infty bdb\, J_0(\beta b) K_0(\alpha b) \qquad (13\text{-}30)$$

Thus we see that Eq. 13-29b has the specific high energy form

$$f(s, t) = \frac{gG\eta(s)}{8\pi\sqrt{s}} \left(\frac{1}{m_e{}^2 + |t|}\right) = \frac{gG\eta(s)}{8\pi\sqrt{s}} \int_0^\infty [bdb\, J_0(b\sqrt{|t|}) K_0(m_e b)] \qquad (13\text{-}31)$$

To obtain the absorption corrected amplitude, we choose for $S_{\text{el}}(s, b)$ a form that will allow us to express the integral in known functions. We use the sharp cutoff model of the diffraction peak described in Sec. 3-3, namely,

$$S_{\text{el}}(s, b) = 0, \qquad b < R; \qquad S_{\text{el}}(s, b) = 1, \qquad b > R \qquad (13\text{-}32)$$

Then we have

$$f_{\text{abs}}(s, t) = \left[\frac{gG\eta(s)}{8\pi\sqrt{s}}\right] \left[\frac{R}{m_e{}^2 + |t|}\right] [m_e J_0(R\sqrt{|t|}) K_1(m_e R) - \sqrt{|t|}\, J_1(R\sqrt{|t|}) K_0(m_e R)] \qquad (13\text{-}33)$$

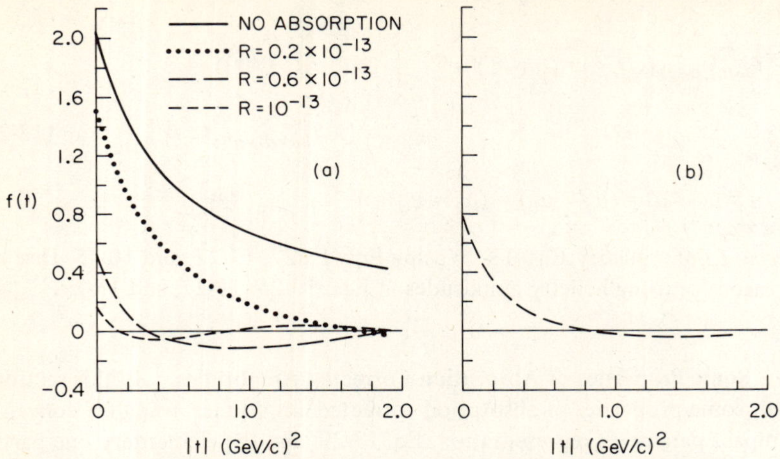

Fig. 13-1 The elementary one-particle exchange amplitude $f(t)$ versus $|t|$. For simplicity the quantity $(gG\eta(s)/8\pi\sqrt{s})$ has been set equal to 1. (*a*) The solid line is the amplitude without an absorption correction. The dashed lines are the amplitudes corrected for absorption using the sharp cutoff model of Eq. 13-32 with various values of R in centimeters. (*b*) The amplitude corrected for absorption using the Gaussian model of Eq. 13-34 with $R = 1.0 \times 10^{-13}$ cm.

Exercise 13-7. Derive Eq. 13-33. Compare Ref. 13DAR.

As an example we choose $m_e = 0.7$ GeV and $R = 10^{-13}$ cm $= (0.2 \text{ GeV})^{-1}$. Figure 13-1*a* compares $f(s, t)$ and $f_{\text{abs}}(s, t)$. We make two observations.

1. The absorption correction narrows the forward peak. This is one of the classic effects of the absorption correction. For the models that ignored vertex form factors or their equivalent, this was a very important effect, because except for pion exchange, the elementary one-particle exchange model usually gives too broad a forward peak. But the effect is not as important when we recognize that vertex form factors in some guise may be used.

2. The second observation is that the absorption correction produced zeros in the amplitude at larger t values. These zeros are directly caused by the sharp cutoff model for $S_{\text{el}}(s, b)$. A smooth cutoff model, such as our old friend from Chs. 2 and 3, the Gaussian function

$$S_{\text{el}}(s, b) = 1 - C e^{-b^2/R^2} \tag{13-34}$$

leads to fewer zeros. Figure 13-1*b* shows $f_{\text{abs}}(s, t)$ using Eq. 13-34 with $C = 1$ and $R = (0.2 \text{ GeV})^{-1}$. Unfortunately, the use of Eq. 13-34 rather than Eq. 13-32 leads to an integral for f_{abs} that is rather intractable analytically. Figure 13-1*b* was obtained by numerical computation.

The second observation is most important, for it shows that a suitably chosen absorption correction can produce zeros in the amplitude. Such zeros can in turn cause dips in the differential cross section—dips of the type discussed in Ch. 18. Thus the absorption correction can provide an explanation of the occurrence and position of these dips. We return to this subject in Ch. 18.

Exercise 13-8. In the foregoing discussion we noted that the absorption correction is usually assumed to remove completely the lowest partial wave amplitudes. Thus in Eq. 13-34 we set $C = 1$. Give a reason for this based on the discussion in the last paragraph of Sec. 13-3. Also consult the literature (13DO, 13GO, 13RO, 13DU). Show that this idea of complete absorption is only in fair agreement with the optical model of elastic scattering. Identify C in Eq. 13-34 with β_0 in Eqs. 2-44 and 2-45, thus showing $C = \beta_0 = 4\sigma_{el}/\sigma_{tot}$. For hadronic systems at high energy, β_0 is 0.4 to 0.8. Unfortunately, there is a more serious problem. As discussed in Ch. 18, values of $C > 1$ are required in some absorption models.

Exercise 13-9. For a neat and simple example with future applications, suppose

$$f(s, t) = A(s)\, e^{-a|t|} \tag{13-35}$$

Use Eqs. 13-29a, 13-34 and B-8 to show

$$f_{abs}(s, t) = A(s)\, e^{-a|t|} \left[1 - C\left(\frac{R^2}{R^2 + 4a}\right) \exp\left(\frac{4a^2}{R^2 + 4a}\right)|t|\right] \tag{13-36}$$

For the remainder of this section we assume that some of the particles in the reaction $a + b \rightarrow c + d$ have nonzero spin. Examination of Eq. 13-29 indicates that the absorption correction will change the different helicity amplitudes by *different* amounts. This will *change the differential cross section* because the relative amounts of the different helicity amplitudes will change. Also, the *density matrices* of the final state particles will *change*. These changes have been extensively discussed (13DO, 13GO, 13DU). In many cases the absorption corrections bring the density matrix predictions of the one-particle exchange model closer to the experimental results.

In spite of some success with the density matrices, the absorption modified one-particle exchange model is not, in general, correct. There are differential cross sections and density matrices that cannot be easily fit by the model. But a much more important failure of the model is that like the uncorrected one-particle exchange model, it gives the wrong energy dependence. This is easily seen by using Eqs. 13-29 with the absorption correction of Eq. 13-34. The form and meaning of Eq. 13-34 come directly from elastic scattering, as discussed in Sec. 3-4, and we know that at high energies the elastic scattering cross section changes very slowly, if at all, with energy. At high energies, therefore, C and R must be constant or possess a very slight energy dependence. Furthermore, in Eq. 13-29a, S_{el} is actually energy *independent* at high energy. Thus the energy dependence of $f_{abs,\mu_c,\ldots}(s, t)$ comes only from $T_{\mu_c\ldots}(s, b)$, and is the *same* as the energy dependence of $f_{\mu_c\ldots}(s, t)$. Hence the absorption corrected one-particle exchange model possesses the same energy dependence as the elementary one-particle exchange model. And we saw in Ch. 12 that in general the latter predicts a completely wrong energy dependence.

The basic difficulty associated with the absorption corrected one-particle exchange model is not the concept of absorption corrections. The problem is that we have been attempting to correct elementary one-particle exchange amplitudes that are themselves very wrong. Therefore, the uncorrected amplitude must be drastically altered. How this might be done is discussed in Chs. 17 and 18. The single apparent exception to this situation—namely, one-pion exchange in the very small t region—is discussed in Sec. 13-8.

13-7 Exceptional Terms and Small t Behavior. When the particles have non-zero spin, the absorption correction calculation is tedious, even if the impact parameter representation is used. But in the small t region, all conventional absorption calculations modify the uncorrected amplitude in a way that can be described rather simply. We present this description in this section, because it is not only convenient but also sheds light on the fundamental nature of the absorption correction. Here by the small t region we mean $|t| \lesssim 0.2\ (\text{GeV/c})^2$, and of course high energy. We avoid the larger t region both because the arguments we present below would not be valid and because the dips observed in the differential cross section in that region require more complicated explanations.

An uncorrected one-particle exchange amplitude has the general form

$$f_{\mu_c\ldots}(s\theta\phi) = [e^{i(\mu_i-\mu_f)\phi}] \left[\frac{1-x}{2}\right]^{|\mu_i-\mu_f|/2} \left[\frac{1+x}{2}\right]^{|\mu_i+\mu_f|/2} \left[\frac{P(s,x)}{z-x}\right] \tag{13-37}$$

where

$$x = \cos\theta; \qquad \frac{1}{z-x} = \frac{2p_i p_f}{m_e^2 - t} \tag{13-38}$$

z is defined in Eq. 13-6, and $P(s, x)$ is a polynomial in x with coefficients that are functions only of s. That the $(1-x)/2$ and $(1+x)/2$ factors can be extracted follows from Eq. 10-35. We can rewrite Eq. 13-37 in the form

$$f_{\mu_c\ldots}(s\theta\phi) = [e^{i(\mu_i-\mu_f)\phi}] \left[\frac{1-x}{2}\right]^{|\mu_i-\mu_f|/2} \left[\frac{1+x}{2}\right]^{|\mu_i+\mu_f|/2} \times \left[\frac{C(s)}{z-x} + P'(s,x)\right] \tag{13-39}$$

where $C(s)$ is a function only of s, and $P'(s, x)$ is a polynomial in x with coefficients that depend only on s. The order of the polynomials P and P' in Eqs. 13-37 and 13-39 will be approximately equal to the magnitude of the spins of the particles in the reaction. Thus the order will be less than 3 or 4. The $F'(s, x)$ term in Eq. 13-39 is called the *exceptional term.*

The absorption correction does two things. First, the lower partial waves are completely absorbed. Therefore, the *exceptional term* in Eq. 13-39, which consists entirely of low order partial waves, is completely absorbed. Thus the exceptional term should be set equal to zero! Second, the absorption correction narrows the forward peak. But in doing this it must preserve the minimum $\cos\theta$ dependence explicit in the $(1 \pm x)/2$ factors. A simple way to accomplish this empirically is to multiply the first term in Eq. 13-39 by the exponential function $(a\, e^{-b|t|})$. Thus the absorption corrected amplitude [writing a new function of s, $C'(s)$, returning to the t notation, and setting $1 + x \approx 2$] is

$$f_{\text{abs},\mu_c,\ldots}(s, t, \phi) = [e^{i(\mu_i - \mu_f)\phi}] \left[\frac{C'(s)[\sqrt{|t - t_{\min}|}]^{|\mu_i - \mu_f|}\, e^{-b|t|}}{m_e^2 - t}\right] \tag{13-40}$$

This absorption correction method emphasizes the importance of the exceptional terms. Their removal can cause a drastic change in the small t behavior of the amplitude, and this is usually the major effect of the absorption correction in the small t region. In the next section we present a dramatic example of this effect. For further discussion the reader is referred to Refs. 13DO, 13GO, 13DU and, in particular, Ref. 13WI, which also has an interesting discussion of the interaction of vertex form factors with the absorption correction.

13-8 One-Pion Exchange in the Small t Region. In this section and its exercises we apply the absorption correction, using the method of the previous section, to the reactions

$$\gamma + p \to \pi^+ + n \tag{13-41a}$$

$$\pi^- + p \to \rho^0 + n \tag{13-41b}$$

$$n + p \to p + n \qquad (n + p \to n + p \text{ at } 180°) \tag{13-41c}$$

We assume that one-pion exchange is the dominant mechanism for these reactions in the small t region, $|t| < 0.2\ (\text{GeV/c})^2$. The reason the one-pion exchange amplitude can be believed in this region is simply the very small mass of the pion. Using the analyticity ideas introduced in Ch. 15, the pole in the pion propagator $(m_\pi^2 - t)^{-1}$ occurs at $t = m_\pi^2$, which is *very close* to the physical region. Therefore, it is reasonable that the one-pion exchange amplitude dominates the part of the physical region in which $|t|$ is less than $5m_\pi^2$ or $10m_\pi^2$. If some other particle (e.g., a ρ) is exchanged, then the pole in the propagator is relatively far from the physical region. And from the analyticity viewpoint it is less likely that such a one-particle exchange amplitude will dominate (Sec. 15-10).

Let us consider the photoproduction reaction of Eq. 13-41a. Figure 7-9 reveals that the small t region shows a sharp forward peak. But at high energies the dominant uncorrected helicity amplitudes are

$$T_{-1/2;+1,+1/2} = T_{+1/2;+1,-1/2} = +\sqrt{2}gG\left[\frac{|t|}{m_\pi{}^2+|t|}\right]$$
$$T_{-1/2;-1,+1/2} = T_{+1/2;-1,-1/2} = -\sqrt{2}gG\left[\frac{|t|}{m_\pi{}^2+|t|}\right] \qquad (13\text{-}42)$$

Here we use the notation $T_{\mu_p,\text{out};\mu_\gamma,\mu_p,\text{in}}$. Also, $\phi = 0$, and $|t_{\text{min}}| = m_\pi{}^4/4E_{\text{lab}}^2$ is ignored.

Exercise 13-10. Derive Eqs. 13-42 using Sec. 12-6 and recognizing that the only formal difference in the calculations is the absence of a zero helicity state for the photon compared to the ρ^0.

Since Eqs. 13-42 will clearly lead to a *dip*, not a *peak*, at $t = 0$, they do not explain Fig. 7-9. But let us apply the absorption correction using the method of Sec. 13-7. Writing Eqs. 13-42 in the form of Eq. 13-39, we obtain

$$|\mu_i - \mu_f| = 0, \quad T_{-1/2;+1,+1/2} = +\sqrt{2}gG\left[-\frac{m_\pi{}^2}{m_\pi{}^2+|t|}+1\right] \qquad (13\text{-}43a)$$

$$|\mu_i - \mu_f| = 2, \quad T_{-1/2;-1,+1/2} = -\sqrt{2}gG\left[\frac{|t|}{m_\pi{}^2+|t|}\right] \qquad (13\text{-}43b)$$

Setting the exceptional term in Eq. 13-43a to zero and introducing a peak narrowing factor, we obtain

$$T_{-1/2;+1,+1/2} = -\sqrt{2}gG\left[\frac{m_\pi{}^2}{m_\pi{}^2+|t|}\right][a\,e^{-b|t|}] \qquad (13\text{-}44a)$$

$$T_{-1/2;-1,+1/2} = -\sqrt{2}gG\left[\frac{|t|}{m_\pi{}^2+|t|}\right][a\,e^{-b|t|}] \qquad (13\text{-}44b)$$

And now the differential cross section will clearly have a peak at $t \approx 0$! This peak is given quite well by the denominator $(m_\pi{}^2 + |t|)$, the exponential $e^{-b|t|}$ is hardly necessary at very small t. The crucial change was made by the removal of the exception term!

Exercise 13-11. Carry out the same procedure for the reaction in Eq. 13-41b. Calculate the differential cross section and the density matrix, and compare your results with the theory and experiment given in Ref. 13BU.

Exercise 13-12. Perform the same procedure for the *np* backward scattering reaction (also called the *np* charge exchange reaction) given in Eq. 13-41c. Show that the small t peak given in Ref. 13DA is obtained.

In this section we have not discussed the one-pion exchange amplitude at larger values of t, $|t| \gtrsim 0.2\ (\text{GeV}/c)^2$, or such competing theories of the small t region as the so-called *electric Born* approximation. A general discussion of these questions is provided by Ref. 13GOT.

Finally, we must note that the arguments presented in this and the previous section for the removal of the exceptional term are not universally accepted, even though the usefulness of the procedure is very generally acknowledged. Thus Fox (13FO) has argued that the removal of the exceptional term has nothing to do with absorption corrections; rather, the removal is necessitated by a smoothness requirement on the amplitudes as they pass from the pion pole to the physical region.

REFERENCES

13BU F. Bulos et al., *Phys. Rev. Lett.* **26**, 1453 (1971).

13DA M. B. Davis et al., *Phys. Rev. Lett.* **29**, 139 (1972).

13DAR A. Dar, *Phys. Rev. Lett.* **13**, 91 (1964).

13DO J. T. Donohue, Ph.D. thesis, University of Illinois, 1962.

12DU L. Durand, III, and Y. T. Chiu, *Phys. Rev.* **139B**, 646 (1965).

13DUR H. P. Dürr and H. Pilkuhn, *Nuovo Cimento* **40**, 899 (1965).

13FE E. Ferrari and F. Selleri, *Phys. Rev. Lett.* **7**, 387 (1961).

13FO G. Fox, Argonne National Laboratory Report ANL/HEP-7208, Vol. II, 1972.

13GO K. Gottfried and J. D. Jackson, *Nuovo Cimento* **34**, 735 (1964).

13GOT E. Gotsman and U. Maor, *Nucl. Phys.* **B46**, 525 (1972).

13PI H. Pilkuhn, *The Interaction of Hadrons* (North-Holland, Amsterdam, 1967).

13RO M. H. Ross and G. L. Shaw, *Phys. Rev. Lett.* **12**, 627 (1964).

13SC N. J. Sopkovich, *Nuovo Cimento* **26**, 186 (1962).

13WI P. K. Williams, *Phys. Rev.* **181**, 1963 (1969).

CHAPTER

14

The Quark Model and $SU(3)$

14-1 Introduction. In most of the book we ignore the differences between the strong interactions of the different hadrons. Our discussions generally assume that the basic dynamics of a strong interaction have nothing to do with the type of hadrons that participate in the interaction. But this proposed picture cannot be complete, for the mass spectrum of the hadrons must be related to the nature of strong interactions, and the internal quantum numbers of the hadrons are conserved in strong interactions. In this chapter we are interested in the differences and similarities between hadrons. And we are interested in how these differences and similarities can be used to classify the hadrons into sets—the $SU(3)$ multiplets—which are generalizations of isospin multiplets. We are also interested in what we call the *relative strong interaction properties* of the hadrons. An example of such a property is the ratio of the pion-nucleon total cross section to the nucleon-nucleon total cross section—about 0.55 at high energy. Why is it 0.55 and not 0.25 or 1 or 2?

There are two different approaches to the material covered in this chapter. One can make full use of group theory, developing the general concept of the Lie group and then constructing the full apparatus of the $SU(3)$ group. This approach illustrates the use of group theoretical methods in elementary particle physics and also provides the reader with a better introduction to the specific calculational methods used in the application of the $SU(3)$ group to elementary particles. The other approach begins with the concept of fundamental quark triplets and constructs the $SU(3)$ multiplets in an ad hoc and unsystematic manner.

The two approaches overlap in most of their predictions, for they are based on the same basic symmetry assumptions. However, additional assumptions are often used in the quark model, and these lead to additional specific predictions by that model. The two approaches tend to emphasize different theoretical beliefs or feelings. The pure group theory approach favors the feeling that elementary particles obey *basic symmetries.* These symmetries are described by group theory, but we may never be able to find more basic reasons for them. The quark approach favors the feeling that the very complicated world of hadrons can be constructed out of a few basic *components* or *constituents* such as quarks. Thus the quark model approach tends to develop insight into *composite* or *constituent* models of the hadrons.

Since we want to develop composite model insight, as well as some understanding of the predictions of $SU(3)$, we have chosen to emphasize the quark model approach. Besides, we could not allot much space to this subject, and the quark model can be developed more rapidly. Fortunately there are a number of books that develop the group theoretical approach starting from simple ideas (14LI, 14LIC, 14CA, 14GO), and a shorter treatment is furnished in Ref. 14GA.

To provide an analogy to use in the construction of the quark model, we consider briefly in Sec. 14-2 how angular momentum multiplets can be constructed out of spin $\frac{1}{2}$ states. This analogy is then used in Sec. 14-3 to build and classify the $SU(3)$ multiplets. The $SU(3)$ octets corresponding to the pseudoscalar mesons, the vector mesons, and the spin $\frac{1}{2}$ baryons; and the $SU(3)$ decuplet corresponding to the spin $\frac{3}{2}$ baryons, are briefly described in Secs. 14-4 and 14-5.

The next steps in most discussions of $SU(3)$ involve the examination of the mass relations inside a multiplet and the strong interaction decay properties of the resonances. The $SU(3)$ model is particularly successful in explaining the relative decay rates of the many different baryon resonances. But we must move on to our main interest—strong interactions at high energy. Thus in Sec. 14-5 we mainly consider the application of the $SU(3)$ symmetry to high energy, two-body reactions. We conclude the chapter with a brief introduction to some specific applications of the quark model (Sec. 14-6), and a general discussion of the significance of composite models of hadrons (Sec. 14-7).

14-2 The Angular Momentum Analogy. We rarely do so, but it is possible to construct any angular momentum state $|j, m\rangle$ out of the direct product of spin $\frac{1}{2}$ angular momentum states $|\frac{1}{2}, \pm\frac{1}{2}\rangle$. To illustrate this idea, we define the spin $\frac{1}{2}$ states

$$|+\rangle \equiv |\tfrac{1}{2}, +\tfrac{1}{2}\rangle, \qquad |-\rangle \equiv |\tfrac{1}{2}, -\tfrac{1}{2}\rangle \tag{14-1}$$

By taking the direct product of two of these states and making suitable combinations, we can construct spin 1 and spin 0 states. Explicitly, the Clebsch-Gordan coefficients yield

$$j = 1 \text{ multiplet} \qquad |1, 1\rangle = |+\rangle|+\rangle \tag{14-2a}$$

$$|1, 0\rangle = \frac{1}{\sqrt{2}}(|+\rangle|-\rangle + |-\rangle|+\rangle)$$

$$|1, -1\rangle = |-\rangle|-\rangle$$

$$j = 0 \text{ multiplet} \qquad |0, 0\rangle = \frac{1}{\sqrt{2}}(|+\rangle|-\rangle - |-\rangle|+\rangle) \tag{14-2b}$$

For later use we introduce a notation that is unconventional in angular momentum work but conventional in *SU*(3) work. An angular momentum multiplet with spin j has $2j + 1$ components, the *multiplicity* of the multiplet being $2j + 1$. The multiplet is then identified by its multiplicity in bold face type, $\mathbf{2j + 1}$. Thus the $j = \frac{3}{2}$ multiplet is denoted by **4**, and we can easily summarize the result of taking the direct product of angular momentum states. For example,

$$\mathbf{2} \otimes \mathbf{2} = \mathbf{3} \oplus \mathbf{1} \tag{14-3}$$

summarizes Eqs. 14-2. The circles around the $\otimes$ and $\oplus$ signs indicate that these are not the ordinary operators of multiplication and addition.

If we take the direct product of three spin $\frac{1}{2}$ states we obtain

$$\mathbf{2} \otimes \mathbf{2} \otimes \mathbf{2} = \mathbf{4} \oplus \mathbf{2} \oplus \mathbf{2} \tag{14-4}$$

That is, the direct product of three spin $\frac{1}{2}$ states yields *one* $j = \frac{3}{2}$, or **4**, multiplet and *two* $j = \frac{1}{2}$, or **2**, multiplets.

Exercise 14-1. Use Clebsch-Gordan coefficients to explicitly construct the eight states that comprise the three multiplets on the right-hand side of Eq. 14-4. The two **2** multiplets are

$$|\tfrac{1}{2}, +\tfrac{1}{2}\rangle = \sqrt{\tfrac{2}{3}}|+\rangle|+\rangle|-\rangle - \frac{1}{\sqrt{6}}|+\rangle|-\rangle|+\rangle - \frac{1}{\sqrt{6}}|-\rangle|+\rangle|+\rangle$$

$$|\tfrac{1}{2}, -\tfrac{1}{2}\rangle = \frac{1}{\sqrt{6}}|+\rangle|-\rangle|-\rangle + \frac{1}{\sqrt{6}}|-\rangle|+\rangle|-\rangle - \sqrt{\tfrac{2}{3}}|-\rangle|-\rangle|+\rangle \tag{14-5a}$$

$$|\tfrac{1}{2}, +\tfrac{1}{2}\rangle = \frac{1}{\sqrt{2}}|+\rangle|-\rangle|+\rangle - \frac{1}{\sqrt{2}}|-\rangle|+\rangle|+\rangle$$

$$|\tfrac{1}{2}, -\tfrac{1}{2}\rangle = \frac{1}{\sqrt{2}}|+\rangle|-\rangle|-\rangle - \frac{1}{\sqrt{2}}|-\rangle|+\rangle|-\rangle \qquad (14\text{-}5b)$$

The two multiplets have the *same* behavior under rotations. They differ only in their *internal* symmetry. The multiplet in Eq. 14-5a is symmetric under the interchange of the first two spin states, counting from the left; the Eq. 14-5b multiplet is antisymmetric under that interchange. Therefore we cannot distinguish the two multiplets until we know how they were constructed.

To achieve an overview, we apply some general group theoretical ideas to the group of rotations in three-dimensional space (14LIC, 14HA). These are the rotations discussed in detail in Chs. 6 and 9, and there are an infinite number of irreducible representations of this group. The $2j + 1$ members of the jth angular momentum multiplet are the basis vectors for the $(2j + 1)$-dimensional representation of the rotation group. Any rotation is then represented by a *unitary* matrix of dimension $(2j + 1)^2$. The lowest dimension nontrivial representation is two-dimensional with $j = \frac{1}{2}$. And the corresponding matrices are 2×2 unitary matrices. This group of matrices also comprises the *unitary* group $U(2)$. We can remove the group redundancy (of some matrices differing from other matrices by an overall constant factor) by demanding that the matrices be unimodular, that is, that their determinant be 1. These 2×2 matrices then comprise the *special unitary* group $SU(2)$. The $SU(2)$ group is locally isomorphic to the group of rotations in three-dimensional space. Therefore, the higher dimension representations of the rotation group are also representations of the $SU(2)$ group. In Eqs. 14-2 to 14-5, we have been constructing the higher dimension multiplets that form the basis vectors of the higher dimension representations of the $SU(2)$ group.

14-3 Quarks and *SU*(3). As discussed in Sec. 4-3, the angular momentum concepts of the previous section can be used by analogy in the construction of a formal theory of isotopic spin. Isotopic spin multiplets are directly analogous to angular momentum multiplets; and one can think of rotations in three-dimensional isotopic spin space as being analogous to rotations in real three-dimensional space. All this is justified by the validity of the predictions of isotopic spin theory in strong interactions.

But isotopic spin theory is only part of a more general theory of the relations between the hadrons. And this more general theory is based on extending and generalizing the analogy between isotopic spin and angular momentum. We consider first a *three*-dimensional hypothetical space with *three* basis vectors,

$$q_1 = \begin{bmatrix} 1 \\ 0 \\ 0 \end{bmatrix}, \quad q_2 = \begin{bmatrix} 0 \\ 1 \\ 0 \end{bmatrix}, \quad q_3 = \begin{bmatrix} 0 \\ 0 \\ 1 \end{bmatrix} \tag{14-6}$$

In this space, the group of all 3 x 3, unitary, unimodular matrices is called the $SU(3)$ group. Just as there are higher dimension representations of the $SU(2)$ group, there are higher dimension representations of the $SU(3)$ group. And in analogy to our use in the previous section of the basis vectors in Eq. 14-1, we intend to use the basis vectors in Eq. 14-6 to construct the basis vectors of the higher dimension representations of $SU(3)$. The set of basis vectors for a particular dimension we call an $SU(3)$ *multiplet.* Thus the higher dimension multiplets will be constructed out of the direct products of the triplet in Eq. 14-6.

But in doing this we make the following audacious demands.

1. We shall assign to each basis vector in Eq. 14-6 the internal quantum numbers of B, S, Q, I, and I_z. The internal quantum numbers of the higher dimension basis vectors will be obtained using the rules for additive quantum numbers and isotopic spin described in Ch. 4.

2. A multiplet shall correspond in a one-to-one relationship to a set of hadrons. All hadrons in the set shall have the same baryon number and intrinsic spin.

3. The internal quantum numbers of the hadrons and the mass relationships in the set shall be explained by the construction.

4. The *relative strong interactions properties* of the hadrons, as defined in Sec. 14-1, shall be explained by the structure of the multiplet and the coupling between multiplets.

It is amazing to think that these demands can be met even approximately, as indeed they are. Let us see how this is done.

Table 14-1 Properties of the quarks and antiquarks

Name	B	I	I_z	Q	S	Y
Quarks						
p	$\frac{1}{3}$	$\frac{1}{2}$	$\frac{1}{2}$	$\frac{2}{3}$	0	$\frac{1}{3}$
n	$\frac{1}{3}$	$\frac{1}{2}$	$-\frac{1}{2}$	$-\frac{1}{3}$	0	$\frac{1}{3}$
λ	$\frac{1}{3}$	0	0	$-\frac{1}{3}$	-1	$-\frac{2}{3}$
Antiquarks						
$\bar{p}$	$-\frac{1}{3}$	$\frac{1}{2}$	$-\frac{1}{2}$	$-\frac{2}{3}$	0	$-\frac{1}{3}$
$\bar{n}$	$-\frac{1}{3}$	$\frac{1}{2}$	$\frac{1}{2}$	$\frac{1}{3}$	0	$-\frac{1}{3}$
$\bar{\lambda}$	$-\frac{1}{3}$	0	0	$\frac{1}{3}$	1	$\frac{2}{3}$

For convenience, we refer to the basis vectors in Eq. 14-6 as the *basic quark triplet* or just the quark triplet. We must assign internal quantum numbers to these quarks. We leave aside the question of intrinsic spin until Sec. 14-6. In Table 14-1 we give the internal quantum numbers of the quarks. The symbol q_i has been replaced by the mnemonic names p, n, and λ. Since p and n have $S = 0$, they are like the proton and neutron in that respect. The λ has $S = -1$, as does the Λ^0. The necessity for fractional baryon numbers and charges will become apparent. But certainly if one wishes to think of these quarks as real particles, they have astonishing properties compared to the elementary particles we know. The properties in Table 14-1 have been selected to fit the Gell–Mann–Nishijima relation (Eqs. 4-2 and 4-3)

$$q = I_z + \tfrac{1}{2}(S + B) = I_z + \tfrac{1}{2}Y \tag{14-7}$$

But if we are going to construct mesons with $B = 0$, we also need the antiparticles of the quarks. These *antiquarks* $q_i = \bar{n}$, $\bar{p}$ and $\bar{\lambda}$ are listed in Table 14-1 also.

Exercise 14-2. Verify Eq. 14-7 for the quarks and antiquarks.

Using the notation developed in the last section, we denote the quark triplet by **3**, indicating that its multiplicity is 3. We use the notation **3*** for the antiquark triplet.

Next we construct multiplets to correspond to the mesons by taking the direct product **3** ⊗ **3***. We obtain nine independent states of the form

$$\bar{q}_i q_j; \qquad i = 1, 2, 3; \qquad j = 1, 2, 3 \tag{14-8}$$

Under a unitary transformation U which is a member of the $SU(3)$ group, we have $q \to q' = Uq$. Therefore, these states transform into each other, but they do *not* transform out of the set of nine states. Thus these nine states comprise in themselves one or more complete $SU(3)$ multiplets.

In our next task, examining the nine states, the ad hoc and unsystematic nature of our approach becomes apparent. Our approach is a justification, not a derivation; and as the dimensions of the multiplets increase, this unsystematic approach becomes increasingly difficult to use. First it is very helpful to make as the completest possible use of isotopic spin considerations; for not only is the conservation of isotopic spin built into this more general theory, but we understand how to work with isotopic spin. Therefore, we construct the following isotopic spin multiplets

$$\begin{aligned} &I = \tfrac{1}{2}, \quad S = +1 \qquad \bar{\lambda}n,\ \bar{\lambda}p \\ &I = 1, \quad S = 0 \qquad \bar{p}n,\ \frac{1}{\sqrt{2}}(\bar{p}p - \bar{n}n),\ \bar{n}p \\ &I = \tfrac{1}{2}, \quad S = -1 \qquad \bar{p}\lambda,\ \bar{n}\lambda \end{aligned} \tag{14-9}$$

Here we have constructed seven independent states, ψ_α, ψ_β, . . ., out of the nine states in Eq. 14-8; there are two more to be constructed. We have used here and will continue to use the orthonormal property

$$\langle \psi_\beta | \psi_\alpha \rangle = \delta_{\beta\alpha} \tag{14-10a}$$

where

$$\psi_\alpha = \sum_{ij} a_{\alpha ij} \bar{q}_i q_j \tag{14-10b}$$

Next we note that the particular state

$$\psi_1 = \frac{1}{\sqrt{3}} (\bar{p}p + \bar{n}n + \bar{\lambda}\lambda) \tag{14-11}$$

is invariant under an arbitrary *SU*(3) transformation, that is, it transforms into itself. Therefore, it is an *SU*(3) *singlet.* This **1** multiplet accounts for the eighth state. Using Eq. 14-10, we are left with the state

$$\psi_8 = \frac{1}{\sqrt{6}} (\bar{p}p + \bar{n}n - 2\bar{\lambda}\lambda) \tag{14-12}$$

Under an arbitrary *SU*(3) transformation, this state will change into some combination of the states in Eq. 14-9 and itself. Therefore the eight states in Eqs. 14-9 and 14-12 comprise an *SU*(3) *octet* or **8** *multiplet.* Thus the direct product of a quark and antiquark triplet is an octet and singlet

$$\mathbf{3^* \otimes 3 = 8 \oplus 1} \tag{14-13}$$

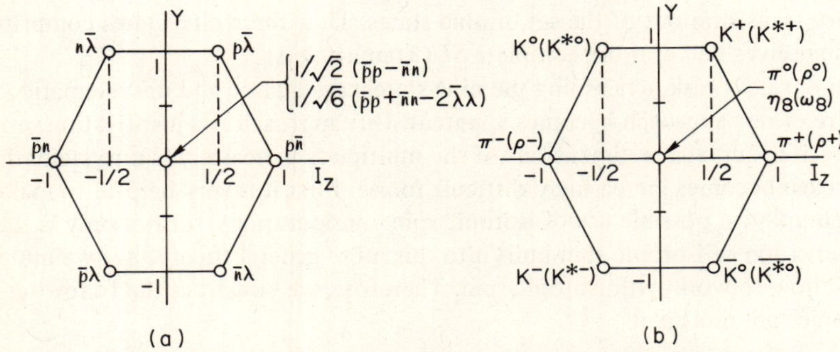

Fig. 14-1 (*a*) The *SU*(3) octet, constructed out of quark-antiquark pairs. But note that an octet can be constructed out of more complicated quark combinations, such as the three-quark states of the baryon octet (Sec. 14-4). (*b*) The *SU*(3) octet classifications of the families of pseudoscalar and vector mesons. Note that in each family the $I = 0$, $Y = 0$ member is a superposition of the octet state ψ_8 (Eq. 14-12) and the singlet state ψ_1 (Eq. 14-11). Only the ψ_8 component is indicated here.

Exercise 14-3. Fill in the steps leading to Eqs. 14-9 through 14-12.

The structure of the octet can be graphically represented in a beautiful way, represented in Fig. 14-1*a*. The x axis is I_z and the y axis is the hypercharge $Y = (S + B)$. We note that the position $I_z = 0$, $Y = 0$ is occupied by two states. Moving to higher $SU(3)$ multiplets, we see that more and more positions are occupied by more than one state. It is also very instructive to use the same graphical scheme for the quark and antiquark triplets, as in Fig. 14-2.

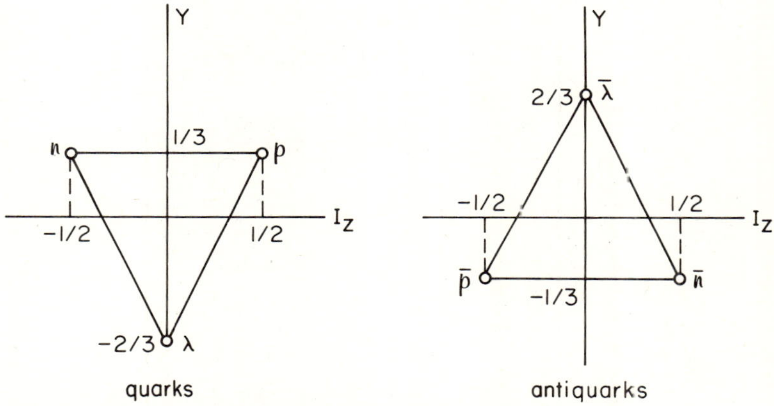

Fig. 14-2 The quark and antiquark $SU(3)$ triplets.

Amazingly, our best-known sets of mesons, the $\pi, K, \bar{K}, \eta$ family of pseudo-scalar mesons and the $\rho, K^*(890), \bar{K}^*(890), \omega, \phi$ family of vector mesons can be exactly fitted into $SU(3)$ octets and singlets. The classification is given in Fig. 14-2*b* and in Table 14-2, which speak for themselves.

Exercise 14-4. Consult the literature to determine the present status of fitting higher mass mesons, such as the spin 2 family (which includes the A_2 and f), into $SU(3)$ octets and singlets.

As we might expect from quantum mechanical principles, the two $I = 0$, $Y = 0$ states can interfere or *mix*. Thus in the vector mesons, the physical particles ϕ and ω are mixtures of the ψ_1 and ψ_8 states of Eqs. 14-11 and 14-12.

Exercise 14-5. Consult the literature for the evidence that ϕ and ω are mixtures of $SU(3)$ octet and singlet states. Because of this mixing, the vector mesons are often said to belong to an $SU(3)$ nonet.

Table 14-2 The quark model content of the octet and singlet multiplets of the pseudoscalar and vector mesons

$SU(3)$ Multiplet	Quark Content	I	I_z	S	Pseudoscalar Name	Mesons Mass (MeV/c^2)	Vector Name	Meson Mass (MeV/c^2)
8	$\bar{n}p$	1	1	0	π^+	140	ρ^+	765
8	$(\bar{p}p-\bar{n}n)/\sqrt{2}$	1	0	0	π^0	135	ρ^0	765
8	$\bar{n}p$	1	-1	0	π^-	140	ρ^-	765
8	$\bar{\lambda}p$	$\frac{1}{2}$	$+\frac{1}{2}$	$+1$	K^+	494	$K^{*+}(890)$	892
8	$\bar{\lambda}n$	$\frac{1}{2}$	$-\frac{1}{2}$	$+1$	K^0	498	$K^{*0}(890)$	892
8	$\bar{n}\lambda$	$\frac{1}{2}$	$+\frac{1}{2}$	-1	$\bar{K}^0$	498	$K^{*0}(890)$	892
8	$\bar{p}\lambda$	$\frac{1}{2}$	$-\frac{1}{2}$	-1	K^-	494	$K^{*-}(890)$	892
8	$(\bar{p}p+\bar{n}n-2\bar{\lambda}\lambda)/\sqrt{6}$	0	0	0	Mixture of $\eta(549)$		Mixture of $\omega(784)$	
1	$(\bar{p}p+\bar{n}n+\bar{\lambda}\lambda)/\sqrt{3}$	0	0	0	and perhaps $\eta'(958)$		and $\phi(1019)$	

If the mesons were completely invariant under $SU(3)$ transformations, all the mesons in an octet would have the same mass. This is not so. Hence there are said to be interactions that break the $SU(3)$ symmetry. Since we do not discuss this question, or the mass relations prevailing inside the octet; we refer the reader to the literature (14LI, 14LIC, 14CA, 14GA).

Mesons that cannot be explained as a quark–antiquark pair are said to be *exotic.* A clear case of an exotic meson is a $B = 0$ hadrons with $|\text{charge}| \geqslant 2$ or $|Y| \geqslant 2$. As this book is being written, there is no sure evidence for such exotic mesons. This then is one of the beautiful predictions of the $SU(3)$ theory of hadron classification.

Exercise 14-6. Justify the definition of exotic mesons and consult the literature for present evidence on their existence.

14-4 The Baryon $SU(3)$ Multiplets. We can only afford the space here to outline the baryon multiplets in $SU(3)$. For help in the exercise and further details, consult Refs. 14KO, 14LI, 14LIC, 14CA, and 14GA. We construct the baryons, the $B = +1$ hadrons, by taking the direct product of an antiquark triplet with *two* quark triplets. Twenty-seven independent states will be produced, and the resolution into $SU(3)$ multiplets is

$$3 \otimes 3 \otimes 3 = 1 \oplus 8_s \oplus 8_a \oplus 10 \tag{14-14}$$

The octet structure has already been discussed. The **10** *multiplet* or *decuplet* appears in Fig. 14-3. The quark configurations which comprise the members of

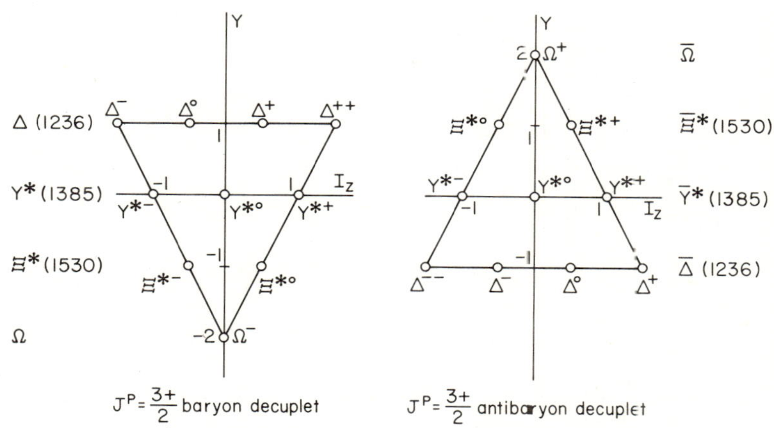

Fig. 14-3 The $SU(3)$ decuplet classification of the $J = \frac{3}{2}$ baryon and antibaryon multiplets.

Table 14-3 The quark model content of the baryon octet. The terms symmetric (s) and antisymmetric (a) refer to the symmetry of the first two quarks in the state function. To emphasize the symmetries, we use the notation $q_1 = p$, $q_2 = n$, $q_3 = \lambda$. (Taken from Ref. 14LIC)

Baryon	I	I_z	S	Mass (MeV/c^2)	Symmetric (s) Quark State Function	Antisymmetric (a) Quark State Function
p	$\frac{1}{2}$	$\frac{1}{2}$	0	938	$(2q_1q_1q_2 - q_1q_2q_1 - q_2q_1q_1)/\sqrt{6}$	$(q_1q_2q_1 - q_2q_1q_1)/\sqrt{2}$
n	$\frac{1}{2}$	$-\frac{1}{2}$	0	940	$(q_1q_2q_2 + q_2q_1q_2 - 2q_2q_2q_1)/\sqrt{6}$	$(q_1q_2q_2 - q_2q_1q_2)/\sqrt{2}$
Σ^+	1	1	-1	1189	$(q_1q_3q_1 + q_3q_1q_1 - 2q_1q_1q_3)/\sqrt{6}$	$(q_3q_1q_1 - q_1q_3q_1)/\sqrt{2}$
Σ^0	1	0	-1	1192	$(q_1q_3q_2 + q_3q_1q_2 + q_2q_3q_1$ $+ q_3q_2q_1 - 2q_1q_2q_3 - 2q_2q_1q_3)/\sqrt{12}$	$(q_3q_1q_2 - q_1q_3q_2 + q_3q_2q_1$ $- q_2q_3q_1)/2$
Σ^-	1	-1	-1	1197	$(q_2q_3q_2 + q_3q_2q_2 - 2q_2q_2q_3)/\sqrt{6}$	$(q_3q_2q_2 - q_2q_3q_2)/\sqrt{2}$
Λ^0	0	0	-1	1116	$(q_1q_3q_2 + q_3q_1q_2 - q_2q_3q_1$ $- q_3q_2q_1)/2$	$(2q_1q_2q_3 - 2q_2q_1q_3 + q_3q_2q_1$ $- q_2q_3q_1 + q_1q_3q_2 - q_3q_1q_2)/\sqrt{12}$
Ξ^0	$\frac{1}{2}$	$\frac{1}{2}$	-2	1314	$(2q_3q_3q_1 - q_1q_3q_3 - q_3q_1q_3)/\sqrt{6}$	$(q_3q_1q_3 - q_1q_3q_3)/\sqrt{2}$
Ξ	$\frac{1}{2}$	$-\frac{1}{2}$	-2	1321	$(2q_3q_3q_2 - q_2q_3q_3 - q_3q_2q_3)/\sqrt{6}$	$(q_3q_2q_3 - q_2q_3q_3)/\sqrt{2}$

Table 14-4 The quark model content of the baryon decuplet: $q_1 = p$, $q_2 = n$, $q_3 = \lambda$ (Taken from Ref. 14LIC)

Baryon	I	I_z	S	Mass (MeV/c²)	Quark State Function
Δ^{++}	$\frac{3}{2}$	$\frac{3}{2}$	0	1236	$q_1q_1q_1$
Δ^{+}	$\frac{3}{2}$	$\frac{1}{2}$	0	1236	$(q_1q_1q_3 + q_1q_2q_1 + q_2q_1q_1)/\sqrt{3}$
Δ^{0}	$\frac{3}{2}$	$-\frac{1}{2}$	0	1236	$(q_1q_2q_2 + q_2q_1q_2 + q_2q_2q_1)/\sqrt{3}$
Δ^{-}	$\frac{3}{2}$	$-\frac{3}{2}$	0	1236	$q_2q_2q_2$
Y^{*+}	1	1	-1	1385	$(q_1q_1q_3 + q_1q_3q_1 + q_3q_1q_1)/\sqrt{3}$
Y^{*0}	1	0	-1	1385	$(q_1q_2q_3 + q_1q_3q_2 + q_2q_1q_3$ $+ q_2q_3q_1 + q_3q_1q_2 + q_3q_2q_1)/\sqrt{6}$
Y^{*-}	1	-1	-1	1385	$(q_2q_2q_3 + q_2q_3q_2 + q_3q_2q_2)/\sqrt{3}$
Ξ^{*0}	$\frac{1}{2}$	$\frac{1}{2}$	-2	1530	$(q_1q_3q_3 + q_3q_1q_3 + q_3q_3q_1)/\sqrt{3}$
Ξ^{*-}	$\frac{1}{2}$	$-\frac{1}{2}$	-2	1530	$(q_2q_3q_3 + q_3q_2q_3 + q_3q_3q_2)/\sqrt{3}$
Ω^{-}	0	0	-3	1672	$q_3q_3q_3$

the multiplets are given in Tables 14-3 and 14-4. In Table 14-3 we see the origin of the subscripts s and a on the two octets in Eq. 14-14. In analogy to the discussion accompanying Eqs. 14-5, the two octets differ in the symmetry of the first two quarks.

Exercise 14-7. With the aid of the literature, justify Eq. 14-14. Show that the assumption that *all* baryons can be constructed using **3* ⊗ 3 ⊗ 3** explains the

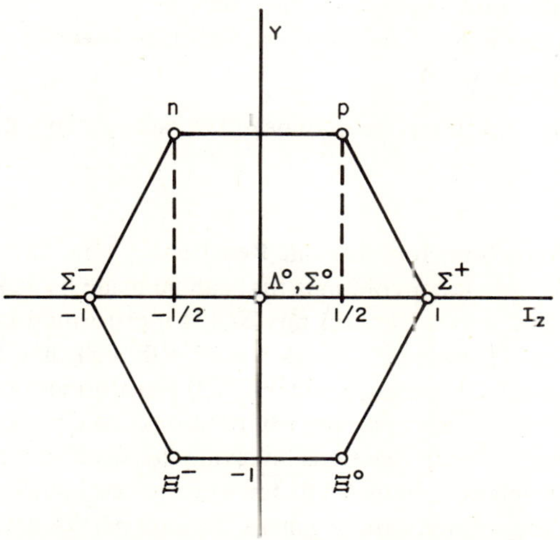

Fig. 14-4 The $SU(3)$ octet classification of the $J = \frac{1}{2}$ baryons.

absence of nucleon–nucleon, hyperon–nucleon, and K^+-nucleon resonances, as mentioned at the end of Ch. 5.

Once again, nature does not disappoint $SU(3)$. As Figs. 14-4 and 14-3 indicate, the spin $\frac{1}{2}$ baryons we know best form an octet; and the spin $\frac{3}{2}$ baryon resonances, which we know almost as well, form a decuplet.

A major difference between the meson and the baryon multiplets is that the former include in the same multiplet the particle and its antiparticle. But antibaryons are in a separate multiplet from the corresponding baryons. For example, the antiparticles of the spin $\frac{3}{2}$ decuplet form the decuplet **10*** (Fig. 14-3). This is in agreement with the requirement that a baryon–antibaryon pair have opposite intrinsic parity, hence be in different $SU(3)$ multiplets.

As with the meson multiplets, the baryons inside a multiplet have different masses; this is again ascribed to interactions that break $SU(3)$ symmetry.

We conclude this section with a few general remarks on $SU(3)$ multiplets.

1. Although we have developed the multiplet structure with the aid of the quark model, that model is not necessary. The entire $SU(3)$ multiplet structure can be developed just using the properties of the $SU(3)$ group.

2. The multiplet structure on the Y, I_z diagrams have beautiful geometric symmetries. We have not proved this, but they always are symmetric under 120° rotations and under a reflection about a vertical axis.

3. Compared to isotopic spin—namely, $SU(2)$—multiplets, it is more difficult to describe a member of an $SU(3)$ multiplet. We must use *four* quantities:

(a) The multiplicity of the multiplet—**M**.
(b) The total isotopic spin of the member—I.
(c) The z component of the isotopic spin of the member—I_z.
(d) The hypercharge—Y.

For isotopic spin multiplets, we only need the multiplicity, $\mathbf{M} = 2I + 1$, or I itself, and I_z.

14-5 *SU*(3) Predictions for Hadronic Reactions. The discussion in this section concerns only the predictions that can be made by requiring that strong interactions be invariant under arbitrary $SU(3)$ transformations. These predictions do not depend on the use of the quark model, although they can of course be derived by this means. Invariance under $SU(3)$ transformations is a generalization of the invariance under isotopic spin rotations discussed in Ch. 4. The latter gave relations between different reactions involving the same isotopic spin multiplets. The former gives relations between different reactions involving the same $SU(3)$ multiplets; and the former is more general because particles with different strangeness are included in the same $SU(3)$ multiplet.

As a first example of an $SU(3)$ prediction for a hadronic reaction, we consider the decay of a particle

$$a \to b + c \qquad (14\text{-}15)$$

through the strong interactions. In particular, the decay of a spin $\frac{3}{2}$ baryon to a spin $\frac{1}{2}$ baryon and a pseudoscalar meson has been extensively studied (14TR). Table 14-5 lists these decays. The relative decay rates are fixed completely, within invariance under $SU(3)$, by the $SU(3)$ Clebsch–Gordan coefficients for the coupling of a decuplet to two octets.

The calculation of the coupling of $SU(3)$ multiplets through the direct use of $SU(3)$ Clebsch–Gordan coefficients is tedious (14LIC). It is usually easier to use a scheme involving what is called *U-spin* (14LI, 14LIC, 14GA). Since we cannot

Table 14-5 The $SU(3)$ Clebsch–Gordan coefficients for the strong interaction decay of the $J = \frac{3}{2}^+$ baryon resonances to a $J = \frac{1}{2}^+$ baryon plus a pseudoscalar meson.[a]

Decay	$SU(3)$ Coefficient (Isoscalar)
$\Delta \to N + \pi$	$-\frac{\sqrt{2}}{2}$
$\Delta \to \Sigma + K$	$+\frac{\sqrt{2}}{2}$
$Y^* \to N + \bar{K}$	$-\frac{\sqrt{6}}{6}$
$Y^* \to \Xi + K$	$+\frac{\sqrt{6}}{6}$
$Y^* \to \Sigma + \pi$	$+\frac{\sqrt{6}}{6}$
$Y^* \to \Sigma + \eta_8$	$+\frac{1}{2}$
$Y^* \to \Lambda + \pi$	$-\frac{1}{2}$
$\Xi^* \to \Xi + \pi$	$+\frac{1}{2}$
$\Xi^* \to \Sigma + \bar{K}$	$+\frac{1}{2}$
$\Xi^* \to \Xi + \eta_8$	$+\frac{1}{2}$
$\Xi^* \to \Lambda + \bar{K}$	$-\frac{1}{2}$
$\Omega^- \to \Xi + \bar{K}$	$+1$

[a] The coefficients are given for the entire isotopic spin multiplets; they are the so-called *isoscalar* $SU(3)$ coefficients. They must be multiplied by the ordinary isotopic spin $SU(2)$ Clebsch–Gordan coefficients to calculate the complete $SU(3)$ coefficients (14DE, 14PDG).

take the space to describe that method here, however, we use instead $SU(3)$ multiplet coupling tables that have already been calculated. This is certainly the easiest and safest procedure. Fortunately, tables exist (14DE, 14PDG, 14GO) for most situations that commonly arise in the coupling of $SU(3)$ multiplets, and Table 14-5 gives the relative decay amplitudes for the decay of a decuplet to two octets.

In comparing these predictions with experiments one must take into account the large mass differences within an $SU(3)$ multiplet. This problem does not arise when we deal only with isotopic spin multiplets, because the masses within such a multiplet are almost equal. In the reaction of Eq. 14-15 the center of mass momentum p of the decay products depends on the mass difference $(M_b + M_c) - M_a$. If l is the orbital angular momentum involved in the decay, the decay rate will depend on the centrifugal barrier penetration factor $F_l(p)$, where very roughly (Eq. 5-41) we have

$$F_l(p) \sim p^{2l+1} \tag{14-16}$$

Once the measured decay rates are corrected for this factor—usually a more sophisticated form is used—the relative rates agree fairly well with the $SU(3)$ predictions. There predictions are simply the square of the numbers listed in Table 14-5. To quote Rosenfeld (14RO) on these decay rate predictions: "$SU(3)$ succeeds, on the average, to better than $\pm 50\%$."

We consider next a two-body elastic or inelastic reaction in which a member of a meson octet interacts with a member of a baryon octet to yield a member of a meson octet and a member of a baryon octet

$$\begin{aligned} &a + b \rightarrow c + d \\ &8 \otimes 8 \rightarrow 8 \otimes 8 \end{aligned} \tag{14-17}$$

The intrinsic spins of the particles are not intrinsically involved. Thus both the initial and final meson may be pseudoscalar

$$\begin{aligned} &0^- + \tfrac{1}{2}^+ \rightarrow 0^- + \tfrac{1}{2}^+ \\ &\pi^- + p \rightarrow \pi^- + p \\ &K^- + p \rightarrow \pi^+ + \Sigma^- \end{aligned} \tag{14-18}$$

or the final meson may be vector

$$\begin{aligned} &0^- + \tfrac{1}{2} \rightarrow 1^- + \tfrac{1}{2}^+ \\ &\pi^- + p \rightarrow \rho^- + p \end{aligned} \tag{14-19}$$

In analogy with isotopic spin, we have to resolve the direct product $8 \otimes 8$ into $SU(3)$ multiplets. This cannot be done automatically, as in $SU(2)$; one must know or look up the answer. In this case we have

$$\mathbf{8} \otimes \mathbf{8} = \mathbf{1} \oplus \mathbf{8}_a \oplus \mathbf{8}_s \oplus \mathbf{10} \oplus \mathbf{10}^* \oplus \mathbf{27} \tag{14-20}$$

The two octets $\mathbf{8}_a$ and $\mathbf{8}_s$ have different internal symmetry, as discussed in the previous section, but $SU(3)$ allows the transition $\mathbf{8}_a \leftrightarrow \mathbf{8}_s$.

Exercise 14-8. Consult the literature for the structure of the **27** multiplet.

If the reaction conserves $SU(3)$, there will be an amplitude for each multiplet on the right-hand side of Eq. 14-20 and we can write

$$\begin{aligned}\langle cd\,|\,T\,|\,ab\rangle = \langle cd\,|[T(1) + T(8_{a\leftrightarrow a}) + T(8_{s\leftrightarrow s}) + T(8_{a\leftrightarrow s}) + T(8_{s\leftrightarrow a})\\ + T(\mathbf{10}) + T(\mathbf{10}^*) + T(\mathbf{27})]\,|\,ab\rangle\end{aligned} \tag{14-21}$$

Here time reversal yields $T(8_{s\leftrightarrow a}) = T(8_{a\leftrightarrow s})$. Equation 14-21 relates 64 possible $|\,ab\rangle$ states to 64 possible $\langle cd\,|$ states through a handful of amplitudes. Thus as a restriction on the behavior of the class of reactions represented by Eq. 14-18, it is more powerful than isotopic spin conservation by itself. We rewrite Eq. 14-21 in the form

Table 14-6 The $SU(3)$ Clebsch-Gordan coefficients for the set of reactions in Eq. 14-18, taken from Ref. 14LE. The subscripts 1 and 2 refer to the internal symmetry of the octet. Data illustrate how $SU(3)$ invariance connects together a large set of two-body reaction amplitudes.

	Intermediate State	27	8_1	8_2	10	$\overline{10}$	1	$8_1/8_2$
1.	$(K^-N\,|\,K^-N)$	$\frac{1}{5}$	$\frac{3}{10}$	$\frac{1}{6}$	$\frac{1}{6}$	$\frac{1}{6}$	0	$-1/\sqrt{5}$
2.	$(\pi^-P\,|\,\pi^-P)$	$\frac{1}{5}$	$\frac{3}{10}$	$\frac{1}{6}$	$\frac{1}{6}$	$\frac{1}{6}$	0	$1/\sqrt{5}$
3.	$(K^-N\,|\,K^0\Xi^-)$	$\frac{1}{5}$	$\frac{3}{10}$	$-\frac{1}{6}$	$-\frac{1}{6}$	$-\frac{1}{6}$	0	0
4.	$(\pi^-P\,|\,K^+\Sigma^-)$	$\frac{1}{5}$	$\frac{3}{10}$	$-\frac{1}{6}$	$-\frac{1}{6}$	$-\frac{1}{6}$	0	0
5.	$(\pi^+P\,|\,\pi^+P)$	$\frac{1}{2}$	0	0	$\frac{1}{2}$	0	0	0
6.	$(\pi^+P\,|\,K^+\Sigma^+)$	$\frac{1}{2}$	0	0	$-\frac{1}{2}$	0	0	0
7.	$(K^+N\,|\,K^0P)$	$\frac{1}{2}$	0	0	0	$-\frac{1}{2}$	0	0
8.	$(K^+N\,|\,K^+N)$	$\frac{1}{2}$	0	0	0	$\frac{1}{2}$	0	0
9.	$(K^-P\,|\,K^-P)$	$\frac{7}{40}$	$\frac{1}{5}$	$\frac{1}{3}$	$\frac{1}{12}$	$\frac{1}{12}$	$\frac{1}{8}$	0
10.	$(K^-P\,|\,\bar{K}^0N)$	$\frac{1}{40}$	$\frac{1}{10}$	$-\frac{1}{6}$	$\frac{1}{12}$	$\frac{1}{12}$	$-\frac{1}{8}$	$-1/\sqrt{5}$
11.	$(K^-P\,|\,\pi^-\Sigma^+)$	$\frac{1}{40}$	$\frac{1}{10}$	$-\frac{1}{6}$	$\frac{1}{12}$	$\frac{1}{12}$	$-\frac{1}{8}$	$1/\sqrt{5}$
12.	$(K^-P\,|\,\pi^+\Sigma^-)$	$\frac{1}{40}$	$\frac{1}{10}$	$\frac{1}{6}$	$-\frac{1}{12}$	$-\frac{1}{12}$	$-\frac{1}{8}$	0
13.	$(K^-P\,|\,K^0\Xi^0)$	$\frac{1}{40}$	$\frac{1}{10}$	$\frac{1}{6}$	$-\frac{1}{12}$	$-\frac{1}{12}$	$-\frac{1}{8}$	0
14.	$(K^-P\,|\,\pi^0\Sigma^0)$	$\frac{1}{40}$	$\frac{1}{10}$	0	0	0	$-\frac{1}{8}$	$-1/2\sqrt{5}$
15.	$(K^-P\,|\,\pi^0\Lambda^0)$	$\sqrt{3}/10$	$-\sqrt{3}/10$	0	$1/4\sqrt{3}$	$-1/4\sqrt{3}$	0	$1/2\sqrt{15}$
16.	$(\pi^-P\,|\,K^0\Lambda^0)$	$-\frac{1}{10}\sqrt{\frac{3}{2}}$	$\frac{1}{10}\sqrt{\frac{3}{2}}$	$1/2\sqrt{6}$	0	$-1/2\sqrt{6}$	0	$\sqrt{\frac{2}{15}}$
17.	$(\pi^-P\,|\,\pi^0N)$	$\frac{3}{10}\sqrt{\frac{1}{2}}$	$-\frac{3}{10}\sqrt{\frac{1}{2}}$	$-1/6\sqrt{2}$	$1/3\sqrt{2}$	$-1/6\sqrt{2}$	0	$-\sqrt{\frac{1}{10}}$

$$\langle cd\,|\,T\,|\,ab\,\rangle = \sum_n c_n T(\mathbf{n}) \tag{14-22}$$

The c_n coefficients (Table 14-6) predict relations between sets of reactions such as those in Eq. 14-18. The testing of these predicted relations is not easy because amplitudes not cross sections are involved, and usually many amplitudes must be measured to test a relation. Thus in terms of the reactions of Eqs. 14-18, there are only a few simple relations such as

$$\langle \pi^+\Sigma^-\,|\,T\,|\,K^-p\,\rangle = \langle K^0\Xi^0\,|\,T\,|\,K^-p\,\rangle \tag{14-23a}$$

$$\langle K^+\Sigma^-\,|\,T\,|\,\pi^-p\,\rangle = \langle K^0\Xi^-\,|\,T\,|\,K^-n\,\rangle$$

$$\langle \pi^+p\,|\,T\,|\,\pi^+p\,\rangle + \langle K^+\Sigma^+\,|\,T\,|\,\pi^+p\,\rangle = \langle K^+p\,|\,T\,|\,K^+p\,\rangle \tag{14-23b}$$

Almost all other relations involve more amplitudes.

Exercise 14-9. Use Table 14-6 to verify these relationships.

At the time of writing, we do not know how precisely such *SU*(3) predictions agree with experiment. At low energies the correction in Eq. 14-16 or its equivalent must be used. It also seems that such corrections must be used at high energy (14TRI) where one might have hoped to ignore mass differences.

14-6 The Quark Model.

The full use of the quark model requires the quarks be taken as physical entities, not just as basis vectors for the construction of *SU*(3) multiplets. Once we accept the quarks as physical entities, we can go beyond the pure group theoretical predictions of *SU*(3). In this section we briefly illustrate the two major classes of quark model predictions. Extensive discussions are given in Refs. 14FE, and 14DA.

We consider first the quark model for two-body hadronic reactions

$$A + B \to C + D \tag{14-24}$$

The hadron A consists of quarks or antiquarks $(a_1, \bar{a}_2)$ if it is a meson or quarks (a_1, a_2, a_3) if it is a baryon. Tables 14-2 to 14-4 give these quark combinations. But in what follows we ignore the internal symmetry of the combinations. To calculate the amplitude for the reaction in Eq. 14-24 using a simple quark model (where quark means quark or antiquark), we make the following assumptions:

1. The quarks act as free particles inside the hadrons.
2. When hadrons interact, the interaction consists solely of the interaction of pairs of quarks $a_i + b_j \to c_k + d_m$. The amplitude for that two-quark reaction is $\langle c_k d_m\,|\,T\,|\,a_i b_j\,\rangle$. Quark–quark interactions in which more than two quarks enter or leave the interaction are ignored. The *impulse approximation* theory of

hadron–hadron interactions through quark-quark interactions leads to the general prediction

$$\langle CD|T|AB\rangle = \sum_{ijkm} \langle c_k d_m|T|a_i b_j\rangle \tag{14-25}$$

The simplest application of Eq. 14-25 is to elastic scattering. Thus for π^+P elastic scattering, using P for the proton to avoid confusion, we have

$$\begin{aligned}\langle \pi^+P|T|\pi^+P\rangle = 2\langle pp|T|pp\rangle + 2\langle \bar{n}p|T|np\rangle + \langle pn|T|pn\rangle \\ + \langle \bar{n}n|T|\bar{n}n\rangle\end{aligned} \tag{14-26a}$$

and for PP elastic scattering

$$\langle PP|T|PP\rangle = 4\langle pp|T|pp\rangle + 4\langle pn|T|pn\rangle + \langle nn|T|nn\rangle \tag{14-26b}$$

where p and n are quarks. Note that internal quantum number conservation does not allow amplitudes like $\langle p\bar{n}|T|pp\rangle$. Using the optical theorem and Eqs. 14-26 for the forward elastic scattering amplitude, we obtain

$$\begin{aligned}\sigma_{\text{tot},\pi^+P} &= 2\sigma_{\text{tot},pp} + 2\sigma_{\text{tot},\bar{n}p} + \sigma_{\text{tot},pn} + \sigma_{\text{tot},\bar{n}n} \\ \sigma_{\text{tot},PP} &= 4\sigma_{\text{tot},pp} + 4\sigma_{\text{tot},pn} + \sigma_{\text{tot},nn}\end{aligned} \tag{14-27}$$

where the right-hand sides of Eqs. 14-27 are hypothetical quark-quark total cross sections. Now if we assume that the quark-quark total cross section is independent of the nature of the quark and whether it is a quark or an antiquark, we obtain the classic prediction

$$\frac{\sigma_{\text{tot},\pi^+P}}{\sigma_{\text{tot},PP}} = \frac{2}{3} \tag{14-28}$$

Comparing this with the observed high energy ratio (Sec. 14-1) of about 0.55, we see that this is quite a good prediction.

Exercise 14-10. Fill in the steps leading to the preceding sets of equations. Assume that the λ quarks act exactly the same as the non-strange quarks. Then show that Eq. 14-28 holds for the ratio of any meson–baryon total cross section to any baryon–baryon total cross section. How does this prediction agree with experimental measurements? Why is it impossible to obtain Eq. 14-28 using $SU(3)$ alone?

As discussed in Refs. 14KO and 14FE, this impulse approximation quark model of hadron–hadron interactions can be made very much more sophisticated.

Certainly the quark-quark amplitudes can be dependent on the type of quarks. In addition, the quarks can be given intrinsic spin, as we discuss next. And finally, the distribution of the quarks inside the hadrons can be described by quark wave functions, this allows the model to be applied to inelastic as well as elastic processes and to nonforward amplitudes. The quark model has its successes and its failures. It is perhaps best regarded as a model that dramatically emphasizes certain aspects of two-body hadronic reactions.

To conclude this section we discuss the second major class of predictions that can be made specifically with the quark model—predictions involving the intrinsic angular momentum of the quark. It is customary to assign an intrinsic spin of $\frac{1}{2}$ to the quarks. Now if we suppose, and this is very big if, that the intrinsic spin of the hadrons can be constructed *nonrelativistically* from the intrinsic spins of the constituent quarks, we obtain a very beautiful explanation of the intrinsic spin of the hadrons. For if we further suppose that the relative orbital angular momentum of the quarks is zero, we immediately predict that the mesons will have spin zero or 1 and the baryons will have spin $\frac{1}{2}$ or $\frac{3}{2}$. And this is indeed the case for the two lowest mass meson octets and for the lowest mass baryon octet and decuplet. But how are we to explain a higher spin meson such as the A_2 with spin 2? Simply by allowing the relative orbital angular momentum l of the quarks in the meson to be nonzero. Continuing to think nonrelativistically, we would also expect that as the average mass of the multiplet increases, the linear momentum of the quarks inside the hadron increases. Therefore, higher mass multiplets should tend to have higher intrinsic spins. And this is also what we find: quite detailed fits can be made to the intrinsic spins, charge conjugation, and parity properties of the higher mass multiplets using this model (14DA, 14FE).

14-7 An Overview of Composite or Constituent Models of Hadrons. From this chapter we can obtain a feeling for the virtues and the vices of composite or constituent models of hadrons. The virtues derive from our return to a system involving only a *few truly elementary* particles—the constituent particles. We can then derive at least the static properties of the many different hadrons using a few simple rules and these few constituent particles. The vices derive from our ignorance of the fundamental nature of the constituent particles and the dynamics of the interactions between those constituent particles. We discuss the static properties and the dynamic properties of such models separately.

1. *Static Properties.* In these models the hadrons are assumed to be composed of a few, more elementary constituent particles. As we have already discussed, the hadronic internal quantum numbers B, Q, I, I_z, and S are obtained by simply combining the internal quantum numbers of the constituents in the

usual way. We then assume that in strong interactions these constituents can only be produced or destroyed in constituent–anticonstituent pairs. Hence the conservation of these internal quantum numbers in strong interactions is explained. Of course a yet more fundamental question about this explanation is why the restriction, in strong interactions, to creation or annihilation in pairs?

As we have demonstrated with the quark model explanation of $SU(3)$ symmetry, the construction of the internal quantum numbers of the hadrons, and their subsequent classification, can be carried out regardless of whether we believe in the physical existence of the constituents. If we believe in their physical existence, we can calculate such additional hadronic properties as the intrinsic spin and magnetic moment, although by including $SU(3)$ into higher symmetry groups such as $SU(6)$, we may postpone the need to believe in the physical existence of the constituents.

But once we accept their physical existence, we face the grave problem that the constituents have not been detected. We can explain this in two ways. We may assume that the mesons of the constituents are in the range of tens or hundreds of GeV. Then their mutual binding energy must be tens or hundreds of GeV if hadron masses of the order of 1 GeV are to be obtained. This is a very unusual dynamic situation, and it leads to vast uncertainties when we try to understand the dynamic properties of constituent models. Even the situation with respect to the static properties of the hadron, as calculated by the quark model, is uncertain. For to use the nonrelativistic versions of concepts like spin-orbit coupling, one must think about these heavy constituents as moving very slowly and nonrelativistically at the bottom of the deep potential well which is the hadron (14DA, 14LIPK). This deep potential well must be assumed to somehow occur as a result of the strong force between constituents. And the well must somehow replace most of that force, leaving the remaining force relatively weak.

This model of strongly bound, heavy constituents has another problem, which is illustrated by the quark model. If three quarks bind together strongly to form a baryon, why do not four or five quarks also bind together strongly to form other kinds of hadrons? And in the mesons, why is quark–antiquark binding limited to a single pair? There appears to be some mysterious saturation-of-forces effect at work here, and we are not happy with such a mysterious effect.

An alternative to the strongly bound, heavy constituent model is a model in which the constituents have masses on the order of hadronic masses. But to explain our failure to find these constituents, we must invoke some other mysterious principle forbidding the existence of free single constituents. Perhaps such a principle does exist; equally unlikely principles have been discovered in quantum physics. But for the present we are uncomfortable with such a principle because it has no predictive power.

2. *Dynamic Properties.* We have noted that the use of $SU(3)$ invariance has led to reasonably successful predictions of the relative decay properties of the hadron; and where it can be applied, it has been reasonably successful in explaining the relative amplitudes of two-body reactions. We have also noted that some additional success can be achieved if the quark model is used explicitly.

One attractive feature of the quark model in dynamic calculations is the assumption that all interactions can be resolved into sums over interactions between pairs of quarks (rule 2, Sec. 14-6). Using this assumption we can avoid the sums over infinite numbers of Feynman diagrams required in conventional hadron–hadron dynamics. But here again there is a puzzle. The quark–quark interaction must be of the same strength as the strong interaction—why then is it sufficient to consider only interactions between pairs of quarks? A related puzzle is the nature of the quark–quark interaction. Is it a local interaction, or is there some other particle that carries the quark–quark force? We have used the quark model for illustrative purposes, but any constituent model that limits constituent–constituent interactions in the same way poses the same problems.

Next we turn to the assumption, made in most constituent models, that the constituents are *point* particles. This is not a necessary assumption if our only concern is with symmetry questions such as $SU(3)$. But it is a very useful assumption if we wish to do detailed dynamics. Otherwise, the model has too much freedom. The point particle assumption is extremely useful when considering the electromagnetic and weak interactions of the hadrons. For we believe we know how to calculate the electromagnetic interactions of point particles to any order. And we know how to calculate the weak interactions of point particles to first order—which is sufficient for many purposes. Indeed, as we discuss in detail in Ch. 20, a particular point constituent model—the parton model—has been developed in order to use our knowledge of the electromagnetic interactions of point particles. We postpone further detailed discussion of constituent models until that chapter.

We close with two general remarks. First, in principle it is not necessary for a constituent model to explain both the symmetry properties and the dynamic properties of the hadrons. For example, we may postulate a very large number of constituents in a hadron if that very large number is necessary to explain some dynamic property. The symmetry properties of the hadron would then be due to some overall symmetry requirements, not to the two- or three-constituent composition of the hadron. Second, with all their problems, why do we return again and again to constituent, especially point constituent, models of hadrons? We do this because one of the things we understand well is the quantum physics of point particles to lowest order in their interactions.

REFERENCES

14CA P. A. Carruthers, *Introduction to Unitary Symmetry* (Wiley-Interscience, New York, 1966).

14DA R. H. Dalitz, in *Meson Spectroscopy*, C. Baltay and A. H. Rosenfeld, Eds. (Benjamin, New York, 1968).

14DE J. J. de Swart, *Rev. Mod. Phys.* **35**, 916 (1963).

14FE B. T. Feld, *Models of Elementary Particles* (Blaisdell, Waltham, Mass., 1969).

14GA S. Gasiorowicz, *Elementary Particle Physics* (Wiley, New York, 1966).

14GO M. Gourdin, *Unitary Symmetries* (North-Holland, Amsterdam, 1967).

14HA M. Hamermesh, *Group Theory* (Addison-Wesley, Reading, Mass., 1962).

14KO J. J. J. Kokkedee, *The Quark Model* (Benjamin, New York, 1969).

14LE C. A. Levinson et al., *Phys. Lett.* **1**, 44 (1962).

14LI H. J. Lipkin, *Lie Groups for Pedestrians* (North-Holland, Amsterdam, 1965).

14LIC D. B. Lichtenberg, *Unitary Symmetry and Elementary Particles* (Academic Press, New York, 1970).

14LIP H. J. Lipkin, *Phys. Rev.* **183**, 1221 (1969).

14LIPK H. J. Lipkin, *Quarks for Pedestrians*, Weizmann Institute Report, 1973 (unpublished).

14PDG See the yearly *Review of Particle Properties* by the Particle Data Group; for example, P. Söding et al., *Phys. Lett.* **39B**, 1 (1972). Also published in reduced size by CERN, Geneva; and Lawrence Berkeley Laboratory, Berkeley, Calif.

14RO A. H. Rosenfeld, in *Proceedings of the Irvine Particle Physics Conference–1971*, M. Bander, G. L. Shaw, and D. Y. Wong, Eds. (American Institute of Physics, New York, 1972).

14TR R. D. Tripp et al., *Nucl. Phys.* **B3**, 10 (1967).

14TRI G. Trilling, *Nucl. Phys.* **B40**, 13 (1972).

CHAPTER

15

Analyticity and Crossing in the Interaction of Spinless Particles

15-1 Introduction. For the last two decades elementary particle physicists have felt that the application of perturbation theory to strong interactions was an uncertain occupation, possibly a hopeless occupation. The qualitative successes of the one-particle exchange model must be balanced against the many quantitative failures. Therefore, particle physicists have tried to develop more fundamental approaches to strong interaction theory. And most of these approaches are based on or related to the treatment of scattering amplitudes as analytic functions.

As with so many other parts of strong interaction theory, we do not possess a rigorous and complete theory of the analytic properties of strong interaction amplitudes. Rather, we have constructed a loose theory based on different types of arguments and insights. To illustrate some of the origins of the theory, we discuss the analytic properties of nonrelativistic partial wave amplitudes in Sec. 15-2 and perturbation theory and analyticity in Sec. 15-5; in Sec. 15-6 we show how the general unitarity relation relates to the singularities of the amplitude. We do not go into the derivation of analytic properties from axiomatic quantum field theory.

The incompleteness of the theory of the analytic behavior of amplitudes pervades not only the foundations of theory but also the applications of the theory. We cannot calculate an amplitude using the analytic properties alone because in general we do not know the residues and discontinuities at the singularities. We only know the position and nature of the singularities. Therefore the analytic nature of the amplitudes is used primarily as a guide and as a framework for describing the amplitudes. We illustrate this in Secs. 15-2 and 15-11 with the examples of bound states and resonances. Another illustration is provided by the development of the dispersion relation concept in Secs. 15-8 and 15-10.

15-2 Nonrelativistic Partial Wave Amplitudes and Analyticity. To begin our study we consider nonrelativistic elastic scattering, for by starting with nonrelativistic quantum mechanics we can most easily develop an understanding of the significance of the analytic properties of amplitudes. Consider a real, spherically symmetric potential $V(r) = (\hbar^2/2m)U(r)$. The radial Schrödinger equation for partial wave l, with $\psi(\mathbf{r}) = (u_l(k^2, r)/r)Y_l^m(\theta, \phi)$, is

$$\frac{d^2u_l(k^2,r)}{dr^2} + \left[k^2 - U(r) - \frac{l(l+1)}{r^2}\right]u_l(k^2,r) = 0 \tag{15-1}$$

where $k^2 = 2mE_{\text{kinetic}}$. We limit the form of $U(r)$ by requiring that as r approaches zero, $U(r)$ must not go to infinity as fast as $1/r^2$. As $r \to \infty$, however, $U(r)$ must go to zero faster than $1/r^2$. As $r \to 0$, $u_l(k^2, r)$ has two solutions

$$u_l(k^2,r) \underset{r\to 0}{\longrightarrow} A_l r^{(l+1)} \tag{15-2a}$$

$$u_l(k^2,r) \underset{r\to 0}{\longrightarrow} A_l r^{-l} \tag{15-2b}$$

Physically k^2 must be real and positive. But let us consider solutions of Eq. 15-1 for complex values of k^2. If these solutions $u_l(k^2, r)$ obey Eq. 15-2a, they are *regular analytic functions* of k^2 for finite values of r. That is, they are analytic functions without singularities in the complex k^2 plane. We apologize for not giving the proof of such an important statement, but the proof is based on the theory of differential equations and cannot be easily summarized here (15DE).

For scattering problems we are interested in the asymptotic form of $u_l(k^2, r)$. From Eq. 3-5 we have

$$u_l(k^2, r) \underset{r\to\infty}{\to} \frac{B_l}{2ik} [S_l(k^2)\, e^{ikr} - (-1)^l\, e^{-ikr}] \qquad \text{(15-3a)}$$

or the alternate form

$$u_l(k^2, r) \underset{r\to\infty}{\to} [\phi_l^-(k^2)\, e^{ikr} + \phi_l^+(k^2)\, e^{-ikr}] \qquad \text{(15-3b)}$$

We have only passing interest in $\Phi_l^\pm(k^2)$, the Jost functions, for we are really interested in using $\Phi_l^\pm(k^2)$ to study

$$S_l(k^2) = (-1)^{l+1} \frac{\Phi_l^-(k^2)}{\Phi_l^+(k^2)} \qquad \text{(15-4)}$$

Our objective is to determine the analytic nature of $S_l(k^2)$ on the complex k^2 plane. As we shall see the singularities have direct physical significance.

But before we can do this we face an annoying task. Since Eq. 15-3 contains $k = \sqrt{k^2}$, which is a double-valued function of k^2, we must exercise care in associating a particular value of k with k^2. We write $k^2 = \kappa\, e^{i\phi}$ and $k = \kappa^{1/2}(\cos\phi/2 + i \sin\phi/2)$ where κ is real and positive and $0 \leqslant \phi \leqslant 4\pi$. Then we use two Riemann sheets to map k onto the k^2 plane; the sheets are defined below and pictured in Fig. 15-1. The k^2 plane is then cut along the real k^2 axis from zero to infinity.

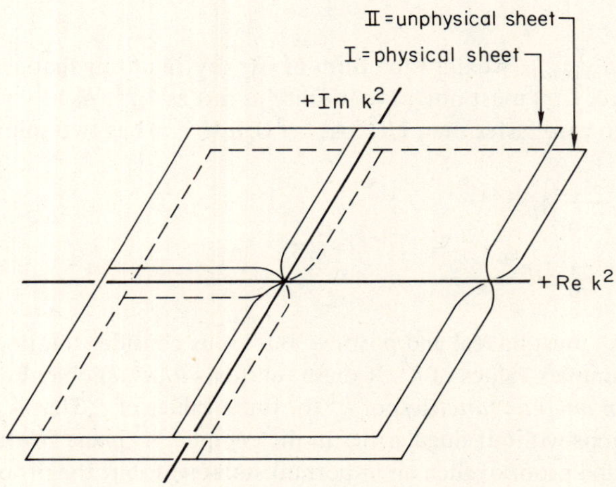

Fig. 15-1 The k^2 analytic plane for S_l, showing the two Riemann sheets.

sheet I, called the physical sheet, $0 \leqslant \phi \leqslant 2\pi$, $\quad \text{Im } k > 0$ (15-5a)

sheet II, called the unphysical sheet, $2\pi \leqslant \phi \leqslant 4\pi$, $\quad \text{Im } k < 0$ (15-5b)

The names physical sheet and unphysical sheet are more conventional than meaningful. In all our discussions of the analytic behavior of amplitudes, we have to describe how amplitudes for complex values of the kinematic parameters are connected with the physical amplitudes for the real values of the kinematic parameters. Here we adopt the convention

$$f(\text{Re } k + i \text{ Im } k) \xrightarrow[\text{Im } k \to 0]{} f_{\text{physical}}(\text{Re } k) \qquad \text{for} \quad \text{Im } k > 0 \tag{15-6}$$

Hence the designation of sheet I as the physical sheet. The information contained on both sheets is redundant, for taking $\Phi_l^{\pm}$ as functions of k we have

$$\Phi_l^+(k) = \Phi_l^-(-k) \tag{15-7a}$$

$$S_{l\text{I}}(k^2) = S_{l\text{II}}^{-1}(k^2) \tag{15-7b}$$

where the subscripts I, II denote the sheets.

Exercise 15-1. Derive Eq. 15-7 using Eq. 15-3b; then derive Eq. 15-8. See Ref. (15OMN).

In addition, the reality of the coefficients in Eq. 15-1 leads to the relations

$$\Phi_l^+(k^{2*}) = [\phi_l^+(k^2)]^*; \qquad \Phi_l^-(k^{2*}) = [\Phi_l^-(k^2)]^* \tag{15-8}$$

Examining the poles of $S_{l\text{I}}(k^2)$ in the complex k^2 plane, we see that $S_{l\text{I}}(k^2)$ will have a pole at k_0^2 if $\Phi_l^+(k^2)$ has a zero at that point. In that case, Eq. 15-3b reduces to

$$u_l(k_0^2, r) \underset{r\to\infty}{\rightarrow} [\Phi_l^-(k_0^2)\, e^{ir\,\text{Re}\,k_0} e^{-br}] \tag{15-9}$$

Here $b = \text{Im } k_0$, and b is positive according to Eq. 15-5a. Therefore, the right-hand side of Eq. 15-9 represents a wave function that is localized in space and normalizable. *If* k_0^2 is real and negative, the pole in $S_{l\text{I}}(k^2)$ at $k^2 = k_0^2$ represents a *bound state* of angular momentum l. We are in a nonrelativistic situation, and the binding energy of this bound state is simply $k_0^2/2m$. If a pole in $S_{l\text{I}}(k^2)$ occurs at a point not on the negative real k^2 axis, we cannot use this bound state interpretation; and there is no other physical significance to ascribe to such a pole. Indeed, conventional potentials do not produce poles in $S_{l\text{I}}(k^2)$ which are off the negative real k^2 axis. (See, however, Refs. 15OMN and 15FR.) Thus a pole in $S_{l\text{I}}(k^2)$ occurs only on the negative real k^2 axis and corresponds to a bound state.

Next we consider the zeros of $\Phi^-(k^2)$. Suppose a zero occurs at $k_0^2 = \text{Re } k_0^2 + i \text{ Im } k_0^2$, with $\text{Re } k_0^2 > 0$ and $\text{Im } k_0^2 > 0$. Then by Eqs. 15-4 and 15-7

$$S_{lI}(k^2) \text{ has zeros at } \mathrm{Re}\, k_0^2 \pm i\, \mathrm{Im}\, k_0^2 \tag{15-10a}$$

$$S_{lII}(k^2) \text{ has poles at } \mathrm{Re}\, k_0^2 \pm i\, \mathrm{Im}\, k_0^2 \tag{15-10b}$$

Let us further suppose that $\mathrm{Im}\, k_0^2 \ll \mathrm{Re}\, k_0^2$. Then the zero of S_{lI} at $\mathrm{Re}\, k_0^2 + i\, \mathrm{Im}\, k_0^2$ and the pole of S_{lII} at $\mathrm{Re}\, k_0^2 - i\, \mathrm{Im}\, k_0^2$ lie close to the point $k^2 = \mathrm{Re}\, k_0^2$ on the real k^2 axis.

Exercise 15-2. Fill in the steps in the arguments just presented. Use Eq. 15-5 to explain which zero and which pole lie close to the real k^2 axis.

The behavior of $S_l(k^2)$ in the vicinity of $k^2 = \mathrm{Re}\, k_0^2$ must be dominated by the nearby zero and pole. Hence for *real* values of k^2 near $\mathrm{Re}\, k_0^2$, the simplest form of S_l is

$$S_l(k^2) = \frac{k^2 - (\mathrm{Re}\, k_0^2 + i\, \mathrm{Im}\, k_0^2)}{k^2 - (\mathrm{Re}\, k_0^2 - i\, \mathrm{Im}\, k_0^2)} \tag{15-11a}$$

Note that Eq. 15-11a must and does fulfill the condition $|S_l^2| = 1$. We can bring Eq. 15-11a into a very familiar form by setting $E_0 = \mathrm{Re}\, k_0^2/2m$ and $\Gamma = \mathrm{Im}\, k_0^2/m$, giving

$$S_l(E) - 1 = \frac{-i\Gamma}{(E - E_0) + i\Gamma/2} \tag{15-11b}$$

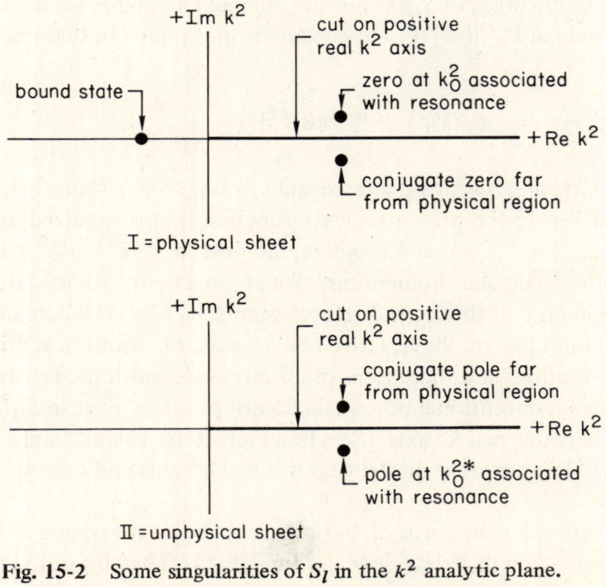

Fig. 15-2 Some singularities of S_l in the k^2 analytic plane.

which is the Breit–Wigner formula for a resonance! In Sec. 5-6 we gave a rather limited derivation of this formula. But the derivation offered here, which is based only on general ideas about the analytic properties of S, indicates the fundamental nature of the resonance concept. A resonance occurs whenever $S_{l\text{II}}(k^2)$ has a pole close to the real positive k^2 axis (also see Sec. 15-11).

As the pole in $S_{l\text{II}}(k^2)$ moves away from the real axis, the influence of the pole and the zero weakens, and other terms begin to dominate $S_l(E)$. From a more physical viewpoint, the Breit–Wigner curve becomes flatter and more difficult to detect. Thus a pole in $S_{l\text{II}}(k^2)$ that is far from the positive real axis cannot be given a direct physical significance.

In Fig. 15-2 we show the singularities discussed so far. Among singularities, not covered here is a very important branch point that occurs on the negative real k^2 axis. The position of this branch point and the discontinuity across the cut depend on the potential $U(r)$ in Eq. 15-1. But we must turn our attention to the relativistic problem, and we refer the reader to Refs. 15DE and 15OMN.

15-3 Two-Body Reactions–Kinematics and Cross Channels. To begin our study of the analyticity of relativistic amplitudes, we must discuss the kinematics that relate two-body crossed reactions. For a reaction $a + b \to c + d$, the *crossed*

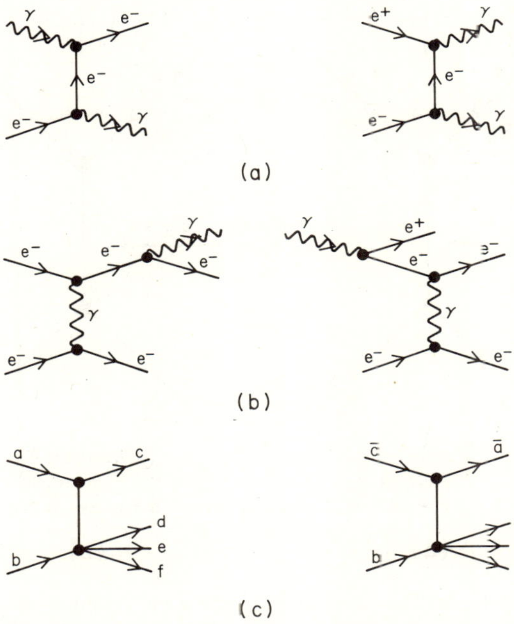

Fig. 15-3 Examples of crossed reactions in perturbation theory.

reactions, or *crossed channels*, are $\bar{c} + b \to \bar{a} + d, a + \bar{c} \to \bar{b} + d, \bar{c} + \bar{d} \to \bar{a} + \bar{b}$, and so forth. Here $\bar{g}$ is the antiparticle of particle g.

In perturbation theory there is a direct connection between particle and antiparticle reactions. For example, the Feynman diagram for the elastic scattering (Compton scattering) of a photon on an electron and the diagram for electron-positron annihilation into two photons are very similar. As Fig. 15-3*a* indicates, the only difference between the diagrams is whether particles are entering or leaving the reaction. Figure 15-3*b* presents a more complicated comparison—electron-electron *bremsstrahlung* versus the photoproduction of an electron pair on an electron (triplet production). For such pairs of amplitudes, the only differences are the signs of kinematic quantities, phases, and those changes necessitated by particle symmetry requirements.

To take yet another example from perturbation theory, there is a direct relation between the reactions

$$a + b \to c + d + e + \cdots \tag{15-12a}$$

$$\bar{c} + b \to \bar{a} + d + e + \cdots \tag{15-12b}$$

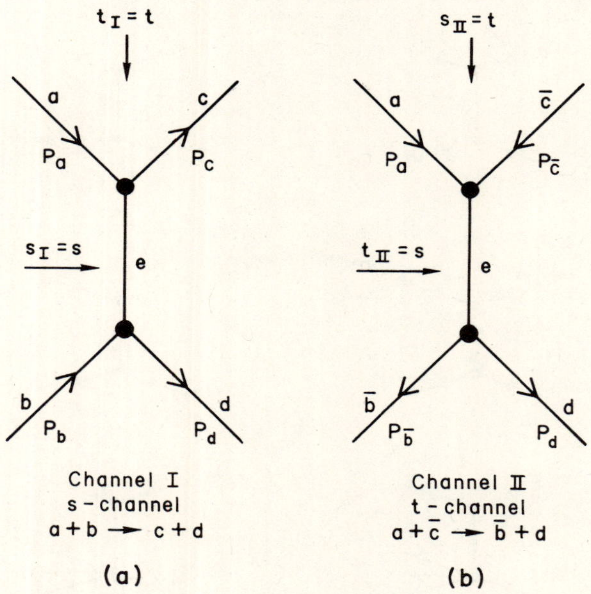

Fig. 15-4 (*a*) The reaction $a + b \to c + d$ and (*b*) the crossed reaction $a + \bar{c} \to \bar{b} + d$, both occurring through the exchange of particle e.

if we assume that these reactions proceed by the elementary one-particle exchange diagram of Fig. 15-3*c*. The pair of reactions in Eqs. 15-12 are said to be related by *line reversal* at a vertex

Let us become more specific and consider in detail the reaction $a + b \to c + d$, which we assume to proceed by the elementary one-particle exchange diagram of Fig. 15-4*a*. For simplicity, all particles are given spin zero. We call this reaction *channel I*, or the *s-channel.* Then we summarize the following information

$$\text{channel I, } s\text{-channel:} \qquad a + b \to c + d, \qquad s_{\text{I}} = (P_a + P_b)^2,$$

$$t_{\text{I}} = (P_a - P_c)^2, \qquad T_{\text{I}}(s_{\text{I}}, t_{\text{I}}) = -\frac{g_{aec}g_{bed}}{m_e^2 - t_{\text{I}}} \tag{15-13a}$$

where P_i is the four-momentum of particle *i.* Next we consider the crossed reaction $a + \bar{c} \to \bar{b} + d$, called *channel II*, or the *t-channel.* As shown in Fig. 15-4*b*, again only particle *e* exchange is allowed, and we have

$$\text{channel II, } t\text{-channel:} \qquad a + \bar{c} \to \bar{b} + d, \qquad s_{\text{II}} = (P_a + P_{\bar{c}})^2,$$

$$t_{\text{II}} = (P_a - P_{\bar{b}})^2, \qquad T_{\text{II}}(s_{\text{II}}, t_{\text{II}}) = -\frac{g_{ae\bar{c}}g_{\bar{b}ed}}{m_e^2 - s_{\text{II}}} \tag{15-13b}$$

where g_{aec} and $g_{ae\bar{c}}$ are vertex coupling constants, hence equal, and

$$T_{\text{II}}(s_{\text{II}}, t_{\text{II}}) = -\frac{g_{aec}g_{bed}}{m_e^2 - s_{\text{II}}} \tag{15-13c}$$

Thus the equations for T_{I} and T_{II} are similar except for the apparently different kinematic variables t_{I} and s_{II}. But by making a more general definition of kinematic variables, we can relate t_{I} to s_{II}. We define $\tilde{P}_i$ as the four-momentum *carried into a reaction* by particle or antiparticle *i*, irrespective of whether *i* is entering or leaving the reaction. Thus

$$\begin{aligned} &\text{for particle } i \text{ actually going into a reaction} && P_i = \tilde{P}_i \\ &\text{for particle } i \text{ actually leaving a reaction} && P_i = -\tilde{P}_i \end{aligned} \tag{15-14}$$

The labeling method of Eq. 15-14 enables us to unify the equations for T_{I} and T_{II} by defining

$$\begin{aligned} s &= (\tilde{P}_a + \tilde{P}_b)^2 = s_{\text{I}} = t_{\text{II}} \\ t &= (\tilde{P}_a + \tilde{P}_c)^2 = t_{\text{I}} = s_{\text{II}} \end{aligned} \tag{15-15}$$

then

$$T_{\text{I}} = T$$
$$T_{\text{II}} = T$$
$$T = -\frac{g_{aec}g_{bed}}{m_e^2 - t} \tag{15-16}$$

And we now have one equation giving the amplitude for the reaction $a + b \rightarrow c + d$ and the crossed reaction $a + \bar{c} \rightarrow \bar{b} + d$. Equation 15-16 illustrates the goal of the application to two-body reactions of *crossing* or, better, *crossing relations.* We would like to be able to write a single function of s and t giving the amplitude for the reaction $a + b \rightarrow c + d$ *and* the crossed reactions. This would mean, for example, that we could use information from one reaction for explaining the behavior of a crossed reaction. In the following sections we develop further the concept of crossing relations and the closely intermeshed concept of the analytic behavior of amplitudes. However, we continue in this section with our discussion of kinematic relationships.

The kinematic relationships we are about to present were greatly emphasized in the early and middle 1960s, when there was hope that a basic and complete understanding of two-body hadronic reactions was at hand. Or at least, it was believed that two-body reactions were the best entrance to the elucidation of hadronic interactions. The hope that complete understanding was at hand turned out to be illusionary, and we are no longer sure that the two-body reactions are the best route to that understanding. Therefore, our treatment of the general kinematics of two-body interactions is shorter than the treatment to be found in older books on strong interactions.

We start with the relationship

$$s + t + u = \sum_{i=1}^{4} m_i^2 \tag{15-17}$$

where m_i is the mass of particle i and

$$s = (\tilde{P}_a + \tilde{P}_b)^2$$
$$t = (\tilde{P}_a + \tilde{P}_c)^2$$
$$u = (\tilde{P}_a + \tilde{P}_d)^2 \tag{15-18}$$

Exercise 15-3. Justify the last definition in Eq. 15-18. See the definition of u in Ch. 7.

Using a theorem in Euclidean geometry—namely, the sum of the perpendicular distances from a point to the sides of an equilateral triangle is a constant—Eq. 15-17 can be represented geometrically (Fig. 15-5*a*). In this representation, called

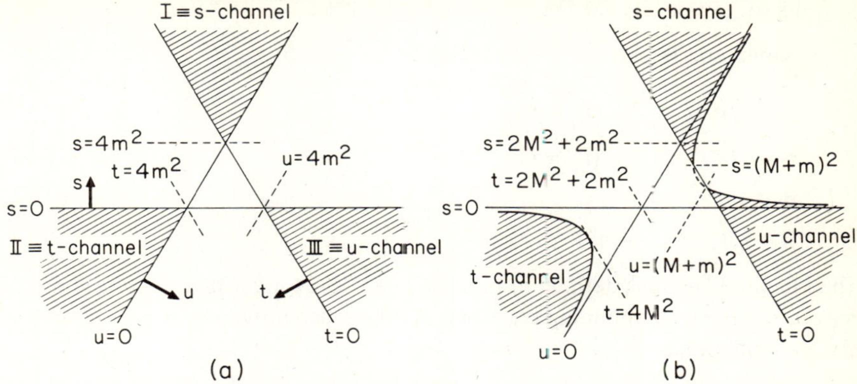

Fig. 15-5 (*a*) Mandelstam diagram for all masses equal to *m*. The shaded areas indicate the physical regions of three crossed reactions. The short heavy arrows indicate the positive directions for each quantity. (*b*) Mandelstam diagram for elastic scattering of unequal mass particles. See text.

a Mandelstam diagram, the quantities s, t, and u can take on all *real* values consistent with Eq. 15-17. Therefore, the diagram includes the regions in which these variables take on the values belonging to the reactions $a + b \to c + d$ and $a + \bar{c} \to \bar{b} + d$, as well as the values corresponding to the reaction $a + \bar{d} \to c + \bar{b}$ (see below).

Exercise 15-4. Justify the position and size of the equilateral triangle in Fig. 15-5*a*.

The kinematic regions belonging to each reaction are most easily determined when all the masses are equal. According to Fig. 15-5*a*, we have

Kinematic Region I ≡ *s*-Channel, $a + b \to c + d$

$$\begin{aligned} s_{\text{I}} &= (P_a + P_b)^2 = (\tilde{P}_a + \tilde{P}_b)^2 = s \\ t_{\text{I}} &= (P_a - P_c)^2 = (\tilde{P}_a + \tilde{P}_c)^2 = t \\ s &\geqslant 4m^2, \\ 0 &\geqslant t \geqslant -[s - 4m^2] \end{aligned} \tag{15-19}$$

Kinematic Region II ≡ *t*-Channel, $a + \bar{c} \to \bar{b} + d$

$$\begin{aligned} s_{\text{II}} &= (P_a + P_{\bar{c}})^2 = (\tilde{P}_a + \tilde{P}_c)^2 = t \\ t_{\text{II}} &= (P_a - P_{\bar{b}})^2 = (\tilde{P}_a + \tilde{P}_b)^2 = s \\ t &\geqslant 4m^2 \\ 0 &\geqslant s \geqslant -[t - 4m^2] \end{aligned} \tag{15-20}$$

Similarly we obtain for the reaction $a + \bar{d} \to c + \bar{b}$

Kinematic Region III $\equiv$ u-Channel, $a + \bar{d} \to c + \bar{b}$

$$
\begin{aligned}
s_{\text{III}} &= (P_a + P_{\bar{d}})^2 = (\tilde{P}_a + \tilde{P}_d)^2 = u \\
t_{\text{III}} &= (P_a - P_c)^2 = (\tilde{P}_a + \tilde{P}_c)^2 = t \\
u &\geqslant 4m^2 \\
0 &\geqslant t \geqslant -[u - 4m^2]
\end{aligned}
\tag{15-21}
$$

The kinematic regions defined in Eqs. 15-19 to 15-21 are called the *physical regions* for the corresponding reactions. We have also noted the conventional and convenient names: s, t, and u-channel.

Exercise 15-5. How should the definitions in Eq. 15-18 be changed if we wish to represent the following set of reactions on a Mandelstam diagram?

$$a + b \to c + d$$

$$\bar{c} + b \to \bar{a} + d$$

$$\bar{d} + b \to c + \bar{a}$$

If the masses of the particles in a two-body reaction are not all equal, the geometry of the boundaries of the physical regions immediately becomes complicated, although we retain the simple concept of representing on a single diagram the kinematic regions belonging to a set of three crossed reactions. For example, in the elastic scattering of unequal mass particles $m_a = m_c = m$, $m_b = m_d = M, M > m$, the physical region for the s-channel $a + b \to c + d$ is defined by

$$s \geqslant (m + M)^2 \tag{15-22a}$$

$$0 \geqslant t \geqslant -\left[s - (2m^2 + 2M^2) + \frac{(M^2 - m^2)^2}{s} \right] \tag{15-22b}$$

We see (Fig. 15-5b) that the right-hand boundary of this region is the curve defined in Eq. 15-22b.

Exercise 15-6. Derive Eqs. 15-22. Then consider the reaction $\pi^- + p \to \pi^0 + n$. Use it to illustrate Eq. 15-22b.

The boundaries of the u-channel $a + \bar{d} \to c + \bar{b}$ are the same type as those of the s-channel, $a + b \to c + d$. But the t-channel $a + \bar{c} \to \bar{b} + d$ has more complicated boundaries, for both are curved.

Although the boundaries of the physical region become more complicated when all the masses are different, the general form of Fig. 15-5a is preserved. In particular, at sufficiently high energy in any of the three physical regions, the boundaries are approximately the straight lines of Fig. 15-5*a*. For a fuller discussion, see Refs. 15OMN, 15MA, and 15ED. Some of these references also show how to represent the three kinematic regions of Eqs. 15-19 to 15-21 using a rectangular formulation. There is nothing magic about the triangular form of Fig. 15-5, although it does have a pleasing symmetry.

15-4 Crossing Relations. To describe the concept of *crossing relations* between crossed channels it is easiest to limit the initial discussion to particles with zero spin. Consider the three crossed reactions defined in Eqs. 15-19 to 15-21. The crossing relations concept assumes that there exists a *single analytic* function $T(s, t, u)$ with the following properties:

in kinematic region I, where $s \geqslant 4m^2$,

$$0 \geqslant t \geqslant -[s - 4m^2],\ T_{ab\to cd}(s, t) \equiv T(s, t, u)$$

in kinematic region II, where $t \geqslant 4m^2$,

$$0 \geqslant s \geqslant -[t - 4m^2],\ T_{a\bar{c}\to\bar{b}d}(t, s) \equiv T(s, t, u)$$

in kinematic region III, where $u \geqslant 4m^2$,

$$0 \geqslant t \geqslant -[u - 4m^2],\ T_{a\bar{d}\to c\bar{b}}(u, t) \equiv T(s, t, u) \tag{15-23}$$

We have written $T(s, t, u)$ to emphasize the symmetry of the three kinematic variables, but we always have the restriction $s + t + u = \Sigma m_i^2$. The key phrase in the definition just given is *single analytic*, the word *analytic* being paramount.

A very simple example is the t-channel, one-particle exchange amplitude of Eq. 15-16. This single analytic function is directly defined in all three kinematic regions, but reality will be much more complicated. When considering real reactions we will know at best approximate amplitudes in the various kinematic regions. To connect those amplitudes, we will have to use the concept of *analytic continuation*.

When the particles have nonzero spins, the simple relations given in Eq. 15-23 cannot be used. As discussed in Sec. 16-3, the crossing between physical regions usually mixes helicity states. The mixing is only a kinematic problem, however, and the basic concept of the crossing relations is retained.

There is no complete proof that the crossing relations can be used in elementary particle physics. But they are consistent with everything we know about elementary particles, and they have been accepted as a framework for all theories of two-body reactions. As we have emphasized, there is a very intimate connection between crossing relations and analyticity; therefore, we turn next to the examination of the analytic behavior of two-body reaction amplitudes.

15-5 Analyticity and Perturbation Theory. Since perturbation theory is one of the main sources of justification and inspiration for the application of the theory of analytic function to reaction amplitudes (15BJ, 15GA), in this section we demonstrate these quantities with the use of the two-body reactions of spinless particles.

A very simple illustration is obtained by considering the one-particle exchange amplitude (Eq. 15-16)

$$T_{ab\to cd,\,\text{pole}} = -\frac{g_{aec}g_{bed}}{m_e{}^2 - t} \tag{15-24}$$

for $a + b \to c + d$. For this reaction the allowed range of values for t is $0 \geqslant t_{\min} \geqslant t \geqslant t_{\max}$. But if we now consider Eq. 15-24 as a function of a complex variable t, the right-hand side is a function that is analytic everywhere except at the point $t = m_e{}^2$, where the function has a simple pole. Thus this amplitude is described by an analytic function. We summarize these observations by the diagram (Fig. 15-6a), representing the complex t-plane at fixed s. Thus the analytic behavior of this amplitude as a function of t is very simple. Of course the analytic behavior as a function of s is even simpler. The amplitude is independent of s, and the analytic function in terms of s is just a constant.

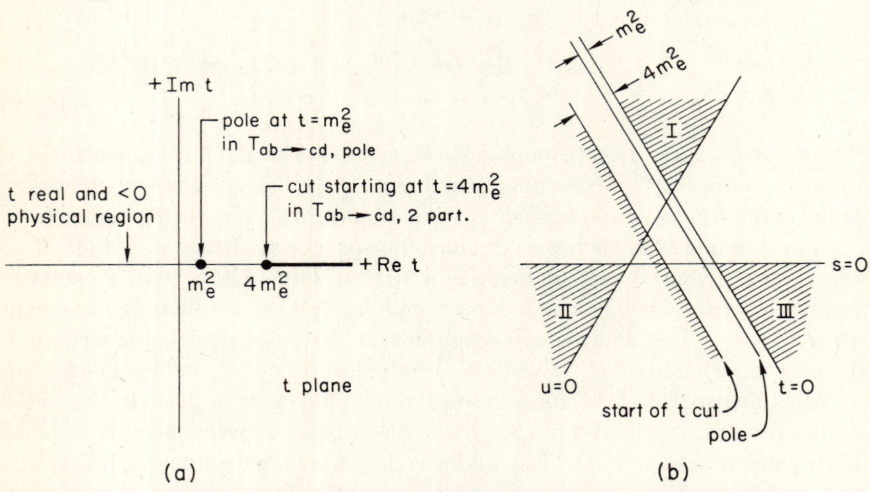

Fig. 15-6 (a) Singularities of the amplitude $T_{ab\to cd}$ in the complex t-plane for s fixed in the $a + b \to c + d$ physical region. (b) The same singularities on an equal mass Mandelstam diagram. Note that these singularities also apply to the other two physical regions.

Exercise 15-7. Suppose that $\pi^- p$ elastic scattering takes place only through the exchange of a neutron in the s-channel. Discuss the amplitude in terms of its analytic behavior. Then suppose that such scattering also takes place through the exchange of a ρ^0 meson in the t-channel. Again discuss the analytic behavior of the amplitude.

Incidentally, the part of the amplitude associated with simple poles produced by elementary particle exchange in the s- or t-channel is often called the *Born term* or the *Born amplitude*.

Next we would like to consider slightly more complicated Feynman diagrams (Fig. 15-7) in which two particles are exchanged in the t-channel. Unfortunately, this slight increase in the complexity of the Feynman diagram greatly increases the complexity of the analysis of the analytic behavior of amplitude represented by this diagram. But the analysis is worth following for the insight it provides. We consider the diagram in Fig. 15-7a in which the lower vertex is a so-called seagull vertex; this seagull vertex simplifies the analysis. All particles are spin zero, the masses and four-momentum vectors are as indicated in the diagram.

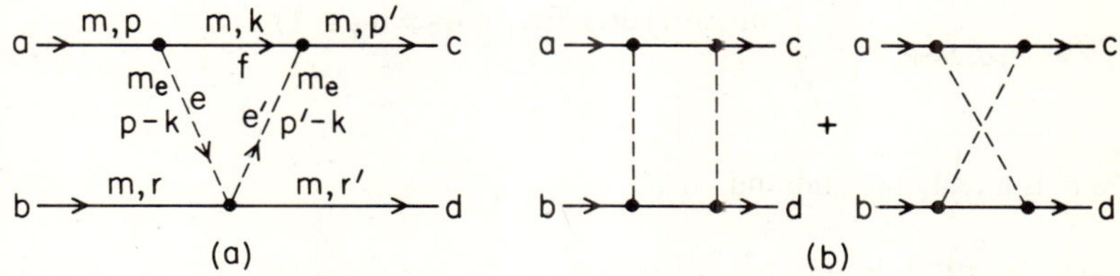

Fig. 15-7 (a) Two-particle exchange diagram with seagull vertex at lower vertex. The mass and four-momentum of each line are indicated. (b) Other two-particle exchange diagrams.

The amplitude is given by

$$T_{ab\to cd,\,2\text{ part}} = ig_{aef}g_{fe'c}G_{ee'bd}$$

$$\times \int \frac{d^4k}{[k^2 - m^2 + i\epsilon]\,[(p-k)^2 - m_e{}^2 + i\epsilon]\,[(p'-k)^2 - m_e{}^2 + i\epsilon]} \tag{15-25a}$$

where the $i\epsilon$'s are exhibited explicitly to indicate the integration contours to be used. The problem—to investigate the analytic properties of the integral in terms of $t = p - p' = r - r$—is traditionally performed in the following way. Equation 15-25a is rewritten in the form (Ex. 15-8)

$$T_{ab\to cd,\,2\text{ part}} = ic \int_0^1 d\alpha_1 \int_0^1 d\alpha_2 \int_0^1 d\alpha_3 \delta(\alpha_1 + \alpha_2 + \alpha_3 - 1)$$

$$\times \int \frac{d^4k}{\{\alpha_1 [k^2 - m^2] + \alpha_2[(p-k)^2 - m_e^2] + \alpha_3[(p'-k)^2 - m_e^2] + i\epsilon\}^3} \tag{15-25b}$$

Exercise 15-8. Look up (15BJ, 15GA) or work out the proof of the following equalities

$$\frac{1}{A_1 A_2 \cdots A_n} = (n-1)! \int_0^1 d\alpha_1 \int_0^1 d\alpha_2 \cdots \int_0^1 d\alpha_n \left[\frac{\delta(\alpha_1 + \alpha_2 + \cdots + \alpha_n - 1)}{\left(\sum_{i=1}^{n} \alpha_i A_i \right)^n} \right] \tag{15-26}$$

$$\int \frac{d^4 k}{(k^2 - L + i\epsilon)^3} = \frac{\pi^2}{2iL} \tag{15-27}$$

Again, using Ex. 15-8, we obtain

$$T_{ab \to cd, 2\text{ part}} = c' \int_0^1 \frac{d\alpha_1\, d\alpha_2 d\alpha_3\, \delta(\alpha_1 + \alpha_2 + \alpha_3 - 1)}{D} \tag{15-28}$$

where c' is a real constant and

$$D = \alpha_1{}^2 m^2 + (1 - \alpha_1) m_e{}^2 - \alpha_2 \alpha_3 t \tag{15-29}$$

The singularities of $T_{ab \to cd, 2\text{ part}}$ occur when D is zero for α_1, α_2, and α_3 in the range of 0 to 1. Using the δ function in Eq. 15-28, we see that the singularities occur when

$$t_{\text{sing}} = \frac{\alpha_1{}^2 m^2 + (1 - \alpha_1) m_e{}^2}{\alpha_2 \alpha_3}; \qquad \alpha_1 = 1 - (\alpha_2 + \alpha_3) \tag{15-30}$$

From the allowed values for the α's, we can deduce that

1. t_{sing} is real and positive.
2. The maximum value of t_{sing} is infinity.
3. The minimum value of t_{sing} is $t_{\text{sing, min}} = 4m_e{}^2$ (15-31)

Exercise 15-9. Prove the last set of observations.

Thus the seagull diagram leads to a branch line, or branch cut, along the positive real t axis. The cut (Fig. 15-6*a*) extends from $t = 4m_e{}^2$ to $t = \infty$. It is instructive to indicate the position of this cut and the pole in Eq. 15-24 on the triangular Mandelstam diagram of Fig. 15-6*b*. Because the position of these singularities is independent of s, the branch point at $t = 4m_e^2$ and the pole at $t = m_e^2$ become straight lines on this diagram.

It would be interesting to continue with this analysis, considering next the two-particle exchange diagram of Fig. 15-7*b*. But the analysis of even this relatively simple diagram is complicated (15BJ), and we cannot spare the space. We only note that this diagram, like the seagull diagram, leads to a cut along the positive, real t axis. Indeed, in perturbation theory, the singularities from higher order diagrams are all on the real t axis when s is real (15BJ, 15ED).

15-6 Singularities in Two-Body Reactions and the General Unitarity Relation.

In the early 1960s a set of rules was developed for the positions of singularities in the amplitudes of two-body reactions. The justification for these rules comes from perturbation theory arguments of the type presented in the previous section, from more sophisticated perturbation theory arguments carried out to all orders in the coupling constants, from axiomatic field theory, and from S-matrix theory. The general unitarity relation Eq. 6-26 also plays an important role in the justification of these rules, and we use that relation in this section. Although the proof of these rules is not complete, the rules are generally accepted. They are presented here because they provide some insight and guidance with respect to the behavior of two-body reaction amplitudes. Unfortunately, as the next section reveals, these rules by themselves do not enable us to calculate the specific form of these amplitudes.

In the previous section we discussed the simplest singularities of the amplitude in the complex t plane for a fixed and real value of s. But now, to broaden our point of view, we consider singularities in the complex s plane for a fixed and real value of t. The reader may wish to translate as he goes along into the t

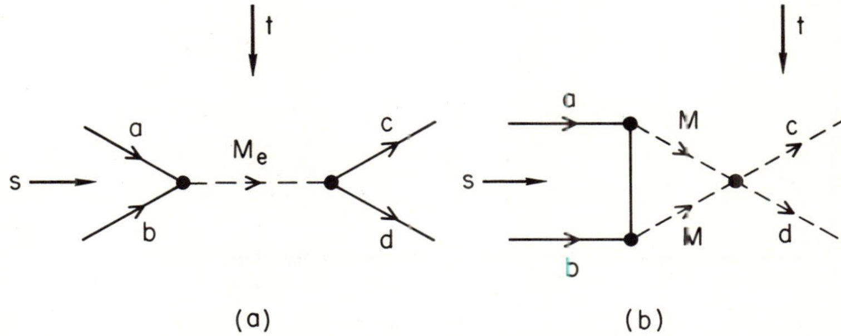

Fig. 15-8 Diagrams leading to (*a*) a pole and (*b*) a cut in the amplitude in the s-channel. To increase the generality, the particles exchanged in (*b*), indicated by the dashed lines, have a mass M that may differ from the mass M_e of the particle exchanged in (*a*).

plane, to develop a feeling for crossing relation ideas. We consider the s-channel reaction $a + b \to c + d$; for simplicity, all spins are taken to be zero, but the particles may have unequal masses. We fix t in the physical region of the s-channel $0 \geqslant t_{\text{min}} \geqslant t \geqslant t_{\text{max}}$.

If the diagrams in Fig. 15-8 can occur in the s-channel, there will be a pole at M_e^2 and a branch point at $s = 4M^2$, both on the positive, real s axis. And as discussed at the end of the previous section, additional Feynman diagrams will lead to additional branch points on the positive, real s axis. However, let us try an alternative way to find the position of these branch points.

We consider the unitarity relation, Eq. 6-26. Let us take the reaction $a + b \to c + d$ to be confined to the x–z plane. This restriction and the zero particle spins combined with Eqs. 10-24 and 6-26 yields the simplified unitarity relation

$$2 \operatorname{Im} T_{ab \to cd}(s, t) = (2\pi)^4 \sum_f \int \left[\prod_{n=1}^{N_f} \frac{d^3 P_n}{(2\pi)^3 2E_n} \times \langle cd | T^\dagger | f \rangle \langle f | T | ab \rangle \delta^4(P_f - P_i) \right] \tag{15-32}$$

We recall that the summation over the intermediate state $|f\rangle$ means that *all* real intermediate states are to be included as long as quantities such as four-momentum, angular momentum, and charge are conserved. This is discussed in Ex. 15-10. Now as s increases along the positive real s axis, starting from $s = 0$, the various thresholds for the intermediate states $|f\rangle$ will be passed. At each of these thresholds we certainly can expect a drastic change to occur in the analytic properties of $T_{ab \to cd}(s, t)$, and this drastic change will be a branch point. The first branch point will occur at the lowest mass intermediate state. All these features appear

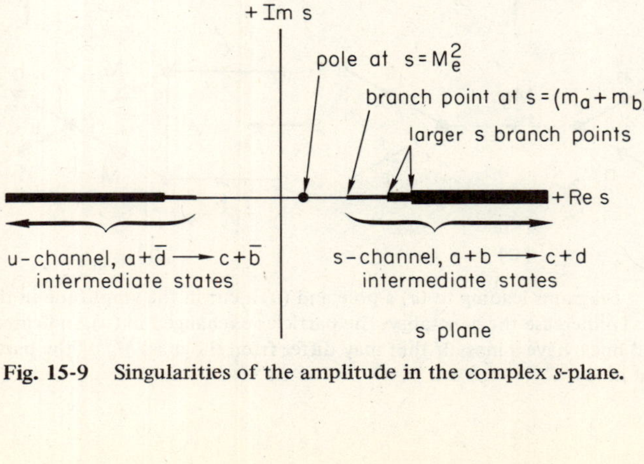

Fig. 15-9 Singularities of the amplitude in the complex s-plane.

in Fig. 15-9 where we have assumed that the lowest mass real intermediate state is $|ab\rangle$, such that the lowest s branch point occurs at $s = (m_a + m_b)^2$.

Every branch point begins a cut extending along the positive, real s axis to $s = \infty$. On most of the positive s axis, therefore, there is a superposition of many branch line singularities coming from different amplitudes; and most of these amplitudes are to intermediate states $|f\rangle$ of many particles. Hence buried in the real s axis singularities is a great amount of very complicated amplitude information —information we rarely possess. We can already see the strength and the weakness of the analytic approach. The overall analytic situation is easily summarized by the position of the singularities. But the details of the analytic behavior usually depend on information on many other amplitudes.

The derivation of the general unitarity relation in Sec. 6-5 assumes $s_{ab} \geqslant (m_a^2 + m_b^2)$ and $s_{cd} \geqslant (m_c^2 + m_d^2)$; therefore, we cannot apply Eq. 15-32, as originally derived, to smaller values of s. However, the significance of the unitarity relation has been expanded by the analyticity concept to allow the application of Eq. 15-32 to lower values of s. For example, suppose there exists a two-particle intermediate state $|f\rangle = |gh\rangle$ such that $s_{gh} = (m_g + m_h)^2$ is less than s_{ab} and s_{cd}. We apply Eq. 15-32, and now the lowest s branch point is at s_{gh}. Of course the amplitude $T_{ab\to cd}(s, t)$ is unphysical when $s_{gh} \leqslant s \leqslant s_{ab}$ or $s_{gh} \leqslant s \leqslant s_{cd}$. We note that the amplitude will also be unphysical for larger values of s when $|t| > 0$. For the amplitude to be physical, s and t must be in the physical region of the reaction $a + b \to c + d$.

Keeping t fixed at the same value, let us consider the situation as we move along the real s axis in the $-s$ direction. As we move below the first threshold, Eq. 15-32 provides no information because it was derived for real initial, intermediate and final states (Sec. 6-5). However, we follow the general philosophy used in analyticity considerations—make the simplest assumption consistent with everything known from perturbation theory, field theory, and non-relativistic quantum mechanics. The simplest assumption in this case is that $\text{Im}\, T_{ab\to cd}(s, t) = 0$ below the first branch point.

As we continue to move along the real s axis in the $-s$ direction, assuming $\text{Im}\, T_{ab\to cd}(s, t)$ is zero except for poles, we soon reach the negative s axis. Remembering that t is fixed, Fig. 15-5 or Eq. 15-17 shows that we will next move into the physical region of the u-channel, $a + \bar{d} \to c + \bar{b}$. Then we can apply Eq. 15-32 to $T_{a\bar{d}\to c\bar{b}}(u, t)$ where $u = \Sigma_i m_i^2 - s - t$, and the singularities of this u-channel will consist of cuts beginning at a u-channel threshold and extending to $u = +\infty$, which is equivalent to $s = -\infty$. Figure 15-9 summarizes the observations on singularities that have been made in this section. A pole could have been indicated in the u-channel, below the u-channel threshold, but we have not bothered to insert one.

So far we have discussed the singularities on the real s axis. What about singularities in the plane which are *off* the real s axis? The simplest assumption, consistent with the requirements imposed by unitarity, is that there are *no singularities off* the real s axis. These considerations, of course, are restricted to real, fixed values of t. This is sometimes called the *maximal analyticity* assumption (15BU). By a maximum amount of analyticity we mean that there are a minimum number of singularities; only those singularities exist which are required by unitarity.

The singularities at fixed s in the complex t–u plane and the singularities at fixed u in the complex s–t plane will have the same structure as those we have just

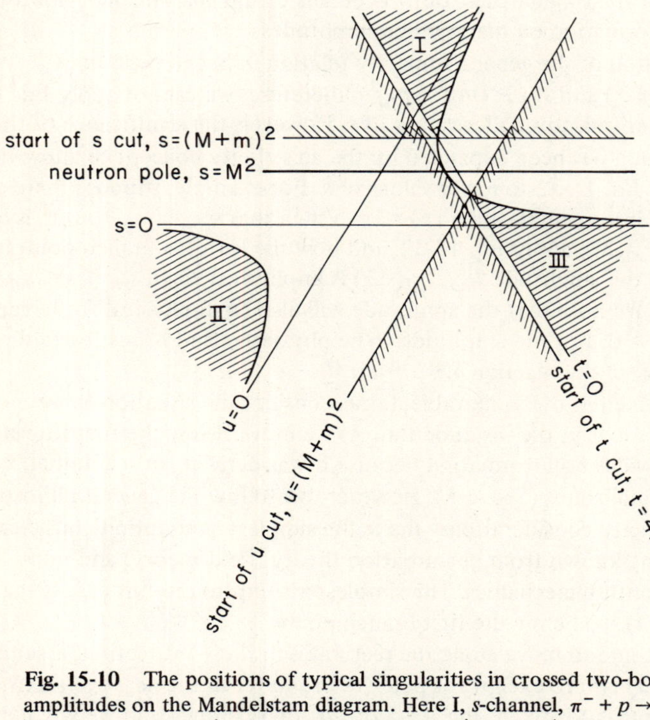

Fig. 15-10 The positions of typical singularities in crossed two-body amplitudes on the Mandelstam diagram. Here I, s-channel, $\pi^- + p \to \pi^- + p$; II, t-channel, $\pi^- + \pi^+ \to \bar{p} + p$; III, u-channel, $\pi^+ + p \to \pi^+ + p$. The mass m of the pion is exaggerated with respect to the mass M of the nucleon.

discussed. It is convenient to summarize all of this information on the Mandelstam diagram of Fig. 15-10. Thus the rules for the position of the singularities are:

1. When one of the variables s, t, u is real and fixed, all the singularities in the complex plane of the other two variables lie on the real axis in that plane.
2. Single-particle exchange or single-particle intermediate states lead to poles.
3. Two or more particle intermediate states, real or virtual, lead to cuts.

Exercise 15-10. Consider the *s*-channel reaction $K^- + p \to K^- + p$ and *u*-channel reaction $K^- + \bar{p} \to K^- + \bar{p}$. Draw a diagram similar to Fig. 15-9 showing the singularities in the *s–u* plane at fixed *t*. Remember that this *s*-channel has single-particle intermediate states like the Λ^0 and below-threshold, several-particle intermediate states like $(\pi^-\Sigma^0)$ and $(\pi^-\pi^0\Sigma^0)$. Draw a Mandelstam diagram similar to Fig. 15-10, in which the *t*-channel is $K^- + K^+ \to \bar{p} + p$.

Thus we have analyzed behavior of the two-body reaction amplitude $T(s, t)$ when one of the kinematic variables is held fixed. This discussion is different from the analysis of the analytic behavior of a *partial wave amplitude* carried out in Sec. 15-2 and 15-11, but the two analytic structures are related. We wish to devote the remainder of this chapter to some illustrations of the uses of the singularity structure of $T(s, t)$. But first we must comment on the distinction between dynamic and kinematic singularities.

15-7 Dynamic and Kinematic Singularities. The singularities we have been considering in this chapter are called *dynamic singularities* because they are produced by the internal dynamics of the reaction process. These dynamic singularities are very important to us because they teach us about the fundamental nature of the reaction process. Unfortunately, in the calculation of specific amplitudes, other singularities can appear that have nothing to do with the internal dynamics. Again it is best to proceed by examples.

Consider the one-particle exchange amplitude of Eq. 15-16

$$T_{ab\to cd}(s, t) = \frac{g_{aec}g_{bed}}{t - m_e^2} \tag{15-33a}$$

Now suppose that we wish to work with

$$f_{ab\to cd}(s, t) = \frac{1}{8\pi\sqrt{s}}\left(\frac{p_f^*}{p_i^*}\right)^{1/2} \times T_{ab\to cd}(s, t) = \frac{1}{8\pi\sqrt{s}}\left(\frac{p_f^*}{p_i^*}\right)^{1/2} \frac{g_{aec}g_{bed}}{t - m_e^2} \tag{15-33b}$$

In addition to the dynamic singularity, a simple pole at $t = m_e^2$, this f amplitude has branch points in s at $p_i^* = 0$, $p_f^* = 0$, and $s = 0$. The branch points have nothing to do with the dynamics of the reaction; they are produced by the definition of the f amplitude. Such singularities, which are one type of *kinematic singularity*, are annoying yet extraneous. The obvious solution is to avoid them by proper definition of the amplitude.

A more subtle form of kinematic singularity occurs when the particles have spin. Remember that we have been considering only spin zero particles in this chapter. As an illustration, let us assume that the reaction $a + b \to c + d$ proceeds by the exchange of a spin zero particle as in Sec. 15-3; but let us suppose that particle c has spin 1 while particles a, b, and d retain their spin zero nature. Then using the methods of Sec. 12-6, the helicity ± 1 amplitude is

$$T_{ab\to cd}^{(\lambda_c=\pm 1)}(s,t) = \frac{\pm\sqrt{2}g_{aec}g_{bed}\sqrt{-t}\sqrt{s+t-4m^2}}{(t-m_e^{\,2})\sqrt{s-4m^2}} \tag{15-34}$$

where we have assumed that all external masses are equal to m. The amplitude in Eq. 15-34 has additional singularities compared with the amplitude in Eq. 15-33a. These singularities (branch points at $s = 4m^2$, $t = 0$, and $s + t - 4m^2 = -u = 0$) are produced by the spin 1 nature of particle c, and they have nothing to do with the dynamics of the reaction, which in this very simple example is elementary one-particle exchange. Therefore, singularities related to the spins of the external particles are also classified as *kinematic singularities.*

To study the dynamic singularities and their effects, it is obviously desirable to use amplitudes having no kinematic singularities. Such kinematic-singularity-free amplitudes have been constructed by a variety of methods, all somewhat complicated (15MA). It would take up too much space to discuss them, and it is sufficient for our purposes if the reader understands the difference between dynamic and kinematic singularities.

15-8 Fixed *t* Dispersion Relations. Once more we consider the reaction $a + b \to c + d$ in the s-channel. And once again for simplicity we take all the external particles to have spin zero. The introduction of nonzero spins greatly complicates the details of the discussion but does not change the essentials. For t fixed, the singularities in the complex s-plane are as discussed in Sec. 15-6 and shown in Fig. 15-9. To increase the generality of our examination, a pole at $s = M^2$ is also included.

Our first order of business, since the s-plane is cut over most of the real s axis, is to define $T_{ab\to cd}(s, t)$, where s is real and in the physical region of the s- or u-channel. This is similar to our task early in Sec. 15-2 when we had to cut the k^2 plane. In making this definition it is useful to be aware of the Schwarz reflection principle. Let c be a closed contour in the upper half of the z complex plane, and let c include a segment AB of the real axis. If

1. A function $f(z)$ is analytic inside c and
2. $\operatorname{Im} f(z) = 0$ on AB (15-35)

then

1. We can reflect c about the real axis obtaining a domain D that is symmetric about the real axis and
2. in $D, f(z^*) = (f(z))^*$ (15-36)

The assumptions on the singularity structure of $T(s, t)$ of course, have made use of Schwarz's principle. Indeed, since the domain D includes all the s–u complex plane except for the cuts and poles on the real s axis, we have

$$T(s^*, t) = (T(s, t))^* \qquad \text{for} \quad \text{Re } s \text{ in physical region, Im } s \neq 0 \qquad (15\text{-}37)$$

Thus in general Im $T(s, t)$ has a discontinuity across the real s axis. Letting s' be *real* and in the s-channel physical region, we define

$$T_{ab\to cd}(s', t) \underset{\epsilon\to 0}{=} T(s' + i\epsilon, t), \qquad \epsilon \geqslant 0 \qquad (15\text{-}38a)$$

Similarly, for the u-channel physical region we define

$$T_{\bar{a}d\to c\bar{b}}(u, t) \underset{\epsilon\to 0}{=} T(u' + i\epsilon, t) \underset{\epsilon\to 0}{=} T(s' - i\epsilon, t), \qquad \epsilon \geqslant 0 \qquad (15\text{-}38b)$$

where $s' = \Sigma_i m_i^2 - u' - t$.

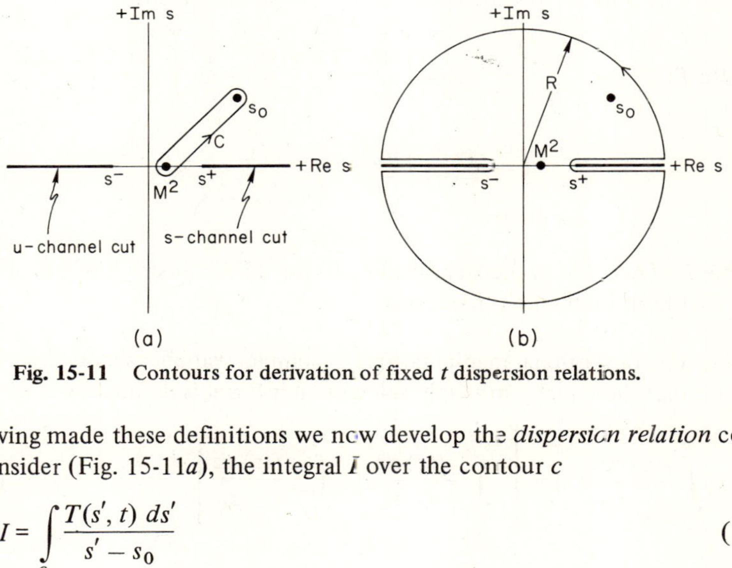

Fig. 15-11 Contours for derivation of fixed *t* dispersion relations.

Having made these definitions we now develop the *dispersion relation* concept. Consider (Fig. 15-11*a*), the integral I over the contour c

$$I = \int_c \frac{T(s', t)\, ds'}{s' - s_0} \qquad (15\text{-}39)$$

where s_0 is a point in the upper half-plane. We use Cauchy's famous theorem of residues to evaluate I and easily obtain

$$T(s_0, t) = -\frac{r}{M^2 - s_0} + \frac{1}{2\pi i} \int_c \frac{T(s', t)\, ds'}{s' - s_0} \qquad (15\text{-}40)$$

where r is the residue of $T(s, t)$ at $s = M^2$.

Note the great importance to this derivation of the rules (Sec. 15-6), limiting the positions of singularities. Next we distort the contour (Fig. 15-11*b*), letting R go to infinity and letting the straight parts of the contour move to within a small distance, $i\delta$ with $\delta \to 0$, of the real axis. Then using Eqs. 15-37 and 15-38, and setting $r = g_{aeb}g_{ced}$, we obtain

$$T(s_0, t) = -\frac{g_{aeb}g_{ced}}{M^2 - s_0} + \frac{1}{\pi}\int_{s^+}^{\infty} \frac{\text{Im}\, T(s', t)\, ds'}{s' - s_0} + \frac{1}{\pi}\int_{-\infty}^{s^-} \frac{\text{Im}\, T(s', t)\, ds'}{s' - s_0} + \int_{\text{semicircles}} \frac{T(s', t)\, ds'}{s' - s_0} \tag{15-41}$$

Here $s^{\pm}$ are defined in Fig. 15-11, and the g's are the vertex coupling constants for the one-particle diagram in the s-channel. We move s_0 to the physical region by writing $s_0 = s + i\epsilon$, where s is *real* and *greater than* s^+ and ϵ is small. As ϵ approaches zero we distort the contour near s to obtain finally

$$\text{Re}\, T(s, t) = -\frac{g_{aeb}g_{ced}}{M^2 - s} + \frac{1}{\pi}P\int_{s^+}^{\infty} \frac{\text{Im}\, T(s', t)\, ds'}{s' - s} + \frac{1}{\pi}\int_{-\infty}^{s^-} \frac{\text{Im}\, T(s', t)\, ds'}{s' - s} + \int_{\text{semicircles}} \frac{T(s', t)\, ds'}{s' - s} \tag{15-42}$$

Exercise 15-11. Fill in the steps leading to Eq. 15-42. Justify the use of the g's in the first term on the right-hand side.

These are conventional manipulations in complex variable theory: recall however, that the Cauchy principle value of an integral is defined by

$$P\int_{z_1}^{z_2} \frac{f(z)\, dz}{z - z_0} = \lim_{\delta \to 0} \left[\int_{z_1}^{z_0 - \delta} \frac{f(z)\, dz}{z - z_0} + \int_{z_0 + \delta}^{z_2} \frac{f(z)\, dz}{z - z_0} \right] \tag{15-43}$$

for $z_1 < z_0 < z_2$.

To make direct use of Eq. 15-42, all the quantities in the equation must be in the physical regions of the s- or u-channels. The first obstacle to direct use of Eq. 15-42 is the integral over the semicircle. The value of this integral is a common problem in complex integration. We recall that the integral over the semicircle will go to zero if

$$|T(s', t)| \to 0 \quad \text{as} \quad R \to \infty \tag{15-44}$$

If the relation of Eq. 15-44 does not hold, the dispersion relation must be changed into a *subtracted dispersion relation* (see next section). Assuming for the remainder of this section that Eq. 15-44 is obeyed, however, we finally obtain the *dispersion relation in s at fixed t:*

$$\operatorname{Re} T_{ab\to cd}(s,t) = \frac{g_{aeb}g_{ced}}{s-M^2} + \frac{1}{\pi} P \int_{s^+}^{+\infty} \frac{\operatorname{Im} T(s',t)\,ds'}{s'-s}$$
$$+ \frac{1}{\pi} \int_{-\infty}^{s^-} \frac{\operatorname{Im} T(s',t)\,ds'}{s'-s} \tag{15-45}$$

This dispersion relation has been derived using very general concepts, hence is a very general and beautiful result. The idea of a dispersion relation extends from classical physics through nonrelativistic quantum mechanics into elementary particle physics (15BJ); and in elementary particle physics, dispersion relations have all sorts of applications. They can be applied not only to two-body reaction amplitudes but to parts of those amplitudes, such as three-particle vertices. They can be employed indirectly in many-body reaction amplitudes. A dispersion relation is often useful in estimating the magnitude or the behavior of an amplitude when a detailed calculation is tedious or impossible.

Of particular interest to us is the ability of dispersion relations to connect lower energies, where measurements can be made, with very high energies, where we may only have speculations about the behavior of amplitudes. Thus it provides a way to test the very high energy aspects of elementary particle theories. On the other hand, there are only a few cases in which the dispersion relation can be used for precise tests, for we rarely possess precise knowledge of an amplitude $T(s, t)$ at fixed t over a large s range. We do possess such information for forward elastic scattering, and this case is discussed next.

15-9 Forward Elastic Scattering Dispersion Relations. A classic application of fixed t dispersion relations in elementary particle physics is the forward elastic scattering dispersion relation, which we derive and discuss in our usual simplified style—there are a number of important points we wish to emphasize without becoming lost in manipulative details.

Let us consider elastic scattering of spin zero particles in the s-channel

$$a + b \to a + b \tag{15-46}$$

Although we have little precise knowledge of the elastic scattering amplitude $T_{el}(s, t)$ at arbitrary t, we do know a good deal about the forward amplitude $T_{el}(s, 0)$; for by the optical theorem of Sec. 6-7

$$\operatorname{Im} T_{el}(s,0) = 2p\sqrt{s}\,\sigma_{tot}(s) \tag{15-47}$$

Therefore, in the integrals of Eq. 15-45 we can replace the Im T by σ_{tot}, wherever σ_{tot} has been measured. This is the fundamental reason for the great utility of forward elastic scattering dispersion relations.

But before proceeding we must note two problems associated with the replacement of Im T_{el} by σ_{tot}. First, the cut on the positive real s axis may begin below the physical threshold for the elastic reaction of Eq. 15-46. This indeed is the situation in K^-p elastic scattering, as discussed in Ex. 15-10, and some theoretical device must be found to estimate Im T_{el} along the unphysical part of the cut below the s-channel physical threshold. Second, the measurements of σ_{tot} cannot be extended to infinite energy; above some s_{max} we have no measurements of σ_{tot}. If we wish to test the analyticity concept by testing the forward dispersion relation, or if we wish to use the forward dispersion relation to evaluate a coupling constant, we must make an assumption about σ_{tot} above s_{max}. We usually assume that σ_{tot} is independent of s above s_{max}.

Returning to the main discussion, we consider the first integral on the right-hand side of Eq. 15-45. Assuming no branch points below $s^+ = (m_a + m_b)^2$, we write

$$J(s) = \frac{2}{\pi} P \int_{(m_a+m_b)^2}^{\infty} \left[\frac{p'\sqrt{s'}\sigma_{tot}(s')}{s' - s} \right] ds' \tag{15-48}$$

Here we have used Eq. 15-47. For $s' \gg s$, the integrand in Eq. 15-48 is a constant if $\sigma_{tot}(s')$ is independent of s', as we have just assumed. Then the integral in Eq. 15-48 is clearly divergent, and the condition of Eq. 15-44 will not be satisfied. Thus we seem to be checkmated in one of the few places that can directly use a dispersion relation. To avoid checkmate, we resort to the trick of constructing a *subtracted* dispersion relation. We select an s_1 in the physical region of Eq. 15-46 but *not* equal to s. Then we have the *once-subtracted* dispersion relation integral

$$J(s) - J(s_1) = \frac{2(s - s_1)}{\pi} P \int_{(m_a+m_b)^2}^{\infty} \left[\frac{p'\sqrt{s'}\sigma_{tot}(s')}{(s' - s)(s' - s_1)} \right] ds' \tag{15-49}$$

where the integrand for $s' \gg s$ and $s' \gg s_1$ now varies as $1/s'$ if we assume that $\sigma_{tot}(s')$ is independent of s'. This is an improvement over Eq. 15-48, but the integral is still divergent. And in general we have to repeat the trick and construct a *twice-subtracted* dispersion relation integral.

Returning to Eq. 15-45, we can evaluate the second integral on the right-hand side if we take the u-channel reaction to be

$$\bar{a} + b \to \bar{a} + b \tag{15-50}$$

hence also elastic. Using Eq. 15-38, with $\Sigma m^2 = 2m_a{}^2 + 2m_b{}^2$, we obtain

$$\frac{1}{\pi}\int_{-\infty}^{s^-} \frac{\text{Im } T(s', 0)\, ds'}{s' - s} = \frac{2}{\pi}\int_{(m_a+m_b)^2}^{+\infty} \left[\frac{p'\sqrt{u'}\sigma_{\text{tot}}(u')}{u' - \Sigma m^2 + s}\right] du' \tag{15-51}$$

where $\sigma_{\text{tot}}(u)$ is the total cross section for $\bar{a} + b$. The same convergence considerations apply to this integral. We have again assumed no branch points in the u-channel below $u = (m_a + m_b)^2$.

For simplicity, let us assume that $\sigma_{\text{tot}}(s)$ decreases as s increases and $\sigma_{\text{tot}}(u)$ decreases as u increases. Let us further assume that both decreases are so rapid that both integrals in Eq. 15-45 will converge after a single subtraction. Then we obtain the once-subtracted, forward dispersion relation

$$\begin{aligned}\text{Re } T_{\text{el}}(s, 0) = \text{Re } T_{\text{el}}(s_1, 0) &+ \frac{g_{aeb}^2}{s - M^2} - \frac{g_{aeb}^2}{s_1 - M_e^2}\\ &+ \frac{2(s - s_1)}{\pi} P \int_{(m_a+m_b)^2}^{+\infty} \left[\frac{p'\sqrt{s'}\sigma_{\text{tot}}(s')}{(s' - s)(s' - s_1)}\right] ds'\\ &+ \frac{2(s - s_1)}{\pi} \int_{(m_a+m_b)^2}^{+\infty} \left[\frac{p'\sqrt{u'}\sigma_{\text{tot}}(u')}{(u' - \Sigma m^2 + s)(u' - \Sigma m^2 + s_1)}\right] du'\end{aligned} \tag{15-52}$$

Exercise 15-12. Fill in the steps leading from Eq. 15-45 to Eq. 15-52.

Exercise 15-13. Evaluate Eq. 15-52 for a fictitious physical system in which both incident particles have 1 GeV/c^2 mass, both total cross sections are 50 mb at the physical threshold, both total cross sections vary as $1/s$, and in which there are no poles or branch points below the physical threshold. Compare Re T_{el} with Im T_{el}.

In actual applications of the forward dispersion relation, there is usually a close relation between the s- and u-channels. For example, if the s-channel is $\pi^- + p \rightarrow \pi^- + p$, the u-channel is $\pi^+ + p \rightarrow \pi^+ + p$. Then the s- and u-channel total cross sections can be resolved into combinations of isotopic spin total cross sections (Sec. 4-5). For some isospin combinations this yields a welcome bonus in that the s-channel and u-channel integrals in Eq. 15-52 are of opposite sign. The divergence appearing in these integrals for constant σ_{tot} is then canceled, and the once-subtracted dispersion relation can be used directly for constant σ_{tot} (15GA).

The forward elastic scattering dispersion relation has many uses.

1. The Re $T_{el}(s, 0)$ can be calculated.
2. The value of $(g_{aeb})^2$ can be determined.
3. The concept of the analyticity of scattering amplitudes can be tested by looking for experimental violations of the dispersion relation. No violations have yet been found.
4. The dispersion relation can be used to examine the consequences of theoretical assumptions about the very high energy behavior of strong interactions. An example is the Pomeranchuk theorem (15ED, 15GA), which states that the total cross section for antiparticle $\bar{a}$ on particle b is equal to the total cross section for particle a on particle b at very high energies.

The applications of the forward elastic dispersion relation comprise a large field in themselves; Refs. 15ED and 15GA can be consulted for more information.

15-10 Singularities in t, the Pole Approximation, and Vertex Form Factors.

We now consider the singularities of $T(s, t)$ in the complex t plane at fixed s, first discussed in Sec. 15-5. The s-channel reaction is again $a + b \to c + d$, and we fix s in the physical region of that reaction. Some of the singularities were illustrated in Fig. 15-6, but we now know that when t becomes sufficiently negative, u-channel singularities will also appear. The complete set of singularities appears in Fig. 15-12.

Now let us suppose that t is in the physical region, but very close to t_{min}. Figure 15-12 indicates that the singularity closest to the s-channel physical region is the pole produced by one-particle exchange in the t-channel. Therefore, we

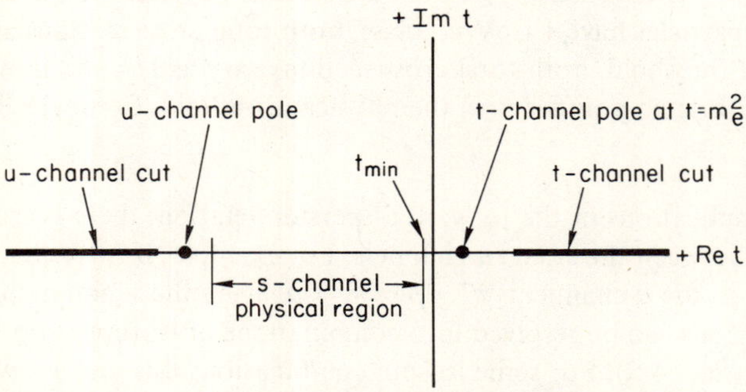

Fig. 15-12 Singularities of $T(s, t)$ in the complex t-plane with s fixed in the physical region of the s-channel.

might hope that in the high energy, very small t region, the amplitude $T_{ab \to cd}(s, t)$ will be dominated by the pole $(m_e{}^2 - t)^{-1}$. This may indeed be the case for one-pion exchange, since the relatively small size of $m_\pi{}^2$ puts the pole close to the physical region of the s-channel. The discussion in Sec. 13-8 was based on this assumption, which is called the *pole approximation.*

But for other particle exchanges, the beginning of the cut on the positive t axis is not much further away than the pole from the s-channel physical region. Thus there is no reason to ignore the influence of the cut, and this produces a much more complicated situation. Extensive attempts have been made to evaluate the influence of the cut, but there has been no broad success. Basically we know almost nothing about $T(s, t)$ in the unphysical regions containing the cuts. For example, if one writes a dispersion relation in t, he does not know how to estimate the Im T along the cuts. He does not even know how many subtractions would be necessary to obtain convergence.

A similar discussion applies when t is in the physical region and close to the u-channel singularities. The pole approximation may be valid in that region if there is a u-channel pole that is much closer to the physical region than any u-channel cut.

To summarize: the cuts in the t plane, which come from two-or-more-particle exchange diagrams are very important singularities. The elementary one-particle exchange model takes account only of the poles singularities, hence it must be a very incomplete model. Indeed, looking at Fig. 15-12, it is remarkable that such a model is even qualitatively successful.

To conclude this section, we turn from two-body reactions to three-particle vertices (Fig. 15-13a). If particles a and c are real, the form factor for this vertex (Sec. 13-2 and Ch. 19) is a function of a *single* Lorentz scalar $t = (P_a - P_c)^2$. The analytic properties of such a vertex form factor are much simpler than the analytic properties of the amplitude of a two-body reaction. But with one exception it is very difficult in strong interactions to acquire experimental information on a vertex form factor. This is because it is difficult to isolate the contribution of the form factor to the reaction amplitude. The exception

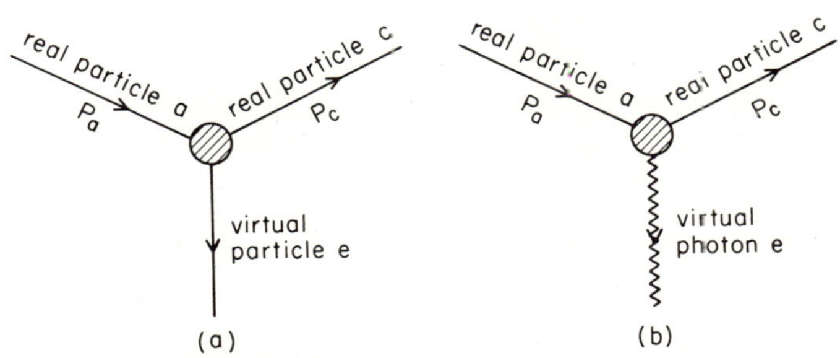

Fig. 15-13 Three particle vertices.

occurs when the virtual particle is a photon (Fig. 15-13*b*). Then we are studying the electromagnetic form factor, which can be directly measured (see Chs. 19 and 21, especially Sec. 21-3).

15-11 Relativistic Partial Wave Amplitudes and Analyticity. We conclude this chapter with a very brief section on the application of analyticity considerations to relativistic partial wave amplitudes. Thus we are returning to the subject introduced nonrelativistically in Sec. 15-2. This again is a very large field (15GA, 15ED, 15MA, 15BU) and we cannot begin to treat it adequately. For example, relativistic partial wave dispersion relations can be constructed and can be used at asymptotically high energies to set bounds on the behavior of scattering amplitudes and cross sections (15ED, 15MA). A classic example is the Froissart upper bound on the total cross section (15ED, 15MA)

$$\sigma_{\text{tot}}(s) < \text{constant} \times (\ln s)^2 \qquad \text{as} \quad s \to \infty \tag{15-53}$$

Restricting our discussion to the elastic scattering of spin zero particles, we have from Eq. 10-9 the basic relationship

$$T_{\text{el}}(s, t) = \frac{8\pi\sqrt{s}}{p} \sum_l (2l + 1) T_l(s) P_l(\cos\theta) \tag{15-54}$$

Using Eq. 15-54, we can relate the analytic properties of the partial wave amplitude $T_l(s)$ in the complex s-plane to the analytic properties of $T_{\text{el}}(s, t)$, since the analytic properties of $P_l(z)$ as a function of s and t are known. We do not carry through this analysis (15GA, 15ED, 15MA) but only sketch a bit of the results. We obtain the same general picture of the singularities of $T_l(s)$ that we obtained in the nonrelativistic situation in Sec. 15-2. The complex k^2-plane is replaced by the complex s-plane. And that s-plane is cut along most of the positive real s axis. Poles in $T_l(s)$, which lie on the real s axis but below the threshold $s = (m_a + m_b)^2$, correspond to bound states of angular momentum l. Or, using elementary particle language, these poles correspond to particles with intrinsic spin l.

Now just as we discussed in Sec. 15-2, poles in $T_l(s)$ that occur on the second sheet and near the positive real s axis correspond to strong interaction resonances. The only change is that Eq. 15-11a takes the form

$$S_l(s) = \frac{s - (\text{Re}\, s_0 + i\, \text{Im}\, s_0)}{s - (\text{Re}\, s_0 - i\, \text{Im}\, s_0)} \tag{15-55a}$$

Then defining $m_0^2 = \text{Re}\, s_0$, $m_0\Gamma = \text{Im}\, s_0$, we obtain a more general justification of the relativistic Breit–Wigner formula first derived in Sec. 5-9:

$$T_l(s) = \frac{-m_0\Gamma}{(s - m_0^2) + im_0\Gamma} \tag{15-55b}$$

Exercise 15-14. Fill in the steps leading to Eq. 15-55b.

Exercise 15-15. We can write Eq. 15-55a, and analogously Eq. 15-11a, in the more general form

$$S_l(s) = \left[\frac{s - (\text{Re}\, s_0 + i\, \text{Im}\, s_0)}{s - (\text{Re}\, s_0 - i\, \text{Im}\, s_0)}\right] A(s) \tag{15-56}$$

where $A(s)$ is a slowly varying function of s with the restriction

$$|A(s)|^2 = 1 \qquad \text{or} \qquad A(s) = e^{2i\delta_B(s)}$$

Show that this leads to the relativistic form of the Breit–Wigner formula with nonresonant background (Eq. 5-36).

REFERENCES

15BJ J. D. Bjorken and S. D. Drell, *Relativistic Quantum Mechanics* and *Relativistic Quantum Fields* (McGraw-Hill, New York, 1965).

15BU H. Burkhardt, *Dispersion Relation Dynamics* (North-Holland, Amsterdam, 1969).

15DE V. de Alfaro and T. Regge, *Potential Scattering* (Wiley, New York, 1965).

15ED R. J. Eden, *High Energy Collisions of Elementary Particles* (Cambridge University Press, Cambridge, 1967).

15FR S. C. Frautschi, *Regge Poles and S-Matrix Theory* (Benjamin, New York, 1963).

15GA S. Gasiorowicz, *Elementary Particle Physics* (Wiley, New York, 1966).

15MA A. D. Martin and T. D. Spearman, *Elementary Particle Physics* (North-Holland, Amsterdam, 1970).

15OM R. Omnès, *Introduction to Particle Physics* (Wiley, New York, 1970).

15OMN R. Omnès and M. Froissart, *Mandelstam Theory and Regge Poles* (Benjamin, New York, 1970).

CHAPTER

16

Crossing, Intrinsic Spin, and Isotopic Spin

16-1 Introduction. In the last chapter we developed the very general concepts of crossing and analyticity. To simplify the discussion, we considered only particles with zero intrinsic spin and we completely ignored isotopic spin. But once one begins to study real two-body reactions, intrinsic spin and isotopic spin must be dealt with. No fundamentally new ideas are required, but we must develop formal manipulative techniques that are often tedious. In keeping with the spirit of this book, we limit our discussion to the basic ideas behind the techniques and a simple example; hence this is a short chapter.

In Sec. 16-2 we provide the foundation for the example, the reactions $N + \pi \rightarrow N + \pi$ and $N + \bar{N} \rightarrow \bar{\pi} + \pi$, by constructing the general helicity amplitudes for those reactions. An overall view of the crossing of helicity amplitudes is presented in Sec. 16-3, and the crossing matrix is constructed for our example. In the concluding section, the isotopic spin relations between crossed reactions are discussed.

Phases are a real headache in the calculation of the crossing relations of helicity amplitudes. In the classic paper of Trueman and Wick (16TR), much space is devoted to phase questions. To keep from obscuring the main arguments, and also out of cowardice, we avoid phase questions.

16-2 Crossing in the Reactions $N+\pi\to N+\pi$ and $N+\bar{N}\to\bar{\pi}+\pi$. We begin with the s-channel reaction

$$s\text{-channel} \equiv \text{channel I:}\qquad N+\pi\to N+\pi \tag{16-1a}$$

$$1+2\to 3+4$$

$$P_1+P_2=P_3+P_4 \tag{16-1b}$$

The most general form of the amplitude is

$$T^{\mathrm{I}}_{\lambda_3,\lambda_1}(s,t)=\bar{u}^{\lambda_3}(\mathbf{p}_3)T^{\mathrm{I}}u^{\lambda_1}(\mathbf{p}_1) \tag{16-2a}$$

where

$$T^{\mathrm{I}}=f_1+f_2V^{\mu}\gamma_{\mu}+f_3Z^{\mu\nu}[\gamma_{\mu},\gamma_{\nu}]+f_4W^{\mu}\gamma_{\mu}\gamma_5+f_5P\gamma_5 \tag{16-2b}$$

Here the f's are scalar functions of s and t; that is, $f_i\equiv f_i(s,t)$. The quantities V^{μ}, $Z^{\mu\nu}$, W^{μ}, and P are, respectively, a vector, a tensor, a pseudovector, and a pseudoscalar in four-dimensional space; they are to be constructed out of the four-vectors P_i, $i=1,2,3,4$. Using parity conservation (Eq. 16-1b) and $\not{P}_1u_1=Mu_1$, Eq. 16-2b reduces to just two independent terms

$$T^{\mathrm{I}}=\frac{1}{2M}\left[A(s,t)+B(s,t)(\not{P}_2+\not{P}_4)/2M\right] \tag{16-3}$$

where $A(s,t)$ and $B(s,t)$ are Lorentz scalars replacing the $f_i(s,t)$'s in Eq. 16-2b. Equation 16-3 is the conventional form of the general amplitude. The second factor of $2M$ occurs to give A and B the same dimensions; the first $2M$ occurs because of our spinor normalization convention; M is the nucleon mass.

Exercise 16-1. Show that Eq. 16-2b reduces to Eq. 16-3. Use the theorem that the four-dimensional pseudoscalar $\epsilon^{\alpha\beta\gamma\delta}A_{\alpha}B_{\beta}C_{\gamma}D_{\delta}$ (App. A) will be nonzero only if the four-vectors A, B, C, D are all linearly independent.

Equation 16-3 is basic to our considerations, for we assume that the scalar functions $A(s,t)$ and $B(s,t)$ have only the kind of dynamic singularities discussed in the previous chapter. We assume that they are analytic functions that can be extended by analytic continuation from the physical region of the reaction in Eq. 16-1a to the physical region of the crossed reactions. All the spin effects are contained in the spinors and matrices of Eqs. 16-2a and 16-3.

We calculate $T_{++}(s, t)$ and $T_{-+}(s, t)$ in the reaction I center of mass system, taking the nucleons as the first particle in the helicity formalism. Using the spin functions in Eq. 12-6, we obtain

$$T^{\mathrm{I}}_{++}(s, t) = \cos \tfrac{1}{2}\theta_{\mathrm{I}} \left[A(s, t) + \left(\frac{Ee + p^2}{M^2} \right) B(s, t) \right] \tag{16-4a}$$

$$T^{\mathrm{I}}_{-+}(s, t) = -\sin \tfrac{1}{2}\theta_{\mathrm{I}} \left[\left(\frac{E}{M} \right) A(s, t) + \left(\frac{e}{M} \right) B(s, t) \right] \tag{16-4b}$$

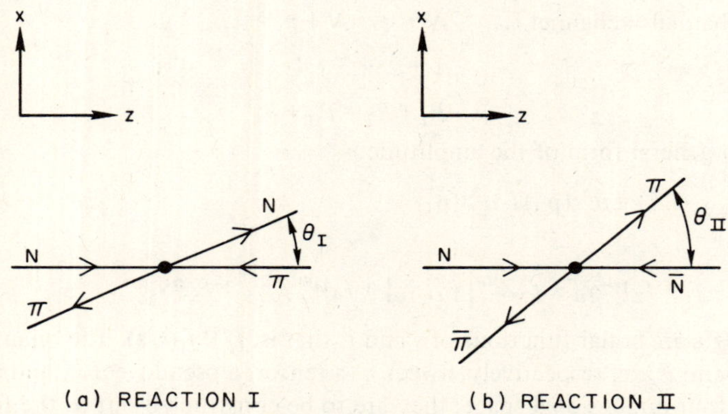

Fig. 16-1 The kinematic situation for (*a*) reaction I, $N + \pi \to N + \pi$; (*b*) reaction II, $N + \overline{N} \to \overline{\pi} + \pi$; in their respective centers of mass.

All quantities are in the center of mass of reaction I (see Fig. 16-1*a*); E and e are the energies of the nucleon and pion, respectively; M and m are the respective masses of the nucleon and pion; and p is the magnitude of the momentum in the center of mass. We restrict the reaction to the x-z plane to keep ϕ at zero and simplify phase considerations.

Exercise 16-2. Derive Eqs. 16-4 and 16-5.

In terms of s, t, and u, we have

$$E = \frac{s + M^2 - m^2}{2\sqrt{s}}, \qquad e = \frac{s + m^2 - M^2}{2\sqrt{s}} \tag{16-5a}$$

$$p^2 = \frac{[s - (M + m)^2]\,[s - (M - m)^2]}{4s} \tag{16-5b}$$

$$\cos \tfrac{1}{2}\theta_{\mathrm{I}} = \frac{[(M^2 - m^2)^2 - su]^{1/2}}{\{[s - (M + m)^2][s - (M - m)^2]\}^{1/2}} \tag{16-5c}$$

$$\sin \tfrac{1}{2}\theta_{\mathrm{I}} = \frac{[-st]^{1/2}}{\{[s-(M+m)^2][s-(M-m)^2]\}^{1/2}} \tag{16-5d}$$

Before considering the crossed reaction, let us study the kinematic singularities in the amplitudes of Eq. 16-4. The $\cos \frac{1}{2}\theta_{\mathrm{I}}$ term in T^{I}_{++} has branch points at $s = (M+M)^2$, $s = (M-m)^2$, and $su = (M^2 - m^2)^2$. There are no other kinematic singularities in T^{I}_{++}. Hence these are the kinematic singularities in T^{I}_{++}. In T^{I}_{-+} the $\sin \frac{1}{2}\theta_{\mathrm{I}}$ term introduces kinematic singularities at $s = 0$ and $t = 0$, as well as at $s = (M+m)^2$ and $s = (M-m)^2$. The E and e factors also introduce a kinematic singularity at $s = 0$.

Exercise 16-3. Verify the foregoing singularity statements. Show that the $su = (M^2 - m^2)^2$ singularity corresponds to $t = -4p^2$. Show that $(Ee + p^2)$ does not introduce a singularity.

To construct amplitudes that are free of kinematic singularities, we define

$$F^{\mathrm{I}}_{++}(s,t) = (\cos \tfrac{1}{2}\theta_{\mathrm{I}})^{-1} T^{\mathrm{I}}_{++}(s,t) \tag{16-6a}$$

$$F^{\mathrm{I}}_{-+}(s,t) = [s^{1/2}](\sin \tfrac{1}{2}\theta_{\mathrm{I}})^{-1} T^{\mathrm{I}}_{-+}(s,t) \tag{16-6b}$$

Note the necessity for the $s^{1/2}$ factor in the second amplitude.

Now we turn to the crossed reaction

$$t\text{-channel} \equiv \text{channel II:} \qquad N + \bar{N} \to \bar{\pi} + \pi \tag{16-7a}$$

$$1 + \bar{3} \to \bar{2} + 4$$

$$Q_1 + Q_{\bar{3}} = Q_{\bar{2}} + Q_4 \tag{16-7b}$$

Here we use Q_i for the four-momenta. In the crossing, P_2 is replaced by $-Q_{\bar{2}}$; hence the most general amplitude is now

$$T^{\mathrm{II}}_{\lambda_3\lambda_1} = \frac{\bar{v}^{\lambda_3}(\mathbf{q}_{\bar{3}})[A(s,t) + B(s,t)(-\not{Q}_{\bar{2}} + \not{Q}_4)/2M]u^{\lambda_1}(\mathbf{q}_1)}{2M} \tag{16-8}$$

The amplitudes in the channel II cms are

$$T^{\mathrm{II}}_{++} = \left[-\left(\frac{q_N}{M}\right)A(s,t) + \left(\frac{q_\pi}{M}\right)\cos\theta_{\mathrm{II}}\, B(s,t)\right] \tag{16-9a}$$

$$T^{\mathrm{II}}_{-+} = -\left(\frac{q_\pi \epsilon}{M^2}\right)\sin\theta_{\mathrm{II}}\, B(s,t) \tag{16-9b}$$

In Fig. 16-1b, which represents the kinematic situation, q_N and q_π are the magnitudes of the center of mass momenta of the baryons and pions, respectively, ϵ is the center of mass energy of one of the particles, and $\bar{N}$ has been taken as the "second particle" in the helicity amplitude.

We next express these quantities in terms of the *channel I s* and *t* kinematic variables:

$$s = (P_1 + P_2)^2 = (Q_1 - Q_{\bar{2}})^2$$
$$t = (P_1 - P_3)^2 = (Q_1 + Q_{\bar{3}})^2 \tag{16-10}$$

Then we have

$$q_N = \left(\frac{t}{4} - M^2\right)^{1/2}, \qquad q_\pi = \left(\frac{t}{4} - m^2\right)^{1/2}, \qquad \epsilon = \left(\frac{t}{4}\right)^{1/2} \tag{16-11a}$$

$$q_\pi \cos\theta_{\text{II}} = \frac{s + t/2 - m^2 - M^2}{[t - 4M^2]^{1/2}} \tag{16-11b}$$

$$q_\pi \sin\theta_{\text{II}} = \frac{[su - (m^2 - M^2)^2]^{1/2}}{[t - 4M^2]^{1/2}} \tag{16-11c}$$

There are kinematic singularities at $t = 0$, $t = 4M^2$ and $su = (M^2 - m^2)^2$, and the singularity-free amplitudes are

$$F^{\text{II}}_{++} = q_N T^{\text{II}}_{++} \tag{16-12a}$$
$$F^{\text{II}}_{-+} = [\epsilon q_\pi \sin\theta_{\text{II}}]^{-1} T^{\text{II}}_{-+} \tag{16-12b}$$

Exercise 16-4. Fill in the steps leading to Eqs. 16-9 through 16-12.

16-3 Crossing Relations for Helicity Amplitudes. In Sec. 15-5 we observed that the crossing relation was very simply stated for two-body reactions involving only particles with zero spin. But when some of the particles have nonzero spins, the crossing relation for helicity amplitudes is not simple. This is to be expected, for helicities are defined according to the direction of motion of the particles; and these directions change drastically when crossing from one physical region to another physical region. As an aside, we note that there is another way to describe spins in a relativistic, two-body, situation—the *transversity formalism*—in which spins are quantized along an axis perpendicular to the reaction plane (16CO). This leads to much simpler crossing relations, but it has other complications and is rarely used.

We are about to remove the reaction II center of mass, helicity amplitudes $T^{\text{II}}_{\lambda_{\bar{3}}\lambda_1}$ (Eq. 16-9) from the reaction II center of mass frame and transform them to the reaction I center of mass frame. Therefore, these amplitudes will be taken from the physical region of reaction II to the physical region of reaction I. The amplitudes $T^{\text{II}}_{\lambda_{\bar{3}}\lambda_1}$ will thereby become unphysical. We then write the reaction I center of mass amplitudes $T^{\text{I}}_{\lambda_3,\lambda_1}$ (Eq. 16-4) in terms of these unphysical amplitudes. For very detailed discussion of the general case, refer to Refs. 16TR and 16MA.

Using Eqs. 16-5 and 16-11, we eliminate $A(s, t)$ and $B(s, t)$ between Eqs. 16-4 and 16-9. The relations thus obtained between the reaction I amplitudes and the reaction II amplitudes are most conveniently expressed by a *crossing matrix C* where

$$C = \begin{bmatrix} C_{11} & C_{12} \\ C_{21} & C_{22} \end{bmatrix} \tag{16-13a}$$

and

$$\begin{aligned} T^{\mathrm{I}}_{++} &= C_{11}T^{\mathrm{II}}_{++} + C_{12}T^{\mathrm{II}}_{-+} \\ T^{\mathrm{I}}_{-+} &= C_{21}T^{\mathrm{II}}_{++} + C_{22}T^{\mathrm{II}}_{-+} \end{aligned} \tag{16-13b}$$

We will find that C is actually a rather simple matrix because it is *orthogonal.* Therefore, it is conventional to write

$$C = \eta \begin{bmatrix} \sin\chi & \cos\chi \\ -\cos\chi & \sin\chi \end{bmatrix} \tag{16-14}$$

where η is some phase and $\sin^2\chi + \cos^2\chi = 1$, *although* χ is *not* a physical angle. By proper definition of the relative phase between the two sets of amplitudes, we can obtain $\eta = 1$. We have not bothered to do this, however; hence we have $|\eta|^2 = 1$ but $\eta \neq 1$.

To find χ is tedious, but we can simplify the calculation by setting $M = m$. Then we write

$$\sin\chi = \frac{2m\sqrt{u}}{[(t - 4m^2)(s - 4m^2)]^{1/2}} \tag{16-15a}$$

$$\cos\chi = \frac{\sqrt{st}}{[(t - 4m^2)(s - 4m^2)]^{1/2}} \tag{16-15b}$$

$$\begin{aligned} T^{\mathrm{I}}_{++}(s, t) &= \frac{\eta}{[(t - 4m^2)(s - 4m^2)]^{1/2}} \\ &\quad \times [2m\sqrt{u}T^{\mathrm{II}}_{++}(s, t) + \sqrt{st}\, T^{\mathrm{II}}_{-+}(s, t)] \\ T^{\mathrm{I}}_{-+}(s, t) &= \frac{\eta}{[(t - 4m^2)(s - 4m^2)]^{1/2}} \\ &\quad \times [-\sqrt{st}\, T^{\mathrm{II}}_{++}(s, t) + 2m\sqrt{u}\, T^{\mathrm{II}}_{-+}(s, t)] \end{aligned} \tag{16-16}$$

Thus the $T^{\mathrm{I}}_{\lambda_3\lambda_1}(s, t)$ amplitudes in the physical region I center of mass frame are written in terms of the $T^{\mathrm{II}}_{\lambda_{\bar{3}}\lambda_1}(s, t)$ amplitudes. But the $T^{\mathrm{II}}_{\lambda_{\bar{3}}\lambda_1}(s, t)$ amplitudes are unphysical for the values of t and s in Eq. 16-16. We can only use them explicitly, if we know enough about the Lorentz scalar amplitudes $A(s, t)$ and $B(s, t)$ (Sec. 16-2). In any case, we see that the effect of the crossing is to *mix* helicity states.

Exercise 16-5. Derive Eqs. 16-15 and 16-16. Show that $\cos \chi$ can vary from zero to infinity. Hence show that χ is not a physical angle.

We conclude this section with the following observations.

1. The dimension of the crossing matrix C is equal to the number of independent helicity amplitudes. Thus in the elastic scattering of one spin $\frac{1}{2}$ particle on a different spin $\frac{1}{2}$ particle, as discussed in Sec. 10-5, C will be a 6×6 matrix. In general, all the helicity amplitudes will mix on crossing.

2. The transformation of the T^{II} amplitudes from the center of mass frame of reaction II to the center of mass frame of reaction I can be interpreted as a generalized Wigner rotation (Secs. 9-8, 9-9; Refs. 16TR, 16MA).

3. The orthogonality of the crossing matrix leads to the following relation between the center of mass helicity amplitudes of crossed reaction I and II:

$$\sum_{\mu_1\mu_2\mu_3\mu_4} |T^{\mathrm{I}}_{34,12}(s,t)|^2 = \sum_{\mu_1\mu_{\bar{2}}\mu_{\bar{3}}\mu_4} |T^{\mathrm{II}}_{\bar{2}4,1\bar{3}}(s,t)|^2 \tag{16-17}$$

Exercise 16.6. What does Eq. 16-17 imply about the cross sections of crossed reactions?

16-4 Crossing and Isotopic Spin. Finally, let us discuss the isotopic spin relations between *crossed* reactions. Consider

$$\begin{aligned} &s\text{-channel reaction:} \qquad a + b \to c + d \\ &t\text{-channel reaction:} \qquad a + \bar{c} \to \bar{b} + d \end{aligned} \tag{16-18}$$

Ignoring all parameters but the isotopic spin state $|I_i, m_i\rangle$ of the particles, we can resolve the reaction amplitudes into isotopic spin conserving amplitudes, as discussed in Ch. 4. Explicitly, we have

$$\langle m_c m_d | T^{(s)} | m_a m_b \rangle = \sum_{\mathrm{I}} C^{\mathrm{I}_a \mathrm{I}_b \mathrm{I}}_{m_a m_b M} C^{\mathrm{I}_c \mathrm{I}_d \mathrm{I}}_{m_c m_d M} T^{(s)}(\mathrm{I}) \tag{16-19a}$$

$$\langle m_{\bar{b}} m_d | T^{(t)} | m_a m_{\bar{c}} \rangle = \sum_{J} C^{\mathrm{I}_a \mathrm{I}_{\bar{c}} \mathrm{J}}_{m_a m_{\bar{c}} M} C^{\mathrm{I}_{\bar{b}} \mathrm{I}_d \mathrm{J}}_{m_a m_d M} T^{(t)}(J) \tag{16-19b}$$

where $C^{\mathrm{I}_a \mathrm{I}_b \mathrm{I}}_{m_a m_b M}$ is a Clebsch–Gordan coefficient. Then under crossing we demand

$$\langle m_c m_d | T^{(s)} | m_a m_b \rangle = \eta \langle m_{\bar{b}} m_d | T^{(t)} | m_d m_{\bar{c}} \rangle \tag{16-20}$$

where η is a phase that is independent of I_i, m_i, and I. By $m_{\bar{a}}$ we mean the I_z of antiparticle $\bar{a}$. The *isotopic spin crossing matrix* X is defined by

$$T^{(s)}(I) = \sum_J X_{IJ} T^{(t)}(J) \tag{16-21}$$

Reference 16CA provides a very general treatment and tables of isotopic spin crossing matrices. But our purpose is to demonstrate the basic idea, and we proceed by an example.

The general method is to equate the right-hand sides of Eqs. 16-19 and obtain a set of simultaneous equations by varying the I_z of one or more of the reactions. For example, consider the reactions in Sec. 16-2

$$s\text{-channel reactions:} \qquad \pi^- + p \to \pi^- + p, \qquad \pi^+ + p \to \pi^+ + p \tag{16-22a}$$

$$t\text{-channel reactions:} \qquad \pi^- + \pi^+ \to \bar{p} + p, \qquad \pi^+ + \pi^- \to \bar{p} + p \tag{16-22b}$$

We easily obtain from a table of Clebsch–Gordan coefficients

$$\langle \pi^- p | T^{(s)} | \pi^- p \rangle = (\tfrac{1}{3}) T^{(s)}(\tfrac{3}{2}) + (\tfrac{2}{3}) T^{(s)}(\tfrac{1}{2}); \quad \langle \pi^+ p | T^{(s)} | \pi^+ p \rangle = T^{(s)}(\tfrac{3}{2})$$

$$\langle \pi^\pm \pi^\mp | T^{(t)} | \bar{p} p \rangle = \pm(\tfrac{1}{2}) T^{(t)}(1) - (1/\sqrt{6}) T^{(t)}(0)$$

Therefore we can write

$$X = \eta \begin{bmatrix} -1/\sqrt{6} & -1 \\ -1/\sqrt{6} & +\frac{1}{2} \end{bmatrix} \qquad \begin{matrix} \text{rows: } I^{(s)} = \frac{1}{2}, \frac{3}{2} \\ \text{columns: } I^{(t)} = 0, 1 \end{matrix} \tag{16-23}$$

Note that the signs of the matrix elements depend on which particle is crossed.

In general, the isotopic spin crossing matrix X is *not* diagonal. Therefore, isotopic spin amplitudes *mix* on crossing. The isotopic spin crossing relations are useful when the isotopic spin amplitudes are simpler or more easily understood in one channel than in the crossed channel. For example, suppose we know that the t-channel reaction in Eq. 16-22b proceeds only through the ρ^0 meson. Then $T^{(t)}(0) = 0$ and

$$T^{(s)}(\tfrac{1}{2}) = -2T^{(s)}(\tfrac{3}{2})$$

Of course there is nothing magical about this result; we could have obtained the same thing by using the one-particle exchange model of Ch. 12, assuming only a ρ^0 meson had been exchanged.

Exercise 16-7. Derive the classic result for the isotopic spin crossing matrix in the crossed reactions $\pi + \pi \to \pi + \pi$ and $\bar{\pi} + \pi \to \pi - \bar{\pi}$.

$$X = \eta \begin{bmatrix} \frac{1}{3} & 1 & \frac{5}{3} \\ \frac{1}{3} & \frac{1}{2} & -\frac{5}{6} \\ \frac{1}{3} & -\frac{1}{2} & \frac{1}{6} \end{bmatrix} \tag{16-24}$$

REFERENCES

16CA P. A. Carruthers, *Spin and Isospin in Particle Physics* (Gordon and Breach, New York, 1971).

16CO G. Cohen-Tannoudgi et al., *Ann. Phys. (N.Y.)* **46**, 239 (1968).

16MA A. D. Martin and T. D. Spearman, *Elementary Particle Physics* (North-Holland, Amsterdam, 1970).

16TR T. L. Trueman and G. C. Wick, *Ann. Phys. (N.Y.)* **26**, 322 (1964).

CHAPTER

17

Regge Theory and the Interactions of Spinless Particles

17-1 Introduction. An immense mathematical structure is associated with the applications of Regge theory to potential scattering (17DE, 17NE) and to strong interactions (17COL, 17CO). To discuss this structure in detail would require a book as long as this one. Obviously we have not written it. Moreover, we have allotted a relatively small amount of space to the foundations of Regge theory and to the complications introduced into the theory by nonzero spin particles, for in our view it is at present impossible to assess the validity of many details of the theory. Regge theory is based on a beautiful idea—poles in the S-matrix in the complex angular momentum plane—and it is rigorous in nonrelativistic potential scattering. When applied to strong interactions it provides the only general explanation of the energy behavior of two-body inelastic cross sections. It also represents a very convenient way to parameterize high energy, two-body data. But we do not know how seriously to take the detailed predictions of the theory, particularly those produced by the spin-induced modifications of the theory (see Ch. 18). For example, we do not understand the comparative importance of these spin-induced modifications relative to absorptive corrections. Therefore our plan in this chapter is to develop as quickly as possible the major features of the Regge theory as applied to the interactions of spin zero particles. In Ch. 18 we discuss briefly how nonzero spin modifies the theory and consider the validity of some detailed predictions of the theory.

In Sec. 17-2 we develop the fundamental thesis of the Regge theory; partial wave amplitudes may be studied as analytic functions of the angular momentum treated as a complex variable. These amplitudes exhibit simple poles, called Regge poles, in the complex angular momentum plane. The position of a Regge pole changes with energy, tracing out a Regge trajectory, which may lead to s-channel resonances. The nonrelativistic potential scattering justification for Regge theory is presented in Sec. 17-3. Some justification for the use of Regge theory in strong interactions is offered in Sec. 17-4, where the Regge trajectory classification of hadrons is discussed. This is one very important application of Regge theory.

In Secs. 17-5 through 17-7 we derive another very important application of Regge theory to strong interactions. We show that Regge poles in the t-channel can be used to predict the form of the amplitude in a high energy, small t, two-body reaction in the s-channel. Thus we develop the Regge model for high-energy, two-body, inelastic reactions. We investigate the connections between this Regge model and one-particle exchange models in Secs. 17-7 and 17-8. In Sec. 17-8 we also briefly discuss a field theory interpretation of the Regge trajectory.

The chapter concludes with two special sections. Section 17-9 is a concise treatment of the problem of applying the Regge model to high energy diffractive elastic scattering. Section 17-10 is a crude derivation of the multiperipheral model of many-body reactions using Regge phenomenology.

17-2 Regge Poles and Resonances in the s-Channel. We begin at once with our old friend, the partial wave analysis of the two-body reaction, elastic or inelastic,

$$a + b \to c + d \tag{17-1}$$

$$P_a + P_b = P_c + P_d$$

$$s = (P_a + P_b)^2, \qquad t = (P_a - P_c)^2$$

We have noted the four-momenta and the usual kinematic variables s and t. We have learned from dealing with the analytic behavior of reaction amplitudes that it is very useful to simplify the exposition by assuming that particles a, b, c, and

d have zero spins and equal masses m, and indeed we make this assumption. The discussion is relativistic, for as we have also learned, one great advantage of the partial wave analysis is that it is easily applied to relativistic two-body reactions. Consider in the center of mass

$$T(s, t) = \sum_{l=0}^{\infty} (2l + 1)T_l(s)P_l(\cos \theta) \qquad (17\text{-}2)$$

In this chapter the factor $(8\pi\sqrt{s}/\sqrt{p_i^* p_f^*})$ (Eq. 10-9b) is included in $T_l(s)$, and we can consider the analytic properties of the entire expression.

In Ch. 15 we studied the analytic properties of $T_l(s)$ as a function of *complex* s. We briefly demonstrated that the singularities of $T_l(s)$ on the complex s-plane were fairly simple in nature and position; we showed that these singularities had direct physical significance, and we indicated that they can be used as guides to the general behavior of $T_l(s)$. The Regge pole hypothesis is also based on a study of the analytic properties of $T_l(s)$. But instead of allowing s to be complex, we keep s real and do something more drastic. We allow l—the angular momentum—to be a *continuous and complex variable*. This is certainly more drastic because physically l is restricted to integral and half-integral values. Our approach to introducing such an unusual idea is itself drastic. We immediately assume that the idea is physically reasonable, and we deduce some of its consequences. Then we try to justify the assumption using (1) nonrelativistic potential scattering theory and (2) the regularities found in the masses and internal quantum numbers of strong interaction resonances.

Following a common convention, we use the symbol α to denote complex values of the angular momentum. For clarity, in this chapter we reserve the symbol l for integral values of the angular momentum.

Then we assume that in the α-plane, the complex angular momentum plane, there exists an analytic function $T(\alpha, s)$ with the property

$$T(\alpha, s)_{\alpha=l} = T_l(s), \qquad l = 0, 1, 2, \ldots \qquad (17\text{-}3)$$

Then Eq. 17-2 becomes

$$T(s, t) = \sum_{l=0}^{\infty} (2l + 1)T(\alpha, s)_{\alpha=l}P_l(\cos \theta) \qquad (17\text{-}4)$$

To completely define $T(\alpha, s)$ we must know the nature and position of the singularities of T in the α-plane. In strong interaction theory we do not know how to rigorously calculate these singularities; therefore, we simplify the problem for ourselves. For this entire chapter we assume that the only singularities are *simple poles*, called *Regge poles*. Indeed, this was assumed when Regge theory was being applied to strong interactions in the early and middle 1960s. Although it was

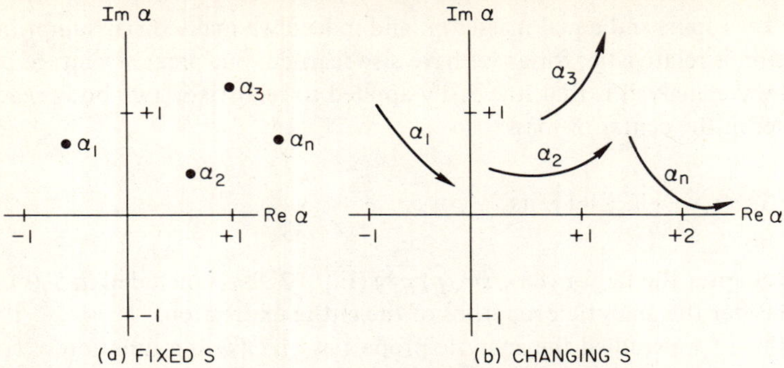

Fig. 17-1 The points $\alpha_1, \alpha_2, \ldots\ldots \alpha_n$ indicate the position of simple poles in the amplitude T in the complex angular momentum plane – the α-plane. (*a*) The positions at fixed value of s. (*b*) The positions change continuously as s changes.

known that branch points could also exist, we had hoped that their physical effects could be ignored. As we discuss in Ch. 18, this was a false hope; and branch points of $T(\alpha, s)$ in the α-plane must also be considered. These are the *Regge cuts*. But let us limit our discussion for the present to Regge poles.

Figure 17-1*a* indicates that at a *fixed* value of s there are a number of simple poles at the points $\alpha_1(s), \alpha_2(s), \ldots \alpha_n(s), \ldots$. If we change s, $\alpha_n(s)$ will in general also change. (We keep s real in this discussion, although the more general case of complex s might also be considered). Figure 17-1*b* illustrates how $\alpha_n(s)$ might change as s changes. These paths traced out by the *moving Regge poles* are called *Regge trajectories.* There is a trajectory for each pole, and there is no a priori relation between the trajectories. Remembering that α is complex, we can also represent these trajectories by the diagrams in Fig. 17-2. And there is no a priori relation between Re $\alpha(s)$ and Im $\alpha(s)$.

Consider a particular partial wave L. When all the Regge poles are relatively far from L, $|\alpha_n(s) - L| \geqslant 1$, we can say nothing specific about $T(\alpha, s)_{\alpha=L}$. But

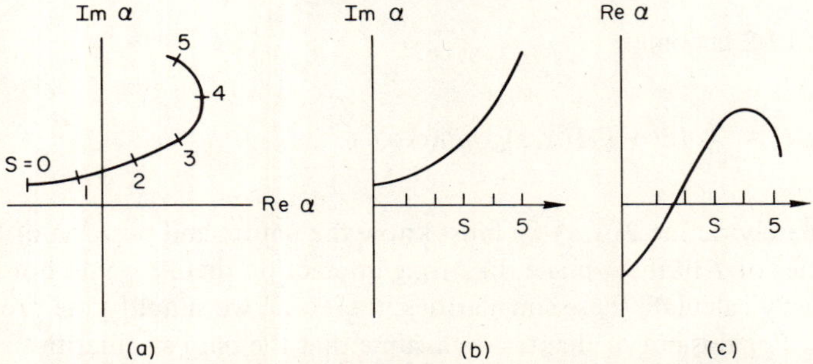

Fig. 17-2 Two ways to indicate the path of a Regge trajectory. Use either (*a*) alone or (*b*) and (*c*) together.

suppose as s changes, one of the trajectories, $\alpha_n(s)$ passes *close* to L. By "close" we mean that if Re $\alpha_n(s_0) = L$, Im $\alpha_n(s_0) \ll 1$. Then when s is in the neighborhood of s_0, $T(\alpha, s)_{\alpha=L}$ will be dominated by the Regge pole. Now for α close to α_n we can always write

$$T(\alpha, s) = \frac{A(s)}{\alpha - \alpha_n(s)} + B(s) \tag{17-5}$$

where $A(s)$ and $B(s)$ are slowly varying functions of s. Also, when α_n is close to $\alpha \equiv L$

$$\text{Re } \alpha_n(s) \approx L + \left(\frac{d \text{ Re } \alpha_n(s)}{ds}\right)_{s=s_0} (s - s_0)$$

since Re $\alpha_n(s_0) = L$.

Hence we obtain

$$T_L(s) = \frac{A'(s)}{(s - s_0) + i\sqrt{s_0}\Gamma} + B(s) \tag{17-6}$$

where $A'(s)$ is another slowly varying function of s, and

$$\sqrt{s_0}\Gamma = \frac{\text{Im } \alpha(s_0)}{\left[\dfrac{d \text{ Re } \alpha_n(s)}{ds}\right]_{s=s_0}} \tag{17-7}$$

But this is just the relativistic Breit–Wigner formula, Eq. 5-38, plus a slowly varying background term. Therefore, as discussed in Secs. 5-7 to 5-9, Eq. 17-5 gives the amplitude for an s-channel resonance. To summarize, a Regge trajectory passing close to an integral value of α describes a strong interaction resonance.

Exercise 17-1. Show that if a Regge pole is not closer to one integer than to all other integers, it is not possible to assign a unique integral spin to the pole. Hence it cannot be interpreted as a resonance with unique spin.

Exercise 17-2. Show that a bound state of intrinsic spin L will occur if a Regge trajectory passes *through* the point $\alpha = L$ for a value of $s = s_B < 4m^2$. Recall that particles a, b, c, and d each have the mass m.

17-3 Regge Poles in Nonrelativistic Potential Scattering. To show that our discussion in the previous section was not pure fantasy, we turn to a realm of physics in which we know how to calculate everything—nonrelativistic elastic scattering

$$a + b \rightarrow a + b \tag{17-8}$$

by a spherically symmetric potential $V(r)$. Regge poles in potential scattering are thoroughly discussed in Refs. 17NE and 17DE.

When using potential scattering to gain insight into strong interaction dynamics, we like to use the Yukawa potential

$$V(r) = \frac{\pm\lambda\, e^{-r/a}}{r} \tag{17-9}$$

Unfortunately, there is no simple expression for $T(\alpha, s)$ for this potential. But $T(\alpha, s)$ indeed has *simple poles*, that is, *Regge poles*. The positions of these poles must be found by numerical calculations. A typical Regge trajectory for one of

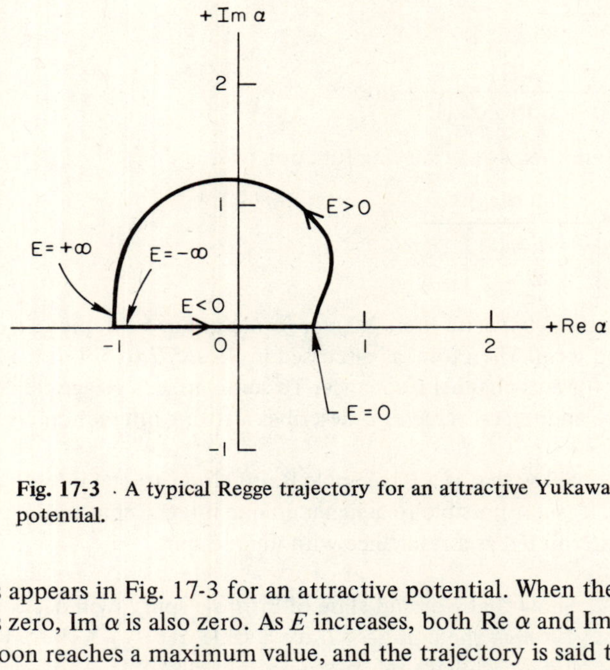

Fig. 17-3 . A typical Regge trajectory for an attractive Yukawa potential.

these poles appears in Fig. 17-3 for an attractive potential. When the total kinetic energy E is zero, Im α is also zero. As E increases, both Re α and Im α increase. But Re α soon reaches a maximum value, and the trajectory is said to *turn over.*

To be more quantitative, let p be the momentum in the center of mass and μ the reduced mass, such that $p^2 = 2\mu E$. Next we define the dimensionless quantities

$$G = \frac{2\lambda\mu a}{\hbar^2}, \qquad k = \frac{pa}{\hbar} \tag{17-10}$$

where a is the range of the potential as given in Eq. 17-9. The shape of the trajectory will, of course, depend on λ and a, and in Fig. 17-4 we show how the shape changes for various values of λ and a for a *single* Regge pole. We emphasize

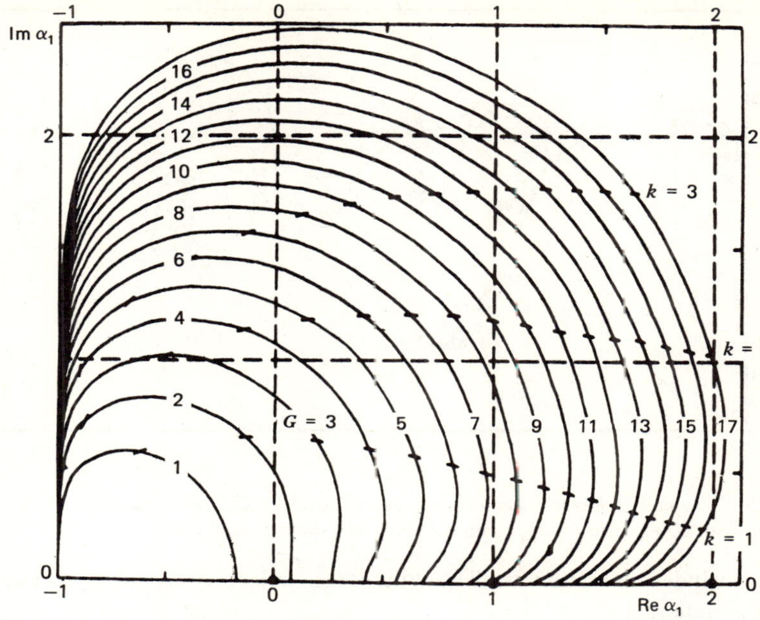

Fig. 17-4 The dependence of the shape of an attractive, Yukawa potential, Regge trajectory on the strength and range of the potential. For the definitions of G and k, see the text (17LO).

that this is not a family of Regge trajectories; it is a single Regge trajectory with different values of λ and a. We also note that a does not appear explicitly but is contained in both G and k. For most combinations of λ and a, the trajectory does not pass very close to α = integer. Hence, applying the discussion of the previous section, we can state that usually no s-channel resonances occur. But consider the $G = 8$ trajectory. When k is slightly greater than zero, that trajectory passes close to $\alpha = 1$. This means that the amplitude will exhibit a spin 1, s-channel resonance when p is small compared with $\hbar/a$.

Exercise 17-3. Carry out a similar discussion when G is about 17. Discuss the significance of the portions of the trajectories in which E is less than zero. Are there bound states?

Thus far we have exhibited the so-called *leading trajectory*, the trajectory having the largest value of Re α at each value of k. In Fig. 17-5 for $G = 2$ and an attractive potential we show the leading trajectory and three more trajectories. The same leading trajectory has already appeared in Fig. 17-4. Clearly, Regge trajectories can have quite complicated shapes.

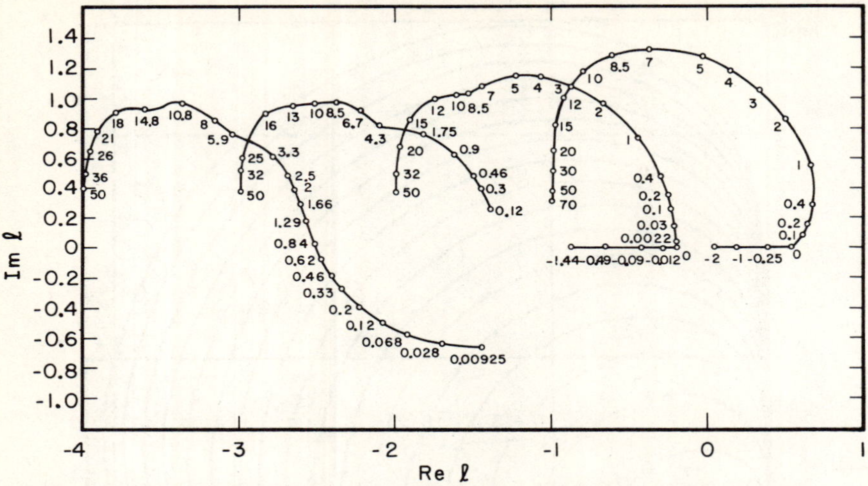

Fig. 17-5 The leading Regge trajectory and other trajectories for an attractive Yukawa potential with $G = 2$. The k^2 values are written along the curves (17AH).

To conclude this section we discuss the Regge trajectory for the attractive Coulomb potential

$$V(r) = \frac{-e^2}{r} \tag{17-11}$$

We did not discuss this potential first because it does not exhibit the resonances that particularly interest us. The S-matrix element is

$$S(\alpha, s) = \frac{\Gamma(\alpha + 1 + i\gamma)}{\Gamma(\alpha + 1 - i\gamma)} \tag{17-12}$$

where

$$\Gamma(z) = \text{gamma function} \tag{17-13}$$

$$\gamma = -\frac{\mu e^2}{\hbar p}, \qquad p^2 = 2\mu E \tag{17-14}$$

Exercise 17-4. Derive Eq. 17-12. This derivation is somewhat difficult, and Ref. 17BAY is particularly useful, as are some other books on nonrelativistic quantum mechanics.

The beautiful thing about Eq. 17-12 is that the analytic properties of $\Gamma(w)$ are simple. On the complex w-plane, $\Gamma(w)$ has no zeros and simple poles at $w = 0, -1, -2, \ldots$. Therefore, in the complex angular momentum plane, the α-plane, $S(\alpha, s)$ has *Regge poles* at

$$\alpha + 1 - i\frac{\mu e^2}{\hbar p} = 0, -1, -2, -3, \ldots \tag{17-15}$$

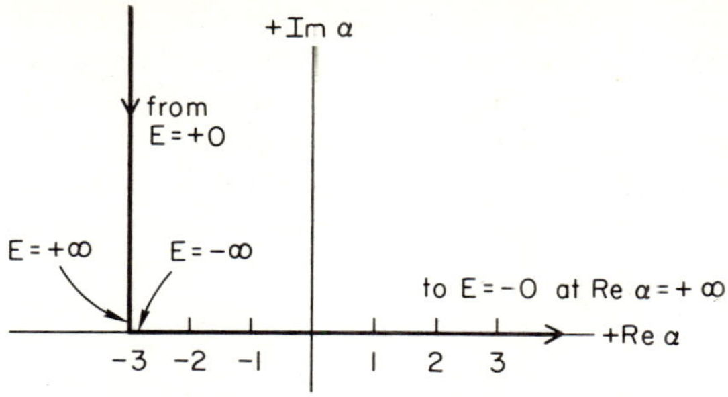

Fig. 17-6 The $N = -3$ Regge trajectory for an attractive Coulomb potential; N is defined in the text.

We make the following observations:

1. There are an infinite number of Regge poles.
2. We may label each trajectory by the integer $N = \alpha(E = \infty)$.
3. The $N = -3$ trajectory is shown in Fig. 17-6.
4. Every trajectory has bound states with $\alpha = 0, 1, 2, \ldots$.
5. There are an infinite number of bound states because of the infinite range of the Coulomb potential. For a short-range potential (e.g., the Yukawa potential) this is not true.

Exercise 17-5. Fill in the steps leading to these observations.

17-4 Regge Trajectories, Strong Interaction Resonances, and Signature.

Consider the Regge trajectory in Fig. 17-7 for the following reaction, elastic or inelastic:

$$a + b \rightarrow c + d \tag{17-16}$$

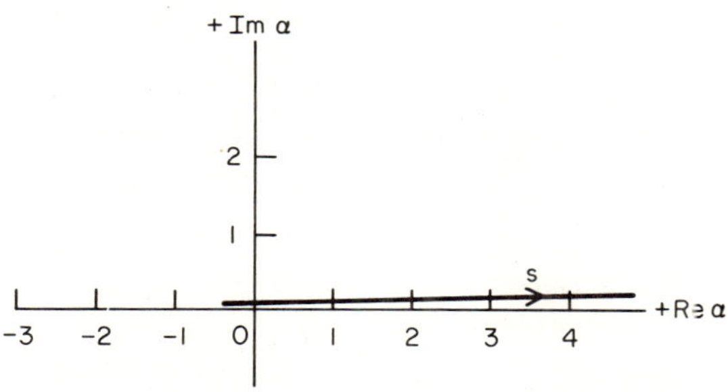

Fig. 17-7 A Regge trajectory predicting a sequence of resonances of intrinsic spin 0, 1, 2, . . ., and increasing mass.

This trajectory predicts a sequence of resonances with intrinsic spin 0, 1, 2, . . . and increasing mass. The resonances will have the total charge, baryon number, and strangeness of the initial and final states in Eq. 17-16. And if we resolve the reaction into separate total isotopic spin channels, as we may do, the resonances will also have the total isotopic spin of the channel. Thus the sequence of resonances will comprise a family of resonances with the same I, I_z, B, and S.

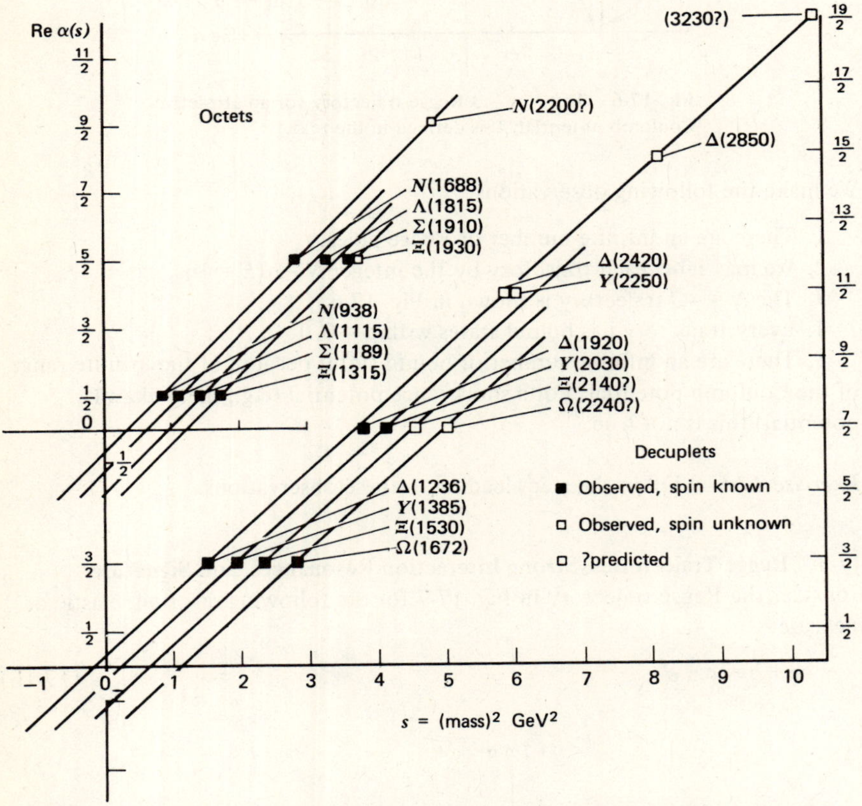

Fig. 17-8 A Chew-Frautschi plot for baryons. The lines, which are assumed to be Regge trajectories, connect baryon families with the same I, B, and S (17CHI). Reproduced with permission from "Evidence for Regge Poles and Hadron Collision Phenomena at High Energies," *Annual Review of Nuclear Science*, **22**, 263 (1972).

The amazing observation, first emphasized by Chew and Frautschi (17CH), is that the hadrons can be arranged in such families. The best examples are provided by the baryons (Fig. 17-8), and the diagram is called a Chew–Frautschi plot. We

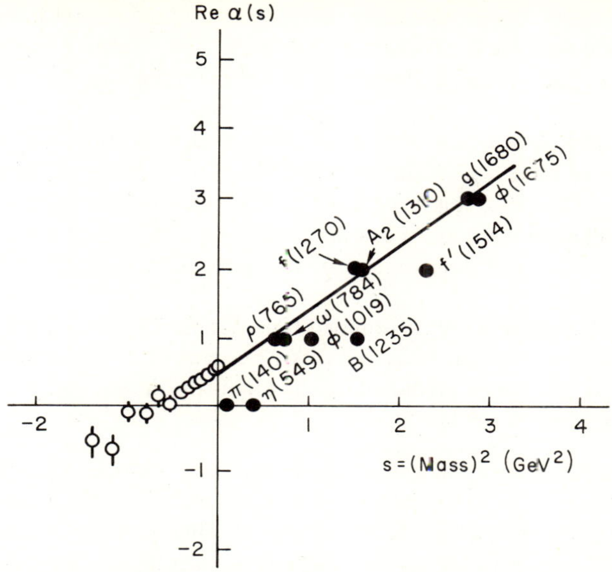

Fig. 17-9 A Chew-Frautschi plot for mesons. The line connects the only two Regge recurrences that are well established at present: the ρ–g and the ω–$\phi(1675)$. The open circles represent an extension of the ρ–g trajectory to $t < 0$, using α_{eff} for the ρ^0 trajectory in the reaction $\pi^- p \rightarrow \pi^0 n$. See the discussion associated with Eq. 17-50.

see that it is nothing more than an application to experimental data of the $s > 0$ part of diagrams such as Fig. 17-2 and 17-3. But what an audacious application. Figure 17-9 is a Chew–Frautschi plot for mesons. Compared with the baryons, there are fewer high mass mesons with established properties. Therefore, the evidence for families of mesons lying on Regge trajectories is a bit weak. Nevertheless, it is conventional to accept such meson families.

We note the following:

1. All the hadrons on a trajectory have the same I, B, and S. We do not bother to exhibit the separate members of the isospin multiplet. The higher mass members of a family on a trajectory are often called *Regge recurrences* of the lowest mass member. Even more generally, we can use the $SU(3)$ classification of hadrons and speak of Regge recurrences of $SU(3)$ multiplets.

2. The trajectories appear, at least roughly, to be straight lines. Their slopes vary somewhat, but a conventional average value that is often used is

$$\frac{d \operatorname{Re} \alpha}{ds} \sim 1 \ (\text{GeV})^{-2} \tag{17-17}$$

We note that these straight line trajectories mean that the masses increase as the square of the intrinsic spin. There is no evidence for the turning over of the trajectories, a phenomenon exhibited by the Regge trajectories of the Yukawa potential (Sec. 17-3).

3. We usually do not exhibit plots of Im α versus s. According to our discussion in Sec. 17-2, the Im α can be calculated by Eq. 17-7 from the measured width of the resonance. Of course this remark only holds when Re α is close to the intrinsic spin. We usually assume that we can interpolate smoothly from one resonance to the next on a trajectory; and we imagine that Im α has the behavior shown in Fig. 17-7.

4. On the experimentally determined trajectories in Figs. 17-8 and 17-9, particles appear at values of Re α that differ by steps of 2, not by steps of 1. This change would not be predicted by the discussion so far, and it calls for a modification of our formulation.

We can cancel every other meson resonance if we multiply $T(\alpha, s)$ by a function, such as $(1 \pm (-1)^l)$ or $(1 \pm (-1)^{-l})$. But we would like to do this in a way that preserves the analytic properties of $T(\alpha, s)$. Therefore, we choose an analytic function that reduces to one of the functions we just mentioned when Im α is small. Such a function is

$$\mathscr{R}^{\pm}(s) = (1 + \eta\, e^{-i\pi\alpha(s)}) \qquad \text{for mesons} \tag{17-18a}$$

where $\eta = \pm 1$ is the *signature* of the Regge trajectory.

The choice of phase here has to do with conventions which we do not discuss. Similarly, we define

$$\mathscr{R}^{\pm}(s) = (1 + \eta\, e^{-i\pi(\alpha(s)\pm 1/2)}) \qquad \text{for baryons} \tag{17-18b}$$

Then we rewrite our basic equation (Eq. 17-4) in the form

$$T^{\pm}_{(s,\,t)} = \tfrac{1}{2} \sum_{l=0}^{\infty} (2l + 1)[T(\alpha, s)\mathscr{R}^{\pm}(s)]_{\alpha=l} P_l(\cos\theta) \tag{17-19}$$

Exercise 17-6. Show that the signature term indeed cancels every other resonance on a trajectory. Use Eq. 17-5. Also confirm the symmetry statements that follow.

This introduction of the signature concept may seem ad hoc, but the concept turns out to have deeper significance. First, the amplitude in Eq. 17-19 is either totally symmetric or totally antisymmetric about $\theta = 90°$. Therefore, we have separated the trajectories and the amplitude into symmetric and antisymmetric parts. This is certainly a natural thing to do. Second, although we do not discuss it, the signature concept is necessary to obtain satisfactory analytic continuation of the amplitudes (17CO, 17COL, 17MA, 17ED). To simplify equations we write the signature term in its meson form (Eq. 17-18a). When baryons are being considered, the $\pm\frac{1}{2}$ in the exponent, Eq. 17-18b, must be included.

Let us see what has been accomplished. The concept of the Regge trajectory has enabled us to connect hadrons into families. Thus it appears very likely that strong interaction dynamics does lead to Regge poles as we have interpreted them. But we should not be misled into thinking that we understand the basis of that dynamics—there is no way known to calculate from first principles these trajectories.

17-5 Regge Poles and High Energy, Two-Body Reactions. We have just applied Regge theory to a low energy phenomenon in strong interaction physics. We now try to understand the dynamics of a high energy, two-body reaction by assuming the existence of Regge poles in the *cross channel*! At this point, we deviate greatly from the conventional mathematical structure used to derive the cross-channel properties of Regge poles. The conventional method uses the Sommerfeld–Watson transform and depends on some intricate discussion of the analytic properties of Legendre function of the first and second kind (17CO, 17COL, 17MA, 17ED). However, since we found it difficult to reproduce this method without using the intricate mathematics that we like to avoid, we were very happy to be given the following method devised by R. Blankenbecler.

We now make a change of nomenclature which we hope is not confusing. We are interested in the high energy dynamics of

reaction I: $$a + b \rightarrow c + d \tag{17-20}$$

$$P_a + P_b = P_c + P_d, \qquad s = (P_a + P_b)^2, \qquad t = (P_a - P_c)^2 \tag{17-21}$$

$$s \geqslant 4m^2, \qquad 0 \geqslant t \geqslant -(s - 4m^2)$$

We assume that there is a Regge pole, denoted by the subscript n, in the cross channel of

reaction II: $$a + \bar{c} \rightarrow \bar{b} + d \tag{17-22}$$

$$P_a + P_{\bar{c}} = P_{\bar{b}} + P_d, \qquad s = (P_a - P_{\bar{b}})^2, \qquad t = (P_a + P_{\bar{c}})^2 \tag{17-23}$$

$$t > 4m^2; \qquad 0 \geqslant s \geqslant -(t - 4m^2)$$

Now we write Eqs. 17-4 and 17-5 in that cross channel in the center of mass frame of the physical region of reaction II. Let $\alpha_n(t)$ be the Regge trajectory. Then, ignoring for a while the signature factor, we have

$$T_n(t, s) = \sum_{l=0}^{\infty} (2l + 1)[T_n(\alpha, t)]_{\alpha=l} P_l(\cos\theta_{\text{II}}) \tag{17-24}$$

$$T_n(\alpha, t) = \frac{A_n(t)}{\alpha - \alpha_n(t)} + B(t) \tag{17-25}$$

where t is the square of the total center of mass energy in reaction II, replacing the s that appeared in Eqs. 17-4 and 17-5. For simplicity, we take all particles to have the same mass and spin zero. And for the present we consider a single Regge pole.

Our object is to continue the amplitude given in Eqs. 17-24 from the physical region of reaction II to the physical region of reaction I. The assumption that this can be done is the essence of the crossing relations discussed in Chs. 15 and 16, and we accept that assumption. But to carry out the crossing we must in principle know the form of the functions $\alpha_n(t)$, $A(t)$, and $B(t)$. Of course we have no detailed rigorous information about the analytic properties of these functions in terms of the *complex variables s and t*. But arguments from nonrelativistic theory, perturbation theory, and field theory make the following simple assumptions plausible (17CO, 17COL, 17MA, 17ED):

1. $\alpha_n(t)$ and $A(t)$ can be continued between physical regions.
2. The continuation of $\alpha_n(t)$ and $A(t)$ leads to reasonably simple functions in the physical region of reaction I.
3. In crossing to the physical region of reaction I, the Regge pole term will dominate over the background term $B(t)$; and we can ignore that term. If we dislike this assumption we can interpret the rest of the discussion as relating only to the effect of the Regge pole term, the effect of $B(t)$ being unknown.

When we carry out the crossing, the relation

$$\cos \theta_{\rm II} = 1 + \frac{2s}{t - 4m^2} \tag{17-26}$$

becomes unphysical. Let us define

$$z = -\cos \theta_{\rm II} = -1 - \frac{2s}{t - 4m^2} \tag{17-27}$$

Now Eq. 17-24 has the form

$$T_n(t, s) = \sum_{l=0}^{\infty} (2l + 1) \left[\frac{A_n(t)}{l - \alpha_n(t)}\right] P_l(-z) \tag{17-28}$$

and our problem is to evaluate the sum on the right-hand side. We remind those who are acquainted with the conventional method that in Eq. 17-28 l remains an integer as we cross, and $P_l(-z)$ remains a simple Legendre polynomial. Thus $P_2(-z) = \frac{1}{2}(3z^2 - 1)$ for any value of z.

Now let us see what happens if we consider the small t, large s region of reaction I. For $s \gg t$, $s \gg m^2$ we have

$$z \to \frac{2s}{m^2 - t} \to \infty \tag{17-29}$$

Every Legendre polynomial in Eq. 17-28 will be dominated by its leading term in the finite series (App. C)

$$P_l(w) = \frac{(2l-1)\,!!}{l\,!}\left\{w^l - \frac{l(l-1)}{2(2l-1)}\,w^{l-2} + \cdots\right\} \tag{17-30}$$

and the leading term in Eq. 17-28 will be

$$T_n(t, s) = A_n(t)\sum_{l=0}^{\infty}\left[\frac{(2l+1)\,!!}{2^l\, l\,!}\ \frac{(-2z)^l}{l - \alpha_n(t)}\right], \quad z \to \infty \tag{17-31}$$

This series diverges for z large, and we must perform the customary maneuver with divergent series. We sum the series when z is small, find the analytic continuation of that sum, and define that as the meaning of the series when z is large. To avoid breaking the flow of the deviation, we leave that process for Exs. 17-7 through 17-9 and write the result

$$T_n(t, s) = -\sqrt{\pi}A_n(t)\left[\frac{\Gamma(\alpha_n(t) + \frac{3}{2})2^{\alpha_n(t)}}{\Gamma(\alpha(t) + 1)}\right]\left[\frac{z^{\alpha_n(t)}}{\sin(\tau\alpha_n(t))}\right], \qquad z \to \infty \tag{17-32}$$

This is the basic equation for the amplitude in the small t, large s region of reaction I when there is a Regge pole in the cross channel–reaction II.

There are a few changes to be made in Eq. 17-32 before it is ready for use. First we must introduce the signature term from Eq. 17-18. We use the meson form, it being understood that the baryon form of Eq. 17-18b should be used for baryons.

$$T_n^{\pm}(t, s) = -\left[\frac{\sqrt{\pi}A_n(t)\Gamma(\alpha_n(t) + \frac{3}{2})2^{\alpha_n(t)}}{\Gamma(\alpha(t) + 1)}\right]\left[\frac{z^{\alpha_n(t)}(1 \pm e^{-i\pi\alpha_n(t)})}{\sin(\pi\alpha_n(t))}\right] \tag{17-33}$$

Since the early days of the Regge theory of strong interactions, there has been a good deal of discussion and calculation aimed at determining whether such terms in Eq. 17-33 as $\Gamma(\alpha(t) + 1)$ have direct physical significance, for such terms produce zeros and poles in $T_s(s, t)$ at particular values of $\alpha_n(t)$. There is still no agreement on this matter, but some of the terms are taken less seriously now than they were five years ago. We return to this problem in the next chapter. For the present we shall sweep it under the rug by combining most functions of t in Eq. 17-33 into the function $\gamma_n(t)$. This function is called the *residue* of the Regge trajectory because it contains the residue of $T(\alpha, t)$ at the Regge pole.

Finally

$$T_n^{\pm}(s, t) = \gamma_n(t)\left[\frac{1 \pm e^{-i\pi\alpha_n(t)}}{\sin(\pi\alpha_n(t))}\right]\left(\frac{s}{s_0}\right)^{\alpha_n(t)} \tag{17-34a}$$

Here we have also introduced an arbitrary constant, conventionally called s_0, for dimensional convenience. We have kept the factor in the square bracket separate to indicate the importance of the *signature factor*

$$\begin{aligned}\mathcal{S}^{\pm} &= \frac{\mathcal{R}^{\pm}(t)}{\sin(\pi\alpha(t))} \\ &= \frac{1 \pm e^{-i\pi\alpha(t)}}{\sin(\pi\alpha(t))} \qquad \text{for mesons}\end{aligned} \tag{17-35}$$

Equation 17-34a is the basic formula for describing the effect of a *t-channel* Regge pole in the *s-channel*. There are two parts to this description.

1. For fixed t, the very high energy behavior of the amplitude is simple: the amplitude is proportional to a power of s. In the next section we develop this very important point.
2. At fixed energy, the behavior as a function of t is unknown and possibly complicated. Regge pole theory itself provides no guide to the function $\gamma_n(t)$ and only sketchy clues to the function $\alpha_n(t)$. Therefore, we have to depend on experiment to determine these functions (see Sec. 17-7).

Equation 17-34a is written for a single Regge pole. But several Regge poles from reaction II may influence the amplitude in reaction I. Therefore, the total amplitude for reaction I is

$$T(s, t) = \sum_n T_n(s, t) \tag{17-34b}$$

Exercise 17-7. To sum the series in Eq. 17-31, first consider the series

$$J(z) = \sum_l^{\infty} \frac{(-z)^l}{l - \alpha} \tag{17-36}$$

To estimate the sum of this series for z sufficiently small to produce convergence, use

$$\frac{1}{l - \alpha} = \int_0^{\infty} dx\, e^{-x(l-\alpha)}$$

to show

$$J(z) = \int_0^\infty dx \left[\frac{e^{x\alpha}}{1 + z\, e^{-x}}\right] \tag{17-37}$$

This can be written as

$$J(z) = z^\alpha \int_0^z \frac{dy}{y^{\alpha+1}(1+y)} \tag{17-38}$$

using $y = z\, e^{-x}$.

Next show that this integral is always finite for

$$-1 < \alpha < 0 \tag{17-39}$$

Hence show that $J(z)$ can be continued outside of the circle of convergence of the series in Eq. 17-36. Thus we have achieved part of our goal. Next show that

$$J(z) = -\frac{z^\alpha \pi}{\sin \alpha\pi} + z^\alpha \int_z^\infty \frac{dy}{y^{\alpha+1}(1+y)} \xrightarrow[z\to\infty]{} -\frac{z^\alpha \pi}{\sin \alpha\pi} \tag{17-40}$$

We could use this result directly in Eq. 17-31, ignoring $(2l+1)!!/(2^l\, l!)$ without too much error. We can do a little better, however, although in the end it will not make any difference.

Exercise 17-8. Use

$$\frac{\pi P_\alpha(z)}{\sin \pi\alpha} = \sum_{l=0}^\infty (-1)^l \left(\frac{1}{\alpha - l} - \frac{1}{\alpha + l + 1}\right) P_l(z) \tag{17-41}$$

or another method to show

$$\sum_{l=0}^\infty \left[\frac{(2l+1)!!\,(-2z)^l}{2^l\, l!\,(l-\alpha)}\right] \xrightarrow[z\to\infty]{} -\frac{\sqrt{\pi}}{\sin \pi\alpha}\,\frac{\Gamma(\alpha + 3/2)2^\alpha z^\alpha}{\Gamma(\alpha+1)} \tag{17-42}$$

Insert the last result in Eq. 17-31 to obtain Eq. 17-32.

Exercise 17-9. Show that we can obtain Eq. 17-34a for values of α outside the range specified in Eq. 17-39 by writing

$$\sum_{l=0}^\infty \frac{(-z)^l}{l - \alpha} = -z \sum_{l=1}^\infty \left[\frac{(-z)^{l-1}}{(l-1) - (\alpha - 1)}\right] - \frac{1}{\alpha} \tag{17-43}$$

Exercise 17-10. In this section we have shown how a Regge pole in the t-channel predicts the form of the amplitude in the small t region of the s-channel. Show that a Regge pole in the u-channel will predict a similar form for the amplitude in the small u region of the s-channel.

Exercise 17-11. In obtaining Eq. 17-32 we used the leading term in each $P_\alpha(z)$. Show that as one sums over the next to leading terms in each $P_\alpha(z)$, a $z^{(\alpha-2)}$ correction term to be added to Eq. 17-32 is obtained.

17-6 Regge Poles and the Energy Dependence of Two-Body Inelastic Reactions.

In this and the next section we use Eq. 17-34 to tell us something about the unknown functions $\alpha(t)$ and $\gamma(t)$. Because we are going in two directions, this discussion must be rather loose, and it should be rather long. Our space is limited, however, and for many details we must refer the reader to the excellent book by Barger and Cline (17BA). In particular, the book discusses baryon trajectories and backward peaks, a subject we must ignore.

In this section and the next section we consider only inelastic reactions. There is a special problem in applying Regge theory to diffractive elastic scattering, and we postpone that discussion to Sec. 17-9.

To discuss the high energy behavior produced by a t-channel Regge pole in the s-channel, we first write Eq. 17-34 in the still simpler form

$$T_{ab\to cd}(s,t) = h(t)\left(\frac{s}{s_0}\right)^{\alpha(t)} = h(t)\, e^{\alpha(t)\ln(s/s_0)} \tag{17-44}$$

where $a + b \to c + d$ is the s-channel reaction. From Eq. 6-44 the differential cross section at very high energies is

$$\left(\frac{d\sigma}{dt}\right)_{ab\to cd} = \frac{|h(t)|^2 s_0^2}{16\pi}\left(\frac{s}{s_0}\right)^{2\alpha(t)-2} \tag{17-45}$$

Therefore, to know the energy dependence of amplitudes or differential cross sections at fixed t we need to know only $\alpha(t)$. Our problem is to find the behavior of $\alpha(t)$ when t is small and negative. In Sec. 17-4 we described the experimentally determined Regge trajectories $\alpha(t)$ when t is greater than zero. In this region the trajectories were approximately straight lines of the form

$$\text{Re}\,\alpha(t) = \alpha_0 + \alpha_0' t, \qquad \alpha_0' = \left(\frac{d\,\text{Re}\,\alpha(t)}{dt}\right)_{t=0} \sim 1\ (\text{GeV/c})^{-2}, \qquad |\text{Im}\,\alpha(t)| \ll 1 \tag{17-46}$$

Returning to our immediate problem, let us assume that Eq. 17-46 still holds if we extend t into the range $0.0 > t > \sim -1(\text{GeV/c})^2$. Thus Eq. 17-44 takes the form

$$T_{ab\to cd}(s, t) = h(t)\left(\frac{s}{s_0}\right)^{\alpha_0+\alpha_0' t} \tag{17-47}$$

In the next section we show that for forward-peak-dominated, two-body reactions $h(t)$ can be approximated in the small t region by the function $g\,e^{bt/2}$. Using this approximation in Eq. 17-47, we obtain

$$T_{ab\to cd}(s, t) = g\left(\frac{s}{s_0}\right)^{\alpha_0} \exp\left[\frac{b}{2} + \alpha_0' \ln\left(\frac{s}{s_0}\right)\right] t \tag{17-48}$$

$$\sigma_{ab\to cd}(s) = \frac{g^2 s_0^{\,2}}{16\pi}\left(\frac{s}{s_0}\right)^{2\alpha_0 - 2}\left[b + 2\alpha_0' \ln\left(\frac{s}{s_0}\right)\right]^{-1} \tag{17-49}$$

To interpret Eq. 17-49, we note that unless $2\alpha_0 \ln(s/s_0) \gg b$, the energy dependence is dominated by the $(s/s_0)^{2\alpha_0-2}$ term. From Fig. 17-9 we recall that for meson trajectories $-0.5 \lesssim \alpha_0 \lesssim 0.5$. Therefore, Eq. 17-49 predicts that s-channel, two-body reactions dominated by t-channel Regge meson trajectories have cross sections whose energy dependence lies between s^{-1} and s^{-3}. And this is just what is found experimentally (Sec. 7-7). Thus one of the earliest and simplest predictions of the Regge theory of strong interactions is the correct high energy dependence of two-body reactions.

Exercise 17-13. Show that the correct energy dependence is also found for s-channel reactions dominated by baryon Regge trajectories in the u-channel. Use Fig. 17-8.

17-7 Differential Cross Sections, Factorization, and the Particle Exchange Analogy. It should be clear from the derivation of the basic equation Eq. 17-34 that Regge theory gives no information on the form of $\alpha(t)$ or $\gamma(t)$ in the $t < 0$ region. Therefore, as we have said before, we must turn to other areas for information on these functions. Let us discuss $\alpha(t)$ first. In the previous section we used a straightforward linear extrapolation (Eq. 17-46) to fix $\alpha(t)$. This is curve I in Fig. 17-10*a*. But such curves as II or III in that figure would be equally reasonable.

There have been many attempts to use the second form of Eq. 17-44 to find $\alpha(t)$, for $\alpha(t)$ fixes the energy dependence of $d\sigma/dt$ at each value of t. In particular, the special form of Eq. 17-48 predicts that the forward peak will narrow as s increases if α_0 is greater than zero. This is the well-known Regge theory prediction

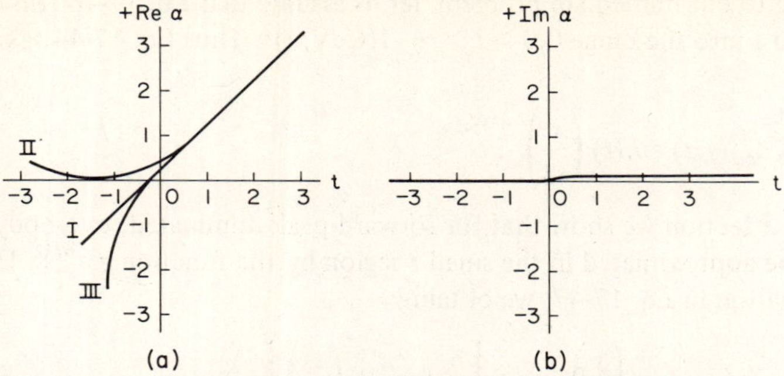

Fig. 17-10 (*a*) Some ways in which the straight line Regge trajectory in the $t > 0$ region may be extended into the $t < 0$ region. (*b*) The assumed behavior of Im α.

of *shrinkage* in forward peaks. Unfortunately, the experimental situation is very complicated:

1. Regge theory predictions are most meaningful at small t, but when shrinkage occurs, and it does occur sometimes, it is often evident at t values of the order of 1 $(\text{GeV}/c)^2$ and larger.

2. The best measurements are in elastic scattering, but this has the special problem discussed in Sec. 17-9.

3. We no longer expect that $d\sigma/dt$ will be determined by a single Regge pole yielding the amplitude in Eq. 17-44. Other Regge poles may also contribute to the amplitude. And most important, absorptive effects can greatly distort $d\sigma/dt$, as we saw in Ch. 13 and will discuss again in Ch. 18.

Therefore we usually take $\alpha(t)$, $t < 0$, to have one of the forms represented by curves such as those in Fig. 17-10*a*. And if we are feeling particularly lazy we use curve 1 and Eq. 17-46. Often, too, we define an $\alpha_{\text{eff}}(t)$ by

$$\frac{d\sigma}{dt} = |F(t)|^2 \left(\frac{s}{s_0}\right)^{2\alpha_{\text{eff}}(t)-2} \tag{17-50}$$

thus burying the various problems. Figure 17-9 contains a classic example. Here Eq. 17-50 has been applied to the charge exchange reaction $\pi^- + p \to \pi^0 + n$ to yield α_{eff} for the ρ^0 Regge trajectory (See Ch. 18).

We have so far said nothing about Im α for t less than zero, except to assume in Eq. 17-46 that $|\text{Im}\,\alpha| \ll 1$. Indeed, since the early days of the application of Regge theory to strong interactions, it has usually been assumed that

$$\text{Im}\,\alpha(t) = 0, \qquad t < 0 \tag{17-51}$$

Although occasionally it has seemed that the data required Im $\alpha \neq 0$, in the end it has always been possible to modify other parts of the theory and to preserve Eq.

17-51. Therefore, on the argument of simplicity, we shall use Eq. 17-51 until compelled by experiment to do otherwise. This assumed behavior of Im $\alpha(t)$ appears in Fig. 17-10*b*.

Exercise 17-13. A word about the mysterious constant s_0, inserted in Eq. 17-34 for dimensional convenience. Show that we cannot determine s_0 unless we know $\alpha(t)$ and $\gamma(t)$ completely. For example, use Eq. 17-48 to show $b_{\text{experimental}} = 2\alpha_0' \ln (s/s_0)$ *if* we know that $h(t)$ is a constant. Show, however, that this excessively simple model disagrees with experiment; use data given in Ch. 7.

Exercise 17-14. Examine what would happen in Eq. 17-34 if Re $\alpha(t)$ passes through the values 0, -1, and -2, as $t \to -\infty$. Show that depending on the signature, $T(s, t)$ may have a pole at these points unless $\gamma(t)$ has a corresponding zero. Now we cannot have poles in $T(s, t)$—a physical amplitude! Therefore $\gamma(t)$ must have the required zeros. Show that the factor $1/\Gamma(\alpha + 1)$ in Eq. 17-32, which we included in $\gamma(t)$, does indeed provide these zeros, except at $\alpha = 0$. Therefore, there will be a so-called *ghost state* at $\alpha = 0$, depending on the signature, unless we require a *dynamic* zero in $\gamma(t)$ at $\gamma = 0$ (17BA). Finally, as an example of the complicated considerations that follow from these arguments, show that the factor $\Gamma(\alpha + 3/2)$ also buried in $\gamma(t)$ cannot be displayed explicitly without producing poles at $\alpha = -1/2, -3/2, \ldots$, depending on the signature. These kinds of problem are another reason for our frequent reluctance to use other than very simple forms for $\alpha(t)$ and $\gamma(t)$.

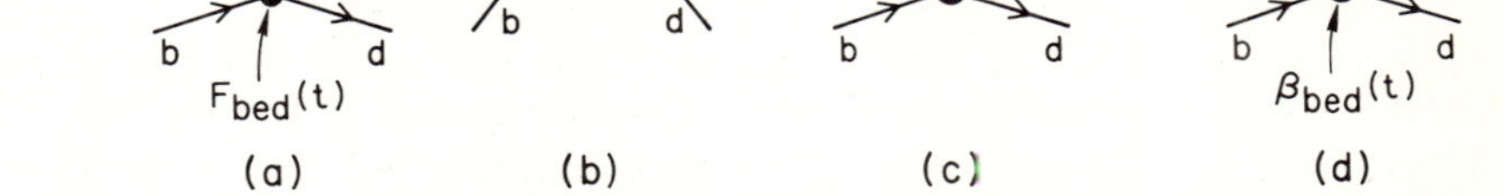

Fig. 17-11 Analogies between the *t*-channel one-particle exchange mechanism and the *t*-channel Regge trajectory exchange mechanism in two-body inelastic reactions.

Our next task is to see if we can construct any analogies or relationships between the Regge model of high energy, two-body reactions and the one-particle exchange model of these reactions. We know that in spite of all its problems (Chs. 12 and 13), the one-particle exchange model provides the basic explanation of the existence of the forward and backward peaks in two-body inelastic reactions. The Regge model must be able to do the same. To begin, we consider the one-particle exchange Feynman diagram (Fig. 17-11*a*) for the reaction

$$a + b \rightarrow c + d \tag{17-52}$$

We take particles *a, b, c,* and *d* to have spin zero, but the exchanged particle *e* may have spin. We have also included vertex form factors of the kind discussed in Sec. 13-2.

Figure 17-11*b* is a diagram for the same reaction in the Regge model. The diagram is not interesting, so we stretch out the cross-hatched blob to emphasize the internal quantum numbers of the Regge trajectory, and we obtain Fig. 17-11*c*. Now this diagram is very interesting, for we see that the Regge trajectory transfers four-momentum t and internal quantum numbers I, I_z, S, and B from the upper vertex to the lower vertex. In this respect it acts just like a virtual elementary particle with that four-momentum and those quantum numbers. Of course the Regge trajectory differs from the virtual particle in that the latter has a *fixed intrinsic* angular momentum j, but the Regge trajectory does not. We can, if we like, say that the Regge trajectory carries a *variable* angular momentum $\alpha(t)$.

This concept of Regge trajectory exchange is very useful, once we begin to think of a Regge trajectory as a substitute for a virtual particle. To discuss this substitution process it is useful to have a convenient way to denote a trajectory. We denote a trajectory by the name of the lowest mass hadron that appears on the $t > 0$ part of the trajectory. Thus we speak of the pion trajectory, the ρ meson trajectory, the proton trajectory. Now the power of the one-particle exchange mechanism is that the same properties of the virtual particle are used to explain many different two-body reactions. For example, in the one-particle exchange model the same virtual ρ meson exchange may be used to explain the reactions

$$\begin{aligned} &\pi^- + p \rightarrow \pi^0 + n \\ &K^+ + n \rightarrow K^0 + p \\ &K^+ + p \rightarrow K^0 + \Delta(1236) \end{aligned} \tag{17-53}$$

The virtual ρ always has the same internal quantum numbers, the same propagator, and the same spin properties. We are only allowed to change, if we wish, the vertex form factors. If we wish to explain the reactions in Eq. 17-53 using the Regge model with a ρ trajectory, we must also treat the ρ trajectory the same way in all these reactions.

As illustrated in Fig. 17-11*c*, we can obviously use the same internal quantum numbers, I, I_z, B, and S for the ρ trajectory in all the reactions in Eq. 17-53. It is also reasonable that we use the same $\alpha(t)$ for the ρ trajectory in all these reactions, for the behavior of $\alpha(t)$ defines the trajectory and fixes the spin properties of the hadrons that lie on the trajectory. Thus $\alpha(t)$ corresponds to the propagator of a virtual particle. In the next section we reinforce this argument.

But what about $\gamma(t)$? If we trace back through the previous sections, we see that the strong interaction dynamical information in $\gamma(t)$ comes from $A(t)$ in Eq. 17-5 or $A'(t)$ in Eq. 17-6. And if we go back even further we see that $A'(t)$ is equivalent to the numerator in Eq. 5-23. Thus for t-channel reactions $\gamma(t)$ depends on the particles a, b, c, and d; as well as on the trajectory. It is reasonable that this is also true in the s-channel. Therefore, $\gamma(t)$ must depend on the nature of all the particles in the reaction, not just on the Regge trajectory.

If there was nothing more we could say about $\gamma(t)$, the predictive powers of the Regge theory with respect to the shape of $d\sigma/dt$ would be destroyed. There are, however, two very useful restrictions on $\gamma(t)$. First we expect $\gamma(t)$ to be *real*. We do not have the space to give the analyticity arguments leading to this conclusion, but we give some weaker arguments in Sec. 17-8. Remembering that $\alpha(t)$ is assumed to be real for $t < 0$, and looking at Eq. 17-34a, we see that the phase of $T(s, t)$ in the Regge model is fixed completely by the signature factor.

The second restriction we impose is called the *factorization hypothesis*. We assume that $\gamma(t)$ factors into a product of two functions, each function dependent only on the particles at one vertex. Explicitly, using Fig. 17-11d, we write

$$\gamma(t) = \beta_{aec}(t)\beta_{bed}(t) \tag{17-54}$$

One way to justify this factorization hypothesis is to argue from the analogy to one-particle exchange. Another way follows.

Exercise 17-15. Use the argument given previously involving Eqs. 17-6 and 5-23 to justify the factorization hypothesis.

One can attempt to test the factorization hypothesis experimentally by fitting several reactions with the same trajectory and with a closed set of $\beta(t)$'s. Unfortunately, such tests are obscured by complications such as: several Regge trajectories contributing to the same reaction, absorption corrections, and the related problem of Regge cuts. Therefore one usually assumes factorization of $\gamma(t)$ to simplify the Regge model and make it useful.

We summarize the analogies between the Regge trajectory exchange model and the elementary one-particle exchange model in Fig. 17-11. From these analogies we see that we can explain the forward and backward peak dominance of two-body reactions using Regge trajectory exchange as well as we can explain it using one-particle exchange (Sec. 7-5 and Ch. 12) *if* $\beta(t)$ or $\beta(u)$ has the right form. The right form for $\beta(t)$ or $\beta(u)$ is simply a function that emphasizes the small t or u values, respectively. We must take this form from experiment. For the purposes of this chapter it is convenient to use

$$\beta_{iej}(t) = g_{iej}\, e^{b_{iej}t/2} \tag{17-55}$$

where g_{iej} and b_{iej} are simply a coupling constant and an exponential slope attached to the *iej* vertex. Spin and absorption corrections will alter this form, but we leave these considerations to the next chapter.

Thus we have constructed a Regge model for high energy, two-body reactions. If it is substituted for the elementary one-particle exchange model, we finally obtain the correct energy dependence of those cross sections. And we can still explain the forward and backward peak dominance in the differential cross section. The price we have paid is the introduction of the functions $\alpha(t)$, $\beta_{aec}(t)$, and $\beta_{bed}(t)$, whose form is not fixed by the model. Table 17-1 lists the most prominent Regge meson trajectories, their properties, and some of the reactions to which they contribute. Information on typical baryon trajectories is provided in Fig. 17-8.

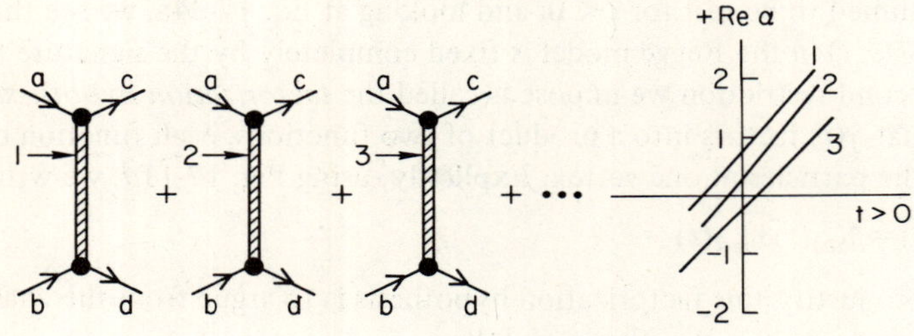

Fig. 17-12 Several Regge trajectories contributing to the same two-body inelastic reaction.

If several trajectories, 1, 2, . . ., N, contribute to a reaction, we have

$$T(s, t) = \sum_{n=1}^{N} \left[\frac{\gamma_n(t)(1 \pm e^{-i\pi\alpha_n(t)})}{\sin(\pi\alpha_n(t))} \left(\frac{s}{s_0}\right)^{\alpha_n(t)} \right] \tag{17-56}$$

as illustrated in Fig. 17-12*a*. Let us suppose that all these trajectories have $\gamma_n(t)$'s of about the same size. Then the relative contributions depend on the $\alpha_n(t)$'s. Suppose that we have ordered the trajectories as in Fig. 17-12*b* such that the $n = 1$ trajectory has the largest Re α in the small t region. As s increases, the relative contribution of the $n = 1$ trajectory will increase; and at very high energy it may be sufficient to consider only the $n = 1$ trajectory. Therefore, the trajectory with the largest Re $\alpha_n(t)$ in the small t region is the dominant trajectory.

Exercise 17-16. Suppose that the π, ρ, and ω trajectories can contribute to a reaction. Use Fig. 17-9 to select the dominant trajectory. What is the relation of this discussion to the lowest mass rule (rule 7 in Sec. 12-3) used to select the dominant one-particle exchange diagram?

We put off until Ch. 18 any detailed discussion of the validity of the model. But we must observe that the general aspects of the model have become imbedded in our thinking about two-body reactions. Even if we do not accept the derivation of the model or its detailed predictive power, we often use the following cruder form of Eq. 17-34a to fit data:

$$T(s,\,t) = h(t)\left(\frac{s}{s_0}\right)^{\alpha_{\text{eff}}(t)} \tag{17-57}$$

We conclude this very long section by emphasizing that the vertex residue functions $\beta(t)$ are simply more sophisticated forms of hadronic vertex form factors. As we discuss in Secs. 13-2 and 19-6, such form factors have to be included sooner or later in any theory of s-channel two-body reactions based on the exchange of a t-channel entity. Now in the Regge model, the shapes of the forward and backward peaks are given almost entirely by $\beta(t)$ and $\beta(u)$, respectively —the effect of the $(s/s_0)^\alpha$ term is usually small. This is different from the one-particle exchange model with vertex form factors, where the shape of the peak comes from both the form factor and the propagator.

17-8 Models for Regge Trajectories.

The analogies between the Regge model and the one-particle exchange model suggest that we should be able to understand the properties of the Regge trajectory using field theory and Feynman diagrams. In this section we describe two methods for doing this. Neither method is rigorous, but they are both enlightening.

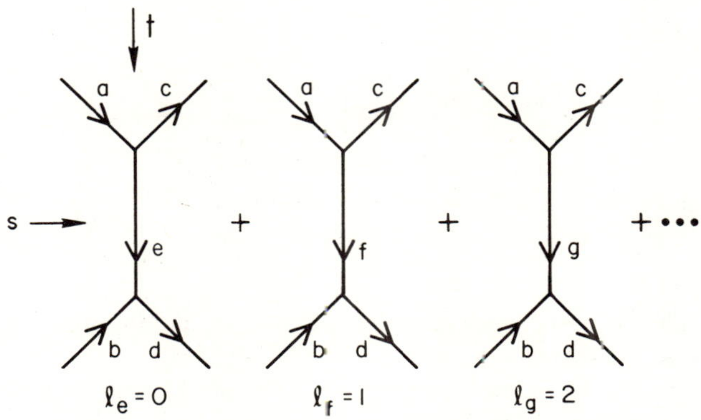

Fig. 17-13 An infinite series of one-particle, t-channel exchange diagrams; the exchanged particles have intrinsic spins 0, 1, 2, ... The external particles have spin 0 and equal masses.

In the first method, we consider the infinite sum of one-particle exchange diagrams in Fig. 17-13. We take each diagram to represent the t-channel exchange of a particle, and we assume that the exchanged particles belong to an infinite family of particles. We take each exchange particle to have the same S, B, and I. But we assume the spin-mass relationship

$$l = aM_l^2 + b, \qquad l = 0, 1, 2, \ldots \tag{17-58}$$

where a and b are constants (straight line Regge trajectories are an example of Eq. 17-58). From Sec. 12-11, in the t-channel physical region we have

$$T(s, t) = C \sum_{l=0}^{\infty} \left[\frac{g_{aec} g_{bed} F_{aec}(t) F_{bed}(t)}{M_l^2 - t} \left(\frac{p_t^*}{p_0} \right)^{2l} (P_l(-z)) \right] \tag{17-59}$$

Here we have inserted vertex form factors; also $-z = \cos\theta_t = 1 + 2s/(t - 4m^2)$ and $p_t^{*2} = (t - 4m^2)/4$. We assume that the external particles have equal masses m. The p_0 is a constant that is required dimensionally, and C is another constant.

Our object is to obtain the amplitude in the s-channel reaction $a + b \to c + d$ as s approaches infinity and t becomes small and negative. Then we have to sum

$$T(s, t) = C \sum_{l=0}^{\infty} \left[\frac{a g_{aec} g_{bed} F_{aec}(t) F_{bed}(t)}{l - (at + b)} \left(\frac{p_t^*}{p_0} \right)^{2l} P_l(-z) \right] \tag{17-60}$$

The problem of defining the analytic continuation of a divergent series has been discussed in Sec. 17-5 and its exercises. If we assume that the coupling constants and vertex form factors are independent of the mass M_l of particle l, we can use the method of that section. In fact, we have a series similar to that in Eq. 17-31. We obtain Eq. 17-32 with

$$\alpha(t) = at + b \tag{17-61a}$$

$$A(t) = C' g_{aec} g_{bed} F_{aec}(t) F_{bed}(t) \left[\frac{s}{s_0} \right]^{(at+b)} \tag{17-61b}$$

Here various constants have been collected in C' and s_0.

Thus we have shown if we sum over the infinite set of diagrams in Fig. 17-13 when the following conditions are satisfied, we obtain a Regge trajectory: $\alpha(t) = at + b$.

1. The exchange particles increase steadily in spin, and the spin is proportional to the square of the mass.

2. The vertex coupling constants and form factors are independent of the mass of the exchange particle.

Thus the Regge model can be interpreted as correcting one of the deficiencies of the one-particle exchange model, for the Regge model replaces a single exchange diagram by an infinite set of exchange diagrams, thus correcting the wrong energy dependence of the one-particle exchange model. For further details see Refs. 17VA, 17BL, 17DU, and 17SU.

Exercise 17-17. Fill in the steps leading to Eq. 17-61. Show that Eq. 17-61 provides an additional argument for the factorization hypothesis and the reality of $\gamma(t)$. When the masses of the external particles are unequal, additional trajectories are created. See the references just given.

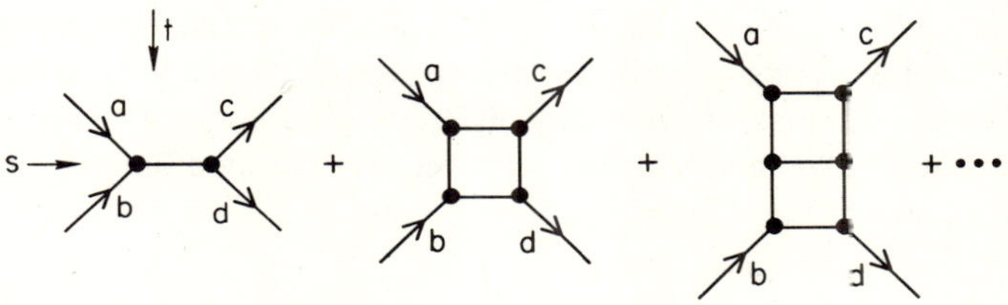

Fig. 17-14 An infinite series of ladder diagrams.

There is another field theory model for Regge trajectories using the ladder diagrams in Fig. 17-14. We first mentioned these diagrams in Sec. 8-9 in connection with the multiperipheral model; an integral can be written down for each diagram. But as we noted in Sec. 15-5, it is very difficult to evaluate these integrals. Thus we did not even attempt in Sec. 15-5 to evaluate the second diagram, which has only one loop. (Even in the limit of large s and small t, where these integrals simplify, it would take several pages to work out these simplified integrals.) Therefore, we must refer the reader to the literature (17ED) and write down the answer. For the infinite set of diagrams in Fig. 17-14, we obtain the infinite series

$$T(s, t) = \frac{g^2}{s}\left[1 + g^2K(t)\ln s + \cdots + \frac{(g^2K(t)\ln s)^n}{n!} + \cdots\right] \tag{17-62}$$

where each term corresponds to a diagram and we have taken all particles to have spin zero; g is the vertex coupling constant and $K(t)$ is a function of t only. The sum of this series is

$$T(s, t) = g^2 s^{[-1+g^2K(t)]} \tag{17-63}$$

which is the Regge form for the two-body reaction amplitude if we set $\alpha(t) = -1 + g^2K(t)$. Thus again we see that the Regge trajectory can be interpreted as a summation over an infinite set of Feynman diagrams.

Exercise 17-18. An interesting question, which we cannot go into here, concerns the relation between the two models just presented. Find out what you can about it. Note that the second model is a summation over a subset of the diffractive elastic scattering Feynman diagrams in Fig. 8-10.

17-9 Regge Trajectories and High Energy, Diffractive Elastic Scattering. The forward peak in diffractive elastic scattering and its connection, through the optical model, with total cross sections, provided one of the inspirations for the application of Regge theory to strong interactions. It is therefore ironic that today we are in doubt about the relation between diffractive elastic scattering and Regge theory. We explained at length in Ch. 2 and Sec. 8-9 that diffractive elastic scattering is closely connected with inelastic processes. Hence in field theory it is explained by the summation over many different infinite sets of complicated Feynman diagrams. We represent that very, very complicated summation by the cross-hatched area in Fig. 17-15*a*.

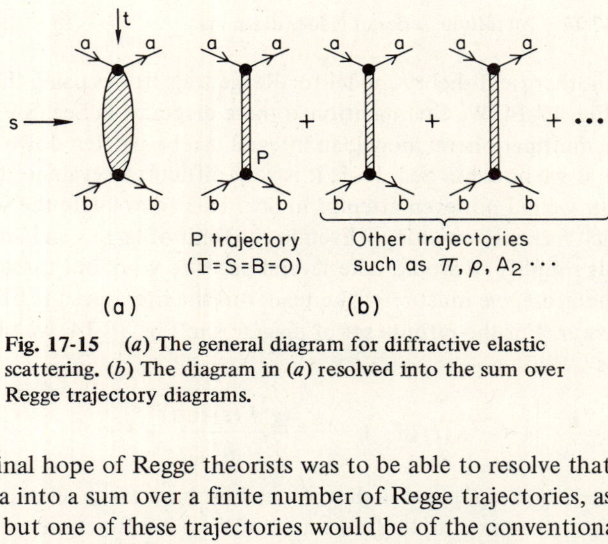

Fig. 17-15 (*a*) The general diagram for diffractive elastic scattering. (*b*) The diagram in (*a*) resolved into the sum over Regge trajectory diagrams.

The original hope of Regge theorists was to be able to resolve that cross-hatched area into a sum over a finite number of Regge trajectories, as in Fig. 17-15*b*. All but one of these trajectories would be of the conventional type—the π trajectory, the ρ trajectory, the A_2 trajectory. But the most important trajectory, shown on the left in Fig. 17-15*b*, would be a special trajectory called the *Pomeranchuk trajectory*, which would contain all the diagrams contributing to the diffractive nature of high energy elastic scattering. The Pomeranchuk trajectory was said to have the internal quantum numbers of the vacuum, $I = S = B = 0$. It also became common to speak of the Pomeranchuk trajectory as representing the exchange of a virtual particle called the *pomeron*.

If we accept the Pomeranchuk trajectory, then at high energies the optical theorem (Eq. 6-46c) and Eq. 17-56 yield

$$\sigma_{\rm tot}(s) = \left(\frac{-1}{2s}\right)\left[\eta_P\gamma_P(0)\left(\frac{s}{s_0}\right)^{\alpha_P(0)} + \sum_{n=2}^{N}\eta_n\gamma_n(0)\left(\frac{s}{s_0}\right)^{\alpha_n(0)}\right] \tag{17-64}$$

Here the subscript P indicates the Pomeranchuk trajectory and the summation is over the other trajectories; $\eta = \pm 1$ is the signature. We can then explain the predominantly imaginary nature of the diffractive scattering amplitude at $t \sim 0$ and the constancy of $\sigma_{\rm tot}(s)$ at large s by the following assumptions:

1. The Pomeranchuk dominates over all the other trajectories at $t = 0$.
2. $\alpha_P(0) = 1$.
3. $\eta_P = +1$; hence $\text{Re } T_P(s, t = 0) = 0$.

Exercise 17-19. Fill in the steps leading to these conclusions.

The concept of a Pomeranchuk trajectory with $\alpha_P(0) = 1$ is attractive; but we have grave doubts about the significance of interpreting diffractive elastic scattering as the result of a normal Regge trajectory. From the field theory point of view, indeed, such a trajectory would have a very different origin from the normal Regge trajectory. And from the Regge point of view it is very difficult to determine the meaning or to fix the shape of the Pomeranchuk trajectory in the $t > 0$ region. (For normal trajectories we use the masses and spins of known hadrons to fix the trajectories in this region.)

Nevertheless, the Regge language is so convenient that we continue to speak of the pomeron and to use such functions as Eqs. 17-44 and 17-56 to describe diffractive elastic scattering data. Of course regardless of whether the application of the Regge language to the Pomeranchuk trajectory has physical significance, we can still use Eq. 17-56 to examine the effects of the normal Regge trajectories in high energy elastic scattering.

As a simple and instructive example, we consider πp elastic scattering. The elastic scattering helicity amplitudes are (see Table 17-1)

$$\begin{aligned} T_{\mu'\mu}(\pi^- p) &= T_{P\mu'\mu} + T_{f\mu'\mu} + T_{\rho\mu'\mu} \\ T_{\mu'\mu}(\pi^+ p) &= T_{P\mu'\mu} + T_{f\mu'\mu} - T_{\rho\mu'\mu} \end{aligned} \tag{17-65}$$

where the relative signs are determined through the use of isotopic spin Clebsch-Gordan coefficients at each vertex. All these quantities are functions of s and t. As we have discussed many times (Ch. 2 and Sec. 11-5), the dominant contribution to the amplitudes will be that part of $\text{Im } T_{++}$ which describes the diffractive mechanism. And this contribution we identify with T_P; even when $t \neq 0$, we assume $|\text{Im } T_P| \gg |\text{Re } T_P|$. We shall examine three elastic scattering phenomena.

1. *Energy Dependence of $d\sigma/dt$.* The energy dependence of $d\sigma/dt$ will depend primarily on $\alpha_P(t)$ in the small t region. There is no agreement at present on the slope of $\alpha_P(t)$ near $t = 0$. It is not as large as the normal Regge trajectory slope

1 $(\text{GeV}/c)^{-2}$, and it might be zero. In any case, in the small $|t|$ region, Re $\alpha_P >$ Re α_f and Re $\alpha_P >$ Re α_ρ; if we use the conventional slopes: $(d\alpha_f/dt) \sim (d\alpha_\rho/dt) \sim$ 1 $(\text{GeV}/c)^{-2}$. Therefore, the contributions of T_f and T_ρ will become increasingly negligible as the energy increases. The high energy shrinkage in $d\sigma/dt$ (Sec. 1-6) depends on $\alpha_P' = d\alpha_P/dt$. Therefore,

$$\text{if } \alpha_P' > 0, \text{ as } s \to \infty, \qquad \text{shrinkage persists,} \qquad \sigma_{\text{el}} \to 0 \tag{17-66a}$$

$$\text{if } \alpha_P' = 0, \text{ as } s \to \infty, \qquad \text{shrinkage stops,} \qquad \sigma_{\text{el}} \to \text{constant} \tag{17-66b}$$

Under the conditions of Eq. 17-66b, shrinkage will be restricted to an intermediate energy phenomenon related to the normal Regge trajectories.

2. *The Crossover Phenomenon.* The contribution of the ρ trajectory to πp elastic scattering can be determined from the relation

$$\frac{d\sigma}{dt}(\pi^- p) - \frac{d\sigma}{dt}(\pi^+ p) \approx \frac{4(\text{Im } T_{P++})(\text{Im } T_{\rho++})}{16\pi s^2} \tag{17-67}$$

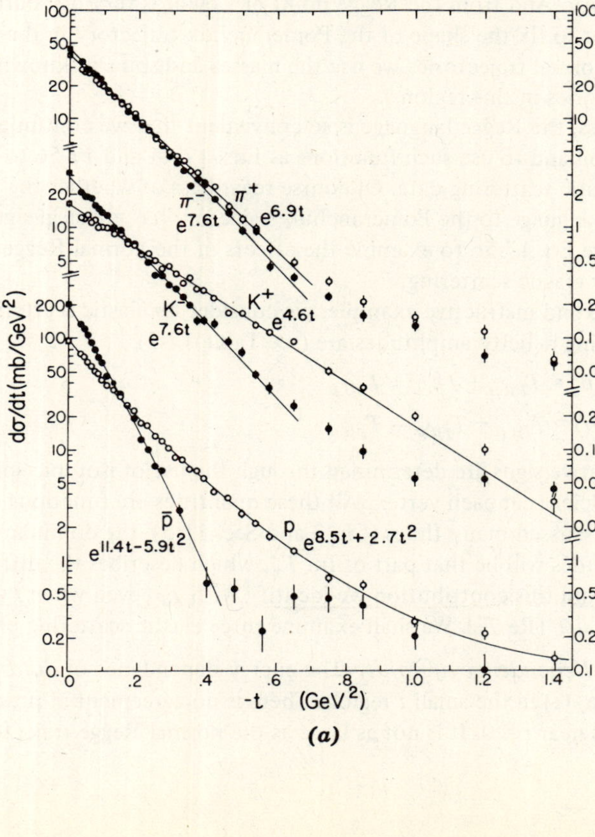

(a)

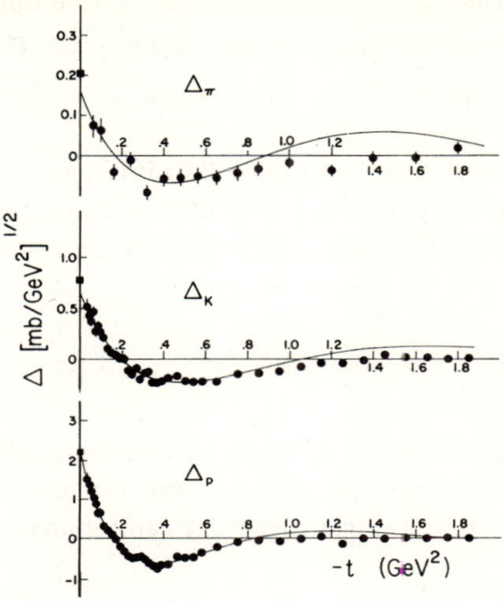

(b)

Fig. 17-16 (*a*) The elastic differential cross section $(d\sigma/dt)_a$ for the scattering of particle *a* on a proton, compared to $(d\sigma/dt)_{\bar{a}}$, for the scattering of the antiparticle $\bar{a}$ on a proton. All at 5 GeV/c. (*b*) The difference between $(d\sigma/dt)_a$ and $(d\sigma/dt)_{\bar{a}}$ as measured by $\Delta_a = [(d\sigma/dt)_a - (d\sigma/dt)_{\bar{a}}] / \{8[d\sigma/dt) + (d\sigma/dt)_{\bar{a}}]\}^{1/2}$ (17AM).

In Eq. 17-67 only the largest term has been retained on the right-hand side. The left-hand side of Eq. 17-67 exhibits experimentally a very interesting effect—the *crossover phenomenon* (Fig. 17-16). At $t = 0$, $d\sigma/dt(\pi^- p) > d\sigma/dt(\pi^+ p)$; but $d\sigma/dt(\pi^- p)$ decreases more rapidly than $d\sigma/dt(\pi^+ p)$ as $|t|$ increases. The two differential cross sections cross each other at $t \sim -0.2\ (\text{GeV/c})^2$. Then from Eq. 17-67

$$\text{Im}\ T_{\rho++} = 0 \qquad \text{at} \quad t \sim -0.2\ (\text{GeV/c})^2 \tag{17-68}$$

Exercise 17-20. The crossover phenomenon occurs in $\bar{K}N$ compared to KN elastic scattering, $\bar{N}N$ compared to NN elastic scattering, and in some pairs of diffractive inelastic reactions. See Fig. 17-16 and consult the literature (17LE, 17BA) on this effect. Extend Eqs. 17-65 and 17-67 to these systems.

3. *Polarization.* Reaching back to Eq. 11-30b, the polarization in πp elastic scattering is given by Eq. 17-69, assuming Re $T_{P+-} \approx 0$.

$$P_0(\pi^\pm p) \approx \frac{-2(\text{Im}\ T_{P++})(\text{Re}\ T_{f-+} \mp \text{Re}\ T_{\rho-+})}{|T_{++}|^2 + |T_{-+}|^2} \tag{17-69}$$

From Fig. 11-2 we see that the $\pi^- p$ and $\pi^+ p$ systems have opposite polarizations and that the polarizations pass through a zero at $t \sim -0.6$ (GeV/c)^2. Therefore,

$$(\text{Im } T_{P++})(\text{Re } T_{f-+}) = 0 \quad \text{in} \quad \text{small } t \text{ region} \tag{17-70}$$

Equation 17-70 will be true if (1) Re $T_{f-+} = 0$ or if (2) T_{f-+} and T_{P++} are in phase or 180° out of phase. In any case, the polarization is caused by the interference of the ρ trajectory with the Pomeranchuk trajectory, and the zero in the polarization requires

$$\text{Re } T_{\rho-+} = 0 \quad \text{at} \quad t \sim -0.6 \text{ (GeV/c)}^2 \tag{17-71}$$

Exercise 17-21. Use Eq. 17-56 to show that according to Regge theory the real and imaginary parts of the amplitude coming from a *single* trajectory have the same phase. Hence show that a single trajectory cannot give a nonzero polarization. Derive Eq. 17-69.

Of course much of the foregoing discussion was based on the quantum numbers of a general t-channel exchange mechanism and had nothing to do with Regge theory. Hence the comments also apply to one-particle exchange. But elementary one-particle exchange cannot provide an amplitude for vector meson exchange which satisfies Eqs. 17-68 and 17-71.

Exercise 17-22. Show that the needed elementary one-particle exchange amplitudes are provided by Eq. 12-54. Then confirm the last statement by showing that these amplitudes do not have the necessary zeros and indeed do not even have the required imaginary part, unless the coupling constant is made complex.

In Sec. 18-5 we inquire whether the Regge ρ trajectory can provide the desired amplitudes.

17-10 Regge Phenomenology in the Multiperipheral Model. In this final section we take up a different topic, the application of Regge phenomenology to the multiperipheral model (Sec. 8-9). Thus we illustrate the use of the Regge formulation in a rather general way, and we complete the discussion of Sec. 8-9. We reproduce Fig. 8-12 in Fig. 17-17 and we refer to Eq. 8-98 for the basic definitions. We assume that *all* particles have spin zero and mass m, and we ignore the discussion associated with Fig. 8-13. The total cross section for the production of N particles is

$$\sigma_N(s) = \frac{(2\pi)^4}{2s(2\pi)^{3N}2^N(N-2)!} \int \left[\prod_{n=1}^{N} \frac{d^3p_n}{E_n} |T_N|^2 \, \delta^4(P_f - P_i) \right] \tag{17-72}$$

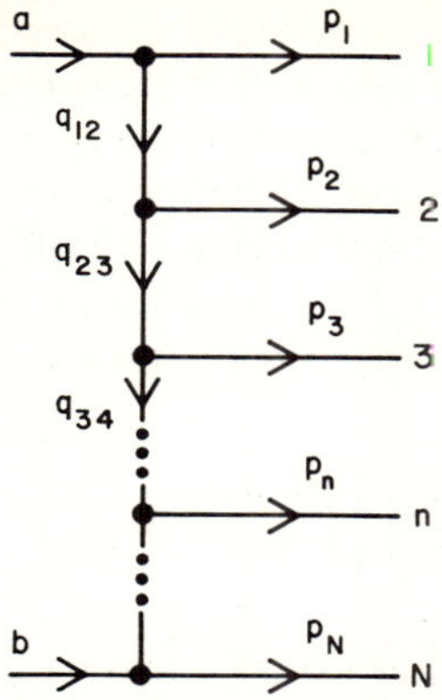

Fig. 17-17 Feynman diagram for the simple multiperipheral model.

from Eq. 8-1. We have $(N-2)!$ rather than $N!$ because we assume particles 1 and N to be leading particles that cannot be interchanged with the other secondary particles. According to Regge theory, the matrix element is

$$T_N = G^N \left(\frac{s_{12}}{s_0}\right)^{\alpha(t_{12})} \left(\frac{s_{23}}{s_0}\right)^{\alpha(t_{23})} \cdots \left(\frac{s_{N-1,N}}{s_0}\right)^{\alpha(t_{N-1,N})} \tag{17-73a}$$

$$\alpha(t) \sim \alpha_0 + \alpha' t, \qquad \alpha' \sim 1\ (\text{GeV/c})^{-2} \tag{17-73b}$$

Here G is the vertex coupling constant, and we have ignored the possibility of vertex form factors. That is, we have assumed that the combined $\beta(t)$ residues and signature factors are simply G. In writing this matrix element we assume that all the subenergies $s_{n,n+1}$ are large, for that is the condition for the validity of this Regge phenomenology. But for this model to be generally applicable to many-body processes, we must certainly allow small subenergies. We can ignore this assumption and hope for the best. Or we can use the duality principle discussed in Ch. 18 and assume that the Regge phenomenology provides an average description, even at low energies.

At this point we have put all the physics into the model and need only calculate. As we noted in Ch. 8, calculations involving phase space integrals can rarely be done analytically. One must use a computer to numerically evaluate the desired quantities. Computer programs (Refs. 17KI and 17FR) that can be used for such calculations now exist. For pedagogical purposes, however, we employ a very simple but instructive method of DeTar (17DET) which evaluates very roughly some quantities of interest. But we must emphasize that the approximations we are about to make are not required by the physics; they are only required by the need to obtain calculational simplicity. And we must not take the detailed results too seriously. Very approximate phase space calculations of the type we are about to perform, can give answers that are wrong in detail.

To proceed, we recognize that the matrix element in Eq. 17-73 will provide *a sharp transverse momentum cutoff.* Therefore we can adopt the calculational philosophy used in the transverse momentum cutoff phase space model of Sec. 8-7. In fact, we have in Eq. 17-73 a physical justification for the transverse momentum cutoff. Then, ignoring the conservation of transverse momentum as in Sec. 8-7, we obtain

$$\sigma_N(s) = \frac{cg^{2N}}{ss_0^{2N\alpha(\bar{t})}(N-2)!} \int \left[\prod_{n=1}^{N} \frac{dp_{\parallel n}}{E_n} (s_{12}s_{23} \cdots s_{N-1,N})^{2\alpha(\bar{t})} \times \delta(P_f^0 - P_i^0)\delta(p_{\parallel f} - p_{\parallel i}) \right] \tag{17-74}$$

where we have replaced each $\alpha(t_{n,n+1})$ by an *average* $\alpha(\bar{t})$ and absorbed factors such as 2^N into G and renamed it g. We have collected other constants into c and continue to do so. It is now convenient to use the rapidity variable defined in Eqs. 8-78 and 8-79. Then we have

$$\frac{dp_{\parallel n}}{E_n} = dy_n, \qquad n = 2, 3, \ldots, N-1$$

We have not changed the variables of particles 1 and N because we are going to use the crude trick, introduced in Sec. 8-7, of allowing two of the particles to take care of the energy and longitudinal momentum conservation. In selecting particles 1 and N for this task, we reason that they are the leading particles and are most likely to carry high energy. Of course in using this trick we are assigning low energies to all the other particles, which is not a good approximation. Finally, we carry out the trick to obtain

$$\sigma_N(s) = \frac{cg^{2N}}{s^2 s_0^{2N\alpha(\bar{t})}(N-2)!} \int \left[\prod_{n=2}^{N-1} dy_n [s_{12}s_{23} \cdots s_{N-1,N}]^{2\alpha(\bar{t})} \right] \tag{17-75}$$

Exercise 17-23. Fill in the steps leading to Eq. 17-75. Show that

$$s_{ij} = 2\omega_i\omega_j \cosh(y_i - y_j) + m_i^2 + m_j^2 - 2\mathbf{p}_{i\perp} \cdot \mathbf{p}_{j\perp}$$
$$\omega_i^2 = m_i^2 + \mathbf{p}_{i\perp}^2 \tag{17-76a}$$

Replacing all ω_i by an average value ω, show that

$$s_{ij} \approx \omega^2 e^{y_i - y_j} \qquad \text{for} \quad y_i \gg y_j \tag{17-76b}$$

Also show that $m^2 e^{y_1 - y_N} \approx s$.

We now make the final calculation approximation, assuming the so-called *strong-ordering* hypothesis

$$y_n \gg y_{n+1}, \qquad n = 1, 2, \ldots, N-1 \tag{17-77}$$

Under this hypothesis the particles are ordered along the multiperipheral string in order of decreasing longitudinal momenta. This is a stronger form of the assumption made in Sec. 8-9 preceding Ex. 8-25. Then, using Ex. 17-23, we obtain the simple and hopefully still useful result

$$\sigma_N(s) = \frac{cg^{2N}}{(N-2)!}\left(\frac{s}{s_0}\right)^{2\alpha(\bar{t})-2} \int \left[\prod_{n=2}^{N-1} dy_n\right] \tag{17-78a}$$

where some constants have been absorbed into c and g. Now we write

$$\int_{y_{\min}}^{y_{\max}} dy_n \sim \ln\left(\frac{s}{s_0}\right) \tag{17-78b}$$

where for simplicity we introduce again s_0. Then

$$\sigma_N(s) = \frac{cg^{2N}}{(N-2)!}\left(\frac{s}{s_0}\right)^{2\alpha(\bar{t})-2}\left(\ln\left(\frac{s}{s_0}\right)\right)^{N-2} \tag{17-79}$$

Equation 17-79 is very similar to Eq. 8-71b; in particular, $g^2 = C$.

We can easily find the total cross section

$$\sigma_{\text{tot}}(s) = cg^4\left(\frac{s}{s_0}\right)^{(2\alpha(\bar{t})-2+g^2)} \tag{17-80}$$

which is constant if

$$2\alpha(\bar{t}) - 2 + g^2 = 0 \tag{17-81}$$

The multiplicity distribution is given by the Poisson formula (Eq. 8-74) with the average multiplicity

$$\langle N\rangle \approx g^2 \ln\left(\frac{s}{s_0}\right) \tag{17-82}$$

Finally, by simply examining Eq. 17-78a, we can see that the one-particle rapidity distribution is flat. And if Eq. 17-81 is true, we obtain the Feynman scaling law for one-particle inclusive distributions.

Exercise 17-24. Verify the last set of conclusions. Use them to confirm the statements made about the multiperipheral model in Secs. 8-9 and 8-10. Also, compare these conclusions with Sec. 8-7, particularly Eqs. 8-71 to 8-74. Which aspects of this multiperipheral model are simply the effect of the sharp transverse momentum cutoff?

Table 17-1 **Some prominent Regge meson trajectories and some of the reactions to which they contribute. The π contribution to high energy KN, NN, and $N\bar{N}$ elastic scattering is taken to be small.**

Lowest Mass Member					Contributes to Forward Elastic Scattering in				
Name	Spin	Mass^2 (GeV^2)	Signature	IPG	πN	KN	NN	$N\bar{N}$	Contributes to Forward Inelastic Reactions
π	0	0.02	+1	1−−	No	(Yes)	(Yes)	(Yes)	$\pi N \to \rho N$, $\gamma N \to \pi N$, $\pi N \to \rho\Delta$ $pn \to np$, $KN \to K^*(890)N$
ρ	1	0.59	−1	1−+	Yes	Yes	Yes	Yes	$\pi^- p \to \pi^0 n$, $pn \to np$, $\pi N \to \pi\Delta$ $\pi N \to \omega N$, $K^- p \to \bar{K}^0 n$
ω	1	0.61	−1	0−−	No	Yes	Yes	Yes	$\pi^{\pm} p \to \rho^{\pm} p$
ϕ	1	1·04	−1	0−−	No	No	No	No	
f	2	1.61	+1	0++	Yes	Yes	Yes	Yes	
A_2	2	1.72	+1	1+−	No	Yes	Yes	Yes	$pn \to np$, $\pi^- p \to \eta^0 n$, $K^- p \to \bar{K}^0 n$ $\pi N \to \rho\Delta$
$K^*(890)$	1	0.79	−1	$\frac{1}{2}$−	No	No	No	No	$\pi^+ p \to K^+\Sigma^+$, $K^- p \to \pi^-\Sigma^+$

REFERENCES

17AH A. Ahmadzadeh et al., University of California Radiation Laboratory Report No. UCRL-10140, Berkeley, Calif., 1962.

17AM I. Ambats et al., *Phys. Rev. Lett.* **29**, 1415 (1972).

17BA V. D. Barger and D. B. Cline, *Phenomenological Theories of High Energy Scattering* (Benjamin, New York, 1969).

17BAY G. Baym, *Lectures on Quantum Mechanics* (Benjamin, New York, 1969).

17BL R. Blankenbecler and R. L. Sugar, *Phys. Rev.* **168**, 1597 (1968).

17CH G. F. Chew and S. C. Frautschi, *Phys. Rev. Lett.* **7**, 394 (1960).

17CHI C. B. Chiu, *Ann. Rev. Nucl. Sci.* **22**, 255 (1972).

17CO P. D. B. Collins, *Phys. Rep.* **1C**, 103 (1971).

17COL P. D. B. Collins and E. J. Squire, *Regge Poles in Particle Physics* (Springer, Berlin, 1968).

17DE V. de Alfaro and T. Regge, *Potential Scattering* (Interscience, New York, 1965).

17DET C. DeTar, *Phys. Rev.* **D3**, 128 (1971).

17DU L. Durand, III, *Phys. Rev.* **154**, 1537 (1967).

17ED R. J. Eden, *High Energy Collisions of Elementary Particles* (Cambridge University Press, Cambridge, 1967).

17FR J. H. Friedman and C. Risk, *Phys. Rev.* **D5**, 2245 (1972).

17KI W. Kittel et al., *Computer Phys. Commun.* **1**, 425 (1970).

17LE D. W. G. S. Leith, in *Proceedings of Sixteenth International Conference on High Energy Physics, 1972* (National Acceleration Laboratory, Batavia, Ill., 1973).

17LO C. Lovelace and D. Masson, in *Proceedings of the 1962 International Conference on High-Energy Physics* (CERN, Geneva, 1962).

17MA A. D. Martin and T. D. Spearman, *Elementary Particle Theory* (North-Holland, Amsterdam, 1970).

17NE R. G. Newton, *The Complex j-plane* (Benjamin, New York, 1964).

17SU R. L. Sugar and J. D. Sullivan, *Phys. Rev.* **166**, 1515 (1968).

17VA L. Van Hove, *Phys. Lett.* **24B**, 183 (1967).

CHAPTER

18

Two-Body Reactions at High Energy

18-1 Introduction. This chapter is concerned with two-body inelastic reactions and the nondiffractive aspects of elastic scattering in the forward peak and backward peak regions. We include the nondiffractive aspects of small angle elastic scattering because the mechanisms used to explain nondiffractive elastic scattering are the same as those used to explain two-body inelastic reactions. Indeed much, of the information on the ρ Regge trajectory, our standard example in this chapter, comes from $\pi^{\pm} p$ elastic scattering.

We found this chapter the most difficult to write, not because so little is known about two-body reactions, but because so much in known. No other aspect of hadron–hadron reactions has been studied more intensely in the past 15 years. However, the large amount of experimental and theoretical knowledge thus acquired simply has not been brought together into a unified theory of two-body interactions. Several times in the past decade we seemed to be on the verge of developing such a theory, particularly when such new theoretical insights as Regge theory were applied to the data. But each time a complete theory eluded us, as new experiments showed that the new insight was useful but not sufficient.

Perhaps the trend we are seeing in two-body inelastic interactions will eventually spread to all of strong interaction physics. The theory of strong interactions may be extremely complicated, and once we make detailed enough measurements those complications are forced on us. But it is also possible that most of our knowledge of two-body inelastic reactions comes from an energy region in which these reactions are particularly complicated—namely, the region of less than 15 or 20 GeV. It may be that the complications that puzzle us will go away as measurements are made in the hundreds of GeV range. At the very least we must reserve judgment until such measurements are made. Unfortunately, they will be very hard to make if, as we expect, the power law energy dependence of individual inelastic cross sections (Sec. 7-7) persists at very high energies. The individual cross sections will be very small. It is with an attitude of reserving judgment that we approach this material.

Section 18-2 summarizes the various types of experimental observations that must be explained by a detailed theory of two-body reactions. In Sec. 18-3 we briefly discuss the extension of Regge pole theory to reactions involving particles with nonzero spin. This used to be a very complicated subject, but in recent years our doubts and uncertainties about the fundamentals of Regge theory, as applied to strong interactions, have led us to ignore most of these complications. In Sec. 18-4 we add to those uncertainties by finally considering the subject of cuts in the complex angular momentum plane.

In Sec. 18-5 we return to the data and briefly discuss some of the mechanisms that have been proposed as explanations of the experimental observations. This section simply describes some of the mechanisms that have been considered, no attempt at a synthesis is made.

In the concluding section, the ideas of Regge trajectory exchange degeneracy and duality are introduced. The importance of the duality concept at energies above 10 or 15 GeV is not clear. Duality seems to be most significant in lower energy range as the transition is being made between low and high energy phenomena. Therefore, our treatment of duality is brief.

The discussion in this chapter is restricted to the regions of the forward or backward peaks $|t| \lesssim 1$ or 2 (GeV/c)^2 or $|u| \lesssim 1$ or 2 (GeV/c)^2. The mechanisms discussed have been extended to larger values of $|t|$ and $|u|$. We feel that such extensions are very provisional at present, but we deal briefly with them in Ch. 22.

To simplify the discussion in this chapter we usually consider only the large s, small t region—the forward peak region. The reader can extend the discussion to the small u region—the backward peak region.

18-2 Detailed Observations on Two-Body Reactions. The detailed observations on two-body reactions, which we must explain, fall in the following classes.

1. *Full Amplitude Analyses.* In the ideal situation, there exists a complete analysis of all the helicity amplitudes that contribute to a reaction. This means that at a fixed s, the real and imaginary parts of every amplitude are known functions of t or u. One of the best examples is πp elastic scattering (Sec. 11-5). Unfortunately, we are rarely able to accomplish the full amplitude analysis of a reaction, particularly at high energies. First, most two-body inelastic reactions have two nonzero spin particles in the final state; hence there are a large number of amplitudes to be determined. Second, it is difficult to make the measurements required to determine the complete joint spin density matrix of both particles in the final state.

2. *Partial Amplitude Analyses.* On the other hand, partial amplitude analyses have been carried out on many reactions in which the spin density matrix of one of the final state particles has been measured. In that case combinations of amplitudes are determined.

3. *Baryon Polarization Measurements.* When the baryon in the final state has spin $\frac{1}{2}$, it is relatively easy to obtain polarization information. For nucleons, the methods mentioned in Sec. 11-5 are used. For the Λ and Σ it is only necessary to observe their decay angular distribution. The polarization observed in πp elastic scattering is shown in Fig. 11-2. In polarization data, we are especially interested in the positions of the zeros, the sign of the polarization, and the energy dependence.

4. *Differential Cross-Section Measurements.* Differential cross sections are most easily measured, but the information they provide is the least restrictive with respect to testing theories of two-body reactions. As discussed in Secs. 17-7 and 13-2, the rough shapes of forward or backward peaks in two-body inelastic reactions are independent of the details of Regge theory or form-factor-modified, one-particle exchange theory. Therefore, models are really tested only by special phenomena in these peaks. One type of phenomenon involves the behavior of $d\sigma/dt$ as t approaches $t_{\min}$; examples are given in Figs. 7-9, 7-10, and 12-2. As discussed in Secs. 12-4 and 13-8, such effects seem to penetrate to the level of the amplitudes themselves; hence their interpretation is significant.

As t or u moves away from its minimum values, we also take as significant: changes in the slope of $d\sigma/dt$, dips in $d\sigma/dt$, or other abrupt changes in the shape of $d\sigma/dt$. One of the most prevalent of such phenomena occurs in the $|t|$ range of 0.5 to 1.0 $(\text{GeV}/c)^2$. As shown in Figs. 17-16 and 1-5, a slope change or dip occurs in this t range in πp, K^-p, and $\bar{p}p$ elastic scattering. Figure 18-1 indicates that some inelastic reactions also show a slope change or dip in $d\sigma/dt$ in this t range.

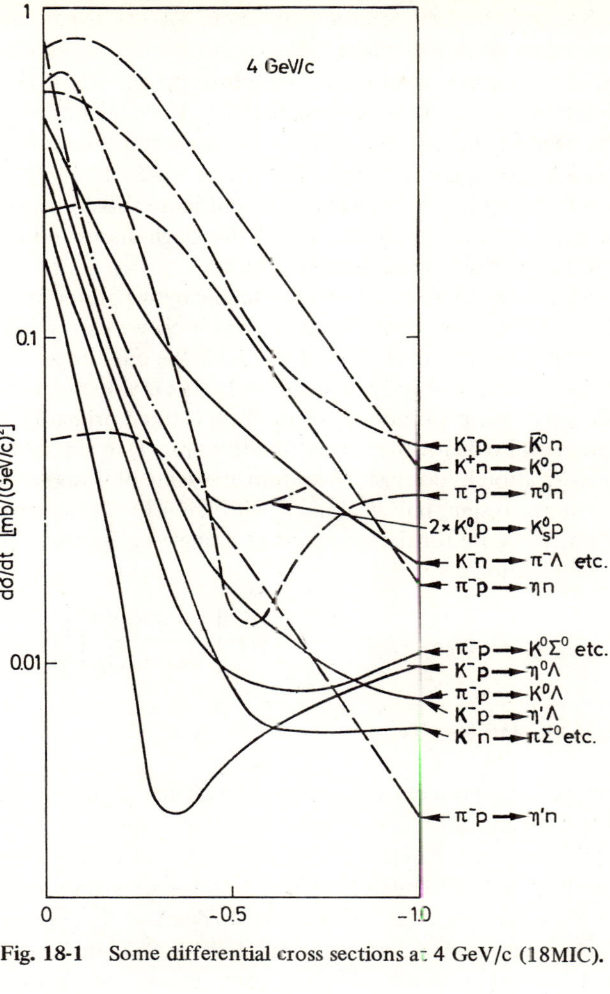

Fig. 18-1 Some differential cross sections at 4 GeV/c (18MIC).

18-3 Regge Trajectories and Nonzero Spin Particles. The entire discussion in Ch. 17 was based on the simplifying assumption that we could ignore the spins of particles a, b, c, and d in the reaction

$$a + b \rightarrow c + d \tag{18-1}$$

Without this assumption, as we remarked in Sec. 17-1, the mathematics becomes very complicated. To see why this is true, we return to our derivation of the Regge behavior in the high energy, small t region of the s-channel by a consideration of Eq. 17-28 in the t-channel (Sec. 17-5). We carried out the summation by using the leading term of the Legendre polynomial $P_l(-z)$. If particles a, b, c, or d have nonzero spin, we must replace $P_l(-z)$ by $d^J_{\mu\mu'}(-z)$ as prescribed by the general partial wave expansion of helicity amplitudes (Eq. 10-9). This will lead to a considerably more complicated version of Eq. 17-32. Other complications arise from the need to change from t-channel helicity amplitudes to s-channel helicity amplitudes using the crossing matrix (Secs. 16-2 and 16-3). Of course kinematic singularities (Sec. 15-7) will occur; and these should be disentangled before the Regge amplitude is constructed. A thorough discussion of how to do all this is provided by Collins and Squires (18CO).

But the modern tendency, or at least the tendency we prefer, is to ignore most of these complications. From Regge theory itself we have no detailed information about the form of $A_n(t)$ in Eq. 17-32. We only hope that it provides $T_n(t, s)$ with the desired analytic properties in t. This is why we went to $\gamma_n(t)$ in Eq. 17-34a, $\gamma_n(t)$ being a function whose form is fixed primarily by experiment. The same statement holds for the vertex functions $\beta_{iej}(t)$ in Eq. 17-54, if we accept the factorization hypothesis. We adopt the same philosophy when the particles have nonzero spin, only making a concession to the nonzero spins by using Eq. 10-35. We write for the exchange of a single t-channel Regge trajectory in the s-channel reaction $a + b \to c + d$

$$T^{\pm}_{\mu_c\mu_d;\mu_a\mu_b}(s, t) = |t_{\min} - t|^{|\mu_i - \mu_f|/2}\gamma(t)\left[\frac{1 \pm e^{-i\pi\alpha(t)}}{\sin(\pi\alpha(t))}\right]\left(\frac{s}{s_0}\right)^{\alpha(t)}$$

$$\mu_i = \mu_a - \mu_b, \mu_f = \mu_c - \mu_d \tag{18-2}$$

The first term on the right-hand side ensures the proper angular momentum dependence as $t \to t_{\min}$, and the last term ensures the energy dependence that is required by experiment. We keep the signature factor because it plays such a crucial role in the $t > 0$ region. (But see Sec. 18-5 for a contrary opinion.) And $\gamma(t)$ takes up the remaining t dependence. We do demand that $\gamma(t)$ be simple–an exponential, for example. In addition, as discussed in Sec. 18-5, $\gamma(t)$ may include factors to cancel the zeros in the $\sin(\pi\alpha(t))$ term. Finally, $\gamma(t)$ must not vary much from reaction to reaction; alternatively, we must use the factorization hypothesis (Eq. 17-54). Otherwise, the model has too much freedom.

Equation 18-2 must not be taken as holy. It is obviously a composite of theory and experiment. Some authors insert additional specific factors from the d^J function in Eq. 10-9 (18CO, 18BA, 18MA). These factors–usually functions of $\Gamma(\alpha(t) \pm r)$, where r is an integer or half-integer–introduce additional zeros and poles. Other authors suppress some of the explicit factors in Eq. 18-2 (Sec. 18-5). Before concluding this section, we note that we have not considered a number of topics: special properties of fermion trajectories; daughter trajectories, and other effects associated with unequal masses; and the construction of helicity amplitudes that preserve parity conservation in the t-channel (18CO, 18BA, 18MA). But in our view none of these topics are of immediate help in penetrating the mysteries of the validity of Regge trajectory theory or how to calculate $\alpha(t)$ and $\gamma(t)$ from first principles. Hence we have not allowed the large amount of space required by such topics.

18-4 Cuts in the Complex Angular Momentum Plane. In nonrelativistic, potential scattering, the only singularities of the scattering amplitude in the complex angular momentum plane are poles—the Regge poles. Since the application of the Regge theory to elementary particle reactions was partially based on analogies to potential scattering, it was natural to try to preserve this restriction. But theoretical models (18CO, 18COL) and experiment (Sec. 18-5) seem to demand the existence of branch point singularities in the complex angular momentum plane. Thus we must modify Fig. 17-1*a* to include branch points and their associated cuts, as illustrated in Fig. 18-2.

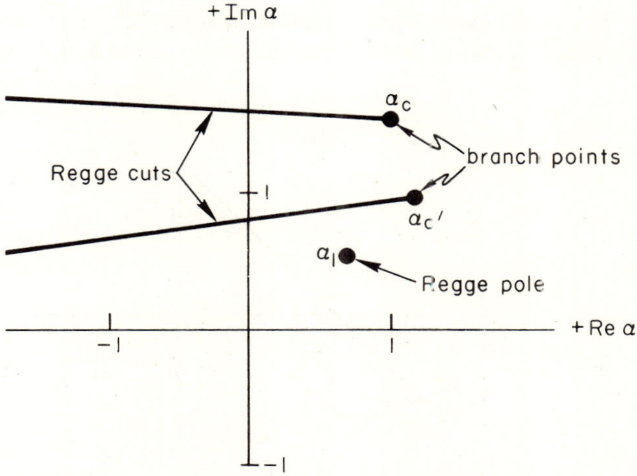

Fig. 18-2 The singularities of the amplitude in the complex angular momentum plane. Regge cuts with branch points at α_c and α_c', as well as a Regge pole at α_1, are shown.

From the viewpoint of simplicity and elegance, the occurrence of Regge cuts is most unfortunate, since a cut singularity is much more complicated than a pole singularity. Assuming that these cuts, like the poles, are functions of t, the full description of the cut requires the specification, as a function of t, of both the position of the entire cut and the discontinuity across the cut. Nevertheless, Regge cuts have been absorbed into the Regge picture of two-body reactions using arguments based partly on experiment, partly on models, and partly on faith. This picture is now sketched.

1. Regge poles are still assumed to be the most important contribution to amplitudes in the large s and small t or small u regions. This assumption is partly justified by the argument that the Regge pole trajectory gives the correct energy dependence of the amplitude. Using models such as the one in Ex. 18-1, the energy dependence of a Regge cut appears to be different from that of a pole; hence a Regge cut alone would not predict the correct energy dependence of the amplitude.

Exercise 18-1. For a simple model of a Regge cut we assume the cut can be described as an integral over a continuous spectrum of Regge poles. We write

$$T_{\text{cut}}(s, t) = \int_{-\infty}^{\alpha_c(t)} \rho(\alpha(t))\gamma(t)\left(\frac{s}{s_0}\right)^{\alpha(t)} d\alpha(t) \tag{18-3}$$

where we have ignored the spin and signature factors of Eq. 18-2. This model is illustrated in Fig. 18-3. We define $\alpha_c(t)$ as the trajectory with the largest value of

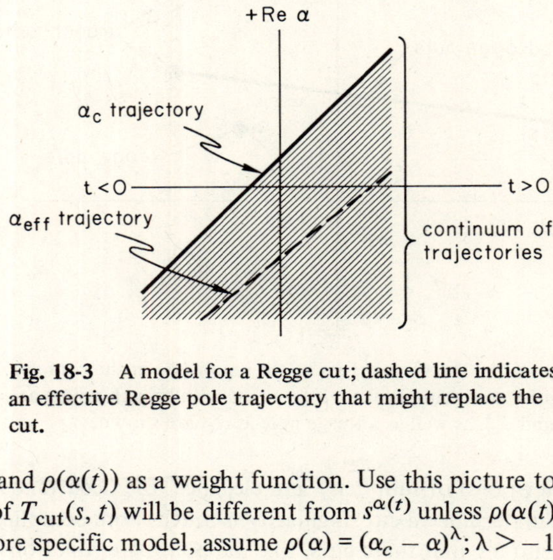

Fig. 18-3 A model for a Regge cut; dashed line indicates an effective Regge pole trajectory that might replace the cut.

Re α at each t and $\rho(\alpha(t))$ as a weight function. Use this picture to show that the s dependence of $T_{\text{cut}}(s, t)$ will be different from $s^{\alpha(t)}$ unless $\rho(\alpha(t))$ is peaked at $\alpha_c(t)$. For a more specific model, assume $\rho(\alpha) = (\alpha_c - \alpha)^\lambda$; $\lambda > -1$, and show

$$T_{\text{cut}}(s, t) = \frac{\gamma(t)\Gamma(\lambda + 1)(s/s_0)^{\alpha_c(t)}}{(\ln (s/s_0))^{\lambda+1}} \tag{18-4}$$

2. There is another argument that partly justifies the assumption that Regge poles remain the most important contribution to the amplitude. Each Regge trajectory $\alpha(t)$ has been connected in the $t > 0$ region with a family of elementary particles with definite S, B, and I. The trajectory $\alpha(t)$ carries these fixed internal quantum numbers into the $t < 0$ region, thus explaining the presence or absence of forward or backward peaks. The Regge cut is not connected with elementary particle families a priori. Therefore, the cut does not necessarily carry specific values of S, B, and I. Hence without further assumptions, the cut does not explain the presence or absence of forward or backward peaks. On the other hand, as discussed in Ex. 18-2, one can construct for the cut models that do assign to it specific values of S, B, and I.

3. What do we do about Regge cuts? They apparently are not the most important contribution to the amplitude, but they cannot be ignored either. The usual solution is to assume that the cuts are intimately associated with the absorption correction to the Regge pole amplitude. One also assumes that the effect of the absorption corrections includes the effects of the Regge cuts. The justification for these assumptions comes from Regge cut models based on multi-Regge trajectory exchange. These models are not really satisfactory, and we discuss them very briefly (18CO, 18COL).

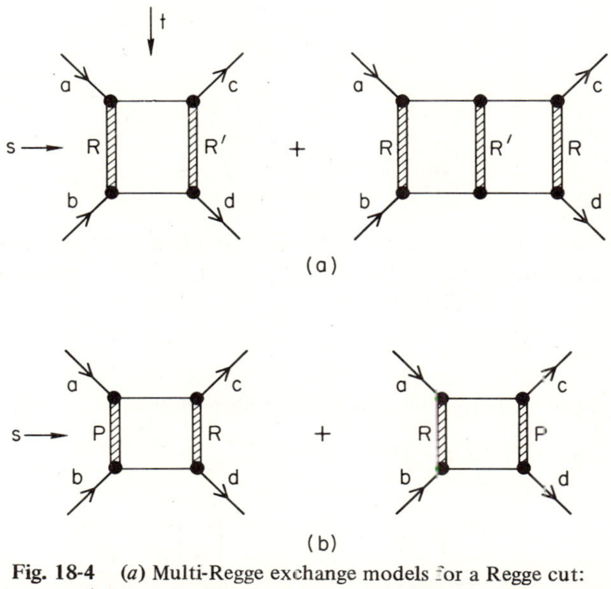

Fig. 18-4 (*a*) Multi-Regge exchange models for a Regge cut: R and R' are normal Regge trajectories. (*b*) Model for a Regge cut in which a pomeron P and a normal Regge trajectory R are exchanged.

In Ch. 15 we showed that cuts in the amplitude $T(s, t)$ in the complex t-plane come from the exchange of two or more elementary particles in the t-channel. Similarly, we might suspect that cuts in $T(s, t)$ in the complex angular momentum plane come from the exchange of two or more Regge trajectories in the same diagram. This is illustrated in Fig. 8-4a, where R and R' denote different Regge trajectories. Unfortunately, the simpler versions of such diagrams do not give Regge cuts with the desired physical properties (18CO, 18COL). Hence the association of multi-Regge trajectory exchange with Regge cuts is not a simple idea—and it could be an incorrect idea.

Nevertheless, the idea has led to the hope that the most important Regge cuts come from the double exchange of a Regge trajectory R and a pomeron P as illustrated in Fig. 18-4b. And if we take the intermediate s-channel particles as nearly real, we can think of these diagrams as simply representing the absorption correction to the Regge trajectory R.

Exercise 18-2. We can make an explicit calculation for this model by using directly the results of Ex. 13-9. We define the Regge trajectory amplitude

$$T_R(s, t) = A\gamma(t)\left(\frac{s}{s_0}\right)^{\alpha(t)} = A\left(\frac{s}{s_0}\right)^{\alpha_0+\alpha_0' t} e^{bt/2}$$

$$= A\left(\frac{s}{s_0}\right)^{\alpha_0} e^{b_R t/2} \tag{18-5a}$$

$$b_R = b + 2\alpha_0' \ln\left(\frac{s}{s_0}\right) \tag{18-5b}$$

where we have ignored the signature factor for simplicity. We assume that the pomeron has a flat trajectory so that $T_P(s, t) \sim (s/s_0)\, e^{B_P t/2}$. Equation 13-34 becomes

$$S_{el} = 1 - C e^{-b^2/2B_P} \tag{18-6}$$

Now simply rewrite Eq. 13-36 in terms of the new parameters and obtain

$$T_{R,\text{modified}}(s, t) = T_R(s, t) + T_{\text{cut}}(s, t) \tag{18-7a}$$

$$T_{\text{cut}}(s, t) = -ACr\left(\frac{s}{s_0}\right)^{\alpha_0} \exp\left(r\left[\frac{b}{2} + \alpha_0' \ln\left(\frac{s}{s_0}\right)\right] t\right) \tag{18-7b}$$

$$r = \frac{B_P}{B_P + b + 2\alpha_0' \ln(s/s_0)} \tag{18-7c}$$

Comparing Eq. 18-7b with Eq. 18-5, show that the forward peak given by T_{cut} is flatter than the forward peak given by T_R, which is usually taken to be one of the characteristics of the cut contribution to the amplitude. Remembering that C is roughly independent of s, discuss the s dependence. Show that in this model the Regge cut has the same B, S, and I as the Regge pole R.

18-5 Absorption and Other Mechanisms. In this section we describe some of the mechanisms that have been proposed as explanations of the detailed behavior of two-body reactions, summarized in Sec. 18-2. Again, we make no attempt to choose between these mechanisms or to present a synthesis using several mechanisms. At present each mechanism has its successes, its failures, and its uncertainties. Most authors begin with a Regge trajectory amplitude like that in Eq. 18-2 with various modifications. They then apply some absorption correction using the method of Sec. 13-5, although they differ widely in the extent of the absorption. We summarize the relevant equations. For convenience, we set $\phi = 0$, take all external masses equal to m, and take $s \gg m^2$.

$$T^{\pm}_{R,\mu_c\cdots}(s, t) = (-t)^{n/2}\gamma(t)\left[\frac{1 \pm e^{-i\pi\alpha(t)}}{\sin(\pi\alpha(t))}\right]\left(\frac{s}{s_0}\right)^{\alpha(t)} \tag{18-8}$$

$$T^{\pm}_{R,\mu_c\cdots}(s, t) = 8\pi s(-1)^n \int_0^\infty \left[b\,db\, J_n(b\sqrt{|t|})T^{\pm}_{\mu_c\cdots}(s, b)\right] \tag{18-9a}$$

$$T^{\pm}_{\text{abs},\mu_c\cdots}(s, t) = 8\pi s(-1)^n \int_0^\infty \left[b\,db\, J_n(b\sqrt{|t|})T^{\pm}_{\mu_c\cdots}(s, b) \times S_{\text{el}}(s, b)\right] \tag{18-9b}$$

Here

$$\mu_{c\cdots} = \mu_c\mu_d;\mu_a\mu_b$$

and

$$n = \mu_i - \mu_f = (\mu_a - \mu_b) - (\mu_c - \mu_d) \geqslant 0 \tag{18-10}$$

is the net s-channel helicity flip. For $n < 0$ use Eq. 10-19 first.

1. *Pion Exchange in the Very Small t Region.* As discussed in Sec. 13-8, when $|t - t_{\min}| \to 0$ sharp peaks or dips can appear in π exchange reactions. We explained these phenomena in Sec. 13-8, using elementary one-particle

exchange with absorption, and the same explanation works using the Regge amplitude, Eq. 18-8. In the very small t region, we have

$$[\sin(\pi\alpha_\pi(t))]^{-1} \sim \frac{\text{constant}}{t - m_\pi^2} \tag{18-11}$$

assuming $d\alpha_\pi(t)/dt \sim 1\ (\text{GeV/c})^{-2}$. Hence the Regge amplitude simply reduces to the elementary one-particle exchange amplitude multiplied by functions of t such as $\gamma(t) = e^{at}$. Thus the Regge trajectory does not add anything very new to our understanding of this very small t region in π exchange.

2. *Amplitude Zeros and Abrupt Changes in $d\sigma/dt$ at $t \neq 0$.* In Sec. 17-9 we showed that πp elastic scattering requires the following zeros in the ρ Regge trajectory amplitudes:

$$\text{Im}\, T_{\rho_{++}} = 0 \quad \text{at} \quad t \sim -0.2\ (\text{GeV/c})^2 \tag{18-12a}$$

$$\text{Re}\, T_{\rho_{-+}} = 0 \quad \text{at} \quad t \sim -0.6\ (\text{GeV/c})^2 \tag{18-12b}$$

These zeros are representative of the amplitude zeros demanded by other two-body reactions in other Regge trajectories. Therefore, we use Eqs. 18-12a and 18-12b to illustrate our discussion.

For further illustration we shall use:

$$\text{change in slope of}\ \frac{d\sigma}{dt}(\pi^\pm p \to \pi^\pm p) \quad \text{at} \quad t = -0.6 \text{ to } -0.8\ (\text{GeV/c})^2 \tag{18-12c}$$

$$\text{minimum in}\ \frac{d\sigma}{dt}(\pi^- p \to \pi^0 n) \quad \text{at} \quad t = -0.5 \text{ to } -0.6\ (\text{GeV/c})^2 \tag{18-12d}$$

as shown in Figs. 17-16 and 18-1.

3. *Nonabsorptive Mechanisms.* A general mechanism for producing phenomena such as those listed in item 2 is that the Regge residue functions [$\gamma(t)$ in Eq. 18-8] have zeros in some of the amplitudes at those t values. The $t \sim -0.6\ (\text{GeV/c})^2$ zero in the ρ trajectory amplitude (Eq. 18-12b) is particularly suggestive in this regard. For if we use $\alpha_\rho(t) = \alpha_{\rho 0} + 0.9t$, we find

$$\alpha_\rho(t) = 0 \quad \text{at} \quad t \sim -0.55\ (\text{GeV/c})^2 \tag{18-13}$$

Therefore, if $\gamma(t)$ contains a factor $\alpha_\rho(t)$ in a particular amplitude, we might expect a zero in that amplitude at $t \sim -0.6\ (\text{GeV/c})^2$.

Exercise 18-3. Use the spins and masses of the ρ and g mesons to derive the formula for $\alpha_\rho(t)$ which we just used.

Exercise 18-4. In the traditional method for generating Regge amplitudes from helicity amplitudes, T_{-+} does indeed have an explicit factor $\alpha(t)$. This factor comes from the function $d^j_{1/2,1/2}(-z)$ which replaces $P_j(-z)$ in Eq. 17-28. Show how the $\alpha(t)$ factor appears by using Eq. C-29 (App. C) and repeating the derivation that follows Eq. 17-28.

The effects illustrated in Eqs. 18-12c and 18-12d have also been attributed to amplitude zeros. It is assumed that in the sum over helicity amplitudes $\Sigma_{\mu_a\mu_b\mu_c\mu_d} |T_{\mu_c\mu_d;\mu_a\mu_b}|^2$, one or more of the Regge trajectory contributions goes to zero.

Exercise 18-5. The ρ trajectory is the only Regge trajectory that contributes appreciably to the charge exchange reaction $\pi^- + p \to \pi^0 + n$. The classic Regge analysis using a pure Regge trajectory was given by Höhler et al. (18HO; see also 18MA). There the $\alpha(t)$ factor of Ex. 18-4 is used to produce the dip at $t \sim -0.6$ (GeV/c)^2 (Fig. 18-1). Since that analysis was performed, it has been definitely established that this reaction exhibits nonzero polarization over most of the small t range. Show that this result invalidates the classic analysis and that we require either absorption corrections or an additional trajectory.

A more general mechanism for production zeros comes from the signature factor

$$\mathscr{S}^{\pm} = \frac{1 + \eta\, e^{-i\pi\alpha(t)}}{\sin(\pi\alpha(t))} \tag{18-14}$$

In the $t < 0$ region this factor causes some embarrassment. As discussed in Ex. 17-14, an unphysical pole or *ghost* state appears whenever there is an infinity in the amplitude in the $t < 0$ region. The $\sin(\pi\alpha(t))$ term in Eq. 8-14 will lead to such ghost states at $\alpha(t) = \pm$ integer, unless there is a corresponding zero in the amplitude. At the *wrong signature points*, defined by Re $(1 + \eta\, e^{-i\pi\alpha}) = 0$, such a zero occurs in the signature factor. But at *right signature points*, defined by Re $(1 + \eta\, e^{i\pi\alpha}) = 2$, there is no corresponding zero. To avoid the infinity at right signature points in the $t < 0$ region, it is customary to assume that $\gamma(t)$ has a zero at those points.

Now the mechanism in $\gamma(t)$ that produces a zero at integers corresponding to right signature points *might* also produce a zero at the other integers, the wrong signature points. And this would produce zeros in the amplitude at wrong signature points. Justification for the "might" comes from the trajectory exchange degeneracy concept (Sec. 18-6).

Exercise 18-6. Apply the foregoing arguments to the ρ trajectory, and show that the first wrong signature zero is at $\alpha(t) = 0$, if it occurs.

We now conclude our examples of nonabsorptive mechanisms that can produce zeros in amplitudes. We do not know if any of these mechanisms have much validity, but we do know that such mechanisms by themselves are not sufficient to explain the many observations on dips and zeros in two-body reactions (18MI).

4. *Absorptive Mechanisms for Amplitude Zeros in the $t \neq 0$ Region.* All present models of two-body reactions accept the presence of absorption corrections. But some models use the absorption corrections to drastically modify the Regge pole amplitude, producing additional zeros, or all the zeros, through the modification. We give two examples.

Several years ago Ross et al. (18RO) developed a model in which all the zeros were produced by absorption. To eliminate any of the mechanisms discussed in item 2, they simplify Eq. 18-8 into the form

$$T_{R,\mu_c,\ldots}(s,t) = |t|^{n/2}\, e^{-(i\pi\alpha(t)/2)} \left[\frac{\gamma(t)}{t - m_e^2}\right] \left(\frac{s}{s_0}\right)^{\alpha(t)} \tag{18-15}$$

where m_e is the mass of lightest particle on trajectory R and $\gamma(t)$ has a simple form (e.g., an exponential in t). To see how a zero is produced by absorption, we simplify a bit further and write (except for π exchange)

$$\frac{\gamma(t)}{t - m_e^2} \sim a\, e^{b_R t/2} \tag{18-16}$$

Now we can use the results of Ex. 18-2 to obtain the absorption corrected T_{R++} amplitude. We simplify the notation by writing

$$B_R = b_R + 2\alpha'_{R0} \ln\left(\frac{s}{s_0}\right), \qquad A = a\left(\frac{s}{s_0}\right)^{\alpha_{R0}} e^{-i\pi\alpha_R/2} \tag{18-17a}$$

$$T_{R++} = A\, e^{B_R t/2} \tag{18-17b}$$

$$r = \frac{B_P}{B_P + B_R} \tag{18-17c}$$

Remember that B_P is the exponential slope of the elastic $d\sigma/dt$. Then we have

$$T_{R,\mathrm{abs}++} = T_{R++}\left[1 - rC\, e^{(rB_R|t|/2)}\right] \tag{18-18}$$

where we have ignored the effect of the absorption correction on the phase factor $e^{-i\pi\alpha/2}$.

Exercise 18-7. We can produce the zero in $T_{\rho,\mathrm{abs}++}$ at $t = -0.2\ (\mathrm{GeV}/c)^2$ by demanding that the factor in the square brackets in Eq. 18-18 be zero at that

t value. Take $B_P = B_\rho = 8\ (\text{GeV/c})^{-2}$. Then show that a zero will occur at $t = -0.2$ when

$$C = 1.35 \tag{18-19}$$

Exercise 18-8. Use Eq. B-8 to show that the absorption correction applied to the amplitude

$$T_R^{(n)} = A|t|^{n/2}\, e^{B_R t/2}; \qquad n = \mu_i - \mu_f \tag{18-20}$$

yields

$$T_{R,\text{abs}}^{(n)} = T_R^{(n)}[1 - Cr^{n+1}\, e^{(rB_R|t|/2)}] \tag{18-21}$$

Then show that the exponential slopes in Ex. 18-7 and the value of C given in Eq. 18-19 yield (with the correction to the phase factor ignored)

$$T_{\rho,\text{abs}-+} = 0 \qquad \text{at} \quad t \sim -0.54\ (\text{GeV/c})^2 \tag{18-22}$$

Hence using this absorption model we have been able to explain the two zeros in the ρ amplitudes in Eqs. 18-12a and 18-12b. But there is a problem. We have used (Eq. 18-19) a value of C greater than 1. Yet in our original definition (Eq. 13-34), C was restricted to the range $0 \leqslant C \leqslant 1$, for C is an amplitude absorption coefficient. We do not go into the question of whether $C > 1$ has physical meaning or of how more sophisticated models may get around this problem (18MI, 18RO, 18TR). Our object has been only to show how the absorption model can produce zeros in amplitudes.

5. *Peripheral Models.* If absorption corrections are believed to produce the zeros in some amplitudes, one might allow the absorption correction to dominate the amplitude completely. In that case the *geometric* impact parameter picture turns out to be very useful. In fact, we have already calculated what we need in Sec. 10-8. We suppose that in Eq. 18-9b the absorption completely removes $T(s, b)$ when b is less than the hadronic radius R. And we may also suppose that $T(s, b)$ itself decreases rapidly when b is greater than R. Then $T(s, b)S_{\text{el}}(s, b)$ is different from zero only in the range

$$R - \delta R < b < R + \delta R \tag{18-23}$$

We may immediately use the δ function approximation of Eq. 10-47 and obtain (with $\phi = 0$ for simplicity)

$$T^{(n)} = \text{constant}\left(\frac{s}{s_0}\right)^{\alpha(t)} J_n(R\sqrt{|t|}), \qquad n = \mu_i - \mu_f, \qquad n \geqslant 0 \tag{18-24}$$

All that we have retained of the Regge input is the energy dependence. We have come far from the Regge philosophy of Ch. 17, and we seem almost back to the simple absorption ideas introduced early in the book. The next exercise shows why we are willing to regress.

Exercise 18-9. Construct the following table for the zeros of $J_n(R\sqrt{|t|})$ where $R = 1$ fermi, the standard hadronic radius and t is in $(\text{GeV}/c)^2$. Show that the table below can explain the zeros in Eq. 18-12.

n	0	1	2
t at first zero	-0.23	-0.57	-1.03
t at second zero	-1.19	-1.92	-2.76

Exercise 18-10. Plot $J_0(R\sqrt{|t|})$ for $R = 1$ fermi. Show that the first algebraic minimum in J_0 occurs in roughly the right t range to explain Eq. 18-12c, if Im T_{f++} and Im $T_{\rho++}$ (Eq. 17-65) have the J_0 behavior.

In impact parameter space, the entire amplitude in Eq. 18-24 comes from the periphery of the hadrons: it comes from glancing collisions. Therefore, this is a peripheral model for inelastic and nondiffractive elastic amplitudes. This peripheral model should be distinguished from the more general application of the description "peripheral" to high energy phenomena (Sec. 7-9). This model predicts a definite relation between the zeros of the amplitude and the net s-channel helicity flip, a relation that is due to the strong absorption of the non-peripheral part of the amplitude.

Of course life among high energy physics models is never simple, and the peripheral model cannot be applied universally. As an illustration, consider the $[F^I_{\mu\mu'}]_J$ amplitudes for πp elastic scattering in Fig. 11-4. Some of the amplitudes are clearly peripheral. But the dominant diffractive amplitude $[\text{Im}\, F^0_{++}]_J$ is clearly not peripheral; and some other amplitudes are also not peripheral.

Exercise 18-11. Why does the peripheral model not apply to the dominant Pomeranchuk contribution in Im T_{++}? For the amplitudes in Fig. 11-4 that demonstrate peripherality, compare their t behavior in Fig. 11-3 with $J_n(R\sqrt{|t|})$.

As has been emphasized by Harari (18HA), only some classes of amplitudes appear to be given by the peripheral model. Others seem to be given by Regge trajectories with only minor absorption corrections (18MI, 18TR). This exposition is intended only as an introduction to the elements of the model. We have ignored considerations that are important in this model such as the phase of the amplitude.

18-6 Exchange Degeneracy and Duality. Throughout this book we have implicitly maintained a separation between the s-channel resonance explanation of s-channel reactions (Sec. 5-7) and the t-channel exchange explanation (one-particle or Regge) of s-channel reactions. We have discussed the s-channel resonance as primarily a low energy mechanism; and we have discussed the t-channel exchange as a high energy mechanism. We have also implied that in the intermediate energy region an amplitude should be written in the form

$$T = T\,(s\text{-channel resonance}) + T\,(t\text{-channel exchange}) \tag{18-25}$$

The first term is given by Eq. 12-62

$$T\,(s\text{-channel resonance}) = \sum_J \frac{A_J(p^*)^{2J}P_J(\cos\theta)}{(s - m_J^2) + im_J\Gamma_J} \tag{18-26}$$

where the sum is over a set of resonances with spins $J_1, J_2, \ldots$. The T (t-channel exchange) has the properties discussed in Ch. 17 and in this chapter.

But is it correct to separate the s-channel resonance mechanism and the t-channel exchange mechanism. Are these really completely separate processes? Table 18-1 provides some reasons for raising this question. There appears to be an intimate relation between the s-channel resonance phenomenon and some t-channel exchange phenomena. Let us explore this in more detail using the Kp system. Our investigation leads to two intertwined concepts–exchange degeneracy and duality, and we consider the exchange degeneracy concept first.

The K^+p and K^-p elastic scattering amplitudes are

$$T_{\text{el}}(K^-p) = T_P + T_f + T_\omega + T_\rho - T_{A_2} \tag{18-27a}$$

$$T_{\text{el}}(K^+p) = T_P + T_f - T_\omega - T_\rho + T_{A_2} \tag{18-27b}$$

Now if we assume that the pomeron has a flat trajectory, $d\alpha_P(t)/dt = 0$, we can make $\sigma_{\text{tot}}(K^+p)$ energy independent by setting

$$\text{Im}\, T_f = \text{Im}\, T_\omega, \qquad \text{Im}\, T_\rho = \text{Im}\, T_{A_2} \tag{18-28}$$

where we have equated amplitudes with the same I but opposite signature. Furthermore, Eq. 18-28 can explain the absence of dips or slope changes in $d\sigma/dt\,(K^+p)$ in the $0.5 \leqslant |t| \leqslant 1.0\ (\text{GeV/c})^2$ region. For those phenomena seem to come from the term in $d\sigma/dt$

$$\left[\sum_{R=f,\omega,\rho,A_2} \text{Im}\, T_P\, \text{Im}\, T_R\right]^2 \tag{18-29}$$

For Eq. 18-28 to be true, we must have

$$\alpha_f(t) = \alpha_\omega(t); \qquad \alpha_\rho(t) = \alpha_{A_2}(t) \tag{18-30}$$

$$\gamma_f(t) = -\gamma_\omega(t); \qquad \gamma_\rho(t) = -\gamma_{A_2}(t) \tag{18-31}$$

where we have used the standard Regge form (Eq. 18-2) for the amplitudes. The relations in Eq. 18-31 are certainly permitted by Regge theory, for we know nothing a priori about the particular form of a residue function. On the other hand, the relations in Eq. 18-30 can be directly tested by looking at the Chew–Frautschi plot (Fig. 17-9). We see that the equalities of the Re $\alpha(t)$ are obvious from the plot, and there is nothing against assuming the equalities of Im $\alpha(t)$, particularly since we assume Im $\alpha(t) = 0$ when $t < 0$. In fact, the near superposition of the meson trajectories on the Chew–Frautschi plot made it necessary for us to introduce the signature idea with theoretical arguments (Sec. 17-4). The signature concept is not at all obvious from the experimental points on the mesonic Chew–Frautschi plot.

Exercise 18-12. Fill in the steps in the foregoing discussion. Show that similar arguments hold for the NN system compared with the $\bar{N}N$ system. Why do both $d\sigma/dt\,(\pi^+p)$ and $d\sigma/dt\,(\pi^-p)$ show a change of slope in this t region?

Exercise 18-13. One argument for the signature factor comes from the *exchange force* concept in nonrelativistic potential scattering. Consult the literature (18BA, 18CO, 18MA) for this argument.

Since the signature factor can be explained using the exchange force concept (Ex. 18-13), when the conditions in Eqs. 18-30 and 18-31 are obeyed, the trajectories are said to be *exchange degenerate.* Thus the f and ω are exchange degenerate, and the ρ and A_2 are exchange degenerate. If only Eq. 18-30 is obeyed there is said to be *weak* exchange degeneracy but not *strong* exchange degeneracy.

Exercise 18-14. Discuss the significance of weak and strong degeneracy on line-reversed two-body reactions (Eq. 15-12). Pairs of line-reversed reactions are used to test the exchange degeneracy concept. Consult the literature (18MI, 18TR, or similar references) for the present status.

Exercise 18-15. Show that if two trajectories of opposite signature have strong exchange degeneracy; a zero in $\gamma(t)$ used to suppress a right signature pole in one trajectory will also lead to an additional zero at a wrong signature point of the other trajectory. This is the justification (mentioned just before Ex. 18-6) for a zero in $\gamma(t)$ at every integral value of $\alpha(t)$ in the $t < 0$ region.

Like the other mechanisms discussed in this chapter, the concept of exchange degenerate Regge trajectories seems to be valid in some cases but not in all. There are also uncertainties about how the concept is to be used. For example, is it to be applied to the Regge trajectories before or after the application of the absorption correction?

To conclude this chapter we return to the other concept suggested by the data of Table 18-1—the duality concept. If we completely accept the philosophy behind Eq. 18-25, we cannot explain the correlation (Table 18-1), between s-channel resonances and t-channel exchange-related properties. But suppose we assume that some part of the basic dynamics in two-body reactions can be interpreted as either an s-channel resonance mechanism *or* as a t-channel exchange mechanism. More specifically, we assume that the s-channel resonances in the πp, $K^- p$, and NN systems are closely connected with the t-channel ρ, ω, f, and A_2 Regge trajectories. Then the absence of s-channel resonances in the $K^+ p$ and NN systems is explained by the exchange degeneracy of these trajectories (in pairs). As given in Eqs. 18-27 and 18-28, the imaginary parts of the amplitudes of these trajectories cancel; and this cancellation is correlated with the absence of s-channel resonances. Thus we seem to have alternate or *dual* ways of viewing the same basic dynamics. The s-channel resonance mechanism is more appropriate at low energies, the t-channel Regge trajectory exchange mechanism is more appropriate at high energies; but both mechanisms should *not* be used simultaneously, as we did in Eq. 18-25, *if* one accepts the duality concept. Generalizing from this example, we say that some part of a two-body reaction amplitude T is dual if that part can be explained using either s-channel mechanisms or t-channel exchange mechanisms. Just what part of an amplitude is dual is an open question. One formulation applies only the duality idea to the imaginary part of the amplitude (18HAR).

$$\text{Im } T(s\text{-channel resonances}) \text{ dual to Im } T(t\text{-channel Regge trajectory exchange}) \tag{18-32}$$

Then for two-body inelastic reactions, we replace Eq. 18-25 partly by

$$\text{Im } T = \text{Im } T \text{ dual } (s\text{-channel resonances or } t\text{-channel Regge exchange}) \tag{18-33}$$

but assume that duality says nothing about Re T. The term "Regge exchange" in Eqs. 18-32 and 18-33 does *not* include Pomeranchuk exchange. Conventional duality assumes that the s-channel resonances are only dual to the normal Regge trajectories, the ρ, f, A_2, and so forth. Now in elastic scattering, the dominant amplitude is diffractive and not s-channel resonant, even at intermediate energies of a few GeV laboratory energy. In elastic scattering, therefore, the dominant amplitude is not included in Eqs. 18-32 or 18-33.

Exercise 18-16. Note that we did not write Im T (s-channel resonances) = Im T (t-channel Regge exchange), for these two amplitudes cannot be exactly equal. If T(s-channel resonances) is computed by summing over a few resonances, T(s-channel resonances) will fluctuate with energy; but T(t-channel Regge exchange) has a smooth energy dependence. Therefore, the two amplitudes can be taken to be dual only when averaged over some energy region sufficiently wide to smooth over the resonant behavior in s. This is best discussed using the concept of finite energy sum rules. Consult the literature (18TR, 18GO, 18MA) for the use of these sum rules in developing the duality concept.

A vast amount of work has been done on applying the duality concept to experiment and on developing models that would explain how the same dynamics can lead to the apparently different behaviors exhibited by s-channel resonance dominated amplitudes as compared to t-channel Regge exchange dominated amplitudes. But this work is difficult to summarize and the reader had best turn to the literature (18GO, 18BAR, and many more).

Table 18-1 General properties of some systems: The high energy properties are observed in the range of 4 or 5 to 20 or 30 GeV (Taken from Ref. 18HAR).

Process	K^+p, pp	K^-p, $\bar{p}p$, π^+p, π^-p
s-Channel resonances	No	Yes
High energy properties related to t-channel exchange mechanisms		
I. Relatively flat total cross section	Yes	No
II. Dip or change of slope in $(d\sigma/dt)_{el}$ in the $0.5 < \lvert t \rvert < 1.0$ (GeV/c)^2 region	No	Yes

REFERENCES

18BA V. D. Barger and D. B. Cline, *Phenomenological Theories of High Energy Scattering* (Benjamin, New York, 1969).

18BAR V. Barger and F. Halzen, *Nucl. Phys.* **B43**, 62 (1972).

18CO P. D. B. Collins and E. J. Squires, *Regge Poles in Particle Physics* (Springer, Berlin, 1968).

18COL P. D. B. Collins, *How Important Are Regge Cuts*? (Springer, New York, 1971).

18GO E. Gotsman, Ed., *Proceedings of the International Conference on Duality and Symmetry in Hadron Physics* (Weizman Science Press, Jerusalem, 1971).

18HA H Harari and A. Schwimmer, *Phys. Rev.* **D5**, 2780 (1972).

18HAR H. Harari, *Ann. Phys.* (*N.Y.*) **63**, 432 (1971).

18HO G. Höhler et al., *Phys. Lett.* **20**, 79 (1966).

18JA J. D. Jackson and A. Roberts, Eds. *Proceedings of the Sixteenth International Conference on High Energy Physics, 1972* (National Acceleration Laboratory, Batavia, Ill. 1973); Dual Models and Theory Session.

18MA A. D. Martin and T. D. Spearman, *Elementary Particle Theory* (North-Holland, Amsterdam, 1970).

18MI C. Michael, in *Proceedings of the Sixteenth International Conference on High Energy Physics, 1972* (National Acceleration Laboratory, Batavia, Ill., 1973).

18MIC C. Michael, in *Proceedings of the Fourth International Conference on High Energy Collisions, 1972* (Rutherford High Energy Laboratory, Chilton, England, 1972).

18RO M. Ross et al., *Nucl. Phys.* **B23**, 269 (1970).

18TR J. Tran Thanh Van, Ed. *Two Body Collisions, Proceedings of the Seventh Rencontre de Moriond* (Paris, 1972).

CHAPTER

19

Photon–Hadron Interactions and Electromagnetic Form Factors

19-1 Introduction. Among the elementary particles, the photon offers unique opportunities for the investigation of the high energy dynamics of hadrons. First except for the difference in the *magnitude* of the cross sections (Sec. 19-3), the real photon–hadron elastic and inelastic reactions behave very similarly to the corresponding hadron–hadron reactions. We have exhibited these similarities in Chs. 1, 2, 7, and 8. But unlike the hadrons, the electromagnetic properties of the photon are completely described by quantum electrodynamics. Thus, although the photon is a particle whose electromagnetic interactions are understood, it can be used to study strong interactions. Second, as discussed in Sec. 19-2, crucial reactions take place by the exchange of a *single virtual* photon. And the effects of higher order virtual photon exchange processes are believed to be less than a few percent. Therefore, the interactions of virtual photons with hadrons can be studied with relative ease and clarity. The study of the interaction of virtual hadrons with hadrons is much more difficult because we rarely can neglect the effects of higher order hadron exchange processes. Indeed this was the lesson of Chs. 13 and 18.

After the overview of real and virtual photon reactions presented in Sec. 19-2, real photon-hadron reactions are summarized in Sec. 19-3. In Sec. 19-4 we begin the study of virtual photon physics by examining charged lepton–nucleon elastic scattering. The significance of electromagnetic form factors and more general vertex form factors is discussed in Secs. 19-5 and 19-6. A phenomenological theory—the vector meson dominance model—which gives some insight into photon-hadron reactions, is discussed in Sec. 19-7.

19-2 Real and Virtual Photons. A major purpose of this chapter is to introduce the concept of virtual photons. But first let us recall the properties of real photons. Since the mass of the real photon is zero, the four-momentum P of the photon obeys the relation

$$P^2 = 0 \tag{19-1}$$

And as discussed in Sec. 12-2, the photon only has two spin states with helicities ± 1.

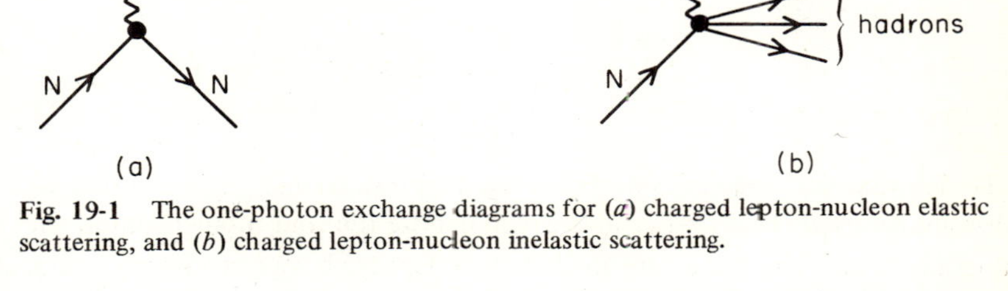

Fig. 19-1 The one-photon exchange diagrams for (*a*) charged lepton-nucleon elastic scattering, and (*b*) charged lepton-nucleon inelastic scattering.

Virtual photons occur in charged lepton–nucleon elastic and inelastic scattering (Fig. 19-1). For brevity we usually use the electron (e) for examples, but the muon (μ) could be used equally well. Indeed, the highest energy studies of charged lepton-nucleon interactions are now carried out by muons. As shown in Fig. 19-1, we always assume that only one virtual photon is exchanged. This assumption is well verified for elastic scattering (Fig. 19-1*a*), where to the few percent level, *no* two-photon exchange effects have been detected. For inelastic reactions (Fig. 19-1*b*), there is little direct experimental verification of this assumption. But we conventionally assume that in inelastic reactions, the effects of two-photon exchange will be less than a few percent. This is based on our knowledge that the coupling of the virtual photon to the charged lepton is of order $\sqrt{\alpha}$ (α = fine structure constant = 1/137) and on the reasoning that the ratio of two-photon exchange to one-photon exchange amplitudes will be of order α.

Exercise 19-1. Discuss the last point in more detail. Also consult the literature (e.g., Ref. 19RU) on two-photon exchange effects in elastic scattering.

Examples of charged lepton–nucleon inelastic reactions are

$$\begin{aligned} &e + p \to e + n + \pi^+ \\ &e + p \to e + p + \rho^0 \\ &e + n \to e + n + K^+ + K^- + \pi^0 \end{aligned} \tag{19-2}$$

Charged lepton–nucleon elastic scattering

$$e + N \to e + N \tag{19-3}$$

can be regarded as a special case of the reactions in Eq. 19-2. As we demonstrate in Secs. 19-4 and 20-2, the charged lepton, virtual photon vertex is completely described by quantum electrodynamics. Therefore we often extract that vertex from the discussion and consider the interaction of a virtual photon γ_V with a nucleon, namely,

$$\gamma_V + \text{hadrons} \to \text{hadrons} \tag{19-4}$$

The reactions in Eq. 19-2 are written

$$\begin{aligned} &\gamma_V + p \to n + \pi^+ \\ &\gamma_V + p \to p + \rho^0 \\ &\gamma_V + n \to n + K^+ + K^- + \pi^0 \end{aligned} \tag{19-5}$$

To describe the kinematics at the charged lepton, virtual photon vertex we follow convention (Fig. 19-1) and use q to denote the four-momentum of the virtual photon. If the initial and final charged leptons have four-momentum p and p', respectively, we can write

$$q = p - p' \tag{19-6}$$

It is also conventional to define ν in the *laboratory frame*

$$\nu = q^0 = p^0 - p'^0 \tag{19-7}$$

If θ is the scattering angle of the charged leptons, we have

$$q^2 = q^2_{\min} - 2|\mathbf{p}||\mathbf{p}'|(1 - \cos\theta) \tag{19-8a}$$

$$q^2_{\min} = 2m^2 - 2p^0p'^0 + 2|\mathbf{p}||\mathbf{p}'| \tag{19-8b}$$

where m is the charged lepton mass. At high energies we can always set $q^2_{\min} = 0$ for electrons, and we generally also do the same for muons. In our metric,

$$q^2 < 0 \tag{19-9}$$

Comparing this with Eq. 19-1, we see that one difference between a real photon and a virtual photon is that the square of the latter's four-momentum is *not* zero. A second difference occurs in the spin properties (see Sec. 20-3). We often speak of the deviation of q^2 from zero as a measure of the *virtuality* of the photon; and the virtual photon that is exchanged in elastic or inelastic charged lepton-nucleon scattering is said to be *spacelike* virtual.

Exercise 19-2. Fill in the steps leading to the various equations and conclusion given in the last paragraph. Also derive Eqs. 19-10 and 19-11.

Finally, in charged lepton-nucleon *inelastic* scattering θ and ν are independent, hence q^2 and ν are independent within the constraint

$$q^2_{\min} > q^2 > -2M\nu \tag{19-10}$$

where M is the nucleon mass. But for charged lepton-nucleon *elastic* scattering, we have

$$q^2 = -2M\nu \tag{19-11}$$

So far we have been discussing *spacelike* virtual photons. But we can also study the reactions

$$e^- + e^+ \rightarrow \text{hadrons} \tag{19-12}$$

by using electron-positron annihilation in electron-positron collisions. As Fig. 19-2 reveals, these reactions take place through the exchange of a single *timelike*

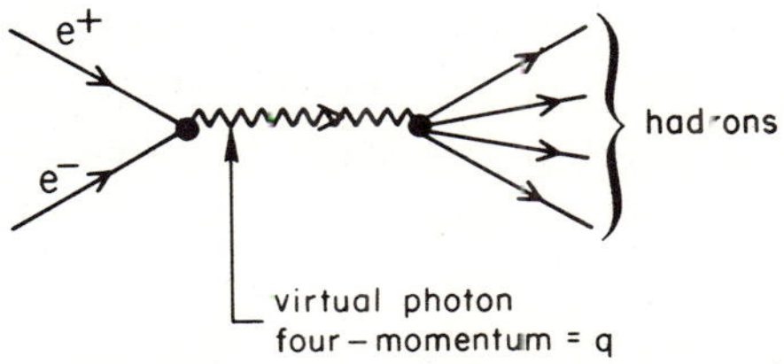

Fig. 19-2 The one-photon diagram for hadron production in electron-positron annihilation.

virtual photon. Again the small size of α leads to the domination of the diagram in Fig. 19-2 over two-photon exchange diagrams. Usually in the laboratory frame the electron and positron each have the same energy E and exactly opposite momenta. In such cases, the four-momentum q of the virtual photon is

$$q^0 = 2E, \qquad \mathbf{q} = 0, \qquad q^2 = 4E^2 \tag{19-13}$$

Thus the virtual photon exchanged here is *timelike*. We discuss hadron production in electron–positron colliding beams in Ch. 21. We only note that by combining the study of charged lepton–nucleon inelastic and elastic scattering with the study of hadron production in electron–positron annihilations, we can compare reactions involving spacelike and timelike virtual photon–hadron interactions.

Exercise 19-3. Show that the colliding beam technique must be used to obtain sufficient center of mass energy for copious production of hadrons and that the annihilation of high energy positrons on electrons at rest cannot be used.

Electron–positron and electron–electron colliding beams can also be used (19BR) to study the production of hadrons by the interaction of two virtual photons. The reaction is

$$e^- + e^+ \to e^- + e^+ + \text{hadrons} \tag{19-14}$$

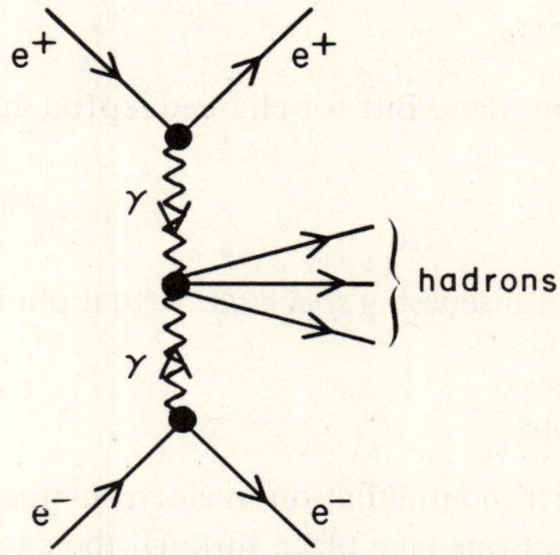

Fig. 19-3 The production of hadrons through the interaction of two virtual photons.

but it is best understood by looking at Fig. 19-3. At the time of writing, very little is known experimentally about this reaction. But it clearly is a reaction of great interest, particularly when *high energy* virtual photons are used.

19-3 The Interaction of Real Photons with Hadrons. Let us summarize the similarities and differences between hadron–hadron and real photon–hadron interactions.

1. *Total Cross Sections.* The energy variations of the total cross sections are very similar (Ch. 1). At low energies, the photon–proton total cross section shows resonance peaks of the same type exhibited by some meson–hadron total cross sections. At high energies the photon–hadron total cross is approximately independent of the energy, as are the hadron–hadron total cross sections. But $\sigma_{\gamma\text{-hadron,tot}}(s)$ is about 1/200 of $\sigma_{\text{hadron–hadron,tot}}(s)$.

2. *High Multiplicity Inelastic Reactions.* The high energy, high multiplicity behaviors of photon–hadron and hadron–hadron reactions are very similar (Chs. 7 and 8). For example, both types of reaction show about the same charged multiplicity at the same energy; and both display rapid cutoff in the transverse momentum distribution.

3. *Low Multiplicity Inelastic Reactions.* Within the restrictions of internal quantum number conservation, both types of reaction exhibit similar inelastic channels. Two-body reactions are dominated by forward and backward peaks. And we can use the same models for explaining two-body photon–hadron and two-body hadron–hadron interactions. Thus in Sec. 13-8 we used the absorption corrected one-pion exchange model for both kinds of reactions. But in agreement with the relative magnitudes of the total cross sections, a real photon–hadron reaction cross section will be 10^{-2} to 10^{-3} of the corresponding hadron–hadron reaction cross section.

4. *High Energy Elastic Scattering.* As discussed in Sec. 2-6, photon–hadron elastic scattering exhibits the same high energy diffraction peak as is exhibited in hadron–hadron elastic scattering. However, $\sigma_{\gamma p,\text{el}}(s)$ is very much smaller than say $\sigma_{\pi p,\text{el}}(s)$. At high energies the former is 0.06 μb and the latter is 4 mb, a difference involving a factor of almost 10^{-5}.

A simple theoretical picture can be used to roughly explain all these relationships. (1) The photon couples to hadrons through the unit electric charge e. This leads to a reduction of the magnitude of the total and inelastic cross sections compared with hadron–hadron cross sections by the factor ($e^2/4\pi = \alpha = 1/137$). This estimate is, of course, very approximate and the reduction factor can easily vary from 10^{-2} to 10^{-3}. (2) In photon–proton elastic scattering, both the initial and final photons must couple to the hadrons. This produces the reduction factor of about α^2. (3) Once the photon is coupled to the hadrons, the photon acts like a hadron!

The last statement, which is descriptive of what appears to occur, is puzzling. In purely electromagnetic interactions we think of the photon as being a point particle, yet if we think of the photon as acting like a hadron, we are ascribing to it, among other properties, the nonzero size of the hadron. We return to this puzzle in Sec. 19-7, where we discuss the vector meson dominance model of photon–hadron interactions.

Recall (Sec. 4-3) that the photon does not have a fixed isotopic spin. In photon-hadron reactions it acts as both an $|I = 0, I_z = 0\rangle$ and an $|I = 1, I_z = 0\rangle$ isotopic spin state.

The general concept that there is a close relation between real photon-hadron reactions and hadron-hadron reactions has been very fruitful. By demanding that strong interaction theories and models explain the observations made in both fields, we get much more incisive and insightful tests of these theories and models. Although we would like to maintain this general concept as we enter the subject of virtual photon-hadron physics, the corresponding subject of virtual hadron-hadron physics is very murky and obscure. Therefore, the field of virtual hadron-hadron physics is a very junior partner in the enterprise of investigating virtual particle physics. This is quite different from real particle physics, where real photon-hadron physics and real hadron-hadron physics are equal partners.

19-4 Charged Lepton–Nucleon Elastic Scattering. We begin our discussion of virtual photon physics with the study of charged lepton-nucleon elastic scattering.

$$(e \text{ or } \mu) + N \to (e \text{ or } \mu) + N \tag{19-15}$$

Here we use N to denote the proton or the neutron. In Sec. 12-8 we laid the foundation for this study; in particular, the most general form of the virtual photon-nucleon vertex is given by Eq. 12-47. We rewrite Eq. 12-49 using the conventional nucleon electromagnetic form factors $F_1{}^N(q^2)$ and $F_2{}^N(q^2)$. We also use the symbolism of the transition current. Thus

$$[j_N^{(\lambda_d \lambda_b)}]^\nu = \bar{u}_d^{\lambda_d}(\mathbf{p}_d) \left[(F_1{}^N(q^2) + \kappa^N F_2{}^N(q^2)\gamma^\nu - \left(\frac{\kappa^N F_2{}^N(q^2)}{2M} \right) (P_b + P_d)^\nu \right] u_b^{\lambda_b}(\mathbf{p}_b) \tag{19-16}$$

where M is the nucleon mass. We can also write Eq. 19-16 in the form

$$[j_N^{(\lambda_d \lambda_b)}]^\nu = \bar{u}_d^{\lambda_d}(\mathbf{p}_d) \left[F_1{}^N(q^2)\gamma^\nu + \left(\frac{\kappa^N F_2(q^2)}{2M} \right) (i\sigma^{\nu\mu} q_\mu) \right] u_b^{\lambda_b}(\mathbf{p}_b) \tag{19-17}$$

Here $q_\mu = (P_d - P_b)_\mu$ and $\sigma^{\nu\mu} = (i/2)[\gamma^\nu \gamma^\mu - \gamma^\mu \gamma^\nu]$.

Exercise 19-4. Show that Eqs. 19-16 and 19-17 are equivalent. Review the significance of the terms in Eq. 19-17 for a nonrelativistic electron in a vector potential **A**.

Recall from Sec. 19-2 that q^2 is simply the quantity we usually call t. The anomalous part of the magnetic moment of the nucleon κ^N is expressed in units of the nucleon magneton $(e\hbar/(2Mc))$

$$\kappa^p = +1.79 \qquad \text{and} \qquad \kappa^n = -1.91 \tag{19-18}$$

From Eq. 19-17 we see that if $\kappa^N = 0$, $F_1^N(t)$ measures the deviation of the t behavior of the nucleon current from that of a pure, charged Dirac point particle; as given in Eq. 12-43. The $\kappa^N F_2^N(t)$ measures the t behavior of that part of the nucleon current which is related to the anomalous magnetic moment. The static properties of the nucleons require at $q^2 = 0$

$$F_1^p(0) = 1, \qquad F_2^p(0) = 1$$

$$F_1^n(0) = 0, \qquad F_2^n(0) = 1 \tag{19-19}$$

From Eq. 19-17 we see that F_1 might be called the *vector* coupling form factor and F_2 might be called the *tensor* coupling form factor. These names are rarely used for the electromagnetic form factors, but the corresponding names are often used in one-particle vector hadron exchange. For example, in Sec. 12-9 G_V (which corresponds to eF_1) is called the vector coupling constant, and G_T (which corresponds to $e\kappa F_2$) is called the tensor coupling constant.

Returning to the electromagnetic form factors of the nucleon, there is another set that is often used because the members simplify the experimental determination of the form factors. And as discussed in Ex. 19-6, they have a neat interpretation in a special Lorentz frame—the nucleon Breit frame. The definitions are

$$G_E^N(q^2) = F_1^N(q^2) + \frac{q^2\kappa^N F_2^N(q^2)}{4M^2}$$

$$G_M^N(q^2) = F_1^N(q^2) + \kappa^N F_2^N(q^2) \tag{19-20}$$

The subscripts E and M stand for electric and magnetic—the motivation for these designations is given in Ex. 19-6. At $q^2 = 0$, we have

$$G_E^p(0) = 1, \qquad G_M^p(0) = 1 + \kappa^p = \mu_p = 2.79$$

$$G_E^n(0) = 0, \qquad G_M^n(0) = \kappa^n = \mu_n = -1.91 \tag{19-21}$$

Here μ_N is the static magnetic moment.

We find experimentally that the form factors have a very strong dependence on q^2. Thus (19RU) with q^2 expressed in $(\mathrm{GeV}/c)^2$ we write

$$G_E^p(q^2) = \frac{G_M^p(q^2)}{\mu_p} = \frac{G_M^n(q^2)}{\mu_n} = \frac{1}{[1 + |q^2|/0.71]^2} \tag{19-22a}$$

$$0 \lesssim G_e^n \lesssim 0.1 \tag{19-22b}$$

The function given in Eq. 19-22a is empirical—we know of no theoretical justification for this simple form. It is often called a *dipole* fit; a *monopole* fit would be $(1 + c|q^2|)^{-1}$. Also we know of no fundamental theoretical reason for $G_E{}^p$, $G_M{}^p$, and $G_M{}^n$ to have the same q^2 behavior, yet they do. This likeness is often expressed by saying these three form factors *scale* together in q^2, or by saying there is a *scaling law* in q^2 for these three form factors. We return to the dipole fit and this scaling law in later sections of this chapter (e.g., see Eq. 19-36). As indicated by Eq. 19-22b, $G_E{}^n$ is approximately zero, but away from $q^2 = 0$ it may not be exactly zero.

Exercise 19-5. Consult the literature (e.g., Ref. 19RU) to obtain more information on $G_E{}^n$.

The empirical relationships given in Eqs. 19-22 are correct to ±10% out to $|q^2| = 25\ (\text{GeV}/c)^2$ for the proton and $|q^2| = 2\ (\text{GeV}/c)^2$ for the neutron. But the separation of $G_E{}^p$ and $G_M{}^p$ for $|q^2| > 4\ (\text{GeV}/c)^2$ depends on *assuming* the validity of the scaling law.

Exercise 19-6. A special Lorentz frame called the *Breit* frame, or the *brick-wall* frame, is defined as follows. It is the Lorentz frame in which

$$E_a = E_b \tag{19-23}$$

Thus there is no energy transferred to the nucleon in this frame, and the nucleon behaves as if it had bounced off a brick wall. Show that in the nucleon Breit frame

$$\begin{aligned} e j_N^{(\lambda'\lambda)0} &= 2M(eG_E)\delta_{\lambda'\lambda} \\ e \mathbf{j}_N^{(\lambda'\lambda)} &= e\bar{u}^{\lambda'}\boldsymbol{\gamma}u^{\lambda}G_M \end{aligned} \tag{19-24}$$

Use these results to justify the designation of G_E as the electric form factor and G_M as the magnetic form factor. Consult Ref. 19ER.

We conclude this section by simply writing down the differential cross section for changed lepton–nucleon elastic scattering—the famous Rosenbluth formula. A rather neat form, with the lepton mass set equal to zero, is

$$\frac{d\sigma}{dq^2} = \frac{4\pi\alpha^2}{q^4(s - M^2)^2}\left[\frac{q^4}{2}G_M{}^2 + \frac{\left(G_E{}^2 + \dfrac{|q^2|}{4M^2}G_M{}^2\right)}{\left(1 + \dfrac{|q^2|}{4M^2}\right)}((s - M^2)^2 + q^2 s)\right] \tag{19-25}$$

A form often seen for the differential cross section in the laboratory frame is

$$\left(\frac{d\sigma}{d\Omega}\right)_{\text{lab}} = \frac{\alpha^2 \cos^2 \frac{1}{2}\theta_{\text{lab}}}{4E_{\text{lab}}^2 \sin^4 \frac{1}{2}\theta_{\text{lab}}} \frac{1}{[1 + (2E_{\text{lab}}/M) \sin^2 \frac{1}{2}\theta_{\text{lab}}]}$$

$$\times \left[\frac{|q^2|}{2M^2} G_M{}^2 \tan^2 \tfrac{1}{2}\theta_{\text{lab}} + \frac{\left(G_E{}^2 + \frac{|q^2|}{4M^2} G_M{}^2\right)}{\left(1 + \frac{|q^2|}{4M^2}\right)} \right] \tag{19-26}$$

Exercise 19-7. The derivation of the Rosenbluth formula is tedious and is given in many books. Those who wish to test their skill and care with trace calculations can compare their derivation with that given, for example, by Ref. 19PI.

We are primarily interested in the high energy approximation, $s \gg M^2$, $s \gg |q^2|$

$$\frac{d\sigma}{dq^2} = \frac{4\pi\alpha^2}{q^4} \left[\frac{G_E{}^2 + \frac{|q^2|}{4M^2} G_M{}^2}{1 + \frac{|q^2|}{4M^2}} \right] \tag{19-27}$$

For the proton we have the simple empirical form, from Eq. 19-22a

$$\left(\frac{d\sigma}{dq^2}\right)_{\text{proton}} = \frac{4\pi\alpha^2}{q^4} \left[\frac{1 + \mu_p(|q^2|/4M^2)}{1 + |q^2|/4M^2} \right] \frac{1}{[1 + |q^2|/0.71]^4} \tag{19-28}$$

We observe that for $s \gg |q^2|$, $d\sigma/dq^2$ is independent of s, in agreement with the discussion of Eq. 12-58. Finally, for future use, we note the high energy, small q^2 form of the Mott differential cross section—the cross section for the electromagnetic scattering of two Dirac point particles—is

$$\left(\frac{d\sigma}{dq^2}\right)_{\text{Mott}} = \frac{4\pi\alpha^2}{q^4}, \qquad s \gg M^2, s \gg |q^2| \tag{19-29}$$

19-5 Nonrelativistic Electromagnetic Form Factors. In Sec. 12-3 we introduced the concept of the vertex form factor $F(t)$. We noted that this fairly arbitrary Lorentz scalar function of t was permitted by a general one-particle exchange model, but the model offers no guidance regarding the specific form that $F(t)$ should take. We have seen in Sec. 19-4 that charged lepton-proton elastic scattering shows a definite need for form factors at the photon-nucleon vertex—those form factors having the specific behavior given in Eq. 19-22. And in Sec.

13-2 we have argued that form factors are also to be expected at purely hadronic vertices. Our purpose in this section is to gain some insight into the meaning of vertex form factors, for until now we treated them in a rather formal way.

We begin with an examination of the form factor concept in atomic structure—the area where the concept originated. The scattering amplitude for the nonrelativistic Coulomb scattering of an electron by a very heavy point source in the first Born approximation is

$$f_{\text{point}}(p, \theta) = -\frac{m}{2\pi} \int \left[\left(\frac{e^2}{4\pi r'} \right) e^{i\mathbf{q}\cdot\mathbf{r}'}\, d^3r' \right] = \frac{2m\alpha}{|\mathbf{q}|^2} = \frac{\alpha}{2vp \sin^2 \frac{1}{2}\theta},$$

$$\mathbf{q} = \mathbf{p} - \mathbf{p}', \qquad |\mathbf{q}^2| = 4p^2 \sin^2(\theta/2) \tag{19-30}$$

Here both the electron and the point source have charge e, the initial and final three-momenta of the electron are $\mathbf{p}$ and $\mathbf{p}'$, the electron mass is m, and the electron velocity is v; θ is the scattering angle. In this and all subsequent formulas in this section we ignore the electron spin.

Exercise 19-8. Derive Eq. 19-30. Remember to use the old trick of replacing the Coulomb potential e^2/r by the shielded Coulomb potential $(e^2/r)(e^{-\lambda r})$ where $\lambda \to 0$ after the integration. Compare Eq. 19-30 with the ultrarelativistic, small angle version of the Mott formula given in Eq. 19-29. Recall that the mass of the electron was neglected in the latter, hence it has $v = c = 1$.

Now we consider a hydrogen atom in which the bound electron has wave functions $\psi_i(\mathbf{r}_1)$. Again we consider the nucleus sufficiently heavy so that we can ignore its motion. Then, in the first Born approximation, the amplitude for the scattering of the free electron on the bound electron is

$$f(p, \theta) = -\frac{m}{2\pi} \int \left[\left(\frac{e^2}{4\pi|\mathbf{r}' - \mathbf{r}_1|} \right) \psi_b^*(\mathbf{r}_1)\psi_a(\mathbf{r}_1)\, e^{i\mathbf{q}\cdot\mathbf{r}'}\, d^3r_1\, d^3r' \right]$$

$$= \frac{2m\alpha}{|\mathbf{q}|^2} \int [\psi_b^*(\mathbf{r}_1)\psi_a(\mathbf{r}_1)\, e^{i\mathbf{q}\cdot\mathbf{r}_1}\, d^3r_1] \tag{19-31}$$

Here we have ignored the scattering of the free electron on the charge of the proton, an amplitude already given by Eq. 19-30. The initial and final state wave functions of the bound electron are $\psi_a(\mathbf{r}_1)$ and $\psi_b(\mathbf{r}_1)$.

Exercise 19-9. Derive Eq. 19-31.

We isolate in the right-hand term of Eq. 19-31 the expression

$$F_{ba}(\mathbf{q}) = \int [\psi_b^*(\mathbf{r}_1)\psi_a(\mathbf{r}_1)\, e^{i\mathbf{q}\cdot\mathbf{r}_1}\, d^3r_1] \tag{19-32}$$

which depends only on $\mathbf{q}$ and the initial and final wave functions of the bound electron; and this expression defines the form factor $F_{ba}(\mathbf{q})$. Clearly we can write Eq. 19-31 as

$$f(p, \theta) = f_{\text{point}}(p, \theta)F(\mathbf{q}) \qquad \text{or} \qquad f(\mathbf{q}) = f_{\text{point}}(\mathbf{q})F(\mathbf{q}) \tag{19-33}$$

Thus $F(\mathbf{q})$ gives the deviation of the scattering amplitude from the point source scattering amplitude.

If the scattering is *elastic*, $b = a$ and $F_{aa}(\mathbf{q})$ is the *elastic form* factor for state a. If $b \neq a$ the scattering is *inelastic* and $F_{ba}(\mathbf{q})$ is an *inelastic* or *transition form factor*. Of course formally $F_{ba}(\mathbf{q})$ is simply the three-dimensional exponential Fourier transform of $[\psi_b^*(\mathbf{r})\psi_a(\mathbf{r})]$.

For further interpretation, we restrict our considerations to elastic scattering. We define

$$\rho_a(\mathbf{r}) = |\psi_a(\mathbf{r})|^2 \tag{19-34a}$$

where $\rho_a(\mathbf{r})$ is simply the charge density distribution in the state a. Usually we assume that $\rho_a(\mathbf{r})$ is normalized such that $\int \rho_a(\mathbf{r})\, d^3r = 1$. Thus $F_{\text{el}}(\mathbf{q})$ is the Fourier transform of the charge density distribution. Assuming spherical symmetry for $\rho(r)$ to further simplify the interpretation, we obtain

$$F_{\text{el}}(|\mathbf{q}|) = \int \rho(r)\, e^{i\mathbf{q}\cdot\mathbf{r}}\, d^3r = 1 - |\mathbf{q}|^2 \frac{\langle r^2 \rangle}{6} + |\mathbf{q}|^4 \frac{\langle r^4 \rangle}{120} + \cdots \tag{19-34b}$$

$$\langle r^n \rangle = \int [\rho(r) r^n]\, d^3r, \qquad n \text{ even} \tag{19-34c}$$

Thus the coefficients in the expansion in Eq. 19-34b measure the dipole, quadrupole, . . ., moments of $\rho(r)$.

Exercise 19-10. Derive the last two equations.

Exercise 19-11. Relate the present discussion to the discussion of Fourier transforms in Sec. 2-4. Note that the $U(r)$ in Eq. 2-15 has been replaced by $\rho(r)$ in Eq. 19-34b. Why has that change occurred?

As we have seen (Ch. 2), a sharp cutoff in a charge distribution, or in a potential, leads to an oscillating transform. On the other hand, a smoothly decreasing charge density distribution, such as the simple exponential

$$\rho(r) = \left(\frac{1}{8\pi b^3}\right) e^{-r/b}, \tag{19-35}$$

leads to a transform that is a monotonic function of $|\mathbf{q}|$. Thus Eq. 19-35 yields

$$F_{\text{el}}(|\mathbf{q}|) = \frac{1}{[1 + b^2|\mathbf{q}|^2]^2} \tag{19-36a}$$

$$F_{el}(|\mathbf{q}|) = 1 - 2b^2|\mathbf{q}|^2 + 3b^4|\mathbf{q}|^4 + \cdots \tag{19-36b}$$

Equation 19-36a is the *dipole* form also given in Eq. 19-22a. Thus we have here a nonrelativistic explanation of that form—it comes from an exponential charge density distribution.

Exercise 19-12. Derive the last two results. An old article by Hofstadter (19HO) gives a very physical and detailed discussion of other forms for nonrelativistic and relativistic form factors.

Equations 19-35 and 19-36 provide a nice example of the relation of the size of the charge distribution to $F_{el}(|\mathbf{q}|)$. When b is large, the charge is spread over a large distance, and $F_{el}(|\mathbf{q}|)$ deviates from 1, even for small $|\mathbf{q}|$ values of the order of $1/b$. On the other hand, if b is very small, the charge is confined to a small region, and $F_{el}(|\mathbf{q}|)$ will deviate from 1 only when $|\mathbf{q}|$ is large. A somewhat more general way to express this important idea is to compare the expansions in Eqs. 19-36b and 19-34b, yielding

$$r_{rms} = \sqrt{\langle r^2 \rangle} = b\sqrt{12} \tag{19-37}$$

Thus we have found two nonrelativistic interpretations of the electromagnetic form factor.

1. Both the elastic and the inelastic form factors can be interpreted as the probability amplitude for going from a particular initial state to a particular final state when a fixed amount of three-momentum $\mathbf{q}$ is transferred. This is the meaning of Eq. 19-32.
2. The elastic form factor can also be interpreted as the Fourier transform of the charge density distribution. This is the significance of Eqs. 19-34.

19-6 Relativistic Electromagnetic and Purely Hadronic Elastic Form Factors. Let us now apply the interpretations developed in the previous section to the elastic form factors of elementary particles. Most of our attention is devoted to the elastic electromagnetic form factors of the nucleon as summarized in Eqs. 19-22, for these are the only elastic form factors that are well measured. In making the transition from nonrelativistic to relativistic considerations, we replace $|\mathbf{q}|^2$ by $(-q^2)$, since the two quantities are equal in the cms. We also write $F_{el}(|\mathbf{q}|)$ as $F_{el}(q^2)$.

As long as

$$|\mathbf{q}^2| \ll M^2 \tag{19-38}$$

(where M is the mass of the nucleon), we can use all the equations developed in the last section.

Exercise 19-13. Justify this last statement.

Therefore under the conditions of Eq. 19-38 we can interpret the nucleon's electromagnetic form factors as follows:

1. $G_E{}^N(q^2)$ is the Fourier transform of the nucleon's charge density distribution.
2. $G_M{}^N(q^2)$ is the Fourier transform of the nucleon's magnetic moment density distribution.

Here we have used the discussion of Ex. 19-6. We can make the formal Lorentz invariant expansion

$$G(q^2) = 1 + a_1|q^2| + a_2|q^2|^2 + \cdots \tag{19-39}$$

and for *small* values of $|q^2|$ we can write, using Eq. 19-34b,

$$a_1 = -\frac{\langle r^2 \rangle}{6}, \qquad a_2 = +\frac{\langle r^4 \rangle}{120} \cdots \tag{19-40}$$

Thus the q^2 dependence of the nucleon form factors in Eq. 19-22a yields

$$r_{\text{nucleon,rms}} = \left[\frac{12}{0.71\ (\text{GeV/c})^2}\right]^{1/2} = 0.81 \times 10^{-13}\ \text{cm} \tag{19-41}$$

Exercise 19-14. Compare the size of the nucleon given in Eq. 19-41 with the size of the hadron obtained by applying the optical model to hadron–hadron elastic scattering (Sec. 2-6). As discussed in Ex. 19-14 the result given in Eq. 19-41 agrees roughly with the size of the proton as measured in other ways. Thus it is assumed that the electromagnetic form factors provide a good measure of the size of the hadrons.

We can also use the interpretation just given to explain the scaling law for $G_E{}^p$, $G_M{}^p$, and $G_M{}^n$, discussed in connection with Eq. 19-22. The scaling law, at small q^2, follows directly from the statement that the charge density distribution of the proton, the magnetic moment density distribution of the proton, and the magnetic moment density distribution of the neutron are very similar. And this is certainly a reasonable situation.

The foregoing discussion was limited to the small q^2 values specified in Eq. 19-38. How then do we interpret the nucleon form factors for larger values of q^2? We have no fundamental answer to this question, but we can gain insight into the significance of large q^2 form factors by using interpretation 1 at the end of the previous section. The $G(q^2)$ is a measure of the probability amplitude that a nucleon will remain a nucleon after absorbing a virtual photon carrying a four-momentum squared of q^2. When q^2 is very small compared with the experimental parameter 0.71 $(\text{GeV/c})^2$, the probability amplitude is close to 1 that the nucleon will remain a nucleon. But as q^2 becomes greater than 0.71, that probability amplitude decreases rapidly.

Exercise 19-15. Use Eq. 19-22a to discuss numerically the behavior of the probability amplitude just mentioned. For $|q^2| \gg 0.71$ the approximation $G(q^2) \sim 1/q^4$ is often used.

Since the probability amplitude at large q^2 for a nucleon to remain a nucleon is very small, what is more likely to happen? Two things. First, when a nucleon absorbs a very virtual photon it is much more likely to change into a state containing two or more hadrons than to remain a nucleon. This is charged-lepton nucleon *inelastic* scattering. Second, given a definition of virtual photon flux, we can say that the total probability for a nucleon to absorb a virtual photon decreases as the photon becomes more virtual. We return to both these points in the next chapter, where we develop the concept of virtual photon–hadron cross sections.

Until this point, we have discussed only the nucleon's elastic electromagnetic form factors. Unfortunately, we have very little knowledge of the elastic electromagnetic form factors of the other hadrons. The only other hadron for which we possess any form factor measurements is the pion. The reasons for the lack of measurements, and the measurement methods used for the pion, are discussed in Ex. 19-16. Nature has been kind in only one respect; the pion, because it has spin zero, has a single form factor $G_\pi(q^2)$.

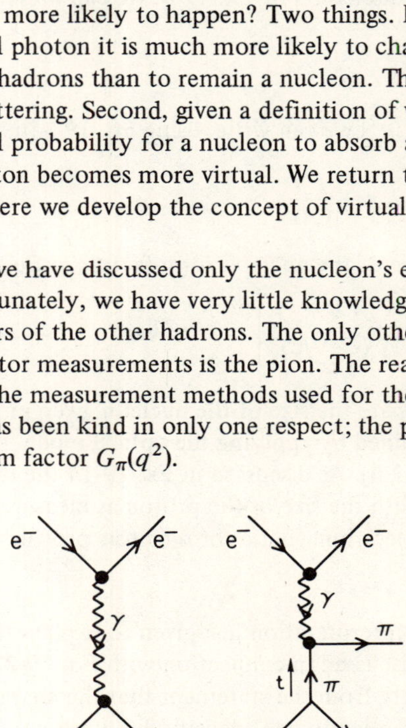

Fig. 19-4 Feynman diagrams for the processes used to measure the pion form factor. (*a*) The direct method. (*b*) The indirect method discussed in Ex. 19-16.

Exercise 19-16. The direct method (Fig. 19-4*a*), for measuring the elastic electromagnetic form factor of a hadron that is not a nucleon requires the scattering of a high energy beam composed of that hadron against the electrons in a target. But to achieve substantial values of q^2, very high energy hadrons are required. Why is this so? Give numerical examples. For the pion an indirect method is also used. This makes use of the electroproduction diagram of Fig. 19-4*b*. For very small values of t, one is indeed measuring the pion form factor at the virtual photon, virtual pion, real pion vertex. One must also use a theory of pion photoproduction of the kind discussed in Sec. 13-8. For more information on both of these methods, see Ref. 19IS.

Using mostly the indirect method described in Ex. 19-16, the pion form factor $G_\pi(q^2)$ has been measured for $0 < |q^2| \lesssim 1.0\ (\text{GeV/c})^2$. For this q^2 range, a fairly good fit is given by the monopole formula

$$G_\pi(q^2) = \left[1 + \frac{|q^2|}{0.59}\right]^{-1}, \qquad q^2 \text{ in } (\text{GeV/c})^2 \tag{19-42}$$

At the higher q^2 values G_π does appear to be falling somewhat faster than Eq. 19-42; but it is definitely not falling as fast as the dipole formula $(1 + |q^2|/0.71)^{-2}$ of Eq. 19-22a. At the end of the next section, we explain the origin of the parameter 0.59 in Eq. 19-42.

With some trepidation, because of the small size of the pion mass, we apply the discussion preceding Eq. 19-41 to the pion. We obtain

$$r_{\text{pion,rms}} \sim 0.6 \times 10^{-13} \text{ cm} \tag{19-43}$$

which says that the pion is slightly smaller than the nucleon. This adds conviction to the assumption we have been making all along—that all hadrons have about the same size. But as this is being written there is little more that can be said about the pion form factor. We cannot answer the crucial question of the behavior of $G_\pi(q^2)$ at large q^2 until we have more experimental data. Thus we do not know whether at large q^2 the monopole behavior of Eq. 19-43 will persist, leading to $G_\pi(q^2) \sim 1/|q^2|$, or whether at large q^2 we will see a $1/q^4$ behavior, as exhibited by the nucleon form factors (Ex. 19-15).

Let us summarize our discussion of the elastic electromagnetic form factor of hadrons. At small q^2, these form factors give us the size of the hadrons, $r_{\text{rms}} \sim 0.5$ to 1.0 fermi. At large q^2, these form factors tell us something about the dynamics at the photon-hadron vertex. They also give us qualitative information about the structure of the hadron—we think of the hadron as being a rather fragile object, easily broken up into two or more hadrons at large q^2 values. This fragility is measured relative to the point particles—the electron and the muon—whose form factors remain 1 even at large q^2 values (19PE). These considerations lead us to relate the fragility of the hadron to its size, even though the limitation imposed by Eq. 19-38 prevents any quantitative considerations at large q^2 values. We think that the hadron is fragile because it is large, because its structure is extended. We think that the electron and muon are not fragile because they are very small.

To conclude this section we broaden the discussion to the consideration of purely hadronic vertices in which a virtual hadron couples to two real hadrons. Although the physics of virtual hadrons is much more obscure than the physics of virtual photons, we can still qualitatively use the two interpretations of vertex form factors given at the end of the previous section.

The form factor $F_{aec}(t)$ must certainly have something to do with the size of hadrons a, c, and e at small values of t. And at large values of t, $F_{aec}(t)$ must be a measure of the probability amplitude for $a + e \to c$. In general, it is certainly wrong to assume that $F_{aec}(t)$ can be a constant at the general hadronic vertex. Hence, as discussed in Sec. 13-2, it should be included in one-particle exchange models of two-body reactions. And as discussed in Chs. 17 and 18, the hadronic vertex form factor appears in modern treatments of two-body reactions in the form of the residue function of the Regge trajectory.

19-7 The Vector Meson Dominance Model of Photon–Hadron Interactions.

This section forms an introduction to the vector meson dominance model of photon–hadron interactions (19SA, 19FE). The treatment is brief; we wish to use the model only as a framework for discussing photon–hadron interactions. For our purposes, the model does not provide special insights into the strong interactions themselves, since it transfers the study of photon–hadron interactions into the study of hadron–hadron interactions. But the model is useful because it provides a fairly specific procedure for making that transfer.

The basic model assumes that photons, real or virtual, interact with hadrons by first changing into the neutral vector mesons ρ^0, ω^0, ϕ^0. Then, as illustrated in Fig. 19-5, the vector meson that is created interacts with the hadronic system. We can temporarily describe the coupling of the photon to the vector meson V by the

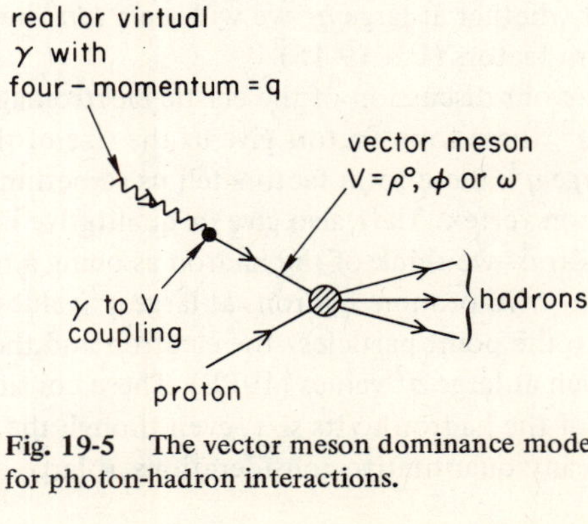

Fig. 19-5 The vector meson dominance model for photon-hadron interactions.

unconventional notation $G_{\gamma V}$. Let $T^{\lambda}_{B,\,VA}$ be the amplitude for a reaction, or for just a vertex, in which the initial state consists of a vector meson V and other hadrons A. The final state contains one or more hadrons, collectively denoted by B. For simplicity, only the helicity λ of V is indicated explicitly. The term $T^{\lambda}_{B,\,\gamma A}$ has the same meaning for a reaction, or vertex, in which a photon, with a four-momentum squared of q^2, replaces the vector meson. The vector meson dominance model then predicts that

$$T^{\lambda}_{B,\,\gamma A} = \sum_{V=\rho^0,\omega^0,\phi^0} \left[G_{\gamma V} \left(\frac{m_V^2}{m_V{}^2 - q^2 - i\Gamma_V m_V} \right) T^{\lambda}_{B,\,VA} \right] \tag{19-44a}$$

for $\lambda = \pm 1$. The term in parentheses is a hybrid between a relativistic Breit–Wigner form (Sec. 5-9) and a simple propagator. The $i\Gamma_V m_V$ term is necessary when $q^2 \sim m_V{}^2$ in electron-positron reactions such as $e^+ + e^- \to \rho^0 \to \pi^+ + \pi^-$, for then we have an s-channel resonance (Sec. 5-7, Ch. 21). When $q^2 \leqslant 0$, which is the case for the reactions discussed in this chapter, we usually neglect the $i\Gamma_V m_V$ term and write

$$T^{\lambda}_{B,\,\gamma A} = \sum_{V=\rho^0,\omega^0,\phi^0} \left[G_{\gamma V} \left(\frac{m_V^2}{m_V{}^2 - q^2} \right) T^{\lambda}_{B,\,VA} \right] \tag{19-44b}$$

Before interpreting Eq. 19-44b we must change to the conventional notations for the coupling $G_{\gamma V}$. First we conceive, in analogy to the photon-hadron charge coupling constant e, a vector meson-hadron coupling constant f_V. We then define f_V by

$$G_{\gamma V} = \frac{e}{f_V} \tag{19-45}$$

There is another somewhat unsymmetric notation:

$$G_{\gamma V} = \frac{\sqrt{\pi\alpha}}{\gamma_V} = \frac{e}{2\gamma_V} \tag{19-46}$$

In using these notations, remember that the strength of the photon-vector meson coupling is *inversely* proportional to γ_V or f_V. The literature contains many slightly different versions of Eqs. 19-44.

To return to the philosophy of the model, in Eqs. 19-44 there is ambiguity about the dependence of $T_{B,\,VA}$ on q^2. If we demand that we know $T_{B,\,VA}$ when $q^2 \neq m_V{}^2$, we are demanding that we know something about the interaction of virtual hadrons—a subject that has been described as rather obscure. Therefore, the simplest form of the model assumes that $T_{B,\,VA}$ is independent of q^2 and that one should use a constant value for $T_{B,\,VA}$. A second problem is that we have no direct experimental information on reactions of the form $V + A \rightarrow B$. In a few cases we can obtain indirect information by using a time-reversed reaction, such as $\pi^- + p \rightarrow \rho^0 + n$. But in general we must rely on more indirect measurements and self-consistent calculations (19WO). In particular, much use is made of the reaction (19WO, 19SA, 19FE).

$$\gamma + p \rightarrow V + p \tag{19-47}$$

This is assumed to be a diffractive inelastic reaction of the type discussed in Ch. 7, and this assumption helps the analysis.

To compare the model with experiment, we first remark that simple models involving $SU(3)$ predict the ratios

$$\frac{1}{\gamma_\rho^{\ 2}} : \frac{1}{\gamma_\omega^{\ 2}} : \frac{1}{\gamma_\phi^{\ 2}} = 9 : 1 : 2 \tag{19-48}$$

More complicated models predict slightly different ratios, but $1/\gamma_\rho^{\ 2}$ is always predicted to be about a factor of 10 larger than $1/\gamma_\omega^{\ 2}$ or $1/\gamma_\phi^{\ 2}$. Experimental measurements are in rough agreement with these predictions, although $1/\gamma_\phi^{\ 2}$ usually turns out to be smaller than predicted.

Exercise 19-17. Using Ch. 14 and the references given in Refs. 19SA and 19WO, derive Eq. 19-48. Use this equation to estimate the ratio of isovector photon to isoscalar photon contributions in some real photon–hadron interactions.

Therefore we simplify the remaining discussion by ignoring the coupling of the ω and ϕ to the photon, and we rewrite Eq. 19-44b in the form

$$T^\lambda_{B,\gamma A} = \frac{e}{2\gamma_\rho}\left(\frac{m_\rho^2}{m_\rho^{\ 2} - q^2}\right) T^\lambda_{B,\rho A} \tag{19-49}$$

It is found experimentally, using real photons, that

$$\frac{\gamma_\rho^2}{4\pi} = 0.3 \text{ to } 0.7 \qquad \text{or} \qquad \frac{f_\rho^2}{4\pi} = 1.2 \text{ to } 2.8 \tag{19-50a}$$

depending on the reaction used to fix the constant value of $T_{B,\,VA}$ (19WO). Hence we have

$$\frac{e^2}{f_\rho^{\ 2}} \sim \frac{1}{160} \text{ to } \frac{1}{380} \tag{19-50b}$$

This, of course, is simply a way to say somewhat more formally what we already said in observation 1 in the middle of Sec. 19-3. The model was designed to fit that observation as well as observation 2.

Exercise 19-18. Draw a Feynman diagram to show how the model fits that second observation.

The model is particularly useful in predicting the more detailed behavior of real photon–hadron reactions. Thus it explains why the $d\sigma/dt$ of the reaction in Eq. 19-47 has roughly the same exponential slope as occurs in hadron–hadron elastic scattering (19WO). In a related area, the model is helpful in thinking about the apparent size of the photon in photon–hadron reactions– a puzzle mentioned at the end of Sec. 19-3. In nonrelativistic quantum mechanics and in quantum electrodynamics, the photon acts as a point particle. But let us consider how the photon appears in high energy photon–proton elastic scattering (Sec. 2-7). Using the Gaussian optical model of Sec. 2-6, the exponential slope b in the elastic scattering of two particles of Gaussian radius R_a and R_b is (Eq. 2-47)

$$b = \frac{{R_a}^2 + {R_b}^2}{2} \qquad (19\text{-}51)$$

Meson–proton elastic scattering in the same energy range has $b = 8$ to $10\ (\text{GeV}/c)^{-2}$. Assuming $R_a \approx R_b$, we obtain for the hadrons ${R_h}^2 = 8$ to $10\ (\text{GeV}/c)^{-2}$. If the photon were a point particle $R_\gamma = 0$, we would find for photon–proton scattering $b_{\gamma p} = ({R_\gamma}^2 + {R_p}^2)/2 = 4$ to $5\ (\text{GeV}/c)^{-2}$. But we find (Sec. 2-7) that $b_{\gamma p} \approx 6$ to $8\ (\text{GeV}/c)^{-2}$. Thus the photon is acting like an extended object, although perhaps not quite as large as a proton. This is the kind of behavior predicted by the vector meson dominance model, in which the photon converts to a vector meson and then scatters as a vector meson.

Having looked at some of the successes of the vector meson dominance model, we conclude this section with a look at one of its tantalizing failures—nucleon form factors. The experimental parameter, $0.71\ (\text{GeV}/c)^2$ is close to square of the mass of the ρ or the other vector mesons. Therefore, one might hope to use the vector meson dominance model to explain the behavior of the nucleon form factors. Indeed, this hope was one of the early reasons for developing the model. We use Eq. 19-49 where $T_{B,\gamma A}$ is now a vertex function. If we follow the simplest philosophy and set $T_{V,\rho A}(q^2) = \text{constant}$, we obtain the monopole function $[\text{constant}/(1 + |q^2|/m_\rho^2)]$, a function which certainly does not fit Eq. 19-22a. On the other hand, if we allow $T_{B,\rho A}(q^2)$ to be dependent on q^2, to fit Eq. 19-22a, we are really begging the question. Thus a first principles application of the vector meson dominance model to the nucleon form factors gives the wrong answer. Nevertheless, the model has such appeal that it is sometimes used for phenomenological inspiration. For example, in the monopole fit to the pion form factor (Eq. 19-42) 0.59 is simply the square of the ρ mass.

Although the vector meson dominance model has other applications, we only consider it once more, when we discuss the reaction $e^- + e^+ \to \pi^- + \pi^+$ in Ch. 21.

REFERENCES

19BR S. J. Brodsky, *Proceedings of the 1971 International Symposium on Electron and Photon Interactions at High Energies* (Laboratory of Nuclear Studies, Cornell University, Ithaca, N.Y., 1971); and references contained in this article.

19ER F. J. Ernst et al. *Phys. Rev.* **119**, 1105 (1960).

19FE R. P. Feynman, *Photon-Hadron Interactions* (Benjamin, New York, 1972).

19HO R. Hofstadter, *Ann. Rev. Nucl. Sci.* **7**, 231 (1957).

19IS See recent volumes of the *Proceedings of the International Symposium on Electron and Photon Interactions at High Energies.*

19PE M. L. Perl, *Phys. Today* **24**, 34 (1971).

19PI H. Pilkuhn, *The Interaction of Hadrons* (North-Holland, Amsterdam, 1967).

19RU R. J. Rutherglen, in *Proceedings of the Fourth International Symposium on Electron and Photon Interactions at High Energies* (Daresbury Nuclear Physics Laboratory, Daresbury, England, 1969).

19SA J. J. Sakurai, in *Proceedings of the Fourth International Symposium on Electron and Photon Interactions at High Energies* (Daresbury Nuclear Physics Laboratory, Daresbury, England, 1969).

19WO G. Wolf, in *Proceedings of the 1971 International Symposium on Electron and Photon Interactions at High Energies* (Laboratory of Nuclear Studies, Cornell University, Ithaca, N.Y., 1971); and references contained in this article.

CHAPTER

20

Lepton–Nucleon Interactions, Hadronic Structure, and the Production of Hadrons

20-1 Introduction. In this chapter we discuss two relatively new methods for investigating strong interactions and the structure of hadrons. In charged lepton–nucleon inelastic scattering, hadrons are produced by the electromagnetic interaction of electrons or muons with nucleons. In neutrino–nucleon inelastic scattering, hadrons are produced by the weak interaction of neutrinos with nucleons, the neutrino turning into an electron or muon in the process. The two methods are at least superficially similar; both utilize a non-strong interaction to study the strong interaction. To the extent to which we understand the non-strong interaction, this gives us new insight into the strong interaction. We believe that we understand the electromagnetic interaction very well, even at very high energies. Hence the first method is certainly fruitful. We know that we do not understand the weak interaction at very high energies, but we believe that we have enough general knowledge about the weak interactions to make the second method useful.

The observations made by these two methods have been very instructive, particularly because they reveal a beautiful regularity called Bjorken scaling. In addition, the observations made by the two methods appear to be closely related both experimentally and theoretically. If we describe these inelastic processes as taking place through current-current interactions, there appears to be strong and profound analogies between the electromagnetic hadronic current and the weak hadronic current in these inelastic processes. Such analogies have already been established in lower energy and simpler reactions.

In Sec. 20-2 we develop the formalism for charged lepton-nucleon inelastic scattering and we introduce the electromagnetic structure functions W_1^{em} and W_2^{em}. Another interpretation of charged lepton-nucleon inelastic scattering using the concept of virtual photon-hadron cross sections is discussed in Sec. 20-3.

In Sec. 20-4 we develop the formalism for charged lepton-nucleon inelastic theory of weak interactions at high energy. And we show why this theory, which is based on Fermi theory, must be wrong at very high energy. In Sec. 20-5 we continue the discussion of the weak interactions by developing the formalism for neutrino-nucleon inelastic scattering. The formal analogy to the material in Sec. 20-2 is emphasized.

Finally in Secs. 20-6 through 20-8 we give the experimental observations. First we present the charged lepton-nucleon inelastic scattering data and illustrate the Bjorken scaling of the structure functions. In Sec. 20-7 we show how the parton model of hadronic structure provides an explanation for this scaling, and in Sec. 20-8 we relate these observations to the data obtained in neutrino-nucleon inelastic scattering.

In this chapter more than any other we have regretted the impossibility of alloting more space to the subject under discussion. We have omitted a great deal of data on the electromagnetic inelastic scattering process, although we feel we have covered the main points. We are most bothered by omissions in theoretical discussion. We have given the parton model explanation of Bjorken scaling because it is simple and provides much physical insight. We have had to omit related theoretical ideas such as calculations based on light cone singularities (20WI). Our greatest omission, current algebra, was necessitated because we could not allow the space required to adequately develop the subject and because most of the development of the subject is based on low energy weak interactions, whereas our interest is in high energies. And as this book is being written, high energy experimental observations on weak interactions are still sparse. We prefer to wait until a fuller experimental foundation has been established.

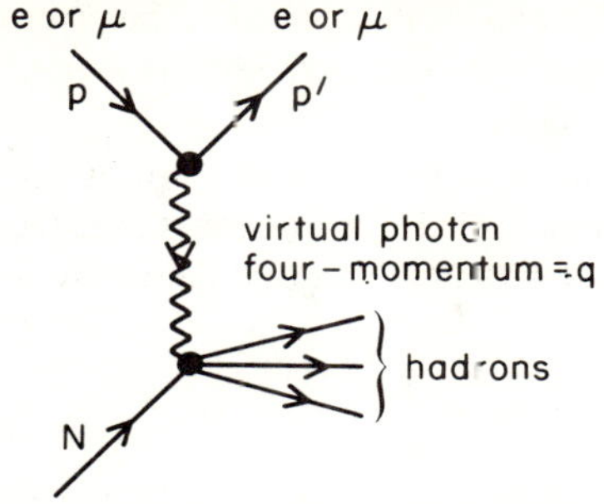

Fig. 20-1 The one-photon exchange diagram for charged lepton-nucleon inelastic scattering.

20-2 Structure Functions in Charged Lepton-Nucleon Inelastic Scattering.

We begin with the one-photon exchange reaction (Sec. 19-2)

$$(e \text{ or } \mu) + \text{nucleon} \to (e \text{ or } \mu) + (2 \text{ or more hadrons}) \tag{20-1}$$

As shown in Fig. 20-1, the initial and final lepton four-momenta are $p = (E, \mathbf{p})$ and $p' = (E', \mathbf{p}')$, the initial nucleon four-momentum is P, the lepton mass is m, and the nucleon mass is M. The four-momentum of the photon is $q = p - p'$. We are concerned with the simplest measurement in which we detect only the final lepton; thus experimentally summing over *all* final hadronic states. The differential cross section (Eq. 6-38) is

$$\frac{d\sigma}{d^3p'} = \frac{(2\pi)^4}{4[(p \cdot P)^2 - m^2M^2]^{1/2}} \frac{1}{(2\pi)^3 2E'}$$

$$\times \sum_{\substack{\text{all hadronic} \\ \text{final states}}} \left(\frac{1}{4}\right) \sum_{\substack{\text{all spin} \\ \text{states}}} \int \left[\prod_{n=1}^{N} \frac{d^3p_n |T_{fi}|^2}{(2\pi)^3 2E_n} \delta^4(P_f - P_i)\right] \tag{20-2}$$

where we have summed or averaged over all spin states, including the lepton spin states. We write the matrix element in the form

$$T_{fi} = \frac{-e^2 j^\mu_{\text{lepton}} J_{\text{hadron},\mu}}{q^2} \tag{20-3a}$$

$$j^\mu_{\text{lepton}} = \bar{u}(p', \lambda')\gamma^\mu u(p, \lambda) \tag{20-3b}$$

$$J_{\text{hadron},\mu} = \langle f | J_\mu | P, \beta \rangle \tag{20-3c}$$

where J is the hadronic current operator, λ and β are lepton and nucleon helicities, respectively, and f denotes a full description of the final hadronic state including the spin states of the hadrons. The electromagnetic coupling constant e has been extracted from the hadronic vertex; and J_{hadron} is defined analogously to j_{lepton}.

Although we know the exact form of the lepton current, we know nothing about the hadronic current. Therefore we must proceed in a very general way, making maximum use of symmetry requirements. We define

$$L^{\mu\nu} = \left[\tfrac{1}{2} \sum_{\lambda,\lambda'} j^{\mu\dagger}_{\text{lepton}} j^{\nu}_{\text{lepton}}\right]$$

$$= 2\left[p^{\mu}p'^{\nu} + p'^{\mu}p^{\nu} + g^{\mu\nu}\,\frac{q^2}{2}\right] \tag{20-4}$$

$$= 2\left[2p^{\mu}p^{\nu} - p^{\mu}q^{\nu} - q^{\mu}p^{\nu} + g^{\mu\nu}\,\frac{q^2}{2}\right]$$

$$W_{\mu\nu} = \left(\frac{(2\pi)^3}{2M}\right)\left(\frac{1}{2}\right)\sum_{\beta}\sum_{f}\int\left[\prod_{n=1}^{N}\frac{d^3p_n}{(2\pi)^3 2E_n}\langle P,\beta|J^{\dagger}_{\mu}|f\rangle\right.$$

$$\left.\times\langle f|J_{\nu}|P,\beta\rangle\,\delta^4(P_f - P - q)\right] \tag{20-5}$$

In Eq. 20-5 the $(\frac{1}{2})$ is for the average over β, and the $(2\pi)^3/2M$ is an old convention justified in Ex. 20-4. Then we can write

$$\frac{d\sigma}{d^3p'} = \frac{\alpha^2 M}{q^4 E'[(p\cdot P)^2 - m^2M^2]^{1/2}}\,L^{\mu\nu}W_{\mu\nu} \tag{20-6}$$

Exercise 20-1. Derive Eqs. 20-2 through 20-6.

Now $W_{\mu\nu}$ must be a tensor constructed out of P and q. There are only two independent four-momenta at the hadronic vertex, and this is the most convenient pair. We also observe that we can construct only two independent Lorentz scalars, other than the nucleon mass. These are taken conveniently to be our old friend q^2, the square of the four-momentum transferred to the hadronic vertex, and the unfamiliar scalar $q \cdot P$. It is conventional to use $q \cdot P$ rather than, say, the square of the invariant mass of the hadronic system

$$s_{\text{hadronic}} = (P + q)^2 = M^2 + 2q\cdot P + q^2 \tag{20-7}$$

With parity conserved at the hadronic vertex, the most general form for $W_{\mu\nu}$ is (cf. Eq. 20-57a)

$$W_{\mu\nu} = Ag_{\mu\nu} + \frac{B}{M^2} P_\mu P_\nu + \frac{C}{M^2} q_\mu q_\nu + \frac{D}{M^2}(q_\mu P_\nu + q_\nu P_\mu)$$

$$+ \frac{E}{M^2}(q_\mu P_\nu - q_\nu P_\mu) \tag{20-8}$$

Here A, B, C, D, and E are Lorentz scalar *functions* of q^2 and $q \cdot P$. Next the hermitian nature of the electromagnetic current leads to

$$W_{\mu\nu} = W^*_{\nu\mu} \tag{20-9}$$

and current conservation (Sec. 12-8), demands

$$q^\mu W_{\mu\nu} = q^\nu W_{\mu\nu} = 0 \tag{20-10}$$

Equation 20-9 tells us A, B, C, and D are real, and E is imaginary. (Also see Ex. 20-14.) Equation 20-10 tells us that $W_{\mu\nu}$ must have the form

$$W_{\mu\nu} = -W_1^{\text{em}}\left[g_{\mu\nu} - \frac{q_\mu q_\nu}{q^2}\right] + \frac{W_2^{\text{em}}}{M^2}\left[\left(P_\mu - \frac{q_\mu q \cdot P}{q^2}\right)\left(P_\nu - \frac{q_\nu q \cdot P}{q^2}\right)\right] \tag{20-11a}$$

W_1^{em} and W_2^{em} are the conventional names for the combinations of A, B, C, and D leading to Eq. 20-11a. Remember

$$W_1^{\text{em}}, W_2^{\text{em}} \text{ are functions of } q^2 \text{ and } q \cdot P \tag{20-11b}$$

The superscript "em" stands for electromagnetic; we soon introduce analogous functions W_1^{weak} and W_2^{weak} for hadronic production through weak interactions.

Exercise 20-2. Derive Eqs. 20-7 through 20-11. Note that unlike the discussion in Sec. 12-8, we have summed over all spin states and no γ four-vectors are available. Incidentally, the use of the M^2 factors and the way the coefficients of D and E are divided are purely conventional. Next derive Eq. 20-12. In doing so, it is best to use the third form of Eq. 20-4 and to recognize that Eq. 20-10 also applies to the lepton vertex. Therefore, all terms in $W_{\mu\nu}$ that contain q_μ can be ignored.

As derived in Ex. 20-2, we write

$$L^{\mu\nu}W_{\mu\nu} = (-2W_1^{\text{em}})[2m^2 + q^2] + \frac{2W_2^{\text{em}}}{M^2}\left[2(p \cdot P)^2 - 2p \cdot Pq \cdot P + \frac{M^2 q^2}{2}\right] \tag{20-12}$$

Finally we can write down the differential cross section. Since we are interested only in high energies and not interested in very small values of $|q^2| \sim m^2$, it is convenient to set $m = 0$. In the *laboratory* frame

$$\frac{d\sigma}{d\Omega'\, dE'} = \frac{\alpha^2}{q^4}\left(\frac{E'}{E}\right) L^{\mu\nu} W_{\mu\nu} \tag{20-13a}$$

$$= \frac{\alpha^2}{q^4}\left(\frac{E'}{E}\right) [2|q^2| W_1^{\text{em}} + (4EE' - |q^2|) W_2^{\text{em}}] \tag{20-13b}$$

One frequently sees

$$\frac{d\sigma}{d\Omega'\, dE'} = \frac{4\alpha^2 E'^2}{q^4} [2 \sin^2 \tfrac{1}{2}\theta\; W_1^{\text{em}} + \cos^2 \tfrac{1}{2}\theta\; W_2^{\text{em}}] \tag{20-14}$$

$$\frac{d\sigma}{dq^2\, dE'} = \frac{\pi\alpha^2}{q^4 E^2} [2|q^2| W_1^{\text{em}} + (4EE' - |q^2|) W_2^{\text{em}}] \tag{20-15}$$

In the last equations, θ is the lepton scattering angle, and E and E' are the energies of the initial and final leptons, all in the laboratory frame. For future use we define in the laboratory system

$$\nu = q^0 = E - E' \tag{20-16a}$$

and we note that

$$q \cdot P = M\nu \tag{20-16b}$$

Exercise 20-3. Derive the last sets of equations.

Thus we have expressed an inelastic differential cross section in terms of two Lorentz scalars $W_1^{\text{em}}(q^2, q \cdot P)$ and $W_2^{\text{em}}(q^2, q \cdot P)$. These are called *structure functions* because they tell us something about the structure of the nucleon. They are generalizations to *inelastic* scattering of the two nucleon electromagnetic *elastic* form factors discussed in Secs. 19-4 through 19-6. But W_1^{em} and W_2^{em} have a broader meaning, for as Eqs. 20-5 and 20-11 indicate, they represent a summation over all final hadronic states. Thus they resemble total cross sections as discussed in the next section.

Exercise 20-4. To illustrate the relation between W_1^{em} and W_2^{em} and elastic form factors, consider the $G_E(q^2)$ and $G_M(q^2)$ form factors given in Eq. 19-20. But for simplicity set $\kappa = 0$, so that $G_E = G_M = G$. Show that in an arbitrary Lorentz frame

$$W^{\text{em}}_{1,\text{elastic}} = \left(\frac{|q^2|}{4M^2}\right) G^2(q^2)\delta\left(\frac{q\cdot P}{M} - \frac{|q^2|}{2M}\right) \tag{20-17a}$$

$$W^{\text{em}}_{2,\text{elastic}} = G^2(q^2)\delta\left(\frac{q\cdot P}{M} - \frac{|q^2|}{2M}\right) \tag{20-17b}$$

In this derivation it is useful to use the trick of writing [with $P = (E, \mathbf{P})$]

$$\iiint \frac{d^3p''\,\delta^4(P''-P')}{2E''} = \iiiint d^3p''\,dE''\,\delta^4(P''-P')\delta(P''^2 - M^2)$$

$$= \delta(P'^2 - M^2)$$

In the literature the δ function in Eqs. 20-17 is usually written $\delta(\nu - |q^2|/2M)$, but ν is simply shorthand for the Lorentz scalar $q\cdot P/M$. The simplicity of Eq. 20-17b is the reason for the convention used to define $W_{\mu\nu}$ in Eq. 20-5. Using Eq. 20-17 and 20-15, derive Eq. 19-27 for $G_E = G_M = G$. Remember that $s \gg M^2$, $s \gg |q^2|$ in Eqs. 19-27.

Exercise 20-5. As another example and for future use, consider the scattering of a charged lepton on a spin zero hadron. Show that

$$W^{\text{em}}_{1,\text{elastic}} = 0 \tag{20-18a}$$

$$W^{\text{em}}_{2,\text{elastic}} = G^2(q^2)\delta\left(\frac{q\cdot P}{M} - \frac{|q^2|}{2M}\right) \tag{20-18b}$$

20-3 Virtual Photon–Nucleon Total Cross Sections. To gain a foothold, we first consider the *real* photon reaction

$$\gamma + \text{nucleon} \to (2 \text{ or more hadrons}) \tag{20-19}$$

which is analogous to the *virtual* photon reaction in Eq. 20-1. For the incident photon in a spin state Φ (Sec. 12-2), the matrix element is

$$T_{fi} = e\Phi^{\mu}J_{\text{hadron},\mu} = e\Phi^{\mu}\langle f|J_{\mu}|P,\beta\rangle \tag{20-20}$$

If Φ represents a transversely polarized photon (Eq. 12-11), we have

$$\Phi_{x\ \text{transverse}} = (0, 1, 0, 0) \qquad \text{and} \qquad \Phi_{y\ \text{transverse}} = (0, 0, 1, 0). \tag{20-21}$$

The total cross section from that spin state Φ is

$$\sigma_{\text{tot}}(s) = \Phi^{\mu\dagger}\Phi^{\nu}\left(\frac{(2\pi^4)e^2}{4Mk}\right)(\tfrac{1}{2})\sum_{\beta}\sum_{f}\int\left[\prod_{n=1}^{N}\frac{d^3p_n}{(2\pi)^3 2E_n}\right.$$

$$\left.\times\langle P,\beta|J_\mu^\dagger|f\rangle\langle f|J_\nu|P,\beta\rangle\,\delta^4(P_f - P_i)\right] \tag{20-22}$$

Here k is the photon energy. Of course we are being unnecessarily fancy here, for we have also summed over the spin states of the initial nucleon. And using the notation defined in Eq. 20-21 for transversely polarized photons, we also have

$$\sigma_{\text{tot}, x \text{ transverse}}(s) = \sigma_{\text{tot}, y \text{ transverse}}(s) = \sigma_{\text{tot}}(s) \tag{20-23}$$

But this elaborateness has a point. Most of the right-hand side of Eq. 20-22 looks just like most of the right-hand side of Eq. 20-5. Thus there is a strong analogy between part of $\sigma_{\text{tot}}(s)$ and $W_{\mu\nu}$. And $\Phi^{\mu\dagger}\Phi^{\nu}$ is a tensor like $L^{\mu\nu}$ in Eq. 20-4; thus there is another analogy here. We now construct that analogy, but we must remember that it is only an analogy. Equations 20-4 and 20-5 are for virtual photons, Eq. 20-22 is for real photons.

To develop the $\Phi^{\mu\dagger}\Phi^{\nu} \leftrightarrow L^{\mu\nu}$ analogy we first consider real photons. Let ρ be the spin density matrix in a Cartesian four-vector basis. Explicitly

$$\rho^{\mu\nu} = \Phi^{\mu\dagger}\Phi^{\nu} \tag{20-24}$$

For $\Phi_{x \text{ transverse}}$ given in Eq. 20-21, we have simply

$$\rho^{11} = 1, \qquad \text{all other} \quad \rho^{\mu\nu} = 0 \tag{20-25}$$

For a circularly polarized photon, that is, a photon with helicity ± 1 (Eq. 12-9), we write

$$\Phi = \frac{1}{\sqrt{2}}(0, \mp 1, -i, 0), \qquad \rho = \begin{bmatrix} 0 & 0 & 0 & 0 \\ 0 & \frac{1}{2} & \pm i\frac{1}{2} & 0 \\ 0 & \mp i\frac{1}{2} & \frac{1}{2} & 0 \\ 0 & 0 & 0 & 0 \end{bmatrix} \tag{20-26}$$

And for an elliptically polarized photon

$$\Phi = \frac{1}{\sqrt{2}}(0, \mp(1+\epsilon)^{1/2}, -i(1-\epsilon)^{1/2}, 0), \qquad -1 \leqslant \epsilon \leqslant 1$$

$$\rho = \begin{bmatrix} 0 & 0 & 0 & 0 \\ 0 & (\frac{1}{2})(1+\epsilon) & \pm(i\frac{1}{2})(1-\epsilon)^{1/2} & 0 \\ 0 & \mp(i\frac{1}{2})(1-\epsilon^2)^{1/2} & \frac{1}{2}(1-\epsilon) & 0 \\ 0 & 0 & 0 & 0 \end{bmatrix} \tag{20-27}$$

To continue the analogy, we interpret j_{lepton} (Eq. 20-3b) as proportional to a spin function Φ for the virtual photon; and $L^{\mu\nu}$ (Eq. 20-4) will be proportional to a spin density matrix ρ_{virtual} for the virtual photon. We gain the most insight if we take the *z axis along the direction of the virtual photon* and set the proportionality factor N by requiring

$$\rho^{11}_{\text{virtual}} + \rho^{22}_{\text{virtual}} = 1 \tag{20-28}$$

This is in analogy to Eqs. 20-25 through 20-27. Then we have

$$j_{\text{lepton}} = \bar{u}(p', \lambda')\gamma u(p, \lambda) = N^{1/2}\Phi(\lambda', \lambda) \tag{20-29a}$$

$$\rho^{uv}_{\text{virtual}}(\lambda', \lambda) = \Phi^{u\dagger}(\lambda', \lambda)\Phi^{\nu}(\lambda', \lambda) \tag{20-29b}$$

where, we recall, j, γ, and Φ are four-vectors.

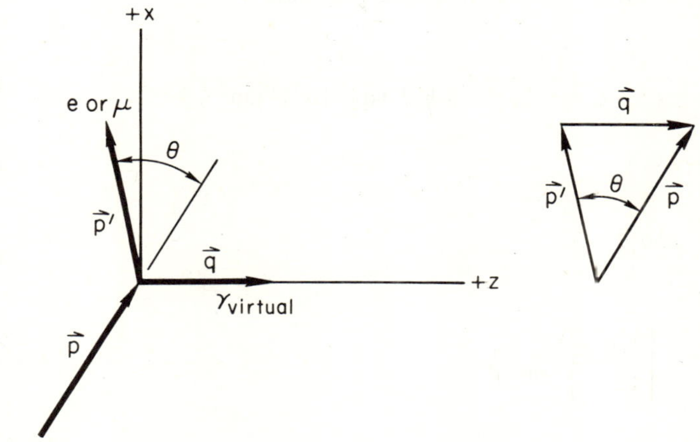

x-z plane in laboratory frame

Fig. 20-2 Kinematics at the charged lepton-photon vertex.

Now in the simpler lepton–nucleon inelastic scattering experiments, and this includes most experiments that have been done, the lepton helicity state is not observed. Then maintaining the normalization of Eq. 20-28, we put

$$\rho_{\text{virtual}} = \tfrac{1}{2}[\rho_{\text{virtual}}(+\tfrac{1}{2}, -\tfrac{1}{2}) + \rho_{\text{virtual}}(+\tfrac{1}{2}, +\tfrac{1}{2})] \tag{20-30a}$$

$$\rho_{\text{virtual}}^{\mu\nu} = \frac{L^{\mu\nu}}{N} \tag{20-30b}$$

where $L^{\mu\nu}$ is given by Eq. 20-4. We can now calculate ρ_{virtual} in any Lorentz frame. A somewhat tedious calculation leads to the following results in the *laboratory* frame. The kinematic quantities are defined in Fig. 20-2. The lepton mass is set to zero, and the z axis is along $\mathbf{q}$.

$$\rho_{\text{virtual}} = \begin{bmatrix} \dfrac{|\mathbf{q}|^2}{|q^2|}\epsilon & \cdot & 0 & \cdot \\ \cdot & \tfrac{1}{2}(1+\epsilon) & 0 & \cdot \\ 0 & 0 & \tfrac{1}{2}(1-\epsilon) & 0 \\ \cdot & \cdot & 0 & \dfrac{\nu^2}{|q^2|}\epsilon \end{bmatrix} \tag{20-31}$$

where the elements not given explicitly are nonzero; we shall not need them. In terms of laboratory kinematic variables, we have

$$\epsilon = \frac{4EE' - |q^2|}{2(E^2 + E'^2) + |q^2|} = \frac{1}{1 + 2(|\mathbf{q}|^2/|q^2|)\tan^2 \tfrac{1}{2}\theta} \tag{20-32a}$$

$$0 \leqslant \epsilon \leqslant 1 \tag{20-32b}$$

$$N = \frac{2|q^2|}{1-\epsilon} \tag{20-33a}$$

$$L^{\mu\nu} = \left[\frac{2|q^2|}{1-\epsilon}\right]\rho_{\text{virtual}}^{\mu\nu} \tag{20-33b}$$

Exercise 20-6. Derive Eqs. 20-30 through 20-33. Compare Eq. 20-31 with Eq. 20-27, showing that the transverse components are elliptically polarized. But note that now there are nonzero z and timelike components of Φ and ρ.

We observe that the net transverse polarization in the x direction is given by

$$\frac{\rho^{11}-\rho^{22}}{\rho^{11}+\rho^{22}}=\epsilon \tag{20-34}$$

When $\theta = 0$, $\epsilon = 1$, and all the transverse polarization is in the x direction, which by Fig. 20-2 is in the lepton scattering plane. We speak of ρ^{11} and ρ^{22} as representing the *transverse component* of the *virtual photon flux*. Similarly we speak of ρ^{33} and ρ^{00} as representing the *longitudinal component* and the *scalar component*, respectively. And even more crudely we talk about *transverse virtual photons, longitudinal virtual photons*, and *scalar virtual photons.*

Exercise 20-7. Lorentz transformations along $\mathbf{q}$ (the z axis) will not change the transverse components of the virtual photon flux, but they will change the scalar and longitudinal components; hence they will change ρ^{00} and ρ^{33}. As an example, calculate ρ_{virtual} in the Breit frame of the photon and the final hadronic system (see Ex. 19-6 for a definition) with $\mathbf{q}$ along the z axis.

Turning to $W_{\mu\nu}$, we are going to calculate $L^{\mu\nu}W_{\mu\nu}$ while interpreting $W_{\mu\nu}$ in terms of virtual photon total cross sections. Two observations simplify the calculation. First, in the laboratory frame with the orientation given in Fig. 20-2, the only nonzero elements of $W_{\mu\nu}$ (Eq. 20-11a) are W_{11}, W_{22}, W_{33}, W_{00}, W_{03}, W_{30}. Second, by current conservation, we have

$$\rho^{33}W_{33}+\rho^{03}W_{03}+\rho^{30}W_{30}+\rho^{00}W_{00}=\left(\frac{|q^2|}{\nu^2}\right)^2\rho^{33}W_{33}=\frac{|q^2|}{\nu^2}\,\epsilon W_{33} \tag{20-35}$$

Next comparing Eqs. 20-5 and 20-22 we define the total cross sections $\sigma_T(q^2,\nu)$ and $\sigma_L(q^2,\nu)$ for *transverse* and *longitudinally polarized virtual photons*, respectively, as they interact with *unpolarized* protons. In the definition of $\sigma_L(q^2,\nu)$ we have absorbed into σ_L the $|q^2|/\nu^2$ factor in the right-hand term of Eq. 20-35. This is a convention, not a necessity.

$$\frac{k_{\text{virtual}}}{4\pi^2\alpha}\,\sigma_T(q^2,\nu)=W_{11}(q^2,\nu)=W_{22}(q^2,\nu)=W_1^{\text{em}}(q^2,\nu) \tag{20-36a}$$

$$\frac{k_{\text{virtual}}}{4\pi^2\alpha}\,\sigma_L(q^2,\nu)=\frac{|q^2|}{\nu^2}\,W_{33}(q^2,\nu)=-W_1^{\text{em}}(q^2,\nu)+\frac{|\mathbf{q}|^2}{|q^2|}\,W_2^{\text{em}}(q^2,\nu) \tag{20-36b}$$

Putting everything together and using Eq. 20-13a, we obtain in the laboratory frame

$$\frac{d\sigma}{d\Omega'\,dE'}=\frac{\alpha k_{\text{virtual}}}{4\pi^2|q^2|}\left(\frac{E'}{E}\right)\left(\frac{2}{1-\epsilon}\right)[\sigma_T(q^2,\nu)+\epsilon\sigma_L(q^2,\nu)] \tag{20-37}$$

There is one more decision to be made before we reach the end of this long road What do we mean by k_{virtual}? At $q^2 = 0$, $k_{\text{virtual}} = k$ = photon energy = photon momentum. But at $q^2 \neq 0$ photon energy $(q^0) \neq$ photon momentum $(|\mathbf{q}|)$. Clearly we can use any definition that yields $k_{\text{virtual}} \to k$ and $q^2 \to 0$. We choose the following convention introduced by Hand (20HA):

$$k_{\text{virtual}} = K = \nu - \frac{|q^2|}{2M} \tag{20-38}$$

in the laboratory frame. By Eq. 20-7, the s of the final hadronic system is

$$s = M^2 + 2MK \tag{20-39}$$

Thus if we keep K fixed as q^2 goes to zero, s is fixed. This is useful when comparing electroproduction and real photoproduction. But other definitions of k_{virtual}, such as $k_{\text{virtual}} = |\mathbf{q}|_{\text{lab}}$ have other virtues. The important point is that in the end $\sigma_T(q^2, \nu)$ and $\sigma_L(q^2, \nu)$ for $q^2 \neq 0$ are arbitrary quantities, not physical quantities, because k_{virtual} is defined arbitrarily.

Finally, we have

$$\frac{d\sigma}{d\Omega'\, dE'} = \Gamma_T(q^2, \nu, E)[\sigma_T(q^2, \nu) + \epsilon(q^2, \nu, E)\sigma_L(q^2, \nu)] \tag{20-40}$$

where Γ_T can be interpreted as the *flux of transverse virtual photons* per unit energy and per unit solid angle of the final lepton; ϵ is sometimes interpreted as the ratio of the flux of longitudinal virtual photons to the flux of transverse virtual photons, but remember that we have defined σ_L to include the contribution of the $\mu = 0$ and $\nu = 0$ components of $\rho^{\mu\nu}$. Writing the longitudinal photon flux

$$\Gamma_L = \epsilon\Gamma_T \tag{20-41}$$

$$\frac{d\sigma}{d\Omega'\, dE'} = \Gamma_T\sigma_T + \Gamma_L\sigma_L \tag{20-42}$$

Thus $d\sigma/d\Omega'\, dE'$ is the sum of a transverse photon contribution and a longitudinal photon contribution. And as we discuss in the next section, the relative size of these contributions is of great interest.

Exercise 20-8. Show that

$$W_1^{\text{em}} = \frac{K}{4\pi^2\alpha}\sigma_T, \qquad W_2^{\text{em}} = \frac{K}{4\pi^2\alpha}\left(\frac{|q^2|}{|q^2| + \nu^2}\right)(\sigma_T + \sigma_L) \tag{20-43}$$

Exercise 20-9. Discuss the $q^2 \to 0$ limits

$$\sigma_T(q^2, K) \underset{q^2\to 0}{\longrightarrow} \sigma_{\text{tot, real } \gamma\text{-nucleon}} \tag{20-44a}$$

$$\sigma_L(q^2, K) \underset{q^2\to 0}{\longrightarrow} 0 \tag{20-44b}$$

Write explicit expressions for Γ_T and Γ_L using Eq. 20-32.

20-4 Weak Interactions at High Energy. Only in this chapter do we consider weak interactions. Therefore, we begin this section with a very brief introduction to the part of weak interaction theory of concern. We are interested in high energy interactions of the type

$$\text{neutrino} + \text{nucleon} \to (e \text{ or } \mu) + (2 \text{ or more hadrons}) \tag{20-45}$$

$$(e \text{ or } \mu) + \text{nucleon} \to \text{neutrino} + (2 \text{ or more hadrons}) \tag{20-46}$$

The reactions in Eq. 20-46 have not been observed because it is now experimentally impossible to detect such reactions in the presence of the much more copious charged lepton–nucleon reactions of Eq. 20-1. We remind the reader that lepton number (Sec. 4-2) is conserved in these reactions.

To orient ourselves, we consider first the purely leptonic reaction

$$\nu_\mu + e^- \to \mu^- + \nu_e \tag{20-47}$$

(muon neutrino) + electron → muon + (electron neutrino)

$$p_{\nu_\mu} + p_e \to p_\mu + p_{\nu_e}$$

At the time of writing we have little direct experimental information of this reaction, but it provides a clear illustration of what our present theory of weak interactions predicts at high energies. Actually, the theory we are about to use is incorrect at very high energies in its simplest form (see next paragraph). The simplest form of the theory, the Fermi theory, explains purely leptonic weak processes through the local interaction of two *weak currents.* The weak current for a neutrino described by the spinor $u(p, \lambda)$ going to a charged lepton described by the spinor $\bar{u}(p, \lambda)$ is

$$j_{\text{lepton}}^{\text{weak},\mu} = \bar{u}(p, \lambda')\gamma^\mu(1 - \gamma_5)u(p, \lambda) \tag{20-48a}$$

If the initial state is an antineutrino $v(p, \lambda)$, we have

$$j_{\text{lepton}}^{\text{weak},\mu} = \bar{v}(p', \lambda')\gamma^\mu(1 + \gamma_5)v(p, \lambda). \tag{20-48b}$$

We assume the reader knows that the factor $(1 \mp \gamma_5)$ is required by the two-component neutrino theory. The factor $(1 \mp \gamma_5)$ limits the helicity of the neutrino to $-1/2$ and that of the antineutrino to $+1/2$.

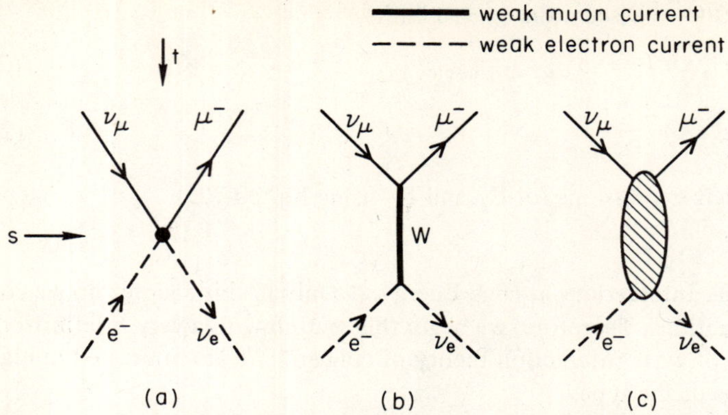

Fig. 20-3 Diagrams for the reaction $\nu_\mu + e^- \to \mu^- + \nu_e$: (*a*) according to the Fermi theory of the local interaction of weak currents, (*b*) if the reaction proceeds through the exchange of the hypothetical vector boson W, (*c*) if at very high energies the reaction proceeds through the summation over many different Feynman diagrams.

Exercise 20-10. If necessary, review the Fermi theory of purely leptonic weak interactions (20GA, 20BJ, 20BE). Derive Eq. 20-52.

According to the Fermi theory (Fig. 20-3*a*), the matrix element for the reaction in Eq. 20-47 is the contraction of the electron and muon weak currents. Explicitly

$$T^{\text{weak, Fermi}}_{(\nu_\mu e^- \to \mu^- \nu_e)} = \frac{G}{\sqrt{2}} j^{\text{weak}, \mu}_{\text{lepton}} (\nu_\mu \to \mu^-) j^{\text{weak}}_{\text{lepton}, \mu} (e^- \to \nu_e) \tag{20-49}$$

Here G is the weak interaction coupling constant. From low energy measurements it has the value

$$G = \frac{1.02 \times 10^{-5}}{M^2_{\text{proton}}} \tag{20-50}$$

Exercise 20-11. Comparing Eqs. 20-48 and 20-49 with Eqs. 20-3a and 20-3b, show that aside from the $(1 \mp \gamma_5)$ factors

$$\frac{T^{\text{weak, Fermi}}(\nu_\mu e \to \mu \nu_e)}{T^{\text{em}}(\mu e \to \mu e)} = \frac{G/\sqrt{2}}{e^2/|q^2|} \approx \frac{0.8 \times 10^{-4} |q^2|}{M_p{}^2} \tag{20-51}$$

This shows the relative scale of the weak to electromagnetic interaction

$$\frac{\sigma^{\text{weak}}}{\sigma^{\text{em}}} \sim 10^{-8} \qquad \text{for} \quad |q^2| \sim M_p{}^2$$

The differential and total cross sections for the reaction in Eq. 20-47 are

$$\frac{d\sigma}{dt} = \frac{G^2}{\pi}, \qquad \sigma = \frac{sG^2}{\pi} \tag{20-52}$$

with the charged lepton masses set equal to zero. We note that only s-wave scattering is present in Eq. 20-52. Therefore, unitarity requires

$$\sigma(\nu_\mu e \to \mu\nu_e) \leqslant \frac{2\pi}{s} \tag{20-53}$$

Equations 20-52 and 20-53 are contradictory when s exceeds $\pi\sqrt{2}/G$; therefore, the Fermi theory must be wrong at very high energies.

Exercise 20-12. Of course $s = \pi\sqrt{2}/G$ occurs at very, very high energy for the reaction in Eq. 20-47; it is an energy which we do not know how to obtain. We do not even know how to obtain that energy in the reaction $\nu_\mu + n \to \mu^- + p$. What laboratory energy is required for the ν_μ? Show, however, that the nucleon elastic form factor will obscure the contradiction. Show that the exchange in the t-channel of an intermediate vector boson (see below) replaces the σ in Eq. 20-52 by an s-independent σ.

One way to postpone the contradiction to higher energies is to assume that the weak interactions occur through the exchange of a spin 1 boson usually designated by the symbol W (20BE). As illustrated in Fig. 20-3b, this is in analogy to electrodynamics, but the W, unlike the photon, has a nonzero mass M_W, usually taken to be of the order of many GeV. Sometimes W is called the *intermediate* (for weak interactions) *vector boson* or *meson.* At the time of writing there is *no* evidence to substantiate this hypothesis. The intermediate vector boson hypothesis is useful as a way of defining the deviation of the weak interaction from a point interaction. If the boson existed, the range of the weak interactions would be about $1/M_W$ according to our usual ideas. Thus to the extent that we take the weak interactions to have zero range—that is, to be a point interaction—we take M_W to be very large compared with typical hadronic masses.

But the intermediate vector meson hypothesis is not the only way in which the point interactions can deviate from a point interaction. We must allow the possibility that the weak interactions can become quite complicated at very high energy; for the t-channel exchange of a single W does not eliminate the unitarity violation as s goes to infinity. The weak interaction may proceed through the exchange of many particles; and as in strong interactions, many complicated Feynman diagrams may contribute. This possibility is suggested by the cross-hatched area in Fig. 20-3c, the cross-hatching representing our ignorance, as in strong interactions.

20-5 Neutrino–Nucleon Inelastic Scattering. We now return to the description of the production of hadrons in the weak interaction of a neutrino with a nucleon given by Eq. 20-45

neutrino + nucleon → (e or μ) + (2 or more hadrons)

We write the interaction as the product of two weak currents

$$T_{fi}^{\text{weak}} = \frac{G}{\sqrt{2}} j_{\text{lepton}}^{\text{weak},\,\mu} J_{\text{hadron},\,\mu}^{\text{weak}} \tag{20-54}$$

Equation 20-54 is an extrapolation to many-body production of the two-body reaction amplitude of Eq. 20-49, and it is analogous to Eq. 20-3a. We take $j_{\text{lepton}}^{\text{weak},\mu}$ to be given by Eq. 20-48a or 20-48b. Thus we ignore our lack of knowledge of the very high energy behavior of the weak interactions as discussed in the last section. Or we assume that we have buried that ignorance in $J_{\text{hadron}}^{\text{weak}}$. For, as we soon discuss, we know very little a priori about $J_{\text{hadron}}^{\text{weak}}$ in many-body production. As in Eq. 20-3c,

$$J_{\text{hadron},\mu}^{\text{weak}} = \langle f | J_{\mu}^{\text{weak}} | P, \beta \rangle$$

where J^{weak} is the hadronic weak current operator, P and β the initial nucleon's four-momenta and helicity, and f denotes a full description of the final hadronic state, including the spin states of the hadrons.

We now proceed as in the electromagnetic reaction of Sec. 20-2. We define and calculate

$$L^{\text{weak},\mu\nu} = \sum_{\lambda'} j_{\text{lepton}}^{\text{weak},\mu\,\dagger} j_{\text{lepton}}^{\text{weak},\nu} \tag{20-55}$$

$$= 8\left[p^{\mu}p'^{\nu} + p'^{\mu}p^{\nu} + g^{\mu\nu}\left(\frac{q^2 - m^2}{2}\right) \pm i\epsilon^{\mu\nu\alpha\beta}p_{\alpha}p'_{\beta}\right]$$

$$= 8\left[2p^{\mu}p^{\nu} - p^{\mu}q^{\nu} - q^{\mu}p^{\nu} + g^{\mu\nu}\left(\frac{q^2 - m^2}{2}\right) \mp i\epsilon^{\mu\nu\alpha\beta}p_{\alpha}q_{\beta}\right]$$

where p and p' are the four-momenta of the incident neutrino (or antineutrino) and of the final charged lepton, respectively; $q = p - p'$ is the four-momentum transferred to the hadronic system. But unlike the electromagnetic case, q is not carried by a single virtual particle, unless we hypothesize the existence of the intermediate vector boson W (Sec. 20-4). In Eq. 20-55 and subsequent weak interaction equations in this chapter, the upper sign is for an incident neutrino, the lower sign for an incident antineutrino, and m is the charged lepton mass.

Exercise 20-13. Derive Eqs. 20-55. Show that the weak leptonic current does *not* obey current conservation. But show that it does obey current conservation if $m = 0$. The $\epsilon^{\mu\nu\alpha\beta}$ is defined in App. A.

Next we define

$$W_{\mu\nu}^{\text{weak}} = \left(\frac{(2\pi)^3}{2M}\right)\left(\frac{1}{2}\right)\sum_{\beta}\sum_{f}\int\left[\prod_{n=1}^{N}\frac{d^3p_n}{(2\pi)^3 2E_n}\langle P, \beta| J_\mu^{\text{weak},\dagger}|f\rangle \times \langle f|J_\nu^{\text{weak}}|P, \beta\rangle \delta^4(P_f - P - q)\right] \tag{20-56}$$

The most general form for $W_{\mu\nu}^{\text{weak}}$ is similar to the form of $W_{\mu\nu}$ (Eq. 20-8), but the nonconservation of parity allows an additional term. Explicitly, with A^{weak}, $B^{\text{weak}}, \ldots, F^{\text{weak}}$ Lorentz scalar functions of q^2 and $q \cdot P$, we have

$$W_{\mu\nu}^{\text{weak}} = A^{\text{weak}} g_{\mu\nu} + \frac{B^{\text{weak}}}{M^2} P_\mu P_\nu + \frac{C^{\text{weak}}}{M^2} q_\mu q_\nu + \frac{D^{\text{weak}}}{M^2}(q_\mu P_\nu + q_\nu P_\mu) + \frac{E^{\text{weak}}}{M^2}(q_\mu P_\nu - q_\nu P_\mu) - i\left(\frac{F^{\text{weak}}}{M^2}\right)\epsilon_{\mu\nu\alpha\beta}P^\alpha q^\beta \tag{20-57a}$$

The hermitian nature of the weak current (Eq. 20-9) tells us that A, B, C, D, and F are real, but E is imaginary. Unlike the electromagnetic case, we cannot use exact current conservation (see Ex. 20-13) to simplify Eq. 20-57a. But we can use time-reversal invariance. As shown in Ex. 20-14, this requires A, B, C, D, E and F to be real, hence

$$E^{\text{weak}} = 0 \tag{20-57b}$$

But this is as far as we can go.

Exercise 20-14. Verify the statement just made about the requirement imposed by time-reversal invariance. We could have shown $E = 0$ in Eq. 20-8 in the same way, but current conservation also demands $E = 0$.

Thus we are left with five weak interaction structure functions compared to the two, W_1 and W_2, which occurred in the electromagnetic case (Sec. 20-2). These five functions must be determined from experiment; and given the difficulty of carrying out neutrino experiments, this is a most difficult task. Fortunately, this task simplifies at high energy, for we can then set $m = 0$ in Eq. 20-55 and then, as discussed in Ex. 20-13, $q_\mu L^{\text{weak},\mu\nu} = 0$. Hence in the calculation of $L^{\text{weak},\mu\nu}W_{\mu\nu}^{\text{weak}}$ the coefficients of C^{weak} and D^{weak} are zero. Explicitly, we have

$$L^{\text{weak},\mu\nu}W_{\mu\nu}^{\text{weak}} = 8A^{\text{weak}}[q^2] + 8\left(\frac{B^{\text{weak}}}{M^2}\right)\left[2(p\cdot P)^2 - 2p\cdot Pq\cdot P + \frac{M^2q^2}{2}\right]$$
$$\pm 8\left(\frac{F^{\text{weak}}}{M^2}\right)[p\cdot Pq^2 - p\cdot qP\cdot q] \tag{20-58}$$

For $m = 0$, the first two terms are similar to the electromagnetic case (Eq. 20-12); the third term, which is absent in Eq. 20-12, is now allowed by the nonconservation of parity.

To calculate the differential cross section, we continue as in Sec. 20-2. In the laboratory frame we have

$$\frac{d\sigma}{d\Omega'\,dE'} = \frac{G^2}{4\pi^2}\left(\frac{E'}{E}\right)L^{\text{weak},\mu\nu}W_{\mu\nu}^{\text{weak}} \tag{20-59a}$$

It is conventional to define for an incident neutrino (ν) or antineutrino ($\bar{\nu}$)

$$W_1^{\nu,\bar{\nu}} = -A^{\text{weak};\nu,\bar{\nu}},\quad W_2^{\nu,\bar{\nu}} = B^{\text{weak};\nu,\bar{\nu}},\quad W_3^{\nu,\bar{\nu}} = F^{\text{weak};\nu,\bar{\nu}} \tag{20-60}$$

Then we have an equation quite similar to Eq. 20-13b for the electromagnetic case.

$$\frac{d\sigma^{\nu,\bar{\nu}}}{d\Omega'\,dE'} = \frac{G^2}{8\pi^2}\left(\frac{E'}{E}\right)\left[2W_1^{\nu,\bar{\nu}}|q^2| + W_2^{\nu,\bar{\nu}}(4EE' - |q^2|)\right.$$
$$\left.\mp W_3^{\nu,\bar{\nu}}|q^2|\left(\frac{E+E'}{M}\right)\right] \tag{20-59b}$$

Exercise 20-15. Derive Eqs. 20-58 and 20-59.

A priori there is no relation between the electromagnetic structure functions W_1^{em}, W_2^{em}, and $W_1^{\nu,\bar{\nu}}$, $W_2^{\nu,\bar{\nu}}$. But experiments are beginning to show that the two sets of structure functions are related in a profound way. We discuss this relationship in Sec. 20-8. For the present, we regard the inelastic scattering of charged leptons or neutrinos on nucleons as alternative ways to study strong interactions and hadronic structure. Either method is valuable insofar as we understand quite well the leptonic part of the reaction and can concentrate on the hadronic aspects of the process.

20-6 Charged Lepton–Nucleon Inelastic Scattering and Bjorken Scaling. Since W_1^{em} and W_2^{em} are scalar functions of the two independent Lorentz scalar variables q^2 and $q\cdot P$, their experimental description could be quite complicated. But in the high energy, large $|q^2|$ region they exhibit a rather simple and very significant behavior first predicted by Bjorken. The prediction, which we justify on the basis of a particular model in Sec. 20-7, is as follows.

We define the dimensionless variable

$$\omega = \frac{2q \cdot P}{|q^2|} = \frac{2M\nu}{|q^2|} \geqslant 1 \tag{20-61a}$$

Note that $\omega = 1$ for elastic scattering. The prediction is

$$\text{for} \qquad \nu \to \infty \qquad |q^2| \to \infty \qquad \text{with } \omega \text{ fixed} \tag{20-62a}$$

$$MW_1^{\text{em}}(q^2, q \cdot P) \to \text{function of just } \omega \tag{20-62b}$$

$$\nu W_2^{\text{em}}(q^2, q \cdot P) \to \text{function of just } \omega \tag{20-62c}$$

This is a *scaling law* prediction, and we refer to it as *Bjorken scaling*. We also use the dimensionless variable

$$x = \frac{|q^2|}{2q \cdot P} = \frac{1}{\omega} \leqslant 1 \tag{20-61b}$$

And we often write Eq. (20-62) in the form

$$MW_1^{\text{em}}(q^2, q \cdot P) \to F_1^{\text{em}}(x) \tag{20-62d}$$

$$\nu W_2^{\text{em}}(q^2, q \cdot P) \to F_2^{\text{em}}(x) \tag{20-62e}$$

(In the literature various definitions of F_1 and F_2 differ by factors of 2 or M.)

Although Bjorken scaling is a large ν, large $|q^2|$ prediction, we find experimentally that the scaling law is obeyed at quite moderate values of q^2 and ν. Specifically, Eq. 20-62 is obeyed, for both proton and neutron targets, for

$$|q^2| \gtrsim 1 \ (\text{GeV/c})^2 \tag{20-63a}$$

and for s_{hadronic} above the N^* resonance region. That is,

$$(s_{\text{hadronic}})^{1/2} \gtrsim 1.5 \text{ or } 2.0 \text{ GeV} \tag{20-63b}$$

Incidentally, in the literature, it is conventional to use the variable (see also Eq. 20-39)

$$W^2 = s_{\text{hadronic}} = 2MK + M^2 \tag{20-64}$$

Exercise 20-16. Show that Bjorken scaling must be violated as q^2 goes to zero at fixed ω. Use Eqs. 20-43 and 20-44. Consult the literature (e.g., 20GI) for the use of another scaling variable $\omega' = \omega + M^2/|q^2|$.

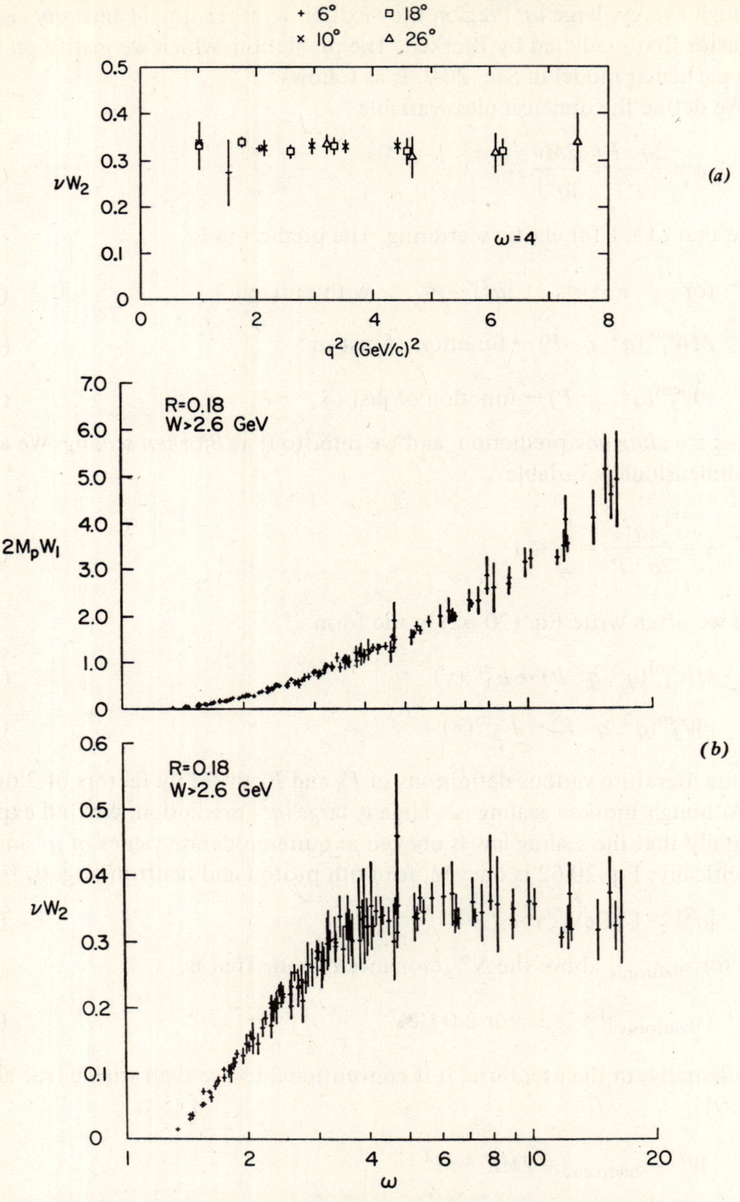

Fig. 20-4 Illustrations of the Bjorken scaling of νW_2 and W_1: (*a*) for a fixed value of ω, and (*b*) for a range of ω values. The points in (*b*) are for many different $|q^2|$ values in the range of 1 to 12 $(\text{GeV/c})^2$. The near overlap of the points at a particular ω indicates scaling (20FR).

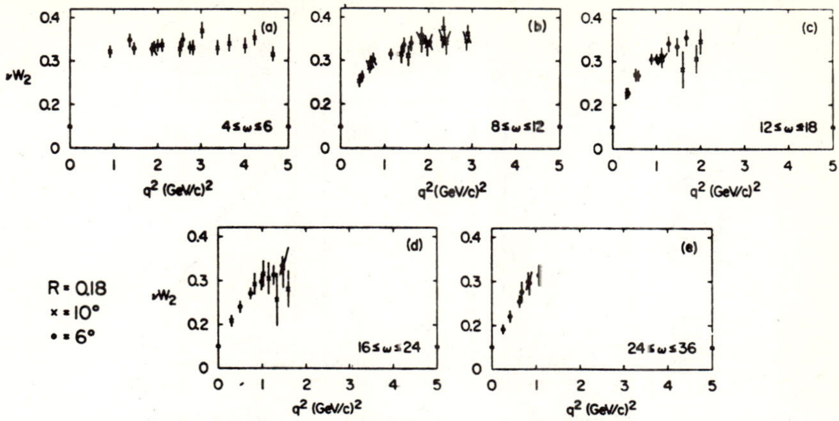

Fig. 20-5 The breakdown of scaling in the $|q^2| \lesssim 1$ (GeV/c)^2 region (20FR).

Figure 20-4*a* is a beautiful illustration of scaling. Another illustration appears in Fig. 20-4*b*. Finally, in Fig. 20-5 we show the breakdown of scaling at low values of q^2. The data are from proton targets, but scaling also holds for neutron targets. The data are less precise because they must be obtained by subtracting proton data from deuteron data. The data discussed here are almost completely from electron accelerators. Data at much higher values of ν are being acquired using muon beams in the several hundred GeV range at very high energy proton accelerators. Therefore we do not know whether the Bjorken scaling law will still be true when $|q^2|$ is of the order of tens of (GeV/c)^2 or ν is of the order of hundreds of GeV.

We shall not discuss in detail the specific forms of $F_1(\omega)$ and $F_2(\omega)$; but it is convenient to remember that for $\omega \gtrsim 4$, $|q^2| \gtrsim 1$ (GeV/c)^2

$$\nu W_2^{\mathrm{em}}(q^2, \nu) \sim 0.35 \tag{20-65}$$

The constancy of νW_2^{em} over a wide range of ω has nothing to do with Bjorken scaling. It represents an additional simplicity in the behavior of the structure functions.

Before discussing one beautifully simple explanation of the remarkable scaling behavior of W_1^{em} and W_2^{em}, we turn to the other way to describe inelastic scattering, σ_T and σ_L. Let us first consider σ_T. We have defined $\sigma_T(q^2, K)$ such that $\sigma_T(q^2, K) \to \sigma_{\mathrm{tot},\gamma N}(K)$, as q^2 goes to zero, where $\sigma_{\mathrm{tot},\gamma N}(K)$ is the photon-nucleon total cross section for a real photon of laboratory energy K. As shown in Fig. 20-6, σ_T decreases rather smoothly but rapidly as $|q^2|$ increases. Thus the dependence on q^2 is quite strong. On the other hand, the dependence on K is rather mild (Fig. 20-7).

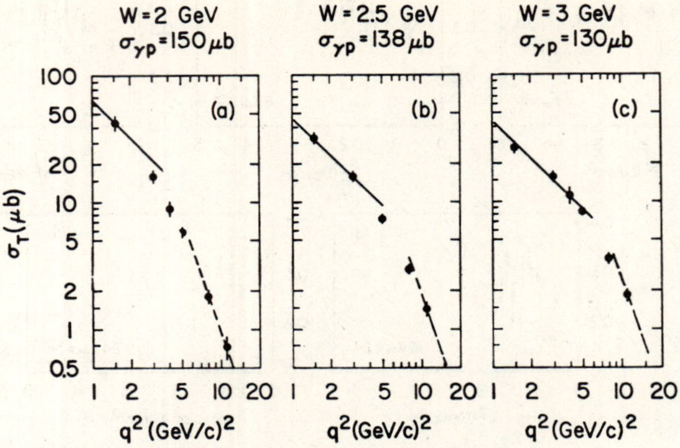

Fig. 20-6 The dependence of σ_T on q^2 (20FR).

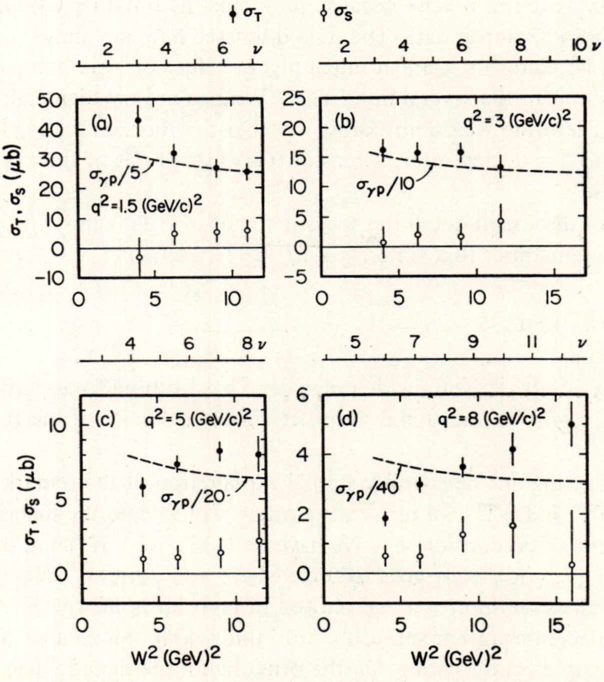

Fig. 20-7 The dependence of σ_T and σ_L on $W^2 = 2MK + M^2$ (20FR).

We do not understand the basic reason for the decrease in σ_T as the photon becomes more virtual. But a superficial explanation is that it is difficult for the target nucleon to absorb a large amount of q^2, even though the nucleon is allowed to break up into any energetically permitted final hadronic state. As $|q^2|$ increases, the difficulty increases. Even so, σ_T does not decrease with $|q^2|$ nearly as fast as the square of the electromagnetic elastic form factors of the nucleon (Sec. 19-4 and Eq. 19-22a).

Exercise 20-17. Discuss this comparison in more detail.

Before we consider σ_L, we must remark that we have ignored the experimental difficulty of separating W_1^{em} from W_2^{em} or σ_T from σ_L. The separation must be done by measuring $d\sigma/d\Omega'\,dE'$ (Eqs. 20-13) at two different values of E for the same values of q^2 and ν. At present, the separation has been carried out only with rather large errors for $|q^2| \gtrsim 1\ (\text{GeV}/c)^2$. We define the ratio

$$R(q^2, \nu) = \frac{\sigma_T(q^2, \nu)}{\sigma_L(q^2, \nu)} \tag{20-66}$$

It is found that R can be fit by the constant 0.18 ± 0.10. Of course $R \to 0$ as $q^2 \to 0$ according to Eqs. 20-44. Finally, R may increase with $|q^2|$; for example, present measurements may be fit by $R \sim |q^2|/\nu^2$ as well as by $R \sim 0.18$. The important conclusion is that $\sigma_L(q^2, \nu)$ is small compared with $\sigma_T(q^2, \nu)$. It is difficult to interpret this conclusion in a directly physical way because of the way σ_L is defined in Eq. 20-36b. And in the end it is probably easier to work directly with W_1^{em} and W_2^{em}. Indeed, for a crude description of these quantities we often set $\sigma_L = 0$.

We could describe many more experimental observations on charged lepton–nucleon inelastic scattering. Thus we have said very little about the results from scattering on neutrons. But we have covered the main points: the Bjorken scaling of W_1^{em} and W_2^{em}, the rather strong dependence of σ_T on q^2, and the small size of R. Our task is now to produce an explanation for these phenomena, particularly for the scaling behavior.

20-7 The Parton Model and Charged Lepton–Nucleon Inelastic Scattering. When we attempt to find a fundamental explanation of these inelastic processes, we usually run into the basic problem that we do not have a fundamental theory of strong interactions. For example, we may use vector meson dominance concepts (Sec. 19-7) or Regge theory concepts (Chs. 17 and 18) to examine these processes. But such examinations cannot be more fundamental than the concepts themselves, and we already know that we do not understand those concepts in a fundamental way. On the other hand, the parton model may provide a fundamental explanation of these inelastic processes, as well as insight into the basic nature of the strong interactions. Even if the parton model is wrong, as it may well be, the model provides a simple descriptive framework that can be used as a test for a correct theory.

The *parton model* is a *constituent* or *composite* model of the hadrons. Such models were briefly discussed in Ch. 14, particularly in Sec. 14-7. We take the partons to be *point* particles with fixed mass and fixed internal quantum numbers. They are bound together by forces at least as strong as the strong interactions. And when two or more of these partons are bound together, they form a hadron. Thus the quark model discussed in Ch. 14 is a particular form of the parton model in which the partons have been assigned a particular set of internal quantum numbers. According to the model, the interaction among hadrons can be resolved into the interaction among partons. Such interactions may be intrinsically simpler than direct hadron–hadron interactions because the partons are point particles. And most important for the electromagnetic inelastic processes under discussion, the interaction of a virtual photon with a hadron can be resolved into virtual photon–parton interactions. But before discussing this point in detail, let us face the fundamental problem in the parton model.

Free partons have never been observed. Therefore, the partons must be bound together by forces that produce binding energies of many GeV, perhaps tens of GeV; or some new physical principle holds the partons together. Yet in applying the parton model we speak of the interactions of individual partons, and we apply the impulse approximation to the partons, treating the partons as quasi-free. We can get around this problem in a somewhat formal way by working in a Lorentz frame in which the hadron has very high momentum. This so-called *infinite momentum frame* has been discussed by many authors (20BJ, 20DR, 20FE), and it is very useful for calculations. Indeed we employ it also, although the use of the impulse approximation in the parton model involves a yet unresolved paradox.

Nevertheless, we shall proceed bravely. In Fig. 20-8*a* we picture a hadron as a volume, indicated by the circle, containing a number of partons. These partons may be of several types; type i has mass M_i and charge $Q_i e$, where Q_i is the ratio of the parton's charge to the unit charge e. To add some generality, they may have spin zero or $\frac{1}{2}$.

The parton model assumes that when a virtual photon of sufficiently large q^2 is absorbed by a hadron, the photon is absorbed *incoherently* by *one* of the partons. The incoherence assumption is crucial. The parton that absorbs the virtual photon is then excited with reference to the remainder of the hadron. But as schematically represented in Fig. 20-8*d*, the parton cannot escape in the end. And finally, to conserve energy and momentum the initial hadron breaks up into two or more hadrons. The same process is illustrated by the Feynman diagram

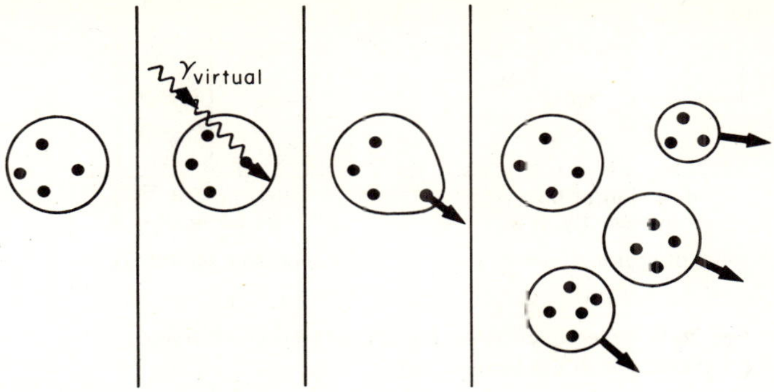

Fig. 20-8 The parton model for a virtual photon-nucleon interaction.

of Fig. 20-11*a*. Again we emphasize that the virtual photon is absorbed by only one of the partons—this is the incoherence assumption.

The nucleon of mass M is at rest in the laboratory frame. But to use the impulse approximation, we apply a Lorentz velocity transformation so that the nucleon velocity becomes β, with β very close to 1. This is a *very high momentum frame*; we need not go to an infinite momentum frame. Now the nucleon has four-momentum $P = (E, \mathbf{P})$

$$E = \gamma M, \qquad |\mathbf{P}| = \beta\gamma M, \qquad \gamma = (1 - \beta^2)^{-1/2}, \qquad \beta \to 1 \tag{20-67}$$

Let us return to the nucleon rest frame for a moment; in that frame each parton will have some variable momentum $\mathbf{p}_{\text{rest}}$. Now we assume that $\mathbf{p}_{\text{rest}}$ is bounded such that

$$|\mathbf{p}_{\text{rest}}| < p_{\max} \tag{20-68}$$

Then we set γ in Eq. 20-67 high enough that in the very high momentum frame the $\mathbf{p}$ of every parton is in the *direction* of $\mathbf{P}$. We also set γ high enough that relativistic time dilation slows down the rate at which partons interact with one another in the nucleon. In this way the partons appear quasi-free to us. We return to this point in Ex. 20-18. In the very high momentum frame, let an individual parton have

$$\mathbf{p}_{\text{parton}} = x\mathbf{P}, \qquad E_{\text{parton}} \approx xE, \qquad 0 \lesssim x < 1 \tag{20-69}$$

Here we have used Eq. 20-68 and taken $|\mathbf{p}_{\text{parton}}|$ very large compared with $p_{\max}$ and with any parton mass m_{parton}. We do not discuss the region of x close to zero. The ratio of the parton's momentum to the nucleon's momentum, x, will turn out to be the x defined in Eq. 20-61b!

So far we have not chosen the direction of the boost. Any direction will do, but it is most convenient to choose a direction such that the three-momentum of the

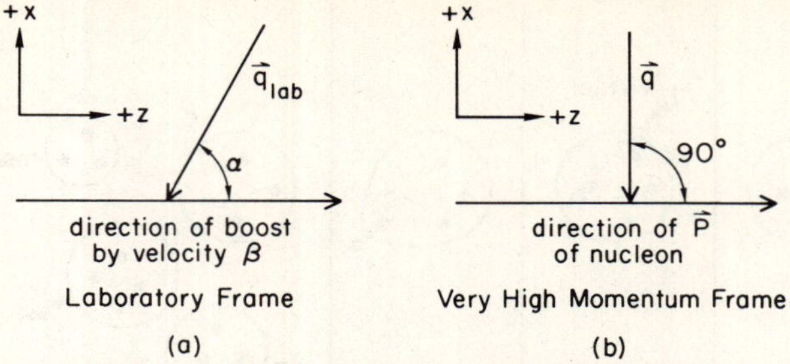

Fig. 20-9 The direction of **q** in (*a*) the laboratory frame and (*b*) the very high momentum frame.

virtual photon **q** is perpendicular to **P** in the *very high momentum* frame. Using the angle α defined in Fig. 20-9*a* we find

$$\beta = \left(\frac{|\mathbf{q}|_{\text{lab}}}{q^0_{\text{lab}}}\right) \cos\alpha, \qquad q^0_{\text{lab}} = \nu \tag{20-70a}$$

in very high momentum frame
$$\begin{cases} q^0 = \dfrac{\nu}{\gamma} & \text{(20-70b)} \\ q_x{}^2 \approx |q^2| & \\ q_y = q_z = 0 & \text{(20-70c)} \end{cases}$$

Exercise 20-18. Derive Eqs. 20-70.

Exercise 20-19. We can now make the impulse approximation argument (20DR, 20FE). Use perturbation theory to show that the lifetime τ_{life} of a state of N partons in the very high momentum frame is given by

$$\tau_{\text{life}} \sim \frac{1}{\sum_{n=1}^{N} [x_n{}^2|\mathbf{P}|^2 + p_{\perp n}^2 + m_{\text{parton},n}^2]^{1/2} - E} \approx \frac{E}{M_{\text{eff}}^2} \tag{20-71a}$$

Note that M_{eff}^2 is finite, assuming m_{parton} is finite, since $p_\perp$ is bounded by Eq. 20-68. Next show that the duration of the interaction is given by

$$\tau_{\text{int}} \sim \frac{E}{M\nu} \tag{20-71b}$$

Thus by going to sufficiently high ν (cf. Eq. 20-62a)

$$\tau_{\text{life}} \gg \tau_{\text{int}} \tag{20-71c}$$

when

$$\nu \gg p_{\text{max}}, \qquad \nu \gg m_{\text{parton},n} \tag{20-72a}$$

Exercise 20-20. The interaction of the virtual photon with the partons will be incoherent if in a virtual photon-nucleon interaction, the virtual photon is completely absorbed by a *single* photon. Give a rough argument to show that this requires

$$|q_x| \gg p_{\max} \qquad \text{or} \qquad |q^2| \gg p_{\max}^2 \tag{20-72b}$$

where we have used the very high momentum frame.

Using the very high momentum frame and the conditions of Eqs. 20-72, we are finally ready to make the calculation. Charged lepton-nucleon inelastic scattering consists simply of a summation, over all partons in the nucleon and over

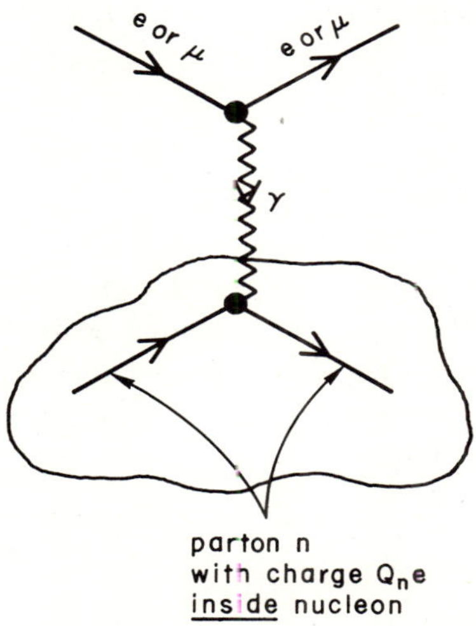

Fig. 20-10 The Feynman diagram for charged lepton-parton elastic scattering.

all values of x, of charged lepton-parton elastic scattering (Fig. 20-10). To begin, let us assume that

1. There are N partons in the nucleon.
2. Parton n has charge eQ_n, where e is the unit electric charge.
3. The probability of finding a parton with parameter x is $f(x)$, normalized such that

$$\int_0^1 f(x)\, dx = 1 \tag{20-73}$$

Here we have assumed, although it is not necessary, that $f(x)$ is the same for all partons. Then we have

$$\langle x \rangle = \int_0^1 x f(x)\, dx = \frac{1}{N} \tag{20-74}$$

Exercise 20-21. Prove Eq. 20-74. Devise some $f(x)$ for $N = 3$ and N arbitrary.

4. The partons are spin $\frac{1}{2}$ particles

The contribution to the structure functions W_1^{em}, W_2^{em} from the elastic scattering on a parton n with parameter x can now be calculated. As described in Ex. 20-22, we can use the results of Ex. 20-4, with $G = 1$, to obtain

$$W_{1,\text{parton elastic}}^{\text{em}} = \frac{1}{2M}\,\delta\left(x - \frac{|q^2|}{2M\nu}\right) \tag{20-75a}$$

$$W_{2,\text{parton elastic}}^{\text{em}} = \frac{x}{\nu}\,\delta\left(x - \frac{|q^2|}{2M\nu}\right) \tag{20-75b}$$

Exercise 20-22. The derivation of Eqs. 20-75 is a bit tricky. It seems easiest to redo Ex. 20-4, changing M and P in the δ function. Thus M^2 is to be replaced by $x^2M^2 \approx E_{\text{parton}}^2 - |\mathbf{p}_{\text{parton}}|^2$ in the very high momentum frame. Also in that frame, $q \cdot p_{\text{parton}} = (\nu/\gamma)(xE)$. Also remember that P_μ, P_ν, and M in Eq. 20-11a refer to the nucleon, not to the parton.

Next we integrate Eqs. 20-75 over all x and sum over the N partons to obtain the very beautiful result

$$MW_1^{\text{em}} = \left(\sum_{n=1}^{N} Q_n{}^2\right)\left[\frac{f(x)}{2}\right] = F_1^{\text{em}}(x) \tag{20-76a}$$

$$\nu W_2^{\text{em}} = \left(\sum_{n=1}^{N} Q_n{}^2\right)[xf(x)] = F_2^{\text{em}}(x) \tag{20-76b}$$

$$x = \frac{|q^2|}{2M\nu} \tag{20-76c}$$

We have obtained Eq. 20-62—Bjorken scaling! Furthermore, $F_1^{\text{em}}(x)$ and $F_2^{\text{em}}(x)$ are simply related to the momentum distribution of the partons in the very high momentum frame (see, however, Ex. 20-24). Thus we have shown how Bjorken scaling is simply explained by the presence of point constituent particles in the nucleon. We have also demonstrated that for this to be true, ν and $|q^2|$ must exceed the lower limits given by Eqs. 20-72. Of course we do not know what

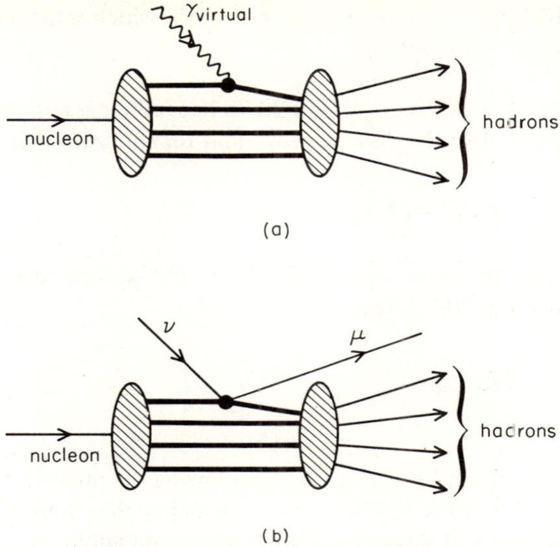

Fig. 20-11 The Feynman diagram for the parton explanations of (*a*) charged lepton-nucleon inelastic scattering, and (*b*) neutrino-nucleon inelastic scattering.

these limits (which depend on the possibly hypothetical parameters $p_{\max}$ and m_{parton}) are. Therefore, the derivation is formally correct only as $|q^2| \to \infty$, $\nu \to \infty$ at fixed x. But experimentally those limits are quite low, as discussed in the previous section. Finally we note that the Feynman diagram for the process we have been describing is simply that given in Fig. 20-11*a*. We observe again that the virtual photon is absorbed by a single parton.

The parton model of charged lepton–nucleon inelastic scattering has been very much elaborated in an attempt to fit all the experimental data on charged lepton–nucleon inelastic scattering. Since a full discussion of these developments and elaborations would be out of place here, we end this section with a few disconnected observations and derivations, chosen to give the flavor of some of these developments.

From Eq. 20-76 we write

$$F_2^{\text{em}}(x) = 2xF_1^{\text{em}}(x) \tag{20-77a}$$

$$\nu W_2^{\text{em}} = 2xMW_1^{\text{em}} \tag{20-77b}$$

Then Eq. 20-43 gives

$$\frac{\sigma_L}{\sigma_T} = \frac{|q^2|}{\nu^2} = x\left(\frac{2M}{\nu}\right) \tag{20-78}$$

Since $\nu \gg M$, this predicts σ_L/σ_L will be very small, which is indeed true, as discussed in the previous section.

Exercise 20-23. Equations 20-77 and 20-78 depend on assumption 4—that the partons have spin $\frac{1}{2}$. Use Ex. 20-5 to prove that for spin zero partons

$$W_1^{\text{em}} = 0, \qquad \nu W_2^{\text{em}} = F_2^{\text{em}}(x) \tag{20-79a}$$

Then show that a substantial contribution from spin zero partons is excluded by experiment, since Eq. 20-79a leads to

$$\sigma_T = 0, \qquad \frac{\sigma_L}{\sigma_T} = \infty \tag{20-79b}$$

The parton model can be generalized by allowing the nucleon to have a variable number of partons, $\mathscr{P}_N$ being the probability that it contains N partons. This allows us to think of the continuous creation and annihilation of parton-antiparton pairs in the nucleon. We may also allow the different partons to have different distributions in x, designated by $f_N(x_n)$ for the nth parton, when there are N partons in the nucleon. Then Eq. 20-76b becomes, for example,

$$F_2^{\text{em}}(x) = x\left[\sum_{N=1}^{\infty} \mathscr{P}_N\left(\sum_{n=1}^{N} Q_n{}^2 f_N(x_n)\right)\right] \tag{20-80a}$$

with

$$\sum_{N=1}^{\infty} \mathscr{P}_N = 1 \tag{20-80b}$$

Exercise 20-24. Use $x = \mathbf{p}_{\text{parton}}/\mathbf{P}$ to show that in the nucleon rest frame

$$x = \frac{E_{\text{parton,lab}} + p_{z,\text{parton,lab}}}{M}$$

where we have taken $\beta \approx 1$ and $\mathbf{P}$ is along the $+z$ axis. Thus it is difficult to attach a direct physical meaning to the x distribution in the nucleon rest frame. Of course if we make the unlikely assumption that the partons are at rest in the nucleon rest frame, we obtain $x = m_{\text{parton}}/M$, and x is simply the distribution of the parton masses.

Continuing with the parton model, we return to our simple assumption that $f_N(x_n)$ is independent of n. Consider the integral

$$Z = \frac{|q^2|}{2M} \int_0^\infty W_2^{\text{em}}(\nu, q^2) \frac{d\nu}{\nu} = \int_0^1 F_2^{\text{em}}(x)\, dx$$

$$= \sum_N \mathscr{P}_N \left(\frac{\sum_{n=1}^{N} Q_n^2}{N} \right) \tag{20-81}$$

If all the partons have unit charge, $Z = 1$. If we use the most naive *quark–parton* model (Ch. 14), in which the partons are simply the three quarks in the quark model of the nucleon, we have

$$Z_{\text{proton}} = \tfrac{1}{3}, \qquad Z_{\text{neutron}} = \tfrac{2}{9} \tag{20-82a}$$

But experimentally (20GI)

$$Z_{\text{proton}} \sim 0.14, \qquad Z_{\text{neutron}} \sim 0.10 \tag{20-82b}$$

This naive quark-parton model is not too bad, but it is not precise either. One must decrease the theoretical value of Z by model changes such as: decreasing Q_n, allowing $f_N(x_n)$ to depend on n, and assuming the presence of neutral constituents carrying momentum but not charge.

Exercise 20-25. Fill in the steps leading to Eqs. 20-81 and 20-82a.

The parton model must be elaborated to give good fits to the data. For example, the creation of parton–antiparton pairs must be considered. The basic reference for much of this is the neat and simple paper by Bjorken and Paschos (20BJ). And it appears that we must also consider the absorption of the virtual photon by the virtual particles carrying the forces between the partons (20KU). But for a book of this nature such considerations are premature, for we have already demonstrated the crucial idea in the parton model. The scaling law and the magnitudes of the structure functions in charged lepton–nucleon inelastic scattering can be explained by the assumption that the nucleon is composed of charged point particles.

To conclude this section we consider Bjorken scaling from a very general viewpoint. The *dimensionless* Lorentz scalars MW_1 and νW_2 are functions of q^2 and $q \cdot P$, each of which has the dimensions of energy squared. To construct a dimensionless function we can start from a dimensionless ratio such as $\omega = 2q \cdot P/|q^2|$. Or we must introduce a mass or a length that will allow us to construct dimensionless quantities such as $|q^2|/M^2$. We should not introduce any mass or any length; it should be a mass or length intimately connected with the *dynamics* of virtual photon–hadron interactions. The significance of Bjorken scaling is that *no* such mass or length is *necessary*. The parton model is an illustration of this, for we observe that the parton mass does not appear in the final equations, and the proton mass enters only in a trivial way.

On the other hand, if at higher energies or q^2 values some mass or length does enter the dynamics in a fundamental way, the scaling law could break down. For example, suppose the interaction among partons is carried by a virtual particle of mass M'. Then according to our standard ideas, there will be propagators of the form $1/(M'^2 - t)$, and the scaling law might break down at $|q^2| \sim M'^2$. Thus we return to the discussion at the end of Ch. 1. Our projection to very high energy of regularities observed at lower energies can be wrong if particles of very large mass are involved in the dynamics.

Exercise 20-26. We have applied the parton model to charged lepton-nucleon inelastic scattering only. But it has many applications in hadron–hadron interactions, although these applications are not nearly as clean as is the use of partons in lepton–nucleon inelastic scattering. The most spectacular application is probably the prediction by Feynman of a scaling law in inclusive hadron–hadron reactions. Rather than reproduce that application, we recommend a full discussion by Feynman (20FE).

20-8 Hadron Production Through the Weak Interactions. As this book is being completed the experimental study of hadron production in neutrino-nucleon interactions is just beginning. Therefore we shall only indicate some directions of interest. Equation 20-59b is the basic equation for the process. In a slightly different form

$$\frac{d\sigma^{\nu,\bar{\nu}}}{dq^2\, dE'} = \frac{G^2}{8\pi E^2}\left[2W_1^{\nu,\bar{\nu}}|q^2| + W_2^{\nu,\bar{\nu}}(4EE' - |q^2|) \mp W_3^{\nu,\bar{\nu}}|q^2|\frac{E+E'}{M}\right] \tag{20-83}$$

There are two crucial differences between this equation and that for charged lepton–nucleon inelastic scattering, Eq. 20-15.

1. There are three structure functions instead of two.
2. The $1/q^4$ factor in the electromagnetic case is not present in this weak interaction case. Hence the fraction of large $|q^2|$ cross section to the total cross section will be much larger in neutrino–nucleon inelastic scattering compared to

charged lepton–nucleon elastic scattering, if the structure functions are not too different. We summarize this by the replacement rule

$$(e \text{ or } \mu + N \to e \text{ or } \mu + \text{hadrons}) \leftrightarrow (\nu + N \to e \text{ or } \mu + \text{hadrons})$$
$$\frac{e^2}{|q^2|} \leftrightarrow G\sqrt{2} \tag{20-84}$$

if we ignore the structure function differences.

1. What are the connections, if any, between $W_{1,2}^{\text{em}}$ and $W_{1,2,3}^{\nu,\bar{\nu}}$?
2. In particular, does Bjorken scaling occur in $W_{1,2,3}^{\nu,\bar{\nu}}$?

We briefly consider these questions in terms of the parton model and in terms of electromagnetic and weak currents.

Figure 20-11*b* is the Feynman diagram for the parton model explanation of neutrino–nucleon inelastic scattering. We simply assume that the coupling of the leptons to the partons is given by the conventional Fermi theory (Sec. 20-4). The coupling constant is taken to be the usual G (Eq. 20-50). Calculations of the same type as in Sec. 20-7 lead to the parton model predictions

$$MW_1^{\nu,\bar{\nu}} = F_1^{\nu,\bar{\nu}}(x), \qquad \nu W_2^{\nu,\bar{\nu}} = F_2^{\nu,\bar{\nu}}(x), \qquad \nu W_3^{\nu,\bar{\nu}} = F_3^{\nu,\bar{\nu}}(x) \tag{20-85a}$$

where

$$2xF_1^{\nu,\bar{\nu}} = F_2^{\nu,\bar{\nu}}, \qquad -xF_3^{\nu,\bar{\nu}} = F_2^{\nu,\bar{\nu}} \tag{20-85b}$$

Therefore, the parton model predicts that $W_1^{\nu,\bar{\nu}}$, $\nu W_2^{\nu,\bar{\nu}}$, and $\nu W_3^{\nu,\bar{\nu}}$ will obey Bjorken scaling.

Exercise 20-27. Derive Eqs. 20-85. Why is the relation of $F_1^{\nu,\bar{\nu}}$ to $F_2^{\nu,\bar{\nu}}$ the same as the relation of F_1^{em} to F_2^{em}?

Exercise 20-28. Show that at high energies one can neglect $|q^2|$ compared to $4EE'$ for most values of q^2 and E'. Then insert the parton model predictions Eq. 20-85b into Eq. 20-83, integrate over the entire x and E' ranges and prove:

1. The total cross section for hadron production $\sigma_{\text{total}}^{\nu,\bar{\nu}}$ is proportional to the incident energy E.

2. $\sigma_{\text{total}}^{\bar{\nu}}/\sigma_{\text{total}}^{\nu} = 1/3$.

Early experiments are in agreement with this predicted 1/3 ratio.

The parton model is a very specific way to connect charged lepton–nucleon inelastic scattering. Here we briefly introduce a much more general way to make such a connection, based on the concept of currents.

As observed in Secs. 4-3 and 19-3, isotopic spin is not conserved in electromagnetic interactions; and the photon has no definite isotopic spin. But to explain existing data it is sufficient to allow the photon to act as the $I_z = 0$ component of an isoscalar or an isovector multiplet. Hence the electromagnetic hadronic current $J^{\text{em}}_{\text{hadron}}$ (Eq. 20-3c) can be divided into isoscalar and isovector parts

$$J^{\text{em},3}_{\text{hadron}} = J^{\text{em},3}_{\text{hadron}}(I = 0) + J^{\text{em},3}_{\text{hadron}}(I = 1) \qquad (20\text{-}86)$$

Here J is a four-vector in real space, but the superscript 3 indicates the third component ($I_z = 0$) in *isotopic spin space.*

Exercise 20-29. Show that $J^{\text{em},3}_{\text{hadron}}(I = 0)$ and $J^{\text{em},3}_{\text{hadron}}(I = 1)$ can be separated by comparing $e + p$ and $e + n$ reactions.

The weak hadronic current is much more complicated (20BE, 20TR). First, as we have already discussed, under spatial reflection the current has both vector V and axial vector A parts. Explicitly

$$J^{\text{weak}}_{\text{hadron}} = V^{\text{weak}}_{\text{hadron}} + A^{\text{weak}}_{\text{hadron}} \qquad (20\text{-}87\text{a})$$

where these quantities are four-vectors or pseudo-four-vectors.

Second, from weak interactions at low energy we know that these currents can be divided into a part that acts as a strangeness conserving ($\Delta S = 0$), $I = 1$ entity, and a part that acts as a strangeness changing ($\Delta S = \pm 1$), $I = \frac{1}{2}$ entity. No additional currents are required at low energy, and we do not know yet what is required at high energy. Assuming that our experience with low energy weak interactions carries over to high energies, we write

$$V^{\text{weak}}_{\text{hadron}} = \cos\theta_c\, V^{\text{weak}}_{\text{hadron}}(\Delta S = 0, I = 1) + \sin\theta_c\, V^{\text{weak}}_{\text{hadron}}(\Delta S = \pm 1, I = \tfrac{1}{2}) \qquad (20\text{-}87\text{b})$$

$$A^{\text{weak}}_{\text{hadron}} = \cos\theta_c A^{\text{weak}}_{\text{hadron}}(\Delta S = 0, I = 1) + \sin\theta_c A^{\text{weak}}_{\text{hadron}}(\Delta S = \pm 1, I = \tfrac{1}{2})$$

Here θ_c, the Cabibbo angle, at low energies has the value

$$\theta_c \sim 0.2 \text{ rad} \qquad (20\text{-}88)$$

Of the four currents in Eq. 20-87b, only $V^{\text{weak}}_{\text{hadron}}(\Delta S = 0, I = 1)$ has the same invariance properties as part of $J^{\text{em}}_{\text{hadron}}$, namely, $J^{\text{em}}_{\text{hadron}}$ ($I = 1$). Therefore, one might hope for a general connection between these two currents. Such a connection, indeed a very simple connection, was postulated by Feynman and Gell-Mann and by Gershtein and Zeldovitch. We write the charge raising and lowering weak current (V^+ and V^-, respectively) in the form

$$V^{\text{weak},\pm}_{\text{hadron}}(\Delta S = 0, I = 1) = V^{\text{weak},1}_{\text{hadron}} \pm i V^{\text{weak},2}_{\text{hadron}} \qquad (20\text{-}89\text{a})$$

Here $V_{\text{hadron}}^{\text{weak},\pm}$ are the charge raising and lowering operators; analogous to the angular momentum operators $J_\pm$ (App. B) for $J = 1$. Then the *conserved vector current* hypothesis is that

$$V_{\text{hadron}}^{\text{weak},1},\ V_{\text{hadron}}^{\text{weak},2},\ J_{\text{hadron}}^{\text{em},3} \tag{20-89b}$$

are the 1, 2, 3 components of an isotopic spin triplet.

This hypothesis, which has been verified at low energies, says that the isovector part of the electromagnetic hadronic current is simply related through Clebsch–Gordan coefficients to the $\Delta S = 0, I = 1$, four-vector part of the weak hadronic current. Remember that we have removed the coupling constant e from the electromagnetic current (Eq. 20-3c); and correspondingly G, or more precisely $\sqrt{G}$, has been removed from the weak current. Therefore, in Eq. 20-89b we are connecting not only the form but also the *magnitude* of the currents.

Returning to the comparison of neutrino–nucleon and charged lepton–nucleon inelastic scattering, we note that if Eqs. 20-87b and 20-88 are true in their high energy reactions, we can neglect the $\Delta S = \pm 1$ currents at the 10% level. Then Eq. 20-89b applies to the entire four-vector part of the weak current.

There is no relation between the electromagnetic current and the axial-vector weak current $A_{\text{hadron}}^{\text{weak}}$ unless one also assumes a relation between $A_{\text{hadron}}^{\text{weak}}$ and $V_{\text{hadron}}^{\text{weak}}$. Indeed there is such a relation, verified at low energies—the *partially conserved axial-vector current* hypothesis (20TR, 20BE). And if this relation is verified in high energy inelastic scattering, $J_{\text{hadron}}^{\text{em},3}$ and almost all $J_{\text{hadron}}^{\text{weak}}$ will be simply related. But to discuss this subject any further would take us beyond the point we are prepared to reach at this time. The recent literature will enable the reader to follow the progress of the comparative study and interpretation of neutrino–nucleon and charged lepton–nucleon inelastic scattering.

Exercise 20-29. Show that the parton model provides an example of a model that obeys the general relationships we have been discussing.

REFERENCES

20BE J. Bernstein, *Elementary Particles and Their Currents* (Freeman, San Francisco, 1968).

20BJ J. D. Bjorken and E. A. Paschos, *Phys. Rev.* **D1**, 3151 (1970).

20DO N. Dombey, *Rev. Mod. Phys.* **41**, 236 (1969); and "Introduction to Photoproduction and Electroproduction", in *Hadronic Interactions of Electrons and Photons*, J. Cumming and H. Osborn, Eds. (Academic Press, New York, 1971).

20DR S. D. Drell and T. M. Yan, *Ann. Phys. (N.Y.)* **66**, 578 (1971).

20FE R. P. Feynman, *Photon–Hadron Interactions* (Benjamin, New York, 1972).

20FR J. I. Friedman and H. W. Kendall, *Ann. Rev Nucl. Sci.* **22**, 203 (1972).

20GA S. Gasiorowicz, *Elementary Particle Physics* (Wiley, New York, 1966).

20GI F. J. Gilman, *Phys. Rep.* **4C**, 95 (1972).

20HA L. Hand, *Phys. Rev.* **129**, 1834 (1963).

20KU J. Kuti and V. F. Weisskopf, *Phys. Rev.* **D4**, 3418 (1971).

20TR See, for example, the articles by S. B. Treiman and by D. J. Gross in *Lectures on Current Algebra and its Applications* (Princeton University Press, Princeton, N.J., 1972).

20WI K. Wilson, in *Proceedings of the 1971 International Symposium on Electron and Photon Interactions at High Energies* (Laboratory of Nuclear Studies, Cornell University, Ithaca, N.Y., 1972).

CHAPTER

21

Hadrons and Timelike Virtual Photons

21-1 Introduction. At this writing, the experimental study of hadron production in electron-positron annihilations is just beginning. We are in the same position with respect to this field that we were in the late 1950s with respect to the entire field of high energy physics. A few relatively low or intermediate energy electron-positron colliding beam facilities have been operating for a few years. Higher energy colliding beam facilities are just now going into operation. Since we have some low and intermediate energy data, but almost no relatively high energy data, this chapter is quite short. Although this new area of physics is of great interest and deserves a chapter of its own; there are few data to describe, if we are primarily interested in high energies. And we are reluctant to discuss in detail theoretical speculations before they have been tested against high energy data.

In Sec. 21-2 we provide some general orientation. The existing data on the total cross section for hadron production are then briefly described. In Sec. 21-3 we discuss the reaction $e^+ + e^- \to \pi^+ + \pi^-$ and the pion electromagnetic form factor. The chapter concludes with a description of a related experimental observation, the indirect study—through charged lepton pair production—of time-like virtual photons produced in hadronic collisions. Some references to Secs. 21-2 through 21-4 are 21PIS, 21DR, and an old classic 21CA.

21-2 General Orientation and Total Hadron Production. We are interested in the reaction

$$e^+ + e^- \to \text{one virtual } \gamma \to 2 \text{ or more hadrons} \tag{21-1}$$

which takes place through the exchange of a single virtual photon (Fig. 21-1). The reaction was discussed in Sec. 19-2: remember that the square of the four-momentum carried by the virtual photon is

$$q^2 = (2E)^2 = E_{\text{cms}}^2$$

where E is the laboratory energy of *each* beam (Sec. 1-2). The virtual photon is timelike and in our metric $q^2 > 0$.

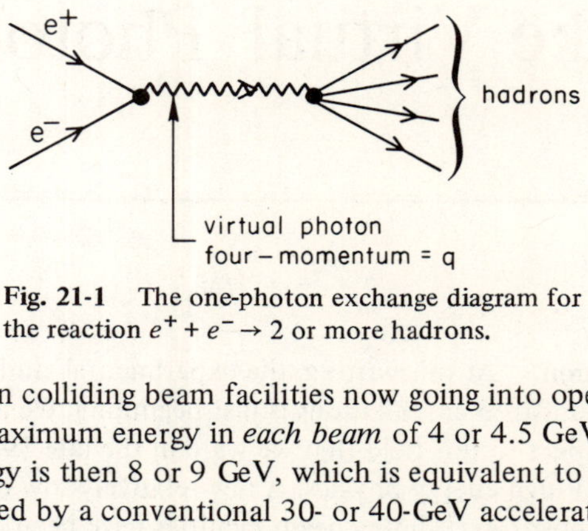

Fig. 21-1 The one-photon exchange diagram for the reaction $e^+ + e^- \to 2$ or more hadrons.

Electron-positron colliding beam facilities now going into operation will probably attain a maximum energy in *each beam* of 4 or 4.5 GeV. The maximum center of mass energy is then 8 or 9 GeV, which is equivalent to the center of mass energy produced by a conventional 30- or 40-GeV accelerator. Thus in terms of a naive comparison with proton acceleration, we are about to enter the region of high energy, but not the region of very high energy, with respect to this field. Nevertheless, the energy contained in an e^+e^- annihilation is in a sense more available to study hadronic production than is the same energy in a hadron–hadron collision. For as discussed at the end of Ch. 8, in the latter case much of the energy is usually carried through the reaction as kinetic energy. But in e^+e^- annihilations, all the energy is contained in a single virtual photon with zero three-momentum.

Since this is such a new field, we have a great many questions and few answers. For example, we have no idea how the total cross section for hadron production $\sigma_{e^+e^- \to \text{hadrons}}(q^2)$ or the average multiplicity $\langle N(q^2)\rangle$ will behave as q^2 increases. The size and q^2 dependence of $\sigma_{e^+e^- \to \text{hadrons}}$ is of particular interest because we have no previous physical intuition of how to calculate that cross section. For example, the optical model explanation of the hadron-hadron total cross section is simply that the total cross section is determined by the size and opacity of the hadrons. The observation that total cross sections are roughly independent of energy at high energy is explained by saying that the size and opacity of hadrons is roughly independent of energy. Of course this is not a fundamental explanation, it is only a useful picture. But for $\sigma_{e^+e^- \to \text{hadrons}}$ we do not even have this picture. There is only one particle entering the hadronic vertex—the virtual photon; and what is its size, anyway?

Another set of questions concerns the relation of the reactions in Eq. 21-1 to the charged lepton-nucleon inelastic scattering reactions in Eq. 20-1. The former reactions involve the interaction of a timelike virtual photon with hadrons and the latter involve the interaction of a spacelike virtual photon with hadrons.

Exercise 21-1. Show that the two sets of reactions just named are related by crossing (Ch. 15; 21DR).

Figure 21-2 gives $\sigma_{e^+e^- \to \text{hadrons}}$. Below $E_{\text{cms}} = 1$ GeV, the cross section is mostly due to the reaction $e^+ + e^- \to \pi^+ + \pi^-$; the peak at $q^2 \sim 0.6 \text{ GeV}^2$ is due to the production of the $\rho^0 \to \pi^+ + \pi^-$ (Sec. 21-3). At the higher energies, $\sigma_{e^+e^- \to \text{hadrons}} \sim 10^{-32} \text{ cm}^2$, which is very small compared to hadron-hadron total cross sections, namely, 10^{-26} cm^2.

Above $E_{\text{cms}} = 1$ GeV, $\sigma_{e^+e^- \to \text{hadrons}}$ decreases as E_{cms} increases, although there may be a second peak at the mass of the ρ' resonance. We don't know how to characterize this decrease because we don't know what to expect. That is, we don't know whether to describe the decrease as fast or slow.

Therefore, it has become conventional, and we have conventions even in this new field, to compare $\sigma_{e^+e^- \to \text{hadrons}}$ with the total cross section for the production of two point particles of unit charge. The term *point particles* is a descriptive way to indicate that the form factor at the photon-hadron vertex is 1 at all q^2 values. A physical example is the reaction

$$e^+ + e^- \to \mu^+ + \mu^- \tag{21-2}$$

with the cross section

$$\sigma_{e^+e^- \to \mu^+\mu^-}(q^2) = \frac{2\pi\alpha^2}{q^2}\,\beta\left[\frac{1-\beta^2}{3}\right] \xrightarrow[q^2 \to \infty]{} \frac{4\pi\alpha^2}{3q^2} = \frac{2 \times 10^{-32}}{E^2}\text{ cm}^2 \tag{21-3}$$

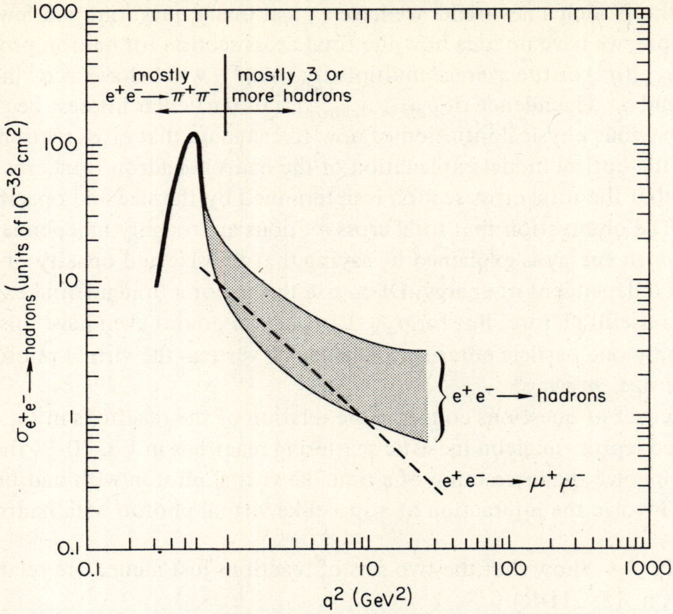

Fig. 21-2 The total cross section $\sigma_{e^+e^-\to\text{hadrons}}$, for the reaction $e^+ + e^- \to 2$ or more hadrons; taken partly from Ref. 21SI. The crosshatched area represents the uncertainties in these early measurements. The dashed line gives $\sigma_{e^+e^-\to\mu^+\mu^-}$. As discussed in connection with Eqs. 21-7, the ratio $R = \sigma_{e^+e^-\to\text{hadrons}}/\sigma_{e^+e^-\to\mu^+\mu^-}$ is significant.

Here β is the muon velocity; and in the extreme right-hand term, E is in GeV. If the particles a^+ and a^- have spin zero, the reaction

$$e^+ + e^- \to a^+ + a^- \tag{21-4}$$

has the cross section

$$\sigma_{e^+e^-\to a^+a^-}(q^2) = \frac{\pi\alpha^2\beta^3}{3q^2} \tag{21-5}$$

Exercise 21-2. Derive Eqs. 21-3 and 21-5.

Returning to Eqs. 21-3 and 21-5, we observe that the cross section goes as $1/q^2$; and this behavior has become the standard of comparison for the energy dependence of cross sections. One reason for this is that a parton model (21BJ) of electron–positron annihilation into hadrons predicts this dependence. The model (Fig. 21-3) is

$$e^+ + e^- \to \text{parton} + \text{antiparton} \to \text{hadrons} \tag{21-6}$$

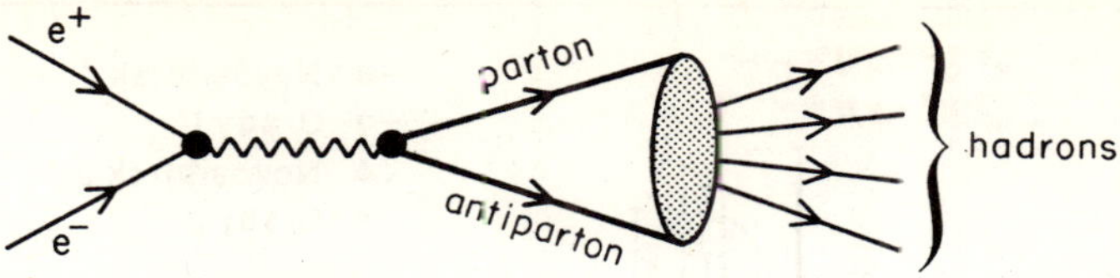

Fig. 21-3 The simple parton model for the reaction $e^+ + e^- \to 2$ or more hadrons.

Since the partons are point particles, the first part of the reaction gives the $1/q^2$ dependence; and the second part has unit probability, since the partons must become hadrons.

Explicitly, if there are N different kinds of spin $\frac{1}{2}$ partons with charge eQ_n, $n = 1, 2, \ldots, N$, the parton model predicts

$$\sigma_{e^+e^- \to \text{hadrons}} \leqslant \frac{4\pi\alpha^2}{3q^2}\left(\sum_{n=1}^{N} Q_n{}^2\right) \tag{21-7a}$$

or

$$\sigma_{e^+e^- \to \text{hadrons}} \leqslant R\sigma_{e^+e^- \to \mu^+\mu^-} \tag{21-7b}$$

where

$$R = \sum_{n=1}^{N} Q_n{}^2 \tag{21-7c}$$

Exercise 21-3. Derive Eqs. 21-7. Show that $R = \frac{2}{3}$ for the three quark-parton model. The upper limit allows for the cross section being lower if the partons have large mass. Early data (Fig. 21-2) show that R has the order of magnitude predicted by this model, but it may be too large for the naive forms of the model. Also, at large q^2, $\sigma_{e^+e^- \to \text{hadrons}}$ decreases slower than $1/q^2$; indeed it may be almost independent of q^2 (Fig. 21-2 and Ref. 21PIS).

21-3 The Reaction $e^+ + e^- \to \pi^+ + \pi^-$ and the Pion Electromagnetic Form Factor. The only $e^+ + e^- \to$ hadron + antihadron reaction for which we have substantial measurements at present is

$$e^+ + e^- \to \pi^+ + \pi^- \tag{21-8}$$

We generalize Eq. 21-5 to the nonpointlike and complicated pions by writing

$$\sigma_{e^+e^- \to \pi^+\pi^-}(t) = \frac{\pi\alpha^2\beta^3}{3t}\,|F_\pi(t)|^2 \tag{21-9}$$

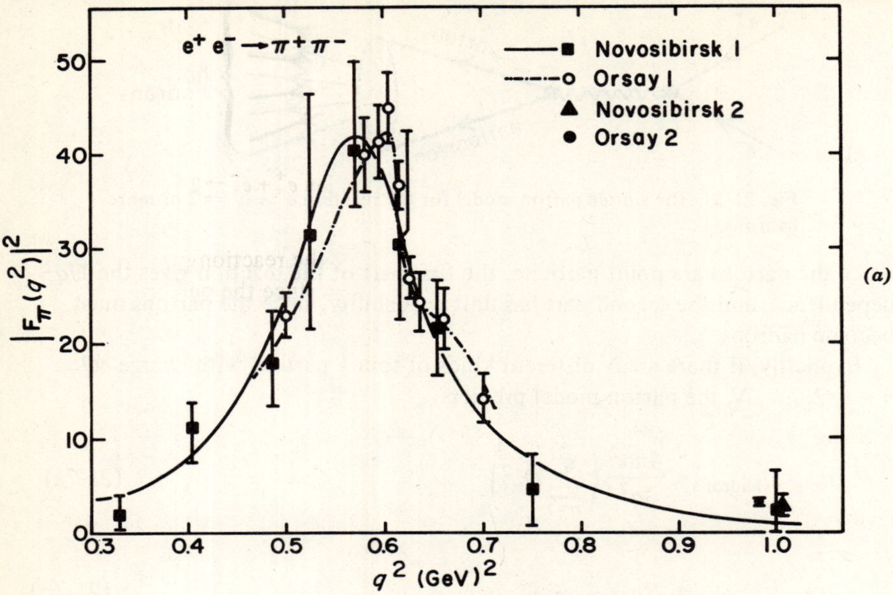

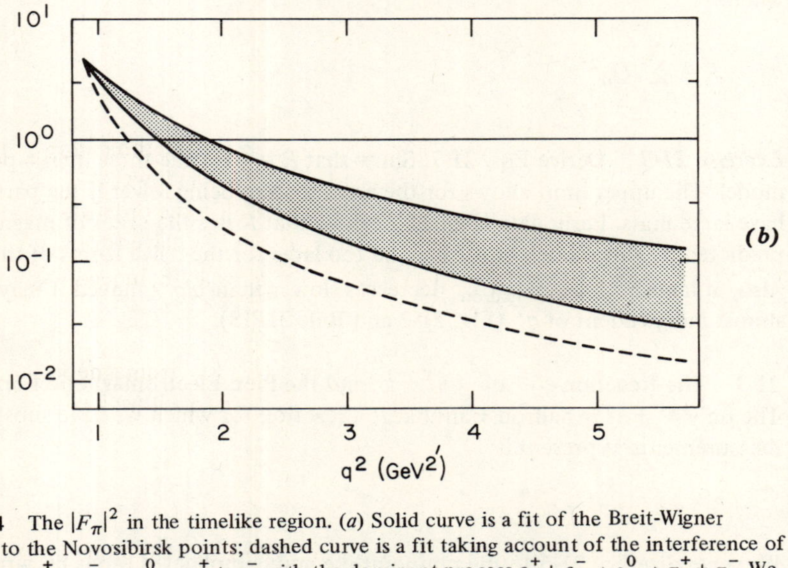

Fig. 21-4 The $|F_\pi|^2$ in the timelike region. (*a*) Solid curve is a fit of the Breit-Wigner formula to the Novosibirsk points; dashed curve is a fit taking account of the interference of the process $e^+ + e^- \to \omega^0 \to \pi^+ + \pi^-$ with the dominant process $e^+ + e^- \to \rho^0 \to \pi^+ + \pi^-$. We have not discussed this interference. Taken from Ref. 21SI. (*b*) Cross-hatched area represents the uncertainties in these early measurements; dashed line is the tail of the ρ-associated form factor, namely, Eq. 21-17, taken from Ref. 21BA. In both figures $q^2 \equiv t$.

(To simplify the appearance of the formulas in this section, we write $t = q^2$.) Here $F_\pi(t)$ is the pion electromagnetic form factor. It is most instructive to present the data in terms of $|F_\pi|^2$, rather than $\sigma_{e^+e^-\to\pi^+\pi^-}$, and we do just that in Fig. 21-4.

The electromagnetic form factor of the pions provides such a simple and fundamental example of the use of dispersion relations that we spend a few pages on the subject, although the data are rather provisional. Thus we take up again our discussion of the vertex form factor, dispersion relation (Sec. 15-10), using for purposes of illustration in the spacelike region the pion form factor.

$$F_\pi(t) \sim \frac{1}{1 - t/a^2}, \qquad t < 0 \tag{21-10}$$

as discussed in Sec. 19-6.

On the complex t-plane (Fig. 21-5), there is a cut on the Re t axis for $t > 4m_\pi{}^2$, where m_π is the pion mass. For real values of t

$$\begin{aligned} &t \geqslant 4m_\pi{}^2, \qquad F_\pi(t) = F_\pi(t + i\epsilon)_{\epsilon\to 0}, \qquad F_\pi(t) \text{ is complex} \\ &0 < t < 4m_\pi{}^2, \qquad \text{unphysical region}, \qquad F_\pi(t) \text{ is real} \\ &t \leqslant 0, \qquad F_\pi(t) \text{ is real} \end{aligned} \tag{21-11}$$

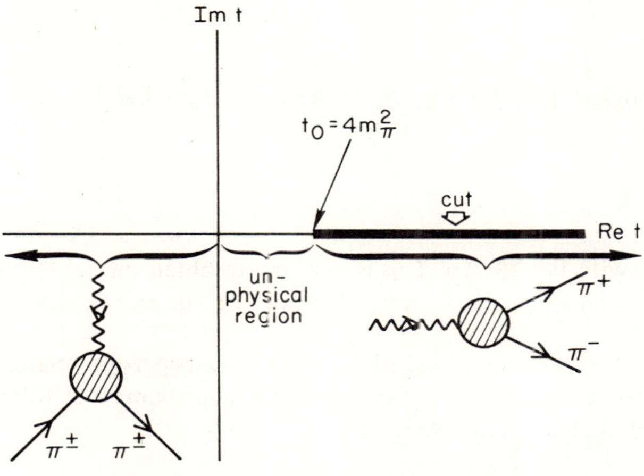

Fig. 21-5 The only singularity of F_π on the complex t-plane is the cut from $t = t_0 = 4m_\pi^2$ to $t = \infty$.

The dispersion relation is

$$\text{Re } F(t) = \frac{1}{\pi} P \int_{t_0}^{\infty} \frac{\text{Im } F(t')}{t' - t}\, dt', \qquad t_0 = 4m_\pi{}^2 \tag{21-12}$$

We begin with the $t > 4m_\pi^2$ in the neighborhood of the ρ^0 peak. The shape of $|F_\pi|$ in this region inspires us to use the vector meson dominance concept (Sec. 19-7) for F_π. Explicitly, we use the Breit–Wigner form

$$F_\pi = \frac{g_{\pi\pi\rho}}{f_\rho}\left(\frac{m_\rho^2}{m_\rho^2 - t - i\Gamma'}\right), \qquad \Gamma' = m_\rho \Gamma_\rho \tag{21-13a}$$

or

$$\text{Re } F_\pi = \frac{g_{\pi\pi\rho}}{f_\rho}\left(\frac{-m_\rho^2(t - m_\rho^2)}{(t - m_\rho^2)^2 + \Gamma'^2}\right) \tag{21-13b}$$

$$\text{Im } F_\pi = \frac{g_{\pi\pi\rho}}{f_\rho}\left(\frac{m_\rho^2 \Gamma'}{(t - m_\rho^2)^2 + \Gamma'^2}\right) \tag{21-13c}$$

Of course, experiment yields only $|F_\pi|$; we rely on the vector meson dominance concept and the Breit–Wigner formula for the phase of F_π. The $\pi\pi\rho$ coupling constant $g_{\pi\pi\rho}$, is discussed in Sec. 12-5. The $1/f_\rho$ comes from Eqs. 19-44 and 19-45, the e having been absorbed into the α^2 term in Eq. 21-9.

Exercise 21-4. Show that Eqs. 21-13 give approximately the observed shape of $|F_\pi|$ in the ρ region (Fig. 21-4*a*). Show that if we set

$$\frac{g_{\pi\pi\rho}}{f_\rho} \approx 1 \tag{21-14a}$$

we obtain roughly the right magnitude of $F_\pi(t = m_\rho^2)$. Use Eqs. 21-14a and 12-30 to obtain

$$\frac{f_\rho^2}{4\pi} \approx 2.6 \tag{21-14b}$$

and compare with Eq. 19-50a. This is one way to obtain these vector meson dominance coupling constants. A justification for Eq. 21-14a appears below.

Exercise 21-5. Verify that Eqs. 21-13 satisfy the dispersion relation (Eq. 21-12) if $t_0 \to -\infty$. Show that Eqs. 21-13 constitute an approximate solution to the dispersion relation when $t_0 = 4m_\pi^2$ if

$$m_\rho^2 - 4m_\pi^2 \gg \Gamma' \tag{21-15}$$

If we accept the approximation that we can use Eqs. 21-13, and if we assume that Eq. 21-13c gives the entire magnitude of Im F_π, we immediately know Re F_π for *all t* by the dispersion relation (of course, from Ex. 21-5 we know that we can use Eq. 21-13b directly). We consider first the spacelike region $t < 0$. By definition

$$F_\pi(0) = 1 \tag{21-16a}$$

We can fulfill Eq. 21-16a using the F_π given in Eqs. 21-13 if we use Eq. 21-14a and $m_\rho^2 \gg \Gamma'$. This is the justification for the equality in Eq. 21-14a. To lowest order in $(t - m_\rho^2)^{-1}$

$$\left.\begin{aligned} \text{Re } F_\pi &\approx \frac{1}{1 - t/m_\rho^2} \\ \text{Im } F_\pi &= 0 \end{aligned}\right\} |t - m_\rho^2| \gg \Gamma', \qquad t < 0 \tag{21-16b}$$

Thus for the spacelike region Eq. 21-16b gives the monopole form factor formula, Eqs. 21-10 and 19-42. Hence using the formalism of vector meson dominance, the ρ^0 resonance explains F_π not only in the t-close-to-m_ρ^2 timelike region but also in the small t spacelike region.

For the larger t timelike region, Eqs. 21-13 and 21-14a predict

$$F_\pi \approx \frac{-m_\rho^2}{t}, \qquad t \gg m_\rho^2 \tag{21-17}$$

This is the type of behavior observed in the higher t region of Fig. 21-4*b*. But the magnitude of F_π in Eq. 21-17 turns out to be too small.

Exercise 21-6. Verify the last statement using Eq. 21-17 and Fig. 21-4b.

There must be some additional contribution to Im F_π in the larger t region beyond that given by ρ production. This is certainly reasonable on general physical arguments, and we write

$$\text{Im } F_\pi = \text{Im } F_{\pi,\rho} + \text{Im } F_{\pi,\text{non-}\rho} \tag{21-18a}$$

where Im $F_{\pi,\text{non-}\rho}$ contributes for

$$t > t_1 \sim 1 \text{ GeV}^2 \tag{21-18b}$$

Exercise 21-7. Consider the following choices for Im $F_{\pi,\text{non-}\rho}$ when $t > t_1$ (*a*) $\mathscr{C}$, (*b*) $\mathscr{C}/t$, (*c*) $\mathscr{C}/t^2$, where $\mathscr{C}$ is a constant. Show that Re F_π = Re $F_{\pi,\rho}$ + Re $F_{\pi,\text{non-}\rho}$ where the corresponding forms for Re $F_{\pi,\text{non-}\rho}$ are

$$\text{Re } F_{\pi,\text{non-}\rho} \to \infty \tag{21-19a}$$

$$\text{Re } F_{\pi,\text{non-}\rho} = \frac{-\mathscr{C}}{\pi t}\left[\ln\left(\frac{|t|}{t_1} \pm 1\right)\right] \tag{21-19b}$$

$$\text{Re } F_{\pi,\text{non-}\rho} = \frac{-\mathscr{C}}{\pi}\left[\frac{1}{t_1 t} + \frac{1}{t^2}\ln\left(\frac{|t|}{t_1} \pm 1\right)\right] \tag{21-19c}$$

and the $\pm$ signs correspond to $t \lessgtr 0$.

Exercise 21-7 indicates that we certainly cannot add a constant to Im F_π and that if we add a (constant/t) term, F_π will fall more slowly then $1/|t|$ as $t \to \pm\infty$. On the other hand, a (constant/t^2) term will simply add

$$\text{Re } F_{\pi,\text{non-}\rho} \approx \frac{\text{constant}}{t}, \qquad t \to \pm\infty \tag{21-19d}$$

to the $m_\rho{}^2/|t|$ contribution from the ρ (Eqs. 21-17).

Exercise 21-8. Show that Im $F_{\pi,\text{non-}\rho} = \mathscr{C}t^{-n}$, $n = 3, 4, 5, \ldots, t > t$, also leads to Eq. 21-19d. We make two observations

1. The (constant/t) behavior of $F_\pi(t \to \pm\infty)$ is produced rather easily.
2. The important equality

$$F_\pi(t \to +\infty) = F_\pi(t \to -\infty) \tag{21-20}$$

is a very general property following from the dispersion relation for well-behaved forms of Im F_π.

Exercise 21-9. Discuss observation 2.

To conclude this section let us see how we might obtain the behavior

$$F_\pi(t) \to \frac{1}{t^2}, \qquad t \to \pm\infty \tag{21-21}$$

First we note that a Dirac δ function

$$\text{Im } F_\pi = \mathscr{C}\delta(t - t_1) \tag{21-22a}$$

inserted in the dispersion relation yields

$$\text{Re } F_\pi = \frac{\mathscr{C}/\pi}{t_1 - t} \tag{21-22b}$$

This is just the monopole form again of the form factor.

Exercise 21-10. Show that Eqs. 21-22 are the $\Gamma \to 0$ limit of the Breit–Wigner formula.

Now suppose there are two pole contributions of opposite sign

$$\text{Im } F_\pi = \mathscr{C}\delta(t - t_1) - \mathscr{C}\delta(t - t_2) \tag{21-23a}$$

Then in the *dipole* limit

$$\mathscr{C}(t_2 - t_1) \to \text{constant}, \qquad \mathscr{C} \to \infty, \qquad t_2 \to t_1 \tag{21-23b}$$

We obtain the *dipole* form factor (Sec. 19-4)

$$\text{Re } F_\pi = \frac{\text{constant}}{(t - t_1)^2} \xrightarrow[|t| \to \infty]{} \frac{\text{constant}}{t^2} \tag{21-23c}$$

This then is the dispersion relations explanation of a dipole form factor. As discussed in Ch. 19, the spacelike form factor of the proton is of the dipole form. But we do not know what basic mechanism leads to that form.

Exercise 21-11. More generally, if Im F_π oscillates about zero, the $1/t^2$ behavior can be obtained. Work out some examples.

Let us return to e^+e^- annihilations. Obviously as the field progresses we will obtain information on other two-body processes such as

$$e^+ + e^- \rightarrow \text{nucleon} + \text{antinucleon} \tag{21-24a}$$

$$e^+ + e^- \rightarrow K^+ + K^- \tag{21-24b}$$

and we will obtain the timelike form factors. The nucleon–antinucleon timelike form factor is of particular interest because we possess so much detailed information on the spacelike form factors.

21-4. Timelike Virtual Photons Produced in Hadronic Collisions. To conclude this chapter we describe briefly yet another way in which the interaction between hadrons and timelike virtual photons can be studied. Consider a reaction of two hadrons (a and b) in which two or more hadrons *and* a charged lepton pair are produced. Explicitly

$$a + b \rightarrow (2 \text{ or more hadrons}) + \mu^+ + \mu^- \tag{21-25a}$$

$$a + b \rightarrow (2 \text{ or more hadrons}) + e^+ + e^- \tag{21-25b}$$

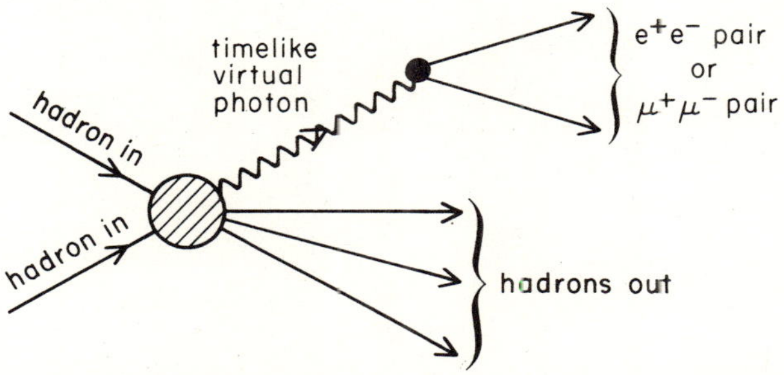

Fig. 21-6 The Feynman diagram for the production and detection of a timelike virtual photon in hadron collisions.

Lepton conservation requires that the dominant mechanism for this reaction be the one shown in Fig. 21-6. A comparison of this diagram and the electron–positron annihilation diagram in Fig. 21-1 reveals similarity. By crossing we can go from Fig. 21-6 to Fig. 21-1. But in more general terms in Eqs. 21-25, we certainly are using a virtual photon to investigate strong interactions. Hence this reaction belongs in the discussion of Sec. 19-2.

Exercise 21-12. For further information consult the original paper (21CH), which also gives references to parton model treatments of this reaction.

REFERENCES

21BA G. Barbiellini, *Lett. Nuovo Cimento* **6**, 557 (1973).

21BJ J. D. Bjorken and S. J. Brodsky, *Phys. Rev.* **D1**, 1416 (1970).

21CA N. Cabibbo and R. Gatto, *Phys. Rev.* **124**, 1577 (1961).

21CH J. H. Christenson et al., *Phys. Rev. Lett.* **25**, 1523 (1970).

21DR For a general discussion of the theoretical significance of this field see S. D. Drell, in *Proceedings of the Amsterdam International Conference on Elementary Particles, 1971* (North-Holland, Amsterdam, 1971).

21PIS For further information and other references, consult recent *Proceedings of the International Symposium on Electron and Photon Interactions at High Energies.*

21SI V. Silvestrini, in *Proceedings of the Sixteenth International Conference On High Energy Physics, 1972* (National Accelerator Laboratory, Batavia, Ill., 1973).

CHAPTER

22

Large Momentum Transfer Collisions

22-1 Introduction. In this, the final chapter of the book, we discuss events involving large momentum transfers between hadrons. In elastic reactions and two-body inelastic reactions we consider events falling into that part of the differential cross section away from the forward peak; that is, we consider events with $|t| > 1$ or 2 $(\text{GeV/c})^2$. But if there is a backward peak we exclude the region near the backward peak; thus we also require $|u| > 1$ or 2 $(\text{GeV/c})^2$. We exclude the backward peak region because we are restricting our attention to those events in which the momentum transferred between hadrons is large. And we believe, as discussed in Chs. 7 and 18, that backward peaks are produced by a u-channel exchange mechanism involving only a small amount of momentum transfer.

In many-body reactions we cannot classify an entire event as either a low momentum transfer event or a high momentum transfer event. But if one or more particles have transverse momenta, $p_\perp$, greater than 1 or 2 GeV/c, we attribute those particles to a large momentum transfer mechanism. The same event can, and usually does, have several low $p_\perp$ particles. (As discussed in Ch. 8, in most many-body events all the particles have small $p_\perp$.)

From the previous two paragraphs it should be clear that we have no rigorous way to define or classify large momentum transfer events. We are seeking insight into a rather vague concept, first mentioned in Sec. 1-4, suggesting that large momentum transfer events result from violent collisions of the elementary particles and that in such collisions we probe most deeply into the internal structure of the hadrons. This is first of all a classical idea—a glancing collision of two classical objects tells us only their size, shape, and mass; but a violent collision breaks open the objects and exposes their internal structure. It is also a quantum mechanical concept, for according to the uncertainty principle, we can detect a spatial structure as fine as Δx only if

$$\Delta p \gtrsim \frac{\hbar}{\Delta x} \tag{22-1}$$

The uncertainty principle says that we can, but it does not say that we will. Therefore we must choose our momentum transfers carefully. Concrete examples of the use of this concept are the interpretation of electromagnetic elastic form factors in Secs. 19-5 and 19-6 and the interpretation of lepton–nucleon inelastic scattering structure functions in Ch. 20.

However, as discussed in Sec. 19-6, it may be difficult to interpret large momentum transfer phenomena in terms of the spatial structure of hadrons because of the problem of constructing a relativistically invariant description. (This problem notwithstanding, most of us like to use spatial structure interpretations.) But it is certainly more rigorous to avoid spatial structure pictures and to try to interpret large momentum transfer phenomena in terms of the interaction between hadrons. Then by the uncertainty principle, large momentum transfers allow us to examine very short range interactions. And this is perhaps the greatest attraction in studying large momentum transfer collisions. We return to this idea in Sec. 22-4, where we discuss the theoretical implications of such collisions.

Before embarking on that theoretical discussion, we describe typical two-body large momentum transfer data in Sec. 22-2. In Sec. 22-3 we make a few remarks on the transverse momentum distribution in many-body reactions. We conclude the chapter with a brief speculative section on the relation of strong, weak, and electromagnetic interactions at large momentum transfers, having already discussed large momentum transfer processes induced by virtual photons in Chs. 19, 20, and 21.

22-2 Large Momentum Transfer Two-Body Reactions. We begin with elastic scattering. The properties of the large momentum transfer region are illustrated

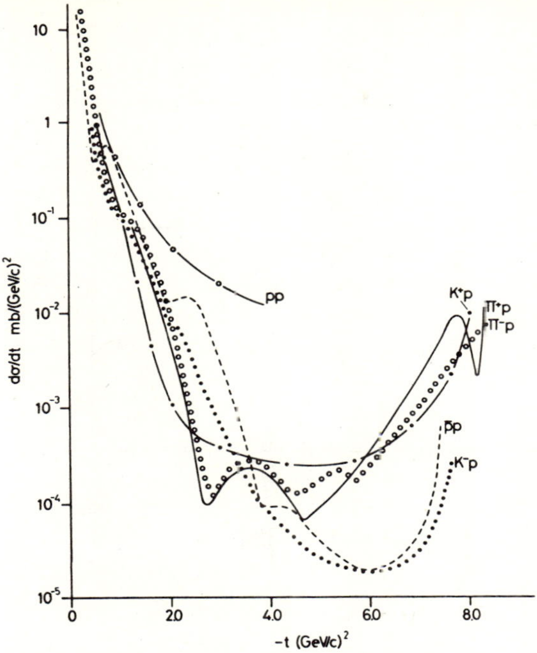

Fig. 22-1 The complete elastic differential cross sections for various systems at 5 GeV/c (22CHA).

by the beautiful composite drawing in Fig. 22-1 and by two of the figures we used very early in the book—Figs. 1-5 and 1-6. We observe the following properties of the elastic cross section in the large t, large u region, which we define as $|t|$ and $|u| > 1$ or 2 $(\text{GeV}/c)^2$.

1. Independent of the energy and the system, the fraction of the total elastic cross section contained in the large momentum transfer region is very small.

2. The differential cross section may be divided into a forward peak region, a backward peak region, and a large t, large u region. But the forward and backward peak regions seem to influence the large t, large u region. For example, in the $p\bar{p}$ and K^-p systems where there is a very small backward peak, $d\sigma/dt$ continues to decrease as t passes through the 90° point. And $d\sigma/dt$ reaches its minimum value quite close to the backward direction. But if there is a relatively strong backward peak, $d\sigma/dt$ reaches its minimum value in the neighborhood of 90°.

Exercise 22-1. Because of its symmetry, we can only examine the pp system in the 0 to 90° region. As we might expect from isotopic spin considerations, the np system has a quite similar $d\sigma/dt$ in the 0 to 90° region. But $(d\sigma/dt)_{np}$ extends to

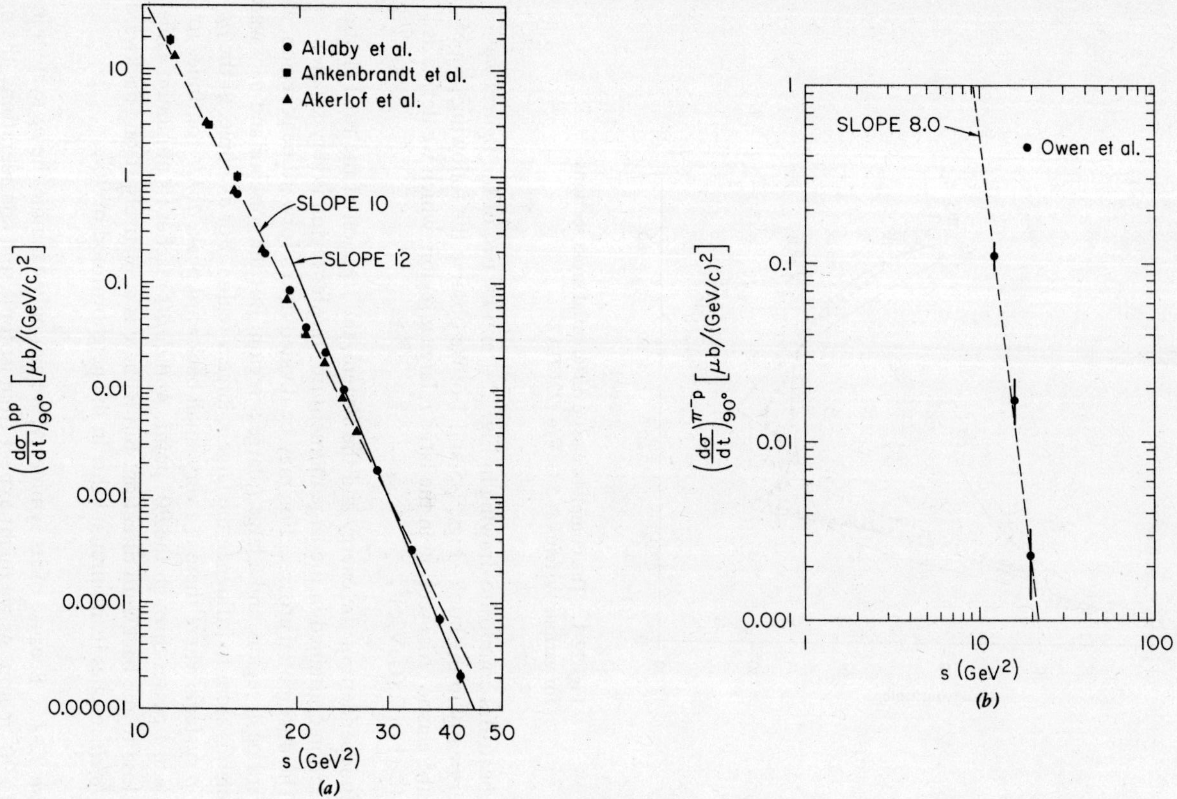

Fig. 22-2 Plot of $1n(d\sigma/dt)_{el,90°}$ versus in s for (*a*) the pp system up to 21 GeV/c and (*b*) the $\pi^- p$ system up to 18 GeV/c. Taken from Ref. 22GUN.

180°. Consult the literature (22DI, 22PE, and recent papers) for the typical behavior of the nucleon–nucleon $d\sigma/dt$ in the 90° region.

3. There are quite large differences between the $(d\sigma/dt)_{el}$ of different systems at the same s and t. But the overall shapes of the entire differential cross sections and their energy dependencies are similar.

4. For example, $(d\sigma/dt)_{el}$ in the large t, large u region always decreases very rapidly with increasing energy (Fig. 22-2). If we use the rough fit

$$\left(\frac{d\sigma}{dt}\right)_{el,90°} \sim cs^{-n} \tag{22-2}$$

we find values of n such as 8 and 12. This is very different from the s dependency of the total elastic cross section, which varies very slowly with s. We note that we tend to use $(d\sigma/dt)_{el,90°}$ as a convenient measure of the magnitude of $(d\sigma/dt)_{el}$ in the large t, large u region.

5. In the large t, large u region, $(d\sigma/dt)_{el}$ often has system-related peaks and valleys. The most impressive example is the sharp dip in $(d\sigma/dt)_{pp}$ at $|t| \sim 1.3$

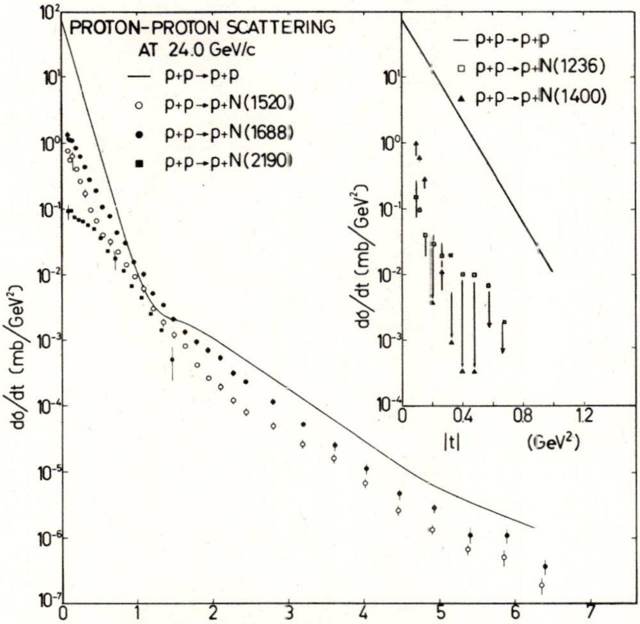

Fig. 22-3 Plot of $d\sigma/dt$ for $p + p \to p + p$ and $p + p \to p + N^*$ at 24.0 GeV/c (22AL).

$(\mathrm{GeV}/c)^2$ (Fig. 1-6). This dip does not occur below $E_{\text{lab}} = 7$ GeV; it is only a break below 20 GeV, but it becomes a clear dip at higher energies. Many of us were surprised to see structure in a differential cross become sharper as the energy increases, for we had felt that $d\sigma/dt$ should become simpler as the energy became very high. This feeling came partly from the observation, illustrated in Fig. 1-5*c*, that the $|t| < 1$ $(\mathrm{GeV}/c)^2$ dips in πp or $K^- p(d\sigma/dt)_{\text{el}}$ disappear as the energy increases.

We now turn our attention to inelastic two-body reactions. There are far fewer data than for elastic reactions because it is difficult to acquire sufficient statistics. In particular, quasi-two-body reactions require very extensive measurements for the resonance to be detected above the background. But Fig. 22-3 contains a beautiful

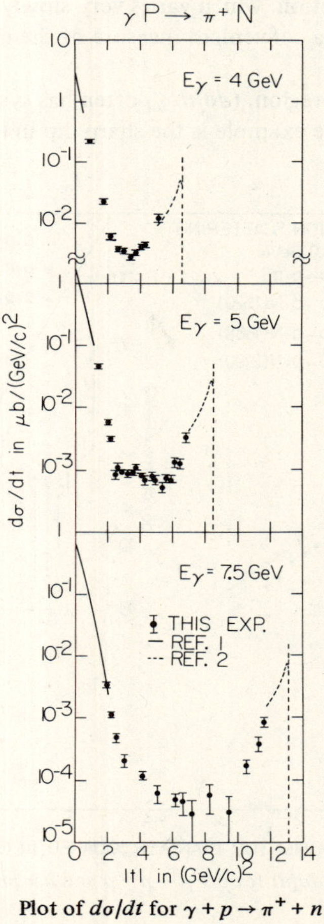

Fig. 22-4 Plot of $d\sigma/dt$ for $\gamma + p \to \pi^+ + n$ (22AN).

example—the large momentum transfer differential cross section for the quasi-two-body reactions $p + p \to p + N^*$, and Fig. 22-4 shows $d\sigma/dt$ for the true two-body reaction $\gamma + p \to \pi^+ + n$. We make two observations:

1. The inelastic two-body $d\sigma/dt$ has an s and t behavior quite similar to the s and t behavior in elastic scattering. For example, in both classes of reactions $d\sigma/dt$ decreases by many orders of magnitude as t increases.
2. Inelastic two-body reactions and elastic reactions from the same initial state may have $d\sigma/dt$'s that are almost equal in magnitude at large t, even though the $d\sigma/dt$'s differ substantially at small t values. This is to be expected. The diffractive elastic scattering mechanism that gives high energy elastic scattering a large forward peak is a small t phenomenon (Ch. 2). At large t there may be no special mechanism leading to elastic scattering; thus inelastic and elastic scattering may proceed through similar mechanisms and the $d\sigma/dt$'s will be of the same size.

Exercise 22-2. Some beautiful regularities have been observed when large angle pp elastic scattering has been studied as a function of $p_\perp{}^2$, not t. Consult some of the original papers (22OR and 22KR).

Exercise 22-3. Use the discussion at the end of Sec. 19-3 or the vector meson dominance model of Sec. 19-7 to relate the large t $d\sigma/dt$ in Fig. 22-4 to the large t pion–proton $d\sigma/dt$ in Figs. 22-1 and 1-5. Compare with Ref. 22AN.

22-3 Large Momentum Transfer in Many-Body Reactions. As we observed in the introduction to this chapter, it is difficult to know what constitutes a large momentum transfer event in many-body reactions.

Exercise 22-4. Suppose that in the fireball model (Ch. 8) a fireball is produced of invariant mass $M^* = E_{\text{cms}}/4$. The production of this fireball certainly involves a large four-momentum transfer. Calculate that four-momentum transfer. But if the fireball decays with high multiplicity, we may see no large $p_\perp$ secondary particles. Devise a numerical example.

If we wish to employ a model independent criterion, we must choose a test such as the requirement of large $p_\perp$. Of course one can devise more sophisticated tests, such as searching for groups of particles, which we might call jets, whose total $p_\perp$ is unusually large. However, we use only the simplest criterion, that of large $p_\perp$; and we give only one example. Although the idea of searching for large momentum transfer events in many-body collisions is old, our understanding of many-body events is in its infancy; and we do not know what data are really significant.

Consider the Lorentz invariant, one-particle distribution $E\, d\sigma/d^3p$ (defined in Sec. 8-6), for

$$p_\perp > 1 \text{ or } 2 \text{ GeV/c} \tag{22-3}$$

There are two general phenomena in the $p_{\text{lab}} \geqslant 100$ GeV region.

1. As $p_\perp$ increases, $E\, d\sigma/d^3p$ decreases more *slowly* than it does in the $p_\perp < 1$ GeV/c region. Thus at fixed x a form like

$$E\frac{d\sigma}{d^3p} = A\, e^{-bp_\perp + cp_\perp{}^2}, \qquad b > 0, c > 0 \tag{22-4}$$

seems appropriate (22ALP). Note that c is greater than zero. This is quite reminiscent of the behavior of $d\sigma/dt$ in elastic and two-body inelastic reactions as t moves out of the forward peak region.

2. As $p_\perp$ increases, the proton/pion ratio increases for a while. The production of other particles relative to pions also increases. Thus pion production does not completely dominate as it does in the small $p_\perp$ region.

The basic, and as yet unanswered, question is whether these effects are merely extensions of the small $p_\perp$ strong interaction dynamics to large $p_\perp$ or whether a new aspect of strong interaction dynamics is manifest here.

22-4 Theoretical Speculations. In Sec. 22-1 we remarked that the hope that large momentum transfer collisions will reveal the fundamental nature of hadron-hadron interactions at very small distances, makes the study of such phenomena very attractive. To be somewhat more quantitative, most strong interaction processes have to do with interaction distances of $\hbar/m_\pi \sim 10^{-13}$ cm. But by studying large momentum transfer strong interaction processes we may penetrate to distances as small as 10^{-14} to 10^{-15} cm.

Theoretical speculations and models for large momentum transfer processes have a long history (22KRZ, 22CH), but in our view we have far to go. First, some of the theories use concepts whose meaning we understand clearly only in the small t region. Second, we must acquire more very high energy, large t, large u data. Third, we are not absolutely sure that the strong interactions will dominate in the very high energy, large t, large u region (Sec. 22-5). Therefore, we make only a few general remarks about some classes of theories of large momentum transfer processes.

One class, which is used for two-body reactions, consists of *multiple-particle exchange* models (22CH). A simple example of this model was given in Ch. 2 in Eqs. 2-71 through 2-74, Fig. 2-7, and the associated discussion. In Eq. 2-74 the $f_n(\theta)$ term in the amplitude may be interpreted as caused by the exchange of n

Pomeranchuk trajectories (Ch. 17). The sum over these $f_n(\theta)$'s will produce a flattening of $d\sigma/dt$ as t increases. Relatively sharp dips can be produced by the interference of the various terms. For example, the sharp dip in $(d\sigma/dt)_{pp,\text{elastic}}$ (Fig. 1-6) may be explained by the interference between $f_1(\theta)$ and $f_2(\theta)$.

Exercise 22-4. Work out a numerical example using Eq. 2-74, inserting reasonable values for the parameters and taking note of the $(-1)^n$ factor.

Clearly such a model is not sufficient, for it does not explain the difference between $a + b$ and $\bar{a} + b$ elastic scattering. Therefore, it is customary to include nonpomeron Regge trajectories and to consider summations over multiple-particle exchange diagrams using both Pomeranchuk and non-Pomeranchuk Regge trajectories. Applying a similar model to the backward peak, without pomeron exchange of course, we can obtain the $d\sigma/dt$'s discussed in Sec. 22-2. This model assumes that there is nothing special about the large t, large u region; it simply consists of the overlap of amplitudes extended from the small t and small u regions. Such models are probably most successful and have most meaning in the *intermediate t region*, $1 \leqslant |t| \leqslant 3$ $(\text{GeV/c})^2$, for these models are based on concepts developed to explain processes involving interaction distances of 10^{-13} cm, and such concepts may not have physical meaning at interaction distances of 10^{-14} to 10^{-15} cm.

There is another class of theories of large momentum transfer, two-body reactions which we call *probability models.* The basic idea is that there is a very low probability for a two-body reaction $a + b \to c + d$ to occur at large momentum transfer because hadrons are fragile objects and break up easily. Therefore when large momentum transfers are involved, the probability is much higher that many particles will be produced in the final state. A very old application of this idea (22FA) used the statistical models of Ch. 8. But such models cannot explain the definite structure seen in $d\sigma/dt$ at large t and u.

Recent forms of the probability model incorporate the form factor concept. The model is simplest if one considers elastic scattering $a + b \to a + b$. Let $F_i(t)$ be the probability amplitude, or form factor, for particle i to remain particle i after it absorbs a four-momentum transfer squared of t. As $|t|$ increases, we expect $F(t)$ to decrease, as discussed in Sec. 19-6. According to the classical interpretation, it is very difficult to accelerate a hadron without breaking it up (22WU). We then expect

$$\frac{d\sigma}{dt} = h(s, t)F_a^{\,2}(t)F_b^{\,2}(t) \tag{22-5}$$

where $h(s, t)$ is involved in the transfer of the four-momentum from particle a to particle b.

Exercise 22-5. As an example, suppose that the four-momentum is transferred by a virtual particle of mass M. Ignoring spin effects, show that

$$\frac{d\sigma}{dt} = \frac{g^4 F_a{}^2(t) F_b{}^2(t)}{16\pi s^2(|t| + M^2)^2} \tag{22-6}$$

Then, as suggested by Ref. 22WU, take $F_i(t)$ to have the same t dependence as the electromagnetic form factor (Secs. 19-4 and 19-6). Finally, show that large s, pp elastic scattering in this model would obey

$$\left(\frac{d\sigma}{dt}\right)_{pp,90°} \sim \frac{s^{-10}}{(s + 2M^2)^2}, \qquad s \to \infty \tag{22-7}$$

Observe that this yields an s^{-10} to s^{-12} dependence, depending on the magnitude of M.

More complicated forms of this model have been devised. For example, the electromagnetic form factor may be used to calculate $\chi(b)$ in the eikonal model; and then the eikonal formula (Eq. 2-69) can be used to calculate the elastic scattering (22CHO). The attraction of models involving the form factor concept is that they inherently incorporate the idea of probing very small interaction distances. Unfortunately, however, they are very phenomenological, since they are based on a form factor whose dynamics we do not understand.

A third class of theories, which is applicable to both two-body and many-body reactions, is based on the constituent model of hadrons (Sec. 14-7 and Ch. 20). Such theories have an appealing simplicity. If we take the constituents to be point particles (e.g., partons), we need only construct a dynamics between these constituent particles, which always have unit form factors and can only be created or destroyed in pairs. The constituent's unit form factor is of particular importance in large momentum transfer processes, for it means that constituent–constituent interactions, if they exist, could dominate conventional strong interaction dynamics in large momentum transfer processes. References 22GU and 22BE are examples of such theories. Of course such theories depend on the validity of the constituent model itself, and one's present view should be that large momentum transfer processes, like the processes discussed in Ch. 20, are tests of the constituent models of hadrons.

22-5 Strong, Electromagnetic, and Weak Interactions at Large Momentum Transfer. We close this long book on strong interactions with a brief speculation. We have assumed throughout that in any process initiated by a hadron–hadron collision, the strong interactions would dominate in magnitude over the electromagnetic and weak interaction. Indeed, the magnitudes of the total hadron production cross sections are in the crude ratio strong hadron production : electromagnetic hadron production : weak hadron production = 10^{-26} cm^2 : 10^{-30} cm^2 : 10^{-38} cm^2 (22-8)

But suppose that in high energy, large momentum transfer processes the strong interactions yield a cross section which decreases as $e^{-b|t|}$ where b is large. The electromagnetic cross sections based on virtual photon exchange may only decrease as $|t|^{-n}$. We do not know how the weak interactions will behave at high energies. But assuming that the low energy behavior will not be modified too much (see, however, Sec. 20-4), we may expect a rather slow dependence on t. Therefore, one may speculate that for very large $|t|$ the electromagnetic or weak interactions may approach or even exceed the strong interactions in cross section. Thus large momentum transfer processes are of great interest in spite of their relative rarity and the obscurity of their dynamics.

REFERENCES

22AL J. V. Allaby et al., *Nucl. Phys.* **B52**, 316 (1973).

22ALP B. Alper et al., *Phys. Lett.* **44B**, 521 (1973).

22AN R. L. Anderson et al., *Phys. Rev. Lett.* **30**, 627 (1973).

22BE S. M. Berman et al., *Phys. Rev.* **D4**, 3388 (1971).

22CH C. B. Chiu, *Rev. Mod. Phys.* **41**, 640 (1969).

22CHA V. Chabaud et al., *Phys. Lett.* **38B** 449 (1972).

22CHO T. T. Chou and C. N. Yang, *Phys. Rev.* **170**, 1591 (1968).

22DI A. N. Diddens, in *Proceedings of Fourth International Conference on High Energy Collision, 1972* (Rutherford High Energy Laboratory, Chilton, England, 1972).

22FA G. Fast et al., *Nuovo Cimento* **27**, 856 (1963).

22GU J. F. Gunion et al., *Phys. Rev.* **D6**, 2652 (1972).

22GUN J. F. Gunion et al., *Phys. Lett.* **39B**, 649 (1972).

22KR A. D. Krisch, *Phys. Rev. Lett.* **19**, 1149 (1967).

22KRZ A. Krzywicki, in *High Energy Phenomenology*, J. Tran. Thanh Van, Ed. (Laboratoire de Physique Théorique et Haute Energies, Orsay, 1971).

22OR J. Orear, *Phys. Lett.* **13**, 190 (1964).

22PE M. L. Perl et al., *Phys. Rev.* **D1**, 1857 (1970).

22WU T. T. Wu and C. N. Yang, *Phys. Rev.* **137**, B708 (1965).

A

Units, Four-Vectors, and Kinematics

A-1 Units. We usually use *elementary particle* or *natural* units in which $\hbar = 1$, $c = 1$; and energy is given in MeV or GeV. Momentum is then in MeV/c or GeV/c, where the c in the denominator is a concise notation for the "c" unit of velocity = 2.998×10^{10} cm/sec. Mass is in MeV/c^2 or GeV/c^2. The "c" unit of velocity is usually ignored in equations, and one writes, for example, E (energy) = M (mass). But in giving numerical values one usually includes the "c" unit: $E = 2$ GeV, but $M = 2$ GeV/c^2.

The following constants are often used. In particular, various powers of $\hbar$ and $\hbar c$ must often be inserted in equations to obtain numerical values. To decide how many powers to use, one must look at the length, time, energy, mass, and momentum quantities in the equation. Then one uses $\hbar$ and $\hbar c$ to give the same units to both sides of the equation.

$$\hbar = 6.58 \times 10^{-25} \approx 6.6 \times 10^{-25} \text{ GeV sec}$$

$$\hbar c = 1.97 \times 10^{-14} \approx 2.0 \times 10^{-14} \text{ GeV cm or (GeV/c) cm} \tag{A-1}$$

Often a length L is given as $L = 1/p$, where p is a momentum. Then

$$L\text{ (cm)} = \frac{2.0 \times 10^{-14}}{p(\text{GeV}/c)} \tag{A-2}$$

Also

$$1 \text{ GeV} = 1.60 \times 10^{-3} \text{ erg}$$
$$\hbar = 1.05 \times 10^{-27} \text{ erg sec}$$
$$c = 2.998 \times 10^{10} \approx 3.0 \times 10^{10} \text{ cm/sec} \tag{A-3}$$

For coupling constants we usually use rationalized units and dimensionless combinations. In particular, for the relation between the fine structure constant α and the unit electric charge e, we write

$$\alpha = \frac{e^2}{4\pi\hbar c} = \frac{1}{137} \text{ cgs units}$$

$$\alpha = \frac{e^2}{4\pi} = \frac{1}{137} \text{ natural units} \tag{A-4}$$

Often useful constants and units are

proton mass $= 0.938$ GeV

$\pi^{\pm}$ mass $= 0.140$ GeV

$K^{\pm}$ mass $= 0.494$ GeV

muon mass $= 0.106$ GeV

electron mass $= 0.51$ MeV

length equivalent to one pion mass $= \hbar c/(m_\pi) = 1.4 \times 10^{-13}$ cm

radius of electron $= r_e = e^2/(m_e c^2) = 2.82 \times 10^{-13}$ cm

Bohr radius $= \hbar^2/(m_e e^2) = 0.53 \times 10^{-8}$ cm

Bohr magneton $= \mu_{\text{Bohr}} = e\hbar/(2m_e c) = 0.58 \times 10^{-14}$ MeV/G

nucleon magneton $= \mu_{\text{nucleon}} = e\hbar/(2M_p c) = 3.15 \times 10^{-18}$ MeV/G

1 fermi $= 10^{-13}$ cm

1 angstrom $= 10^{-8}$ cm (A-5)

A-2 Four-Vectors and Three-Vectors. Three-vectors are in boldface; four-vectors are in lightface and are often uppercase.

The contravariant four-vector is

$$x^\mu = (x^0, x^1, x^2, x^3) = (t, x, y, z) = (t, \mathbf{r})$$

$$P^\mu = (P^0, P^1, P^2, P^3) = (E, p_x, p_y, p_z) = (E, \mathbf{p})$$

The covariant four-vector is

$$x_\mu = (x_0, x_1, x_2, x_3) = (t, -x, -y, -z) = (t, -\mathbf{r})$$

and so forth. The metric tensor is

$$g^{\mu\nu} = g_{\mu\nu} = \begin{bmatrix} 1 & 0 & 0 & 0 \\ 0 & -1 & 0 & 0 \\ 0 & 0 & -1 & 0 \\ 0 & 0 & 0 & -1 \end{bmatrix}$$

The inner product of two four-vectors is

$$a \cdot b = a^\mu b_\mu = a_\mu b^\mu = a^0 b^0 - \mathbf{a} \cdot \mathbf{b}$$

where repeated indices are summed unless otherwise noted. The relation between the mass m and four-momentum $P^\mu = (E, \mathbf{p})$ of a particle is

$$m^2 = P^\mu P_\mu = E^2 - \mathbf{p}^2$$

If in our metric $P^\mu P_\mu > 0$, P is said to be *timelike*; if $P^\mu P_\mu < 0$, P is said to be *spacelike*. The subscripts or superscripts 0, 1, 2, 3 for a four-vector are almost always denoted by a Greek letter. For the subscripts 1, 2, 3 of a three-vector, we usually use a Roman letter. Some useful tensors are:

the Kronecker delta

$$\begin{aligned} \delta^{\mu\nu} &= 1 \quad \mu = \nu \\ &= 0 \quad \mu \neq \nu \end{aligned}$$

the third-rank Levi–Civita symbol

$$\begin{aligned} \epsilon_{ijk} &= +1 \quad ijk = \text{even permutation of } 123 \\ &= -1 \quad ijk = \text{odd permutation of } 123 \\ &= 0 \quad \text{otherwise} \end{aligned}$$

and the fourth-rank Levi–Civita symbol

$$\begin{aligned} \epsilon_{\lambda\mu\nu\sigma} &= +1 \quad \lambda\mu\nu\sigma = \text{even permutation of } 0123 \\ &= -1 \quad \lambda\mu\nu\sigma = \text{odd permutation of } 0123 \\ &= 0 \quad \text{otherwise} \end{aligned}$$

A-3 Kinematics. Consider two particles 1 and 2 with masses m_i and four-momenta P_i. If in the *laboratory frame* $P_1 = (E_1, \mathbf{p}_1)$ and 2 is at rest, we have

$$\beta_{\text{cms}} = \text{velocity of center of mass} = \frac{|\mathbf{p}_1|}{E_1 + m_2} \tag{A-6}$$

$$\gamma_{\text{cms}} = (1 - \beta_{\text{cms}}^2)^{-1/2} = \frac{E_1 + m_2}{(m_1{}^2 + m_2{}^2 + 2E_1 m_2)^{1/2}} \underset{E_1 \to \infty}{=} \left(\frac{E_1}{2m_2}\right)^{1/2}$$

In the center of mass frame of 1 and 2, define

$$P_1 = (E_1^*, \mathbf{p}_1^*), \qquad P_2 = (E_2^*, \mathbf{p}_2^*) \tag{A-7}$$

Then

$$\mathbf{p}_1^* = -\mathbf{p}_2^*, \qquad p^* = |\mathbf{p}_1^*| = |\mathbf{p}_2^*|$$

We define

$$s = (P_1 + P_2)^2 \tag{A-8}$$

$$s = 2m_2 E_1 + m_1{}^2 + m_2{}^2 \underset{s \to \infty}{\longrightarrow} 2m_2 E_1, \qquad \text{laboratory frame}$$

$$s = (E_1^* + E_2^*)^2, \qquad \text{center of mass frame}$$

$$E_1^* = \frac{s + m_1{}^2 - m_2{}^2}{2\sqrt{s}} \underset{s \to \infty}{\longrightarrow} \frac{\sqrt{s}}{2} \tag{A-9}$$

$$E_2^* = \frac{s + m_2{}^2 - m_1{}^2}{2\sqrt{s}} \underset{s \to \infty}{\longrightarrow} \frac{\sqrt{s}}{2}$$

The following symmetric function of three scalars a, b, c is often used.

$$\begin{aligned} \lambda(a, b, c) &= a^2 + b^2 + c^2 - 2ab - 2ac - 2bc \\ &= [a - (\sqrt{b} + \sqrt{c})^2][a - (\sqrt{b} - \sqrt{c})^2] \\ &= (a - b - c)^2 - 4bc \end{aligned} \tag{A-10}$$

$$\text{if} \quad a \gg b, \qquad a \gg c, \qquad \lambda(a, b, c) \to a^2$$

Returning to the two-particle system

$$p^* = \frac{[\lambda(s, m_1{}^2, m_2{}^2)]^{1/2}}{2\sqrt{s}} \underset{s \to \infty}{\longrightarrow} \frac{\sqrt{s}}{2} \tag{A-11}$$

If 2 is at rest in laboratory frame, we write

$$E_2^* = \gamma_{\text{cms}} m_2, \qquad p^* = \beta_{\text{cms}} \gamma_{\text{cms}} m_2 \tag{A-12}$$

Consider the two-body reaction

$$1 + 2 \to 3 + 4$$

The conventional Lorentz kinematic scalars are

$$s = (P_1 + P_2)^2$$
$$t = (P_1 - P_3)^2$$
$$u = (P_1 - P_4)^4 \tag{A-13}$$

All equations for t can be written for u if 3 is replaced by 4. We note

$$s + t + u = \sum_{i=1}^{4} m_i^2 \tag{A-14}$$

Here θ_{13} is the angle between $\mathbf{p}_1$ and $\mathbf{p}_3$; that is, $\cos\theta = \mathbf{p}_1 \cdot \mathbf{p}_3/(|\mathbf{p}_1||\mathbf{p}_3|)$. Then

$$t = m_1^2 + m_3^2 - 2E_1E_3 + 2|\mathbf{p}_1||\mathbf{p}_3| \cos\theta_{13} \tag{A-15}$$

Define the Lorentz scalar

$$t_{\min} = m_1^2 + m_3^2 - 2E_1E_3 + 2|\mathbf{p}_1||\mathbf{p}_3| \tag{A-16}$$

Then

$$t = t_{\min} - 2|\mathbf{p}_1||\mathbf{p}_3|(1 - \cos\theta_{13}) \tag{A-17}$$

In the center of mass

$$t = t_{\min} - 2p^{*2}(1 - \cos\theta^*_{13}) \underset{s\to\infty}{\longrightarrow} -\frac{s}{2}(1 - \cos\theta^*_{13}) \tag{A-18}$$

In our metric, t and u are negative over most of the angular range. Depending on the masses, t may be positive near $\theta_{13} = 0$ and u may be positive near $\theta_{13} = 180°$. There is unfortunately no simple form for Eq. A-16. In terms of s and the masses

$$t_{\min} = m_1^2 + m_3^2 - \frac{(s + m_1^2 - m_2^2)(s + m_3^2 - m_4^2)}{2s} + \frac{[\lambda(s, m_1^2, m_2^2)\lambda(s, m_3^2, m_4^2)]^{1/2}}{2s} \tag{A-19}$$

Equation A-19 is so clumsy that we usually use approximations. For $s \gg m_i^2$, we write

$$t_{\min} \approx -\frac{1}{s}(m_3^2 - m_1^2)(m_4^2 - m_2^2) - \frac{1}{s}(m_3^2 + m_4^2 - m_1^2 - m_2^2) \times \left(\frac{m_3^2 m_4^2}{s - m_3^2 - m_4^2} - \frac{m_1^2 m_2^2}{s - m_1^2 - m_2^2}\right) \tag{A-20}$$

We note that the second term is necessary if $m_1 = m_3$ or $m_2 = m_4$. If $m_2 = m_4$

$$t_{\min} \approx -\frac{m_2^2(m_3^2 - m_1^2)^2}{s^2} \approx -\frac{(m_3^2 - m_1^2)^2}{4E^2_{1,\text{lab}}} \tag{A-21}$$

Just as this manuscript was completed, a book containing a very complete and useful treatment of elementary particle kinematics was published. For details of the kinematics questions discussed in the appendix and throughout the book, one can consult: E. Byckling and K. Kajantie, *Particle Kinematics* (Wiley, New York, 1973).

APPENDIX

B

Bessel Functions

B-1 References. Very useful references with formulas, graphs, and tables are *Handbook of Mathematical Functions*, edited by M. Abramowitz and I. A. Stegun (Washington, D.C., National Bureau of Standards, Government Printing Office, 1965); *Jahnke–Emde Tables of Higher Functions*, new edition (McGraw-Hill, New York, 1960), or old edition (Dover, New York, 1945); I. S. Gradshteyn and I. M. Ryzhik, *Tables of Integrals, Series and Products* (Academic Press, New York, 1965).

B-2 Bessel Functions of Nonnegative Integral Order. Bessel's equation is

$$u'' + \left(\frac{1}{z}\right) u' + \left(1 - \frac{n^2}{z^2}\right) u = 0 \qquad \text{(B-1)}$$

We primarily use *Bessel functions of the first kind* with $n = 0, 1, 2, \ldots$, denoted by $J_n(z)$

$$J_n(z) = \sum_{k=0}^{\infty} \frac{(-1)^k (z/2)^{n+2k}}{k!(n+k)!} \tag{B-2}$$

In particular, we write

$$J_0(z) = 1 - \frac{(z/2)^2}{1} + \frac{(z/2)^4}{4} - \cdots \tag{B-3}$$

$$J_1(z) = \frac{z}{2}\left[1 - \frac{(z/2)^2}{2} + \frac{(z/2)^4}{12} - \cdots\right] \tag{B-4}$$

There are enormous numbers of integrals involving Bessel functions. They are given in the references (Sec. B-1). We list only a few integrals that are used in the book.

$$J_n(z) = \frac{i^{-n}}{\pi} \int_0^{\pi} e^{iz\cos\theta} \cos(n\theta)\, d\theta \tag{B-5}$$

$$\int_0^x J_n(z)\, dz = 2 \sum_{k=0}^{\infty} J_{n+2k+1}(x) \tag{B-6}$$

$$\int_0^x z^{n+1} J_n(z)\, dz = x^{n+1} J_{n+1}(x) \tag{B-7}$$

$$\int_0^{\infty} e^{-a^2 z^2} J_n(bz) z^{n+1}\, dz = \frac{b^n}{(2a^2)^{n+1}} e^{-b^2/4a^2} \tag{B-8}$$

$$\int_0^z J_0(\alpha t) K_0(\beta t) t\, dt = \left(\frac{z}{\alpha^2 + \beta^2}\right) [\alpha J_1(\alpha z) K_0(\beta z) - \beta J_0(\alpha z) K_1(\beta z)] \tag{B-9}$$

$$\int_0^{\infty} J_0(\alpha t) K_0(\beta t)\, t dt = \left(\frac{1}{\alpha^2 + \beta^2}\right) \tag{B-10}$$

Here $K_n(z)$ is a modified Bessel function of the third kind, see references in Sec. B-1 for details.

$$\text{for } z \to 0 \qquad K_0(z) \to \ln \frac{2}{z} \tag{B-11}$$

$$K_n(z) \to \tfrac{1}{2}(n-1)! \left(\frac{2}{z}\right)^n \qquad n = 1, 2, 3; \ldots$$

$$\text{for } z \to \infty \qquad K_n(z) \to \left(\frac{\pi}{2z}\right)^{1/2} e^{-z} \tag{B-12}$$

B-3 Spherical Bessel Functions. Spherical Bessel functions are solutions of the radial Schrödinger equation

$$u'' + \left(\frac{2}{\rho}\right)u' + \left(1 - \frac{l(l+1)}{\rho^2}\right)u = 0 \tag{B-13}$$

Also, $j_l(\rho)$ is the *spherical Bessel function*, $n_l(\rho)$ is the *spherical Neumann function*, and $h_l^{(1)}(\rho) = j_l(\rho) + in_l(\rho)$, $h_l^{(2)}(\rho) = j_l(p) - in_l(\rho)$ are the spherical Hankel functions.

$$\begin{aligned} j_0(\rho) &= \frac{\sin\rho}{\rho} & n_0(\rho) &= -\frac{\cos\rho}{\rho} \\ j_1(\rho) &= \frac{\sin\rho}{\rho^2} - \frac{\cos\rho}{\rho} & n_1(\rho) &= -\frac{\cos\rho}{\rho^2} - \frac{\sin\rho}{\rho} \end{aligned} \tag{B-14}$$

As ρ goes to zero

$$\begin{aligned} j_l(\rho) &\to \frac{\rho^l}{1\cdot 3\cdot 5\cdots(2l+1)} \\ n_l(\rho) &\to -\frac{1\cdot 3\cdot 5\cdots(2l-1)}{\rho^{l+1}} \end{aligned} \tag{B-15}$$

As ρ goes to infinity

$$\begin{aligned} j_l(\rho) &\to \frac{1}{\rho}\cos\left[\rho - (l+1)\frac{\pi}{2}\right] \\ n_l(\rho) &\to \frac{1}{\rho}\sin\left[\rho - (l+1)\frac{\pi}{2}\right] \\ h_l^{(1)}(\rho) &\to \frac{1}{\rho}e^{i[\rho-(l+1)\pi/2]} \\ h_l^{(2)}(\rho) &\to \frac{1}{\rho}e^{-i[\rho-(l+1)\pi/2]} \end{aligned} \tag{B-16}$$

C

Legendre Functions, Angular Momentum Operators, Clebsch–Gordan Coefficients, and Rotation Matrices

C-1 Phase Conventions. The phase conventions we use for spherical harmonics, Clebsch–Gordan coefficients, and $D^j_{m'm}$ functions are those used by M. E. Rose, *Elementary Theory of Angular Momentum* (Wiley, New York, 1957).

C-2 Legendre Functions. The references given in Sec. B-1 are also very useful for Legendre functions.

Legendre functions of the *first kind* $P_\nu^\mu(z)$ and of the *second kind* $Q_\nu^\mu(z)$ are solutions of the equation

$$(1-z^2)u'' - 2zu' + \left(\nu(\nu+1) - \frac{\mu^2}{1-z^2}\right)u = 0 \tag{C-1}$$

where μ and ν are arbitrary complex constants. In general, z is complex, but we usually require only the special range of values, z real and $-1 < z < +1$. When z falls in this range, we replace it by the symbol x. Unless otherwise indicated, x and z mean

x real and $-1 < x < +1$

z = any complex or real number

The general expression for $P_\nu^\mu(x)$ is

$$P_\nu^\mu(x) = \frac{1}{\Gamma(1-\mu)}\left[\frac{1+x}{1-x}\right]^{\mu/2} F\left(-\nu, \nu+1; 1-\mu; \frac{1-x}{2}\right) \tag{C-2}$$

Here F is the hypergeometric function

$$F(a, b; c; z) = 1 + \frac{ab}{c1!}z + \frac{a(a+1)b(b+1)}{c(c+1)2!}z^2 + \cdots$$
$$= \frac{\Gamma(c)}{\Gamma(a)\Gamma(b)}\sum_{n=0}^{\infty}\frac{\Gamma(a+n)\Gamma(b+n)}{\Gamma(c+n)n!}z^n \qquad \text{for } |z| < 1 \tag{C-3}$$

When μ is zero we have

$$P_\nu(x) \equiv P_\nu^0(x) = F\left(-\nu, \nu+1; 1, \frac{1-x}{2}\right) \tag{C-4}$$

The values for P_ν^μ and Q_ν^μ obey the recurrence relations

$$(\nu - \mu + 1)P_{\nu+1}^\mu(z) = (2\nu+1)zP_\nu^\mu(z) - (\nu+\mu)P_{\nu-1}^\mu(z) \tag{C-5}$$

$$(z^2-1)\frac{dP_\nu^\mu(z)}{dz} = \nu z P_\nu^\mu(z) - (\nu+\mu)P_{\nu-1}^\mu(z) \tag{C-6}$$

$$P_{\nu+1}^\mu(x) - P_{\nu-1}^\mu(x) = (2\nu+1)(1-x^2)^{1/2}P_\nu^{\mu-1}(x) \tag{C-7}$$

When $\mu = 0$ we have the useful relation for P_ν or Q_ν

$$(\nu+1)P_{\nu+1}(z) = (2\nu+1)zP_\nu(z) - \nu P_{\nu-1}(z) \tag{C-8}$$

C-3 Legendre Functions for Integral Values of ν and μ. Using n and m to represent integral values of ν and μ, we have the Legendre functions $P_n{}^m(z)$ and $Q_n{}^m(z)$. Some useful relations are

$$P_n{}^m(x) = (1 - x^2)^{m/2} \frac{d^m P_n(x)}{dx^m}; \qquad n, m = 0, 1, 2, 3, \ldots, \qquad n \geqslant m \tag{C-9}$$

$$P_{n-1}^m(z) = P_n{}^m(z) \tag{C-10}$$

$$P_n{}^m(-x) = (-1)^{n+m} P_n{}^m(x) \tag{C-11}$$

Explicit expressions for $P_n(z)$, the *Legendre polynomials*, are

$$\begin{aligned} P_0(z) &= 1 \\ P_1(z) &= z \\ P_2(z) &= \frac{3z^2 - 1}{2} \\ P_3(z) &= \frac{5z^3 - 3z}{2} \\ P_4(z) &= \frac{35z^4 - 30z^2 + 3}{8} \end{aligned} \tag{C-12}$$

Explicit expressions for $Q_n(z)$ are

z not on real axis between +1 and −1	x on real axis between +1 and −1
$Q_0(z) = \left(\frac{1}{2}\right) \ln \left(\frac{z+1}{z-1}\right)$	$Q_0(x) = \left(\frac{1}{2}\right) \ln \left(\frac{1+x}{1-x}\right)$
$Q_1(z) = \left(\frac{z}{2}\right) \ln \left(\frac{z+1}{z-1}\right) - 1$	$Q_1(x) = \left(\frac{x}{2}\right) \ln \left(\frac{1+x}{1-x}\right) - 1$
$Q_2(z) = \left(\frac{3z^2 - 1}{4}\right) \ln \frac{z+1}{z-1} - \frac{3z}{2}$	$Q_2(x) = \left(\frac{3x^2 - 1}{4}\right) \ln \left(\frac{1+x}{1-x}\right) - \frac{3x}{2}$

(C-13)

Special values are

$$\begin{aligned} &P_n(1) = 1, \qquad P_n(-1) = (-1)^n \\ &P_{2n+1}(0) = 0 \\ &P_{2n}(0) = (-1)^n \frac{1 \cdot 3 \cdot 5 \cdots (2n-1)}{2 \cdot 4 \cdot 6 \cdots 2n} \\ &Q_n(1) = \infty \end{aligned} \tag{C-14}$$

For n a positive integer, we have

$$P_n(z) = \frac{2n!}{2^n(n!)^2}\left[z^n - \frac{n(n-1)}{2(2n-1)} z^{n-2} + \frac{n(n-1)(n-2)(n-3)}{2 \cdot 4(2n-1)(2n-3)} z^{n-4} - \cdots\right]$$

$$= \frac{(2n-1)!!}{n!}\left[z^n - \frac{n(n-1)}{2(2n-1)} z^{n-2} + \cdots\right] \quad \text{(C-15)}$$

Useful integrals are

$$\int_{-1}^{+1} P_n{}^m(x) P_{n'}^m(x)\, dx = \frac{2}{2n+1} \frac{(n+m)!}{(n-m)!} \delta_{nn'} \quad \text{(C-16)}$$

$$Q_n(z) = \frac{1}{2} \int_{-1}^{+1} \frac{P_n(t)}{z-t}\, dt, \qquad z \text{ not on real axis between } +1 \text{ and } -1 \quad \text{(C-17)}$$

There is an asymptotic connection between P_n and the Bessel function J_0. For x fixed and > 0

$$\lim_{n\to\infty} P_n\left(\cos \frac{x}{n}\right) = J_0(x)$$

C-4 Spherical Harmonics. The *spherical harmonics* $Y_l^m(\theta, \phi)$ are defined by

$$Y_l^m(\theta, \phi) = \epsilon \left[\frac{2l+1}{4\pi} \frac{(l-|m|)!}{(l+|m|)!}\right]^{1/2} P_l^{|m|}(\cos\theta)\, e^{im\phi} \quad \text{(C-18)}$$

$$\epsilon = (-1)^m \qquad m > 0$$

$$\epsilon = 1 \qquad m \leqslant 0$$

$$\int_0^{\pi} \int_0^{2\pi} Y_l{}^{m*}(\theta, \phi) Y_{l'}{}^{m'}(\theta, \phi) \sin\theta\, d\theta\, d\phi = \delta_{mm'} \delta_{ll'} \quad \text{(C-19)}$$

Examples are given in Table C-1.

C-5 Angular Momentum Operators. The equation for angular momentum operators is

$$[J_x, J_y] = iJ_z; \qquad [J_y, J_z] = iJ_x, \qquad [J_z, J_x] = iJ_y \quad \text{(C-20)}$$

Stepping operators are

$$J_+ = J_x + iJ_y, \qquad J_- = J_x - iJ_y$$

$$J_\pm |jm\rangle = [(j \mp m)(j \pm m + 1)]^{1/2} |j, m \pm 1\rangle \quad \text{(C-21)}$$

Also

$$J_x = \frac{J_+ + J_-}{2}, \qquad J_y = \frac{J_+ - J_-}{2i}$$

C-6 Clebsch–Gordan Coefficients. Clebsch–Gordan coefficients are given in Table C-1. They obey the orthogonality relations

$$\sum_{m_1 m_2} C^{j_1 j_2 J}_{m_1 m_2 M} C^{j_1 j_2 J'}_{m_1 m_2 M'} = \delta_{JJ'} \delta_{MM'} \tag{C-22a}$$

$$\sum_{JM} C^{j_1 j_2 J}_{m_1 m_2 M} C^{j_1 j_2 J}_{m_1' m_2' M} = \delta_{m_1 m_1'} \delta_{m_2 m_2'} \tag{C-22b}$$

Also

$$C^{j_1 j_2 J}_{m_1 m_2 M} \neq 0 \qquad \text{only if} \qquad \begin{array}{l} m_1 + m_2 = M \\ |j_1 - j_2| \leqslant J \leqslant j_1 + j_2 \end{array} \tag{C-23}$$

Some symmetry properties are

$$\begin{aligned} C^{j_1 j_2 J}_{m_1 m_2 M} &= (-1)^{J-j_1-j_2} C^{j_2 j_1 J}_{m_2 m_1 M} \\ C^{j_1 j_2 J}_{m_1 m_2 M} &= (-1)^{J-j_1-j_2} C^{j_1 \ \ j_2 \ \ J}_{-m_1, -m_2, -M} \end{aligned} \tag{C-24}$$

When one of the spins is $\frac{1}{2}$

$$C^{j_1\, 1/2\, J}_{m_1 m_2 M} = \left(\frac{J \pm M}{2j_1 + 1}\right)^{1/2} \qquad \text{for} \quad m_2 = \pm\tfrac{1}{2}$$

C-7 The $D^j_{m'm}$ Functions. We express the $D^j_{m'm}$ function by

$$D^j_{m'm}(\alpha\beta\gamma) = e^{-i\alpha m'} d^j_{m'm}(\beta) e^{-i\gamma m} \tag{C-25}$$

The $d^j_{m'm}(\beta)$ function is given in Table C-2 and by Eqs. 9-8 and 10-33. Some symmetry properties are

$$\begin{aligned} d^j_{m'm}(-\beta) &= d^j_{mm'}(\beta) \\ d^j_{m'm}(\beta) &= (-1)^{m'-m} d^j_{mm'}(\beta) \\ d^j_{m'm}(\beta) &= (-1)^{m'-m} d^j_{-m',-m}(\beta) \\ D^{j*}_{m'm}(\alpha\beta\gamma) &= (-1)^{m'-m} D^j_{-m',-m}(\alpha\beta\gamma) \\ d^j_{m'm}(\pi + \beta) &= (-1)^{j+m'} d^j_{m'm}(\beta) \\ d^j_{m'm}(\pi - \beta) &= (-1)^{j+m'} d^j_{m,-m'}(\beta) \end{aligned} \tag{C-26}$$

Special values are

$$d^j_{m'm}(0) = \delta_{m'm}$$
$$d^j_{m'm}(\pi) = (-1)^{j-m}\delta_{m',-m} \tag{C-27}$$

The connection with the Legendre functions for l = integer is

$$P_l(\cos\beta) = d^l_{00}(\beta) \tag{C-28}$$

$$Y_l^m(\theta,\phi) = \left(\frac{2l+1}{4\pi}\right)^{1/2} D^{l*}_{m0}(\phi,\theta,-\phi)$$

For half-integral values of $j = l + 1/2$, we write

$$d^j_{1/2\,1/2}(\beta) = \frac{1}{l+1}\cos\tfrac{1}{2}\beta\,[P'_{l+1}(\cos\beta) - P'_l(\cos\beta)]$$

$$d^j_{-1/2\,1/2}(\beta) = \frac{1}{l+1}\sin\tfrac{1}{2}\beta\,[P'_{l+1}(\cos\beta) + P'_l(\cos\beta)] \tag{C-29}$$

The $D^j_{m'm}$ functions obey

$$\sum_{m''} D^{j*}_{m''m'}(\alpha\beta\gamma) D^j_{m''m}(\alpha\beta\gamma) = \delta_{m'm} \tag{C-30}$$

and Eq. 9-12.

Table C-1 Clebsch–Gordan coefficients and spherical harmonics

Note: A $\sqrt{\ }$ is to be understood over every coefficient; e.g., for -8/15 read $-\sqrt{8/15}$.

Notation:

m_1	m_2	J, M	J, M	...
m_1	m_2	Coefficients		
.	.			

$$Y_1^0 = \sqrt{\frac{3}{4\pi}} \cos\theta$$

$$Y_1^1 = -\sqrt{\frac{3}{8\pi}} \sin\theta\, e^{i\phi}$$

$$Y_2^0 = \sqrt{\frac{5}{4\pi}} \left(\frac{3}{2}\cos^2\theta - \frac{1}{2}\right)$$

$$Y_2^1 = -\sqrt{\frac{15}{8\pi}} \sin\theta \cos\theta\, e^{i\phi}$$

$$Y_2^2 = \frac{1}{4}\sqrt{\frac{15}{2\pi}} \sin^2\theta\, e^{2i\phi}$$

1/2 × 1/2

m_1	m_2	J=1, M=+1
+1/2	+1/2	1

m_1	m_2	J=1, M=0	J=0, M=0
+1/2	-1/2	1/2	1/2
-1/2	+1/2	1/2	-1/2

m_1	m_2	J=1, M=-1
-1/2	-1/2	1

1 × 1/2

m_1	m_2	J=3/2, M=+3/2
+1	+1/2	1

m_1	m_2	J=3/2, M=+1/2	J=1/2, M=+1/2
+1	-1/2	1/3	2/3
0	+1/2	2/3	-1/3

m_1	m_2	J=3/2, M=-1/2	J=1/2, M=-1/2
0	-1/2	2/3	1/3
-1	+1/2	1/3	-2/3

m_1	m_2	J=3/2, M=-3/2
-1	-1/2	1

2 × 1

m_1	m_2	J=3, M=+3
+2	+1	1

m_1	m_2	J=3, M=+2	J=2, M=+2
+2	0	1/3	2/3
+1	+1	2/3	-1/3

m_1	m_2	J=3, M=+1	J=2, M=+1	J=1, M=+1
+2	-1	1/15	1/3	3/5
+1	0	8/15	1/6	-3/10
0	+1	6/15	-1/2	1/10

m_1	m_2	J=3, M=0	J=2, M=0	J=1, M=0
+1	-1	1/5	1/2	3/10
0	0	3/5	0	-2/5
-1	+1	1/5	-1/2	3/10

m_1	m_2	J=3, M=-1	J=2, M=-1	J=1, M=-1
0	-1	6/15	1/2	1/10
-1	0	8/15	-1/6	-3/10
-2	+1	1/15	-1/3	3/5

m_1	m_2	J=3, M=-2	J=2, M=-2
-1	-1	2/3	1/3
-2	0	1/3	-2/3

m_1	m_2	J=3, M=-3
-2	-1	1

1 × 1

m_1	m_2	J=2, M=+2
+1	+1	1

m_1	m_2	J=2, M=+1	J=1, M=+1
+1	0	1/2	1/2
0	+1	1/2	-1/2

m_1	m_2	J=2, M=0	J=1, M=0	J=0, M=0
+1	-1	1/6	1/2	1/3
0	0	2/3	0	-1/3
-1	+1	1/6	-1/2	1/3

m_1	m_2	J=2, M=-1	J=1, M=-1
0	-1	1/2	1/2
-1	0	1/2	-1/2

m_1	m_2	J=2, M=-2
-1	-1	1

2 × 1/2

m_1	m_2	J=5/2, M=+5/2
+2	1/2	1

m_1	m_2	J=5/2, M=+3/2	J=3/2, M=+3/2
+2	-1/2	1/5	4/5
+1	+1/2	4/5	-1/5

m_1	m_2	J=5/2, M=+1/2	J=3/2, M=+1/2
+1	-1/2	2/5	3/5
0	+1/2	3/5	-2/5

m_1	m_2	J=5/2, M=-1/2	J=3/2, M=-1/2
0	-1/2	3/5	2/5
-1	+1/2	2/5	-3/5

m_1	m_2	J=5/2, M=-3/2	J=3/2, M=-3/2
-1	-1/2	4/5	1/5
-2	+1/2	1/5	-4/5

m_1	m_2	J=5/2, M=-5/2
-2	-1/2	1

3/2 × 1/2

m_1	m_2	J=2, M=+2
+3/2	+1/2	1

m_1	m_2	J=2, M=+1	J=1, M=+1
+3/2	-1/2	1/4	3/4
+1/2	+1/2	3/4	-1/4

m_1	m_2	J=2, M=0	J=1, M=0
+1/2	-1/2	1/2	1/2
-1/2	+1/2	1/2	-1/2

m_1	m_2	J=2, M=-1	J=1, M=-1
-1/2	-1/2	3/4	1/4
-3/2	+1/2	1/4	-3/4

m_1	m_2	J=2, M=-2
-3/2	-1/2	1

3/2 × 1

m_1	m_2	J=5/2, M=+5/2
+3/2	+1	1

m_1	m_2	J=5/2, M=+3/2	J=3/2, M=+3/2
+3/2	0	2/5	3/5
+1/2	+1	3/5	-2/5

m_1	m_2	J=5/2, M=+1/2	J=3/2, M=+1/2	J=1/2, M=+1/2
+3/2	-1	1/10	2/5	1/2
+1/2	0	3/5	1/15	-1/3
-1/2	+1	3/10	-8/15	1/6

m_1	m_2	J=5/2, M=-1/2	J=3/2, M=-1/2	J=1/2, M=-1/2
+1/2	-1	3/10	8/15	1/6
-1/2	0	3/5	-1/15	-1/3
-3/2	+1	1/10	-2/5	1/2

m_1	m_2	J=5/2, M=-3/2	J=3/2, M=-3/2
-1/2	-1	3/5	2/5
-3/2	0	2/5	-3/5

m_1	m_2	J=5/2, M=-5/2
-3/2	-1	1

Table C-2 Values of $d^j_{mm'}(\theta)$ for $j = \frac{1}{2}, 1, \frac{3}{2}$

$j = \frac{1}{2}$		$m' = +\frac{1}{2}$	$m' = -\frac{1}{2}$
m	$+\frac{1}{2}$	$\cos \frac{1}{2}\theta$	$-\sin \frac{1}{2}\theta$
	$-\frac{1}{2}$	$\sin \frac{1}{2}\theta$	$\cos \frac{1}{2}\theta$

$j = 1$		$m' = +1$	$m' = 0$	$m' = -1$
m	$+1$	$\cos^2 \frac{1}{2}\theta$	$-2^{-1/2} \sin \theta$	$\sin^2 \frac{1}{2}\theta$
	0	$2^{-1/2} \sin \theta$	$\cos \theta$	$-2^{-1/2} \sin \theta$
	-1	$\sin^2 \frac{1}{2}\theta$	$2^{-1/2} \sin \theta$	$\cos^2 \frac{1}{2}\theta$

$j = \frac{3}{2}$		$m' = +\frac{3}{2}$	$m' = +\frac{1}{2}$	$m' = -\frac{1}{2}$	$m' = -\frac{3}{2}$
m	$+\frac{3}{2}$	$\cos^3 \frac{1}{2}\theta$	$-3^{1/2} \cos^2 \frac{1}{2}\theta \sin \frac{1}{2}\theta$	$3^{1/2} \cos \frac{1}{2}\theta \sin^2 \frac{1}{2}\theta$	$-\sin^3 \frac{1}{2}\theta$
	$+\frac{1}{2}$	$3^{1/2} \cos^2 \frac{1}{2}\theta \sin \frac{1}{2}\theta$	$\cos \frac{1}{2}\theta(1 - 3 \sin^2 \frac{1}{2}\theta)$	$-\sin \frac{1}{2}\theta(3 \cos^2 \frac{1}{2}\theta - 1)$	$3^{1/2} \cos \frac{1}{2}\theta \sin^2 \frac{1}{2}\theta$
	$-\frac{1}{2}$	$3^{1/2} \cos \frac{1}{2}\theta \sin^2 \frac{1}{2}\theta$	$\sin \frac{1}{2}\theta(3 \cos^2 \frac{1}{2}\theta - 1)$	$\cos \frac{1}{2}\theta(1 - 3 \sin^2 \frac{1}{2}\theta)$	$-3^{1/2} \cos^2 \frac{1}{2}\theta \sin \frac{1}{2}\theta$
	$-\frac{3}{2}$	$+\sin^3 \frac{1}{2}\theta$	$3^{1/2} \cos \frac{1}{2}\theta \sin^2 \frac{1}{2}\theta$	$3^{1/2} \cos^2 \frac{1}{2}\theta \sin \frac{1}{2}\theta$	$\cos^3 \frac{1}{2}\theta$

D

Properties of Hadrons

D-1 Purpose. The accompanying tables are *not* complete; they contain only the most prominent hadron with masses of *less than* 1500 or 1600 MeV. For unstable hadrons we list only the decay modes that contribute at least 10% to the total decay rate. Our purpose is to provide information for use in this book and to enable the reader to refresh his memory about the properties of particular hadrons.

For complete information, the reader should use the tables prepared yearly by the Particle Data Group. These are published, and issued in condensed form, by

Technical Information Division
Bldg. 90
Lawrence Berkeley Laboratory
Berkeley, California 94720

Scientific Information Service
CERN (Comité Européen pour la Recherche Nucléaire)
Geneva 23, Switzerland

In these tables I = isotopic spin, G = G-parity, J = intrinsic spin, P = intrinsic parity, C_n = charge conjugation parity of neutral member, and S = strangeness. For the baryon resonances we also list $L_{2I,2J}$ as discussed in Ch. 5.

D-2 Mesons with $S = 0$

Name	$I^G(J^P)C_n$	Mass (MeV)	Lifetime or Full Width (MeV)	Decay Modes
$\pi^\pm$	$1^-(0^-)$	140	2.60×10^{-8} sec	$\mu\nu$
π^0	$1^-(0^-)^+$	135	8.4×10^{-17} sec	$\gamma\gamma$
η	$0^+(0^-)^+$	549	2.6 keV	38% $\gamma\gamma$, 30% $\pi^0\pi^0\pi^0$ 24% $\pi^+\pi^-\pi^0$
ρ	$1^+(1^-)^-$	770	146	$\pi\pi$
ω	$0^-(1^-)^-$	784	10	90% $\pi^+\pi^-\pi^0$
η' or X^0	$0^+(0^-)^+$	958	<2	72% $\eta\pi\pi$, 26% $\pi^+\pi^-\gamma$
ϕ	$0^-(1^-)^-$	1019	4.4	47% K^+K^-, 35% $K_L{}^0K_S{}^0$ 15% $\pi^+\pi^-\pi^0$
B	$1^+(1^+)^-$	1237	120	$\omega\pi$
f	$0^+(2^+)^+$	1270	163	80% $\pi\pi$
A_2	$1^-(2^+)^+$	1310	100	72% $\rho\pi$, 15% $\eta\pi$
f'	$0^+(2^+)^+$	1516	46	only $K\bar{K}$ seen
$\rho'(1600)$	$1^+(1^-)^-$	~1600	~500	4π
$\omega(1675)$ or $\phi(1675)$	$0^-(3?)^-$	1664	141	$\rho\pi$
g	$1^+(3^-)^-$	1680	160	2π and 4π

D-3 Mesons with $S = \pm 1$

Name	$I(J^P)$	Mass (MeV)	Lifetime or Full Width (MeV)	Decay Modes
$K^\pm$	$\frac{1}{2}(0^-)$	494	1.24×10^{-8} sec	64% $\mu\nu$, 21% $\pi\pi^0$
$K_S{}^0$	$\frac{1}{2}(0^-)$	498	8.9×10^{-11} sec	69% $\pi^+\pi^-$, 31% $\pi^0\pi^0$
$K_L{}^0$	$\frac{1}{2}(0^-)$	498	5.2×10^{-8} sec	39% $\pi e\nu$, 27% $\pi\mu\nu$ 21% $\pi^0\pi^0\pi^0$, 13% $\pi^+\pi^-\pi^0$
$K^*(892)$	$\frac{1}{2}(1^-)$	892	50	$K\pi$
$K^*(1420)$	$\frac{1}{2}(2^+)$	1421	100	$K\pi$ and $K^*\pi$

D-4 Baryons with $I = \frac{1}{2}$, $S = 0$

Name	J^P	Mass (MeV)	Full Width (MeV)	Decay Modes
p	$\frac{1}{2}^+$	938		
n	$\frac{1}{2}^+$	940		
$N'(1470)$	$(\frac{1}{2})^+P'_{11}$	~1470	165–300	60% $N\pi$, 40% $N\pi\pi$
$N'(1520)$	$(\frac{3}{2})^-D'_{13}$	1510–1540	105–150	50% $N\pi$, 50% $N\pi\pi$
$N'(1535)$	$(\frac{1}{2})^-S'_{11}$	1500–1600	50–160	35% $N\pi$, 55% $N\eta$ 10% $N\pi\pi$
$N'(1670)$	$(\frac{5}{2}^-)D'_{15}$	1670–1685	115–175	40% $N\pi$, 60% $N\pi\pi$
$N'(1688)$	$(\frac{5}{2}^+)F'_{15}$	1680–1690	105–180	60% $N\pi$, 40% $N\pi\pi$

D-5 Baryons with $I = \frac{3}{2}$, $S = 0$

Name	J^P	Mass (MeV)	Full Width (MeV)	Decay Modes
$\Delta(1236)$	$(\frac{3}{2}^+)P'_{33}$	1230–1236	110–122	$N\pi$
$\Delta(1650)$	$(\frac{1}{2}^-)S_{31}$	1615–1695	130–200	28% $N\pi$, 72% $N\pi\pi$

D-6 Baryons with $I = 0$, $S = -1$

Name	J^P	Mass (MeV)	Full Width (MeV)	Decay Modes
Λ	$(\frac{1}{2}^+)$	1116	2.5×10^{-10} sec	$N\pi$
$\Lambda'(1405)$	$(\frac{1}{2}^-)S'_{01}$	1405	40	$\Sigma\pi$
$\Lambda'(1520)$	$(\frac{3}{2}^-)D'_{03}$	1518	16	45% $N\bar{K}$, 41% $\Sigma\pi$

D-7 Baryons with $I = 1$, $S = -1$

Name	J^P	Mass (MeV)	Full Width (MeV)	Decay Modes
Σ^+	$(\frac{1}{2}^+)$	1189	0.8×10^{-10} sec	$N\pi$
Σ^0	$(\frac{1}{2}^+)$	1192	$<1 \times 10^{-14}$ sec	$\Lambda\gamma$
Σ^-	$(\frac{1}{2}^+)$	1197	1.48×10^{-10} sec	$N\pi$
$\Sigma'(1385)$	$(\frac{3}{2}^+)P'_{13}$	1383	34	89% $\Lambda\pi$

D-8 Baryons with $I = \frac{1}{2}$, $S = -2$

Name	J^P	Mass (MeV)	Full Width (MeV)	Decay Modes
Ξ^0	$(\frac{1}{2}^+)$	1314	3.0×10^{-10} sec	$\Lambda\pi$
Ξ^-	$(\frac{1}{2}^+)$	1321	1.66×10^{-10} sec	$\Lambda\pi$
$\Xi(1530)$	$(\frac{3}{2}^+)P_{13}$	(0 charge) 1531	9	$\Xi\pi$
		(– charge) 1535	16	

D-9 Baryons with $I = 0$, $S = -3$. In this class we have Ω^-, with $J^P = (\frac{3}{2}^+)$; mass, 1673 MeV; full width, 1.3×10^{-10} sec; and decay modes $\Xi\pi$ and ΛK^-.

E

Dirac Matrices and Spinors

E-1 Dirac Spinors. We follow the notation used by J. D. Bjorken and S. D. Drell, *Relativistic Quantum Mechanics* (McGraw-Hill, New York, 1964). A thorough discussion of an alternative notation is given by J. J. Sakurai, *Advanced Quantum Mechanics* (Addison-Wesley, Reading, Mass., 1967).

We use the slash notation $\gamma^\mu a_\mu = \not{a}$. The Dirac equation for a particle spinor $u(p, \lambda)$ is

$$(\not{p} - m)u(p, \lambda) = 0 \tag{E-1}$$

where p is the four-momentum and λ denotes the helicity or spin state. For an antiparticle spinor $v(p, \lambda)$, we write

$$(\not{p} + m)v(p, \lambda) = 0 \tag{E-2}$$

The adjoint spinors

$$\bar{u} = u^\dagger \gamma^0$$

$$\bar{v} = v^\dagger \gamma^0 \tag{E-3}$$

obey

$$\bar{u}(p, \lambda)(\not{p} - m) = 0$$
$$\bar{v}(p, \lambda)(\not{p} + m) = 0 \qquad \text{(E-4)}$$

Our normalization condition (see Sec. 12-2) is

$$\bar{u}(p, \lambda)u(p, \lambda) = 2m$$
$$\bar{v}(p, \lambda)v(p, \lambda) = -2m \qquad \text{(E-5)}$$

Here and in Eq. E-6 we differ from the Bjorken and Drell convention.

We say that

$$\text{particle projection operator} = \not{p} + m$$
$$\text{antiparticle projection operator} = \not{p} - m \qquad \text{(E-6)}$$

For specific plane wave spinors see Sec. 12-2.

E-2 Matrices. We write

$$\gamma^\mu \gamma^\nu + \gamma^\nu \gamma^\mu = 2g^{\mu\nu} \qquad \text{(E-7)}$$

We use the γ matrix representation

$$\gamma^0 = \begin{bmatrix} I & 0 \\ 0 & -I \end{bmatrix}$$
$$\gamma^j = \begin{bmatrix} 0 & \sigma^j \\ -\sigma^j & 0 \end{bmatrix}, \qquad j = 1, 2, 3 \qquad \text{(E-8)}$$

Where I is the unit matrix and the σ^j's are the Pauli matrices, we have

$$I = \begin{bmatrix} 1 & 0 \\ 0 & 1 \end{bmatrix}, \quad \sigma^1 = \begin{bmatrix} 0 & 1 \\ 1 & 0 \end{bmatrix}, \quad \sigma^2 = \begin{bmatrix} 0 & -i \\ i & 0 \end{bmatrix}, \quad \sigma^3 = \begin{bmatrix} 1 & 0 \\ 0 & -1 \end{bmatrix} \qquad \text{(E-9)}$$

We recall $\sigma^i \sigma^j = i\sigma^k$, ijk cyclic permutation of 123.

$$\gamma^0 \text{ is hermitian} \qquad \gamma^{0\dagger} = \gamma^0$$
$$\gamma^j \text{ is antihermitian} \qquad \gamma^j = -\gamma^j, \qquad j = 1, 2, 3 \qquad \text{(E-10)}$$

We mean the same thing by γ^5 and γ_5, namely,

$$\gamma^5 = \gamma_5 = i\gamma^0\gamma^1\gamma^2\gamma^3$$

$$\gamma^5 = \begin{bmatrix} 0 & I \\ I & 0 \end{bmatrix} \tag{E-11}$$

Where γ^5 is hermitian. We also use

$$\sigma^{\mu\nu} = \frac{i}{2}\,[\gamma^\mu\gamma^\nu - \gamma^\nu\gamma^\mu] \tag{E-12a}$$

For ijk = cyclic permutation of 123 we write

$$\sigma^{ij} = \begin{bmatrix} \sigma^k & 0 \\ 0 & \sigma^k \end{bmatrix} \tag{E-12b}$$

Also

$$\not{a}\not{b} = a \cdot b - i\sigma_{\mu\nu}a^\mu b^\nu \tag{E-13}$$

E-3 Traces. In calculating matrix elements we have to deal with

$$(\bar{u}'\Gamma u)^\dagger = (\bar{u}\bar{\Gamma}u'), \qquad \bar{\Gamma} = \gamma^0\Gamma^\dagger\gamma^0 \tag{E-14}$$

It is convenient to note that

$$\bar{\Gamma} = \Gamma \qquad \text{for} \quad \Gamma = \gamma^\mu, \sigma^{\mu\nu}, i\gamma^5 \tag{E-15}$$

The following trace theorems are useful:

$$\begin{aligned}
&\mathrm{Tr}\ (\text{odd number of } \gamma\text{'s}) = 0 \\
&\mathrm{Tr}\ \gamma^5 = 0 \\
&\mathrm{Tr}\ \gamma^5\not{a}\not{b} = 0 \\
&\mathrm{Tr}\ 1 = 4 \\
&\mathrm{Tr}\ \not{a}\not{b} = 4a \cdot b \\
&\mathrm{Tr}\ \not{a}_1\not{a}_2\not{a}_3\not{a}_4 = 4[a_1 \cdot a_2 a_3 \cdot a_4 - a_1 \cdot a_3 a_2 \cdot a_4 + a_1 \cdot a_4 a_2 \cdot a_3] \\
&\mathrm{Tr}\ \gamma^5\not{a}\not{b}\not{c}\not{d} = 4i\epsilon_{\alpha\beta\gamma\delta}\, a^\alpha b^\beta c^\gamma d^\delta
\end{aligned} \tag{E-16}$$

E-4 Coordinate Transformations. For a spatial rotation about the k axis by an angle θ, the spinor operator is

$$R(\theta, \hat{n}_k) = \exp\left(-\frac{i\theta\sigma^{ij}}{2}\right) \tag{E-17}$$

For a boost along the k axis by a boost parameter ξ (Sec. 9-4), the spinor operator is

$$L(\xi, \hat{n}_k) = \exp\left(-\frac{i\xi\sigma^{0K}}{2}\right) \tag{E-18}$$

where σ^{ij} and σ^{0k} are defined in Eq. E-12.

F

Rules for Feynman Diagrams and Cross Sections

F-1 Cross Sections. The normalization convention, the definition of the T-matrix, and the cross section formulas are given in Ch. 6, particularly in Sec. 6-6. For $a + b \to N$ particles, before summing over spins, we write

$$d\sigma = \frac{(2\pi)^4 |T|^2}{4[(P_a \cdot P_b)^2 - m_a{}^2 m_b{}^2]^{1/2}} \prod_{n=1}^{N} \frac{d^3p_n}{(2\pi)^3 2E_n} \delta^4(P_f - P_i) \tag{F-1a}$$

The density of states term is explicitly

$$\prod_{n=1}^{N} \frac{d^3p_n}{(2\pi)^3 2E_n} = \frac{d^3p_1}{(2\pi)^3 2E_1} \frac{d^3p_2}{(2\pi)^3 2E_2} \cdots \frac{d^3p_N}{(2\pi)^3 2E_N}$$

$$= \frac{1}{(2\pi)^{3N} 2^N} \prod_{n=1}^{N} \frac{d^3p_n}{E_n} \tag{F-1b}$$

But for simplicity, and as a mnemonic device, we always use the left-most form in Eq. F-1b. Because our normalization convention is the same for fermions and bosons, Eq. F-1b holds for both types of particles. For only two particles in the final state, in the center of mass frame, we put

$$\frac{d\sigma}{d\Omega^*} = \frac{1}{64\pi^2 s}\left(\frac{p_f^*}{p_i^*}\right)|T|^2 \tag{F-2}$$

F-2 General Rules for Calculating *T* from Feynman Diagrams. We follow the conventions used by J. D. Bjorken and S. D. Drell, *Relativistic Quantum Mechanics* (McGraw-Hill, New York, 1964), with two minor exceptions. First, our plane wave normalization convention (Eq. 6-15) is the same for fermions and bosons. Second, we use the invariant amplitude $T = -\mathcal{M}$, where $\mathcal{M}$ is their invariant amplitude. Note carefully Rule F.6 below.

A. For each external line entering a diagram with four-momentum p and helicity λ, insert the following spin functions.

spin 0	1
spin 1/2 particle	$u(p, \lambda)$, a four-spinor given in App. E and Sec. 12-2
spin 1/2 antiparticle	$v(p, \lambda)$, a four-spinor given in App. E
spin 1	$\epsilon^{(\lambda)}$, a four-vector given in Sec. 12-2 (F-3)

B. For each external line leaving a diagram, insert the hermitian conjugate of the boson spin functions or the adjoint of the spinor functions given in A.

C. For each internal line with four-momentum p and mass m, insert the following propagators:

spin 0 $\quad \dfrac{i}{p^2 - m^2}$

spin $\frac{1}{2}$ $\quad \dfrac{i(\not p + m)}{p^2 - m^2}$

spin 1 $\quad \dfrac{i(p_\mu p_\nu/m^2 - g_{\mu\nu})}{p^2 - m^2}$

photon $\quad \dfrac{-ig_{\mu\nu}}{p^2} \qquad$ (F-4)

D. For each three-particle vertex with the four-momenta and masses shown in Fig. F-1, insert the following. When a photon is involved, $g = e = \sqrt{4\pi\alpha}$.

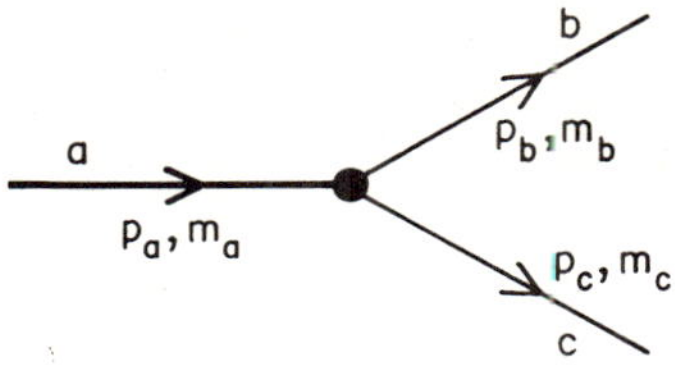

Fig. F-1 Three-particle vertex.

Spin-Parity of Particles

a	b	c	Vertex	
0^+	0^+	0^+	$-ig$	
$0^\pm$	$0^\pm$	1^- or photon	$-ig(p_a + p_b)_\mu$	
0^-	1^-	1^-	$-i(f/m)\epsilon_{\mu\nu\rho\sigma}p_a{}^\nu p_c{}^\rho$	
$\frac{1}{2}^+$	0^-	$\frac{1}{2}^+$	$-ig\gamma_5$	
$\frac{1}{2}^+$	1^- or photon	$\frac{1}{2}^+$, vector coupling	$-ig\gamma_\mu$	
$\frac{1}{2}^+$	1^- or photon	$\frac{1}{2}^+$, tensor coupling	$g\sigma_{\gamma\nu}p_b{}^\nu/(m_a + m_c)$	(F-5)

E. For each four-particle or seagull graph in Fig. F-2, use

Spin-Parity of Particles

a	b	c	d	*Vertex*	
$0^\pm$	$0^\pm$	$0^\pm$	$0^\pm$	$-iG$	
$0^\pm$	$0^\pm$	photon	photon	$2ig^2 g_{\mu\nu}$	(F-6)

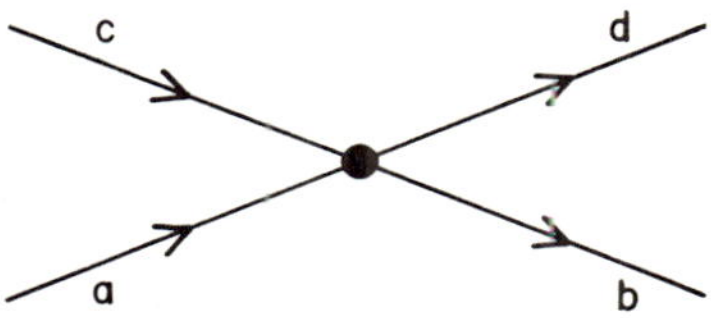

Fig. F-2 "Seagull" graph.

F. There are the following overall factors and considerations.

1. Four momentum is conserved at each vertex.
2. Each internal four-momentum p that is not fixed by momentum conservation requires an integration $\int (d^4p)/(2\pi)^4$.
3. Each closed fermion loop requires a factor of -1.

4. Graphs differing only by the interchange of two identical external fermions require a -1 between them.
5. A loop containing only *two* lines, both photons, or both bosons, requires a factor of $\frac{1}{2}$.
6. To obtain T, the entire amplitude, as calculated here, is finally multiplied by i. (Then a real repulsive potential between spin zero particles–to lowest order–has T real and negative and the phase shifts negative, as it should be.)

INDEX